COMBINATORY LOGIC

PURE, APPLIED AND TYPED

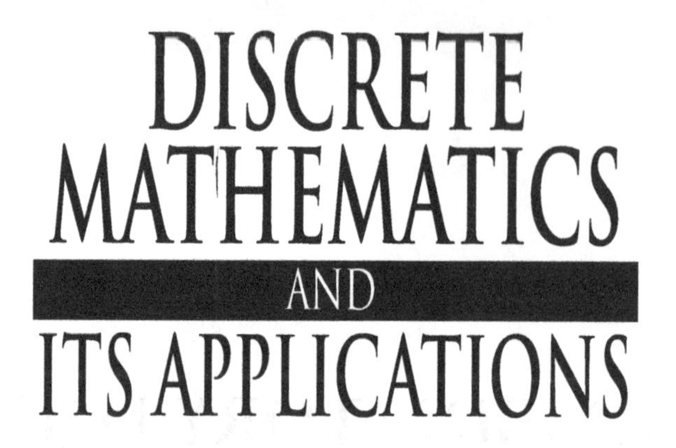

Series Editor
Kenneth H. Rosen, Ph.D.

R. B. J. T. Allenby and Alan Slomson, How to Count: An Introduction to Combinatorics, Third Edition

Juergen Bierbrauer, Introduction to Coding Theory

Katalin Bimbó, Combinatory Logic: Pure, Applied and Typed

Donald Bindner and Martin Erickson, A Student's Guide to the Study, Practice, and Tools of Modern Mathematics

Francine Blanchet-Sadri, Algorithmic Combinatorics on Partial Words

Richard A. Brualdi and Dragoš Cvetković, A Combinatorial Approach to Matrix Theory and Its Applications

Kun-Mao Chao and Bang Ye Wu, Spanning Trees and Optimization Problems

Charalambos A. Charalambides, Enumerative Combinatorics

Gary Chartrand and Ping Zhang, Chromatic Graph Theory

Henri Cohen, Gerhard Frey, et al., Handbook of Elliptic and Hyperelliptic Curve Cryptography

Charles J. Colbourn and Jeffrey H. Dinitz, Handbook of Combinatorial Designs, Second Edition

Martin Erickson, Pearls of Discrete Mathematics

Martin Erickson and Anthony Vazzana, Introduction to Number Theory

Steven Furino, Ying Miao, and Jianxing Yin, Frames and Resolvable Designs: Uses, Constructions, and Existence

Mark S. Gockenbach, Finite-Dimensional Linear Algebra

Randy Goldberg and Lance Riek, A Practical Handbook of Speech Coders

Jacob E. Goodman and Joseph O'Rourke, Handbook of Discrete and Computational Geometry, Second Edition

Jonathan L. Gross, Combinatorial Methods with Computer Applications

Jonathan L. Gross and Jay Yellen, Graph Theory and Its Applications, Second Edition

Titles (continued)

Jonathan L. Gross and Jay Yellen, Handbook of Graph Theory

David S. Gunderson, Handbook of Mathematical Induction: Theory and Applications

Richard Hammack, Wilfried Imrich, and Sandi Klavžar, Handbook of Product Graphs, Second Edition

Darrel R. Hankerson, Greg A. Harris, and Peter D. Johnson, Introduction to Information Theory and Data Compression, Second Edition

Darel W. Hardy, Fred Richman, and Carol L. Walker, Applied Algebra: Codes, Ciphers, and Discrete Algorithms, Second Edition

Daryl D. Harms, Miroslav Kraetzl, Charles J. Colbourn, and John S. Devitt, Network Reliability: Experiments with a Symbolic Algebra Environment

Silvia Heubach and Toufik Mansour, Combinatorics of Compositions and Words

Leslie Hogben, Handbook of Linear Algebra

Derek F. Holt with Bettina Eick and Eamonn A. O'Brien, Handbook of Computational Group Theory

David M. Jackson and Terry I. Visentin, An Atlas of Smaller Maps in Orientable and Nonorientable Surfaces

Richard E. Klima, Neil P. Sigmon, and Ernest L. Stitzinger, Applications of Abstract Algebra with Maple™ and MATLAB®, Second Edition

Patrick Knupp and Kambiz Salari, Verification of Computer Codes in Computational Science and Engineering

William Kocay and Donald L. Kreher, Graphs, Algorithms, and Optimization

Donald L. Kreher and Douglas R. Stinson, Combinatorial Algorithms: Generation Enumeration and Search

Hang T. Lau, A Java Library of Graph Algorithms and Optimization

C. C. Lindner and C. A. Rodger, Design Theory, Second Edition

Nicholas A. Loehr, Bijective Combinatorics

Alasdair McAndrew, Introduction to Cryptography with Open-Source Software

Elliott Mendelson, Introduction to Mathematical Logic, Fifth Edition

Alfred J. Menezes, Paul C. van Oorschot, and Scott A. Vanstone, Handbook of Applied Cryptography

Richard A. Mollin, Advanced Number Theory with Applications

Richard A. Mollin, Algebraic Number Theory, Second Edition

Richard A. Mollin, Codes: The Guide to Secrecy from Ancient to Modern Times

Richard A. Mollin, Fundamental Number Theory with Applications, Second Edition

Richard A. Mollin, An Introduction to Cryptography, Second Edition

Richard A. Mollin, Quadratics

Richard A. Mollin, RSA and Public-Key Cryptography

Titles (continued)

Carlos J. Moreno and Samuel S. Wagstaff, Jr., Sums of Squares of Integers

Dingyi Pei, Authentication Codes and Combinatorial Designs

Kenneth H. Rosen, Handbook of Discrete and Combinatorial Mathematics

Douglas R. Shier and K.T. Wallenius, Applied Mathematical Modeling: A Multidisciplinary Approach

Alexander Stanoyevitch, Introduction to Cryptography with Mathematical Foundations and Computer Implementations

Jörn Steuding, Diophantine Analysis

Douglas R. Stinson, Cryptography: Theory and Practice, Third Edition

Roberto Togneri and Christopher J. deSilva, Fundamentals of Information Theory and Coding Design

W. D. Wallis, Introduction to Combinatorial Designs, Second Edition

W. D. Wallis and J. C. George, Introduction to Combinatorics

Lawrence C. Washington, Elliptic Curves: Number Theory and Cryptography, Second Edition

DISCRETE MATHEMATICS AND ITS APPLICATIONS
Series Editor KENNETH H. ROSEN

COMBINATORY LOGIC

PURE, APPLIED AND TYPED

Katalin Bimbó
University of Alberta
Edmonton, Canada

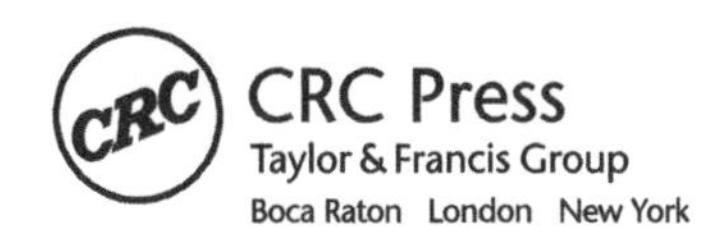

CRC Press is an imprint of the
Taylor & Francis Group, an **informa** business
A CHAPMAN & HALL BOOK

First published in paperback 2024

First published 2012
by Chapman & Hall/CRC
2385 NW Executive Center Drive, Suite 320, Boca Raton FL 33431

and by Chapman & Hall/CRC
4 Park Square, Milton Park, Abingdon, Oxon, OX14 4RN

CRC Press is an imprint of Taylor & Francis Group, LLC

Publisher's Note
The publisher has gone to great lengths to ensure the quality of this reprint but points out that some imperfections in the original copies may be apparent.

ISBN: 978-1-032-92118-1 (pbk)
ISBN: 978-1-439-80000-3 (hbk)
ISBN: 978-0-429-07528-5 (ebk)

DOI: 10.1201/b11046

Visit the Taylor & Francis Web site at
http://www.taylorandfrancis.com

and the CRC Press Web site at
http://www.crcpress.com

Contents

Preface

Combinatory logic was invented in the 1920s and has been developing ever since — often in the company of λ-calculi. Both of these formalisms, together with their variants, can capture the notion of a computable function or algorithm. This places combinatory logic next to recursion theory and computer science. On the other hand, typed combinatory logic straightforwardly connects with intuitionistic logic and other nonclassical logics, for example, relevance logics and linear logic. All this links combinatory logic to mathematical logic, philosophical logic and computational logic.

This book intends to present combinatory logic — starting with the basics — as a self-contained subject. λ-calculi are mentioned, because it seems unavoidable due to the historical development of these areas. However, λ-calculi are not given here as much prominence as is some other texts.

The key themes that are explicit (and sometimes implicit) in the following chapters are the connections to *computability* and to *nonclassical logics*. This implies that some other matters are given less emphasis. Nevertheless, the book aims to give a snapshot of the current state of the development of combinatory logic — together with pointers toward further possible advancements.

There is a comprehensive *two-volume book* on combinatory logic by Haskell B. Curry and his collaborators and students. Those volumes, however, had been completed by the late 1950s and early 1970s, respectively. Of course, mathematical and logical knowledge is typically cumulative, but the way concepts and results are presented might change even when the core results are preserved. Arguably, the role and the perception of combinatory logic within the discipline of logic has changed in the past 30–40 years. During those decades, many new results have been obtained. This creates a need and imparts a justification for a new presentation of the subject with some emphases shifted.

Combinatory logic is briefly mentioned or dealt with in quite a few *books on λ-calculi*, because of the relationship between λ-calculi and combinatory logics. (See, for instance, [10], [79] and [87].) I take the reverse approach here: the book focuses on combinatory logic itself, and I give a concise introduction to λ-calculi (in chapter 4), primarily, for the sake of comparison.

Curry seems to have deemed important to supplement his books with sections on history. (See, for instance, [53], [52] and [54].) Combinatory logic has *matured* since its early days, and it seems to me that brief sections would not suffice to provide an accurate historical account. Recently, separate and detailed publications appeared on the history of combinatory logic and λ-calculus in volumes on the history of logic, which also suggest that the history of the subject is better to be treated in itself.

In sum, I did not aim at providing a historical account of the development of combinatory logic in this book. Instead of trying to follow previous presentations closely, for the sake of historical accuracy, I intended to give an exposition in which the concepts are introduced as the results require them. Although some of the results are labeled by their usual labels, such as "Church–Rosser theorem," I did not try to attach a label to each claim, lemma and theorem in order to point to the first or to a widely accessible publication. An attempt to conduct more vigorous labeling either would have limited the scope of the presentation or would have opened the door for misattributions due to minor (or even more significant) differences in the underlying concepts. An extensive *bibliography* is included at the end of the book, which also includes texts that are either exclusively historical or at least more historically oriented than this book.

I wrote this book to give *a state-of-the-art view of combinatory logic* in the early 21st century. In order to make the book relatively easy to read, I started with an elementary introduction and I placed into the appendix various concepts and definitions that are useful or occasionally presupposed in the main text of the book, but do not belong to combinatory logic itself. I hope that this will help anybody to read the book, who has some aptitude toward formal thinking.

Combinatory logic is not as widely taught as I think it should be. Therefore, a purpose of this book is to serve as a readily accessible source of knowledge for experts in logic, computer science, mathematics, philosophy and certain related disciplines, who are perhaps, less familiar with this branch of logic.

Another potential use of the book, what I kept in mind during the writing, is as a text in a course. The examples and exercises in the text are intended to facilitate learning combinatory logic, alone or in the context of a course. (The starring of the exercises indicates the expected difficulty of their solutions.) The first couple of chapters of the book (perhaps, with a selection of further sections from later chapters) covers topics that can provide the content of an undergraduate course on combinatory logic. The whole book is suitable to be used as the main text for a graduate course — at a beginning or advanced level, depending on the background of the students. Because of the many connections combinatory logic has, I hope that the book will also be used to supplement courses, for example, on philosophical logic, on computability and on the foundations of mathematics.

My first acquaintance with combinatory logic happened in the early 1990s, when I took a course of Raymond Smullyan, at Indiana University in Bloomington. The power and elegance of combinatory logic quickly enthralled me.

The close connection between nonclassical logics and combinatory logic reinforced my interest in the subject. Among relevance logicians, it is common knowledge that Alonzo Church invented the implicational fragment of the logic of relevant implication (what he called "the theory of weak implication"), which corresponds to types for his λI-calculus. J. Michael Dunn and Robert K. Meyer invented structurally free logics and dual combinators in the 1990s. These logics both tighten the connection between relevance logic and combinatory logic, and expand the bond between combinatory logic and a wide range of nonclassical logics.

My own research results in combinatory logic concern some of the newer developments such as dual combinatory logic. The connection between nonclassical logics and combinatory logic means that I always keep in mind combinators when I work on a substructural logic. Some of my work and results in the proof theory of relevance logic (and indirectly in the proof theory of classical logic) had been motivated by combinatory logic and structurally free logics. The splice of combinatory logic and relevance logic has proved fruitful in many ways. For example, J. Michael Dunn and I have solved the famous problem of $T_{\rightarrow}$, which remained open for half a century, by combining insights from proof theory and combinatory logic.

I am indebted to Robert Stern, the acquiring editor for this book, for his patience and for allowing me to ask for a new deadline (or two) for the submission of the manuscript. (The project of writing this book was substantially delayed when — somewhat unexpectedly — I had to teach a course on real ethics, that is, on formal decision theory as part of my regular teaching duties at the University of Alberta.) I am also grateful to Jennifer Ahringer for her help during the production process.

I am grateful to Graham Sullivan, who, as a research assistant, proofread parts of the book and provided helpful comments and corrections. I am indebted to the University of Alberta for awarding a grant from the Endowment Fund for the Future Support for the Advancement of Scholarship Research Fund that allowed me to have a research assistant for a couple of months at the time when the manuscript of the book was nearing its completion.

The book was typeset using the program TEX (by D. Knuth) with the LATEX format. I also utilized various packages that were developed under the auspices of the American Mathematical Society.

The actual writing of this book took place in Edmonton. I always enjoy the prospect of having another snow fall, what can (and sometime does) happen here as early as September and as late as June.

Edmonton, 2011 KB

Chapter 1

Elements of combinatory logic

Functions are everywhere, and combinatory logic is one of the theories that make them their objects of study. To start with, let us consider a simple function such as $+$, the addition of integers. $+$ can be applied to numbers; for instance, $46+68$ is the result of this application, which is also denoted by 114. Then we can view $+$ as a certain map from ordered pairs of integers into integers, or, further, as *a set of ordered triples*. From another perspective, $+$ can be thought of as *a rule to compute* the sum of two integers — indeed, this is how each of us first became acquainted with $+$. In both cases, knowing that $+$ is commutative and associative (and has other properties) is useful in identifying seemingly different versions of $+$. Just as light has a dual character — being a wave and being a beam of particles — functions have two sides to them: they are collections of input–output pairs, and they are rules for computation from inputs to the output.

Combinatory logic (henceforth, CL) is the logic of *functions*, in which the two aspects of functions mentioned above are in constant interplay. Functions are the objects in CL, and every object of CL can be thought of as a function. The functions need not always be thought of as functions on numbers, though we will see in chapter 3 how to represent numerical functions in CL.

1.1 Objects, combinators and terms

The objects in CL are called *terms*, and they stand for functions. Some of the terms cannot be divided into more elemental components: such terms are either *constants* or *variables*. The only way to put terms together is via (function) *application*, and the result of this operation is always a term.

Let us consider some examples from elementary algebra before proceeding to formal definitions. The natural numbers are denoted, in the usual decimal notation, by nonempty strings of digits like $0, 1, \ldots, 11, \ldots, 98765, \ldots$. (In formalized arithmetic, these numbers are denoted, respectively, by 0 and the successor function applied sufficiently many times to 0.) These expressions are constants, although in an elementary exposition they are not thought of as functions. Another kind of constants are functions like $+$ (addition), $\cdot$ (multiplication), $-$ (subtraction), etc. The three functions we listed are *binary* (i.e., each takes two arguments), but there are functions of other *arities* too. The absolute value function has arity 1, that is, it is *unary*, and we can have functions taking n arguments (where n is a positive integer). It is just not that useful in practice to have a *quinary addition*, let us say $+_5$, that produces the sum of five numbers. (A quinary addition is easily emulated by the binary addition, because $+$ is associative.) Variables may enter into algebraic expressions too, and they are often denoted by $x, y, z, x_0, x_1, x_2, \ldots$.

So far we emphasized some *similarities* between algebraic terms and combinatory terms. They both may contain constants and variables. Each function takes a fixed number of inputs (i.e., arguments), but different functions can have more or fewer arguments than others. Also, the use of delimiters (such as parentheses) is common. $5+7\cdot 13$ is ambiguous — unless we have agreed that $\cdot$ takes precedence over $+$. In other words, $(5+7)\cdot 13$ and $5+(7\cdot 13)$ are different terms; moreover, their value is different too: $(5+7)\cdot 13$ is 156, whereas $5+(7\cdot 13)$ is 96.

A notable *difference* between algebraic and combinatory terms is that in elementary algebra functions such as $+$ are applicable to numbers only. That is, we would dismiss as nonsensical (or as "a misprint") something like $+\cdot(5+\cdot)$, where $\cdot$ seems to have $+$ as its argument and vice versa. We do not make any similar assumptions about the nonapplicability of terms in CL. We may reiterate also that variables in CL may stand for functions. This happens much more infrequently in algebra, except perhaps in definitions of properties of functions. For instance, $f(x,y)$ is called a commutative operation when $f(x,y)$ equals $f(y,x)$, for all x and y. This definition applies to *any* binary operation; that is, the whole sentence is understood as if "for all f" were prefixed to it. Lastly, function application is a binary operation in CL that is explicitly introduced (though typically suppressed in the notation). The more explicit and more detailed analysis of function application in CL leads to several notions of computing with terms.

We assume that the language of CL contains *constants*, typically, S and K. However, it may contain other constants too such as B, C, W, I and M — to name some. These constants are called *combinators*. There might be other sorts of constants in the language, including constants that are not definable from S and K (in a sense that we will explain later). B, C, W, M and I are all definable from S and K. For the definition of the set of terms, it is not particularly interesting which constants are included in the language, except that the definition, of course, applies only to those constants that are included. (We focus, especially in the first half of the book, exclusively on combinators.) We also assume that there is a denumerable set of *variables*, that is, there are infinitely many variables, but countably many (i.e., $\aleph_0$ many). The language includes a binary operation called *function application* that is denoted by juxtaposition.

DEFINITION 1.1.1. The *set of* CL-*terms* (or terms, for short) is inductively defined by (1)–(3).[1]

(1) If Z is a constant, then Z is a term;

[1] As usual, we leave tacit and hence omit mentioning that the least set generated by clauses (1)–(3) is the set, that is defined. This will be our practice in what follows.

(2) if x is a variable, then x is a term;

(3) if M and N are terms, then so is (MN).

We use uppercase sans serif letters for combinators. The letters chosen for particular combinators are often standard. To refer ambiguously to some combinator, we use Z; in other words, Z is a meta-variable for combinators.[2] The lowercase letters, $x, y, z, \ldots$, with or without indices, serve as variables (as well as meta-variables for variables). The uppercase letters M, N, P and Q, with or without indices, are meta-variables for CL-terms in general. To put it differently, M can be instantiated with x or I, but also with $(\mathsf{S}(x(xy)))$, etc. The binary application operation figures into the term (MN) in clause (3). We do not add a symbol like $\cdot$ or $+$ to the term that results from M and N, but the resulting term is enclosed in parentheses.

Example 1.1.2. S and K are terms, and so are the variables standing alone. These terms are called *atomic* to distinguish them from the rest of the terms, which are called *complex* or *compound*. Some of the latter are $(\mathsf{I}x)$, $(x(yz))$, $(z\mathsf{S})$ and $((\mathsf{SI})\mathsf{I})$.

There are denumerably many atomic terms and there are denumerably many complex terms; in total, the set of CL-terms is *denumerable* (i.e., the cardinality of the set of CL-terms is $\aleph_0$).

Exercise 1.1.1. Prove informally (or prove using structural induction) that every CL-term either is atomic or starts with a '('.

Exercise 1.1.2. Sort the terms listed into the following categories: "atomic term," "complex term" and "nonterm" (i.e., an expression that is not a term). (x), SKK, x_{12}, $(xx))(((xx)x)x)$, $(((\mathsf{S}(\mathsf{KS}))\mathsf{K})(x(yz)))$, $(((\mathsf{BM})((\mathsf{BW})\mathsf{B}))x)$, $((((\mathsf{W}M)N)N)\mathsf{W})$, C, z, $(((x_1\mathsf{S})((x_2\mathsf{S})x_3))((x_5(x_4\mathsf{K}))y))$.

Exercise* 1.1.3. Prove informally (or by structural induction) for every term that if the term contains parentheses, then the parentheses are well-balanced. (The latter means that there are equal numbers of ('s and) 's, and that scanning the term from left to right, there are never more right parentheses that left ones.)

Exercise 1.1.4. Prove that not all expressions that contain well-balanced parentheses are CL-terms.

We will often use *meta-terms* instead of terms, that is, expressions that contain some of the meta-variables M, N, P, ... By those (meta-)terms, we refer to an arbitrary term that instantiates the expression.

Another notational convention concerns *omitting parentheses*. Delimiters pile up quickly, and then terms soon become practically unreadable. In CL, parentheses are normally omitted from *left-associated* terms, that is, a term of the form $((MN)P)$

[2] Z is a schematic letter that stands for any of the combinators in the meta-language, that is, in the language we use to talk *about* CL. We use 'meta-variable' similarly for other types of expressions.

may be written as (MNP). The outmost parentheses are, as a rule, omitted too; continuing the example, we get MNP.[3]

Example 1.1.3. $(\mathsf{B}xy)z$ is shorthand for $(((\mathsf{B}x)y)z)$. $\mathsf{BWB}x(\mathsf{BWB}x)$ is a result of omitting some parentheses from $((((\mathsf{BW})\mathsf{B})x)(((\mathsf{BW})\mathsf{B})x))$. The former shorthand term could be further abbreviated by omitting the remaining pair of parentheses. The latter, however, cannot be further simplified.

There are terms from which parentheses may be omitted in various ways; hence, we may get more than one abbreviated term. On the other hand, the insertion of all omitted parentheses leads to a unique term, which is important, because we do not want to introduce ambiguity into our notation. In view of this remark, we will call the abbreviated terms simply "terms" — unless the distinction between them is the point, as in the next couple of exercises.

Exercise 1.1.5. Restore all the parentheses in the following abbreviated terms. $\mathsf{K}x(\mathsf{W}yz)y$, xxx, $(\mathsf{SI})y\mathsf{B}$, $x_{14}x_{152}$, $M_1(M_2(M_3xyy)z)$.

Exercise 1.1.6. Omit as many parentheses as possible from the following terms (without introducing ambiguity). $(((yz)(\mathsf{II}))x)$, $((\mathsf{M}x)((\mathsf{B}y)(\mathsf{W}z)))$, $(\mathsf{W}(\mathsf{I}x_{145}(\mathsf{I}x_{72}))x_{58})$, $(((\mathsf{SK})\mathsf{K})x)$, $(((\mathsf{S}M)M)(\mathsf{I}(NP)))$.

Exercise 1.1.7. Is there a term that has at least two (distinct) abbreviated forms such that neither can be obtained from the other by omitting parentheses? (Hint: If the answer is "yes," give an example; if the answer is "no," then prove that such abbreviated terms cannot exist.)

Exercise 1.1.8. Prove informally (or by structural induction) that all terms contain an even number of well-balanced parentheses, and that omitting parentheses preserves this property.

Exercise 1.1.9. Give an example of an expression (from the language of CL) that is not a term or an abbreviated term, but has an even number of parentheses. Give an example of an expression (from the language of CL) that is not a term or an abbreviated term, but contains well-balanced parentheses.

Now that we have a precise definition of CL-terms, it will be helpful to point out a few more similarities and differences between algebraic terms and CL-terms. Arithmetic functions such as multiplication and addition are binary, and the function symbol is usually placed *between* the two arguments. We will see in the next section that the arity of W is 2; nonetheless, the arguments of W *follow* the function symbol (that is, the arguments come after W). In general, putting a function (or relation) symbol between its arguments is called *infix* notation. In CL, a *prefix* notation is used instead, that is, functions precede (or are prefixed to) their arguments.

[3]We parenthesize M, N, P, ..., as if they were atomic terms — as long as they have not been instantiated. This, of course, does *not* imply that they have to be replaced or instantiated by atomic terms.

In mathematics, in general, there is great variation as to where the arguments of a function are located. Consider as examples exponentiation, logarithmic functions, integrals and derivatives. For less frequently used functions, the notation is less idiosyncratic: an n-ary function may be simply denoted by $f(x_1,\ldots,x_n)$. If f were a constant in CL, then the term in which the function f is followed by the arguments $x_1,\ldots,x_n$ would look like $(\ldots(fx_1)\ldots x_n)$. In other words, the commas are all dropped and a pair of parentheses surrounds a function applied to an argument rather than a series of arguments.

Exercise 1.1.10. Assume that # and $*$ are combinatory constants (of arity 2) that stand, respectively, for addition and multiplication in CL. (a) Translate the following arithmetical terms into CL. $z+z$, $x\cdot y$, $(x+y)\cdot x$, $x_1\cdot(x_2\cdot x_3\cdot x_3+x_4\cdot x_5+x_7)$, $x\cdot(x+y)$. (b) Translate the following terms from CL into arithmetics. $\#x(\#xx)$, $(*(\#yz))((\#x)y)$, $*(*(*(*(*xz)y)y)x)(yx)$, $\#M(N\#P)$, $\#\#(*\mathsf{KI})(*\mathsf{K}*)$.

We have not exhausted the range of investigations of syntactic properties of terms.[4] However, we introduce here yet another syntactic concept. Informally, M is a subterm of N if M is itself a term and M is part of N.

DEFINITION 1.1.4. The *subterm of* relation on the set of CL-terms (i.e., M is a subterm of N) is defined inductively by (1)–(5).

(1) x is a subterm of x;

(2) Z is a subterm of Z;

(3) M is a subterm of the terms (NM) and (MN);

(4) (NP) is a subterm of (NP);

(5) if M is a subterm of N and N is a subterm of P, then M is a subterm of P.

All subterms of a term are themselves terms. Terms have more that one subterm — unless they are atomic; and all terms are subterms of infinitely many terms. The joint effect of (1), (2) and (4) is that every term is a subterm of itself, which may appear strange at first (given the ordinary meaning of "sub-"). However, the *reflexivity* of the "subterm of" relation is intended, because it simplifies some other definitions. (The notion of a *proper subterm* is also definable — see section A.1.) Clause (5) means that the "subterm of" relation is *transitive*; indeed, it is a partial order on the set of terms.

Exercise 1.1.11. List all the subterms of each of the following terms. $(x(\mathsf{W}x)y)$, $yy(y(yy))$, $(\mathsf{S}(\mathsf{KW})\mathsf{S})(\mathsf{BWW})$, $z\mathsf{S}(y\mathsf{S})(x\mathsf{WSK})$, $PM((PN)(PP))M$, $\mathsf{W}Mx(NNPPy)$, $(\mathsf{B}M(x)(NM))$.

[4] Some further notions and definitions concerning terms, such as the definition of left-associated terms, the definition of occurrences of terms, the definition of free occurrences of a variable, etc., may be found in section A.1 of the appendix.

We stated but did not prove that the "subterm of" relation is a partial order. Granted that reflexivity and transitivity are established, it remains to show that the relation is *antisymmetric*. This is the content of the next exercise.

Exercise 1.1.12. Prove that if M is a subterm of N and N is a subterm of M, then M is the same term as N.

Exercise 1.1.13. Complex CL-terms have more than one term that is their subterm. Define inductively an operation that gives *the set of subterms* for a term.

Exercise 1.1.14. Exercise 1.1.11 in effect asked you to list the elements of the set of subterms of seven terms. Give another definition of the set of subterms of a term using the "subterm of" relation.

1.2 Various kinds of combinators

Combinators are characterized by *the number of arguments* they can have (i.e., their *arity*) and by *the effect* they have when applied to as many arguments as their arity. The arity of a combinator is always *a positive integer*. For example, I is a unary combinator, whereas S takes three arguments.

The effect of the application of a combinator is given by its *axiom*. A combinatory axiom consists of two terms with a symbol (such as $\rhd$) inserted between them. The following table shows the axioms of some well-known combinators.[5]

$\mathsf{I}x \rhd x$	$\mathsf{B}xyz \rhd x(yz)$	$\mathsf{S}xyz \rhd xz(yz)$
$\mathsf{K}xy \rhd x$	$\mathsf{C}xyz \rhd xzy$	$\mathsf{W}xy \rhd xyy$
$\mathsf{M}x \rhd xx$	$\mathsf{B}'xyz \rhd y(xz)$	$\mathsf{J}xyzv \rhd xy(xvz)$

The terms on the left-hand side of the $\rhd$ tacitly indicate the arity of the combinator, because the combinator precedes as many variables as its arity. Notice also that these variables are not assumed to be the same. For instance, we have the term $\mathsf{W}xy$ (not $\mathsf{W}xx$) on the left-hand side of the $\rhd$ on the second line.

The term on the right-hand side in an axiom shows the term that results after the application of the combinator. In the resulting terms, some of the variables may be repeated, omitted, or moved around. All the combinators listed above (but not all combinators in CL) have the property that their application does not introduce new variables and the resulting term is built solely from some of the variables that appeared as their arguments. Such combinators are called *proper*. Some combinators that are *not proper* (i.e., *improper*), are extremely important, and we will return to them later. However, they are the exceptions among the improper combinators.

---

[5] B' is not a single letter, and this combinator is sometimes denoted by Q. B' is closely related to B; this explains the notation, which we retain because it is widely used.

Variables are used in logic when there is an operator binding them or there is a rule of substitution. CL contains *no variable binding operators*, which is one of the best features of CL; however, substitution is a rule. The combinatory axioms above gave a characterization of applications of combinators to variables only, but of course, we would like combinators to be applicable to arbitrary terms. *Substitution* ensures that this really happens, as we illustrate now.

Example 1.2.1. $\mathsf{I}(yz) \rhd yz$ is an *instance* of the axiom for I with yz substituted for x in both terms (that is, in $\mathsf{I}x$ and in x) in the axiom. Another instance of the same axiom is $\mathsf{I}(y(xz)) \rhd y(xz)$. (Notice that the presence of x in the term that is being substituted is unproblematic, and there is no danger of "circularity" here.)

We get yet another example by substituting I for x, M for y and z for z (or leaving z as it is) in the axiom for B. $\mathsf{BIM}z$ is the left-hand side term, and $\mathsf{I}(\mathsf{M}z)$ is the right-hand side term. The latter term can be viewed also as obtained from the left-hand side term in the axiom for I by substituting $\mathsf{M}z$ for x, which then gives $\mathsf{M}z$ on the right.

The last example shows that we may be interested in successive applications of combinators to terms. It also shows that substitution in distinct terms may result in the same term. Furthermore, the same term may be substituted for different variables; hence, the use of distinct variables in the axioms is not a restriction at all, but rather a way to ensure generality. The informal notion of substitution may be made precise as follows.

DEFINITION 1.2.2. (SUBSTITUTION) The *substitution* of a term M for the variable x in the term N is denoted by N_x^M, and it is inductively defined by (1)–(4).

(1) If N is x, then N_x^M is M, (i.e., x_x^M is M);

(2) if N is y and y is a variable distinct from x, then N_x^M is N, (i.e., y_x^M is y);

(3) if N is Z, then N_x^M is N, (i.e., Z_x^M is Z);

(4) if N is (P_1P_2), then N_x^M is $(P_{1x}^{\ M}P_{2x}^{\ M})$, (i.e., $(P_1P_2)_x^M$ is $(P_{1x}^{\ M}P_{2x}^{\ M})$).

Substitution is a *total operation* in CL. There is no restriction on the shape of N or on occurrences of x in N in the definition, though the result of the substitution, in general, will depend on what N looks like. Substitution can be summed up informally by saying that all the x's in N (if there are any) are turned into M's.

Substitution is a much more complicated operation in systems that contain a variable binding operator — such as λ-calculi or logics with quantifiers. Substitution was not very well understood in the early 20th century, and one of the motivations for combinatory logic — especially, in Curry's work in the 1930s — was the clarification of substitution. Another motivation — mainly behind Schönfinkel's work — was the elimination of bound variables from first-order logic. These two problems turn out to be one and the same, and the general solution in both cases leads to a *combinatorially complete* set of combinators. (We return to combinatorial completeness in the next section; see in particular definition 1.3.8 and lemma 1.3.9.)

Sometimes, it is convenient to substitute at once for more than one variable in a term. This operation is called *simultaneous substitution*. A simultaneous substitution can always be emulated by finitely many single substitutions, which means that simultaneous substitutions can be viewed as convenient notational devices. (However, see section A.1 for a more formal view of this operation.) We limit ourselves to a few examples.

Example 1.2.3. $(\mathsf{S}xyz)^{\mathsf{I},yy,\mathsf{J}}_{x,y,z}$ is $\mathsf{SI}(yy)\mathsf{J}$, that is, I is substituted for x, yy for y and J for z. A more revealing case is when one of the substituted terms contains a variable that is one of the variables for which a term is substituted simultaneously. For instance, $(\mathsf{J}xyzv)^{yy,\mathsf{W}}_{x,y}$ is the term $\mathsf{J}(yy)\mathsf{W}zv$ (rather than $\mathsf{J}(\mathsf{WW})\mathsf{W}zv$).

Other notations for the substitution of M for x include $[x/M]$ (and $[M/x]$) placed in front of or after the term on which the operation is performed. For instance, $[x/M]N$ or $[x_1/M_1, x_2/M_2]N$ are such notations, which may be clearer than N^M_x and $N^{M_1,M_2}_{x_1,x_2}$.[6] (We will use the "slash-prefix" notation too.)

Exercise 1.2.1. Write out the details of the following substitutions step-by-step. (a) $(\mathsf{I}x(\mathsf{W}x))^z_x$, (b) $[z/x(xy)]\mathsf{S}(xx)(yy)(xy)$, (c) $[x_1/\mathsf{M}, x_3/\mathsf{B}, x_4/\mathsf{W}](x_3x_1(x_3x_4x_3))$, (d) $(\mathsf{S}xyz)^{\mathsf{K},\mathsf{K},x}_{x,y,z}$, (e) $(\mathsf{C}xyM)^{PPN}_y$.

Substitution affects *all occurrences* of a *variable*. Another operation on terms affects *selected occurrences* of a *subterm* in a term, and it is called *replacement*. The result of replacing some subterms of a term by a term is, of course, a term. For our purposes, the visual identification of an occurrence of a subterm is sufficient — together with the remark that a completely rigorous numbering or identification of the subterms is possible.

Example 1.2.4. The result of the replacement of an occurrence of a subterm $\mathsf{SKK}x$ in $\mathsf{SKK}x$ by $\mathsf{I}x$ is $\mathsf{I}x$. (Recall that every term is its own subterm.) The result of replacement of the first occurrence (from left) of x in $\mathsf{KS}x(\mathsf{S}x)$ by M is $\mathsf{KS}M(\mathsf{S}x)$, whereas the replacement of the second occurrence of x yields $\mathsf{KS}x(\mathsf{S}M)$. This shows that *not all* occurrences of a subterm (even if that term happens to be a variable) need to be replaced. Lastly, the replacement of the second occurrence of BW in $\mathsf{BWBW}(\mathsf{BWBW})$ by SKC gives us $\mathsf{BWBW}(\mathsf{SKCBW})$. (This looks like, but in fact is not, a typo!)

In the first example, the term we started with and the term we ended up with are very closely related in a sense that will soon become clear. A similar relationship holds in the second example, but only accidentally — because of the occurrence of K in a specific spot in the term. The last two examples show no such relationship between the input and the output terms. Generally speaking, replacement induces

[6] There are still other ways to denote substitution. Occasionally, $\leftarrow$ or $\rightarrow$ is used instead of $/$. We do not intend to give an exhaustive list here, but you should be aware that the notation for substitution is anything but unvarying in the literature.

desirable relationships between input and output terms only if such relationships obtain between the replaced subterm and the newly inserted subterm.

The effect a combinator can have on a term may be thought of as *replacing* a term that is on the left-hand side of the $\triangleright$ by a term that has the form of the term on the right-hand side of the $\triangleright$. More precisely, such replacements can be performed *within* any terms according to instances of combinatory axioms.

The change that a combinator produces in a term is one of the most interesting criteria for classifying combinators. (We have already seen another grounds for categorization: the arity of combinators and the comparison of the set of atomic subterms of the terms on the left and on the right.)

Terms are "structured strings," that is, they typically contain parentheses that show grouping. Terms might differ with respect to the number of occurrences of some subterm, the order of subterms or whether a subterm occurs at all in a term.

There are *five salient features* that we use to categorize combinators.[7] The properties that are exemplified by I, B, K, C and M are the ones that are abstracted away and turned into defining properties of classes of combinators.

Identity combinators. The letter I is intended to remind us that I is a unary identity function. As such, the effect of an application of I is that we get back the input term. (To put it more scintillatingly, I simply "drops off" from the term.) Functions like I can be envisioned for any arity, and accordingly we call the n-ary combinator Z an *identity combinator* when its axiom is

$$(\ldots((\mathsf{Z}x_1)x_2)\ldots x_n) \triangleright (\ldots(x_1x_2)\ldots x_n).$$

The binary identity combinator may be taken to be BI, because $\mathsf{BI}yz \triangleright \mathsf{I}(yz)$ and $\mathsf{I}(yz) \triangleright yz$ are instances of the axioms for B and I. (We already hinted at that we might consider series of replacement steps according to combinatory axioms; we will make this precise in the next section when we will define reduction to be a transitive relation.)

Associators. Recall that the only alteration B introduces into a term is association to the right. A term comprising two atoms, such as two variables x, y or two constants S, C, can be grouped only in one way. Hence, B has the least arity (3) among associators. An n-ary combinator Z is an *associator* when Z's axiom is

$$\mathsf{Z}x_1\ldots x_n \triangleright M \text{ and } M \text{ is not a left-associated term.}$$

For example, M may be a term $(x_1\ldots x_n)$ with x_1 through x_n in the same order as on the left-hand side of the $\triangleright$, but with at least one subterm of the form x_iN, where $i \neq 1$. The following Z_1 and Z_2 are both quinary associators.

$$\mathsf{Z}_1x_1x_2x_3x_4x_5 \triangleright x_1x_2x_3(x_4x_5)$$

[7] We follow the usual classification of combinators (that may be found, e.g., in Curry and Feys [53]) with some slight modifications. A main difference is that we do not limit the use of labels such as "permutator" to regular combinators.

$$\mathsf{Z}_2 x_1 x_2 x_3 x_4 x_5 \rhd x_1(x_2 x_3 x_4 x_5)$$

There are five possible ways to insert parentheses into a term that consists of four atomic terms. Curiously, B by itself is sufficient to define three combinators that yield the terms $x_1x_2(x_3x_4)$, $x_1(x_2x_3x_4)$ and $x_1(x_2(x_3x_4))$ (when the arguments are x_1, x_2, x_3 and x_4 in that order), and an application of B yields $x_1(x_2x_3)x_4$. (Although B is a ternary combinator, we may form a term from the terms $\mathsf{B}x_1x_2x_3$ and x_4, that reduces to the desired term.) $\mathsf{B}(x_1x_2)x_3x_4 \rhd x_1x_2(x_3x_4)$ is an instance of B's axiom, and so is $\mathsf{BB}x_1x_2 \rhd \mathsf{B}(x_1x_2)$. Then BB, sometimes denoted as D, associates x_3 and x_4 — leaving x_1 and x_2 grouped together.

$$\mathsf{D}x_1x_2x_3x_4 \rhd x_1x_2(x_3x_4)$$

Exercise 1.2.2. Find combinators Z_3 and Z_4 that are composed of B's and have axioms $\mathsf{Z}_3x_1x_2x_3x_4 \rhd x_1(x_2x_3x_4)$ and $\mathsf{Z}_4x_1x_2x_3x_4 \rhd x_1(x_2(x_3x_4))$.

Cancellators. An application of the combinator K to two arguments, let us say x and y, results in the term x, that is, y gets omitted or cancelled. This feature may seem spurious at first. However, when K and S are chosen as the only constants, the potential to form terms with some subterms disappearing after an application of K is essential. In general, an n-ary combinator Z is called a *cancellator* when the axiom of Z is

$$\mathsf{Z}x_1\ldots x_n \rhd M \text{ with at least one of } x_1,\ldots,x_n \text{ having no occurrences in } M.$$

As another example of a cancellator, we could consider a combinator K^2 such that when it is applied to x and y the term y results. $\mathsf{KI}x \rhd \mathsf{I}$ is an instance of the axiom for K; hence, KI achieves the desired effect. We will see in chapter 3 that the capability to omit the first or the second argument that is exhibited by K and KI, as well as the definability of cancellators that retain only one of their n arguments is paramount, in the representation of recursive functions by combinators.

Permutators. The combinators C and S both change the order of some of their arguments. Recall that the right-hand side terms in their axioms are xzy and $xz(yz)$. In the former, y and z are swapped; in the latter, an occurrence of z precedes an occurrence of y. A combinator taking n arguments is a *permutator* when its axiom is of the form

$$\mathsf{Z}x_1\ldots x_n \rhd M \text{ with } M \text{ containing an } x_j \text{ preceding an } x_i \text{ (for } 1 \leq i < j \leq n).$$

Of course, M may contain several occurrences of either variable, but in the definition we only require the existence of at least one pair of occurrences with x_j coming first.

Exercise 1.2.3. Find a ternary combinator (composed of the constants already introduced) such that if applied to x, y and z it produces the term yzx, that is, "it moves its first argument behind the two others." This combinator is usually denoted by R and its axiom is $\mathsf{R}xyz \rhd yzx$. The effect of R can be produced by combining the combinators that we already have.

Exercise 1.2.4. Recall that the axiom for the combinator B' is $\mathsf{B}'xyz \rhd y(xz)$. First, define the combinator B' using B and C. Next, find a combinator V that "reverses" the effect of R, that is, an application of V to three arguments x, y and z yields zxy. (Hint: B', B and some permutators may be useful in the latter definition.)

Duplicators. The combinators W and M "copy" or "duplicate" some of their arguments. W's application results in two occurrences of its second argument in the output term, and M returns two copies of its (only) input. An n-ary combinator Z is a *duplicator* when the axiom for Z is

$$\mathsf{Z}x_1 \ldots x_n \rhd M \text{ and } M \text{ contains more than one occurrence of an } x_i \ (1 \leq i \leq n).$$

Notice that S is not only a permutator, but a duplicator too, because $xz(yz)$ has two occurrences of z. S is an excellent example of a combinator that combines various kinds of effects. Indeed, in the terminology of [53], the last four categories would be called "combinators with such-and-such effect." (Cf. footnote 7.)

The only category of combinators in which the shape of the resulting term is completely specified by our definitions is that of identities. Combinators that integrate various effects are useful, but it is also helpful to be able to dissect their blended effect into components. This will become transparent when we will delineate a subset of the set of the terms that can model arithmetic functions.

Exercise 1.2.5. Classify all the combinators introduced so far according to their effects. (Hint: Some combinators will appear in more that one category.)

Exercise 1.2.6. Define or find combinators (among the already introduced ones) that demonstrate each pair of properties that can appear together. (Hint: You need 6 combinators in total.)

We quickly mention two other classes of combinators now.

Regular combinators. Combinators may be thought of as affecting the *arguments* of a function. Let us assume that f is a binary function; then fxy is the application of f to x and y. $\mathsf{C}fxy$ takes f, x and y as arguments and yields fyx, that is, f applied to its arguments in reversed order. This motivates singling out combinators that keep their first argument "in the left-most place." These combinators are called *regular*, and the axiom for a regular combinator Z is

$$\mathsf{Z}x_1 \ldots x_n \rhd x_1 M.$$

The combinators I, K, S, W, C, B, J and D are all regular, whereas B' and R are not. Therefore, it is obvious by now that putting together regular combinators may give a combinator that is not regular. Because of the lack of preservation of regularity under application, we will not place an emphasis on regular combinators — despite the fact that they have an easy to understand informal meaning.

Proper combinators. We already mentioned this kind of combinators. Z is a *proper* combinator if its axiom is

$$\mathsf{Z}x_1 \ldots x_n \rhd M \text{ and } M \text{ contains no variables other than } x_1, \ldots, x_n.$$

Informally speaking, combinators that introduce a new variable are hardly sensible (because an arbitrary term can be substituted for that variable). On the other hand, some combinators that introduce a *constant* make perfect sense and are important (even though they are not proper). The archetypical combinator of this kind is the *fixed point* combinator, often denoted by Y.[8] The axiom for Y is $\mathsf{Y}x \rhd x(\mathsf{Y}x)$. If x is a function, then $\mathsf{Y}x$ is its fixed point when the axiom is viewed from right to left (or $\rhd$ is thought of as some sort of equality). That is, by replacing $\mathsf{Y}x$ by z, x by f and $\rhd$ by $=$ we get $fz = z$.

Exercise* 1.2.7. Find a term built from some of the combinators mentioned so far (save Y) that has the effect of Y.

We introduced some combinators as constants at the beginning of this section. These are the *primitive combinators* (also called *undefined combinators*) in our system. However, we have used the term "combinator" informally to refer to complex terms as well. We make this use official now. A complex term that is built entirely from (primitive) combinators is called a *combinator* too.

The solution of the previous exercise shows that combining primitive proper combinators does not always preserve the property of being proper. Indeed, it is quite easy to generate improper combinators this way.

Example 1.2.5. C and I are both proper combinators, but CII is not. This particular combinator is interesting in the context of typed systems, because its principal type schema is $((A \to A) \to B) \to B$ (i.e., *specialized assertion*), which is a theorem of classical propositional logic and a characteristic axiom of the logic of entailment.

1.3 Reductions and combinatory bases

The understanding of the structure of CL-terms as well as of some of their other syntactic features is vital to grasping CL. However, CL is more about *relations* between terms than about the terms themselves. We have used phrases such as "combinators cause changes in terms," and such changes may be characterized by *pairs of terms*: the first term is the term in which the combinator is applied and the second term is the term that results. We can describe the axioms this way too, and then define one-step reduction.

DEFINITION 1.3.1. If Z is an n-ary combinator, then a term of the form $\mathsf{Z}M_1 \ldots M_n$ is a *redex*. The *head* of this redex is Z, and $M_1, \ldots, M_n$ are its *arguments*.

A redex may contain one or more combinators, because the terms that instantiate the meta-variables $M_1, \ldots, M_n$ may or may not have combinators and redexes in

[8]This combinator is sometimes labeled as "fixpoint" or "paradoxical" combinator. We hasten to point out that pure CL is consistent, and so the latter name might seem to cozen.

them. Also, Z may occur several times in the term, but it surely occurs at least once. That is, the definition is to be understood to mean that the particular occurrence of Z that is apparent in the definition of a redex is the head of the redex.

DEFINITION 1.3.2. (ONE-STEP REDUCTION) Let Z be a primitive combinator with axiom $\mathsf{Z}x_1 \ldots x_n \rhd P$. If N is a term with a subterm of the form $\mathsf{Z}M_1 \ldots M_n$, and N' is N with that subterm replaced by $[x_1/M_1, \ldots, x_n/M_n]P$, then N *one-step reduces* to N'. This is denoted by $N \rhd_1 N'$.

The definition is not as complicated as it may first seem. We already (tacitly) used $\rhd_1$ in examples to illustrate various types of combinators. When we said that $\mathsf{I}(yz)$ yields yz, we applied one-step reduction. $\mathsf{I}M$ is a subterm of $\mathsf{I}(yz)$ with M standing for yz. In I's axiom, x (on the right-hand side of the $\rhd$) plays P's role in the above definition, and $[x/yz]x$ is yz. Thus $\mathsf{I}(yz) \rhd_1 yz$ — it is really quite straightforward.

Example 1.3.3. The term $\mathsf{BWB}x(\mathsf{BWB}x)$ contains two redexes. The head of one of them is the first occurrence of B (counting from left to right); the head of the other is the third occurrence of B. Having chosen the second redex for one-step reduction, we get $\mathsf{BWB}x(\mathsf{BWB}x) \rhd_1 \mathsf{BWB}x(\mathsf{W}(\mathsf{B}x))$. The latter term contains one redex only, and after one-step reducing that, we get $\mathsf{W}(\mathsf{B}x)(\mathsf{W}(\mathsf{B}x))$. (This term also has a redex, because the one-step reduction created a new redex.)

Exercise* 1.3.1. Consider the term S(KS)KSKS. Perform as many one-step reductions as possible in every possible order. (Hint: Write out the one-step reductions as sequences of terms.)

One-step reduction may yield a term with no redexes, and it may yield a term with one or more remaining or new redexes. A one-step reduction of a duplicator may increase the number of redexes — including the number of occurrences of some already existing redexes. In other words, the number of successive one-step reductions counts the number of reduced redexes, but this number does not necessarily equal the number of redexes in the starting term minus the number of redexes in the resulting term. The term $\mathsf{BWB}x(\mathsf{BWB}x)$ (from example 1.3.3) has two redexes. Once the first redex from the left is reduced, the resulting term $\mathsf{W}(\mathsf{B}x)(\mathsf{BWB}x)$ contains a brand new redex headed by W.

Exercise 1.3.2. For each of the categories of combinators introduced in the previous section, consider terms in which one of the redexes M has a head belonging to that category, and all the other redexes are wholly inside arguments of the head of M. Work out how the number of occurrences of the already existing redexes changes with the one-step reduction of M.

Exercise* 1.3.3. Example 1.3.3 showed that one-step reduction may create *new redexes* — not merely *new occurrences* of already existing redexes. Can combinators from each category create new redexes? (Hint: If "yes," then describe the shapes of suitable terms before and after $\rhd_1$.)

The sample one-step reductions we performed on $\mathsf{BWB}x(\mathsf{BWB}x)$ show that redexes need not be reduced in the same order as they occur in a term from left to right. This suggests that we may define *reduction strategies* or *reduction orders* to have more control over the reduction process. One special order for reductions — proceeding from left to right — has a theoretical importance in connection to normal forms, as well as a practical significance in thinking about CL as modeling computation. Sequential programming languages force a linear order upon computational steps; hence, the implicitly nondeterministic view projected by definition 1.3.2 may not be always tenable.

One-step reductions surely entice us to contemplate successive one-step reductions. Indeed, given all the combinators with their diverse effects, what we are really interested in is the compound outcome of their repeated applications.

DEFINITION 1.3.4. (WEAK REDUCTION) The reflexive transitive closure of the one-step reduction relation is *weak reduction*, and it is denoted by $\triangleright_{\mathrm{w}}$.

If R is a binary relation on a set A, then the *reflexive closure* of R, contains all pairs of elements of A of the form $\langle a,a\rangle$, for all $a \in A$. Accordingly, the weak reduction relation holds between every CL-term and itself: $M \triangleright_{\mathrm{w}} M$. A term that has no redexes does not one-step reduce to any term, yet it weakly reduces to itself. Terms M such that $M \triangleright_1 M$ are rare and exceptional. One example is MM.

If R is a binary relation on a set A, then R^+, the transitive closure of R includes all pairs $\langle a,c\rangle$ in which the second element is "accessible" from the first one via finitely many R-steps. If $Rab_1,\ldots,Rb_nc$ for some $b_1,\ldots,b_n$, then R^+ac. For CL-terms, this means that $M \triangleright_1 N \triangleright_1 P$ yields that M weakly reduces to P, that is, $M \triangleright_{\mathrm{w}} P$. Zero intermediate steps (i.e., $n=0$) are allowed too, which means that any one-step reduction is a weak reduction: $M \triangleright_{\mathrm{w}} N$ if $M \triangleright_1 N$.

The reflexive transitive closure of R is usually denoted by R^*. Thus $\triangleright_{\mathrm{w}}$ is defined as $\triangleright_1^*$, but the usual notation is $\triangleright_{\mathrm{w}}$.

The idea that every function can be viewed as unary is not reflected by weak reduction, and that is why this relation is called "weak." Neither the axioms for combinators nor one-step reduction gives any hint as to how to compute with a term in which a combinator is followed by fewer terms than its arity. For instance, we have no clue how to apply the axiom for S to a term of the form $\mathsf{S}MN$. We will return to this issue later on and introduce other reductions. Weak reduction is the most naturally arising reduction in CL, and $\triangleright_{\mathrm{w}}$ takes a central place in the theory.

In arithmetic, there is no need to worry about the order in which the arithmetic operations are calculated — as long as the groupings within the term are respected. For instance, $(5+7)\cdot(2+3)$ may be calculated stepwise as $12\cdot(2+3)$ or $(5+7)\cdot 5$, then the next step in both cases is $12\cdot 5$, and the final result is 60. There are many other ways to calculate the final result if one throws in applications of various identities (or "algebraic laws"), such as distributivity of multiplication over addition. Then $(5+7)\cdot 5=(5\cdot 5)+(7\cdot 5)$, and further $25+(7\cdot 5)=25+35$, etc. But our concern was simply the order of performing arithmetical operations that are explicit in an expression.

One might wonder whether the CL-terms exhibit similar behavior. The short answer is that they do. If a term N weakly reduces to P_1 and P_2, then there is always a term M to which P_1 and P_2 both weakly reduce. This property is called the *confluence of weak reduction*. We will return to this result, which is called the Church–Rosser theorem, as well as to its proof, in the next section. The notion of a normal form would make sense without the confluence of weak reduction. However, the Church–Rosser theorem means that CL-terms are well-behaved; hence, normal forms are even more important.

DEFINITION 1.3.5. (WEAK NORMAL FORM) A term M is in *weak normal form*, in *wnf* (or in *nf*, for short), iff M contains no redexes.

Weak reduction is reflexive, that is, $M \rhd_{\mathrm{w}} M$ is always a possible "further" step in a sequence of weak reductions. One-step reduction is obviously not reflexive, and it is useful to have a reduction — weak reduction — that is not just transitive, but reflexive too. However, the steps that are justified by reflexivity may be likened to having a "skip" step that can be inserted into chains of calculations at any point. Thus the above definition delineates a subset of terms that do not yield a term under $\rhd_1$.

Some obvious examples of terms in nf include the variables and the primitive combinators, each considered by itself. Some complex terms are in weak nf too. $x\mathsf{I}$ and SK are in nf, just as $\mathsf{B(WI)(BWB)}$ is. It is immediate that there are denumerably many terms that are in nf, and there are denumerably many combinators that are in nf.

Exercise 1.3.4. Prove — without assuming an infinite set of primitive constants — that there are infinitely many ($\aleph_0$ -many) combinators that are in nf. (Hint: You may assume that at least S and K are among the primitives.)

Exercise 1.3.5. Consider the combinators I, B, S, K, C, W, M, $\mathsf{B'}$, J and D to be primitive. Is there a subset of these combinators that does not yield infinitely many combinators in nf?

There are terms that in one or more steps reduce to themselves. Let us note first that $\mathsf{II} \rhd_{\mathrm{w}} \mathsf{I}$, which is nearly II except that a copy of I is missing from the term. Replacing the identity combinator with the duplicator M we get the term MM, and $\mathsf{MM} \rhd_1 \mathsf{MM}$ — as we already noted above. Another example involves the combinator W in the term WWW. In general, any combinator that produces a left-associated term in which all arguments occur exactly once, except that one of the arguments is duplicated, can be used (possibly, together with I) to create at least one term that reduces to itself. Such a term is $\mathsf{WI(WI)}$, which produces a cycle via one-step reductions: $\mathsf{WI(WI)} \rhd_1 \mathsf{I(WI)(WI)} \rhd_1 \mathsf{WI(WI)}$. We have taken into account all the redexes at each step. Therefore, $\mathsf{WI(WI)} \rhd_{\mathrm{w}} \mathsf{WI(WI)}$ and $\mathsf{WI(WI)} \rhd_{\mathrm{w}} \mathsf{I(WI)(WI)}$, and there is no other term to which $\mathsf{WI(WI)}$ reduces. Of course, $\mathsf{I(WI)(WI)} \rhd_{\mathrm{w}} \mathsf{I(WI)(WI)}$, which is another term built from W and I such that it reduces to itself.

Exercise* 1.3.6. Consider duplicators that yield a left-associated term with one argument being duplicated and all other arguments occurring exactly once. Prove that all such duplicators (just as M and W above) are suitable to generate terms that reduce to themselves. (Hint: First, give a general description of the shape of the resulting term from the axiom of such combinators.)

These examples show that a term may reduce to itself via one or more one-step reductions. To emphasize again, $\triangleright_1$ is not a reflexive relation. However, there is a CL-term M (e.g., when M is in the set of combinatory constants) such that $M \triangleright_1 M$. If the set of combinators is combinatorially complete (or at least includes M or W, etc.), then there is a CL-term M such that $M \triangleright_{\mathrm{w}} M$ via a nonempty sequence of $\triangleright_1$ steps. This means that another — but equivalent — way to characterize weak normal forms would be to say that M is in *weak normal form* iff there is no term N, which may be the same as M, such that $M \triangleright_1 N$.

The above examples showed that MM $\triangleright_1$ MM, but not WI(WI) $\triangleright_1$ WI(WI), though WI(WI) $\triangleright_{\mathrm{w}}$ WI(WI). Of course, M can be defined as WI, and now we make this concept more precise.

DEFINITION 1.3.6. Let Z be a combinator with axiom $\mathsf{Z}x_1 \ldots x_n \triangleright M$. The CL-term Z', which is built from (primitive) combinators, *defines* Z when $\mathsf{Z}'x_1 \ldots x_n \triangleright_{\mathrm{w}} M$.

This definition involves $\triangleright$ and $\triangleright_{\mathrm{w}}$, which suggests that there may be a difference in one-step reduction sequences when we take a combinator as undefined or when we build it up (i.e., define it) from other primitives.

Exercise 1.3.7. (a) Define B and C from S and K (only). (b) Define S from B, W, C and I.

Exercise* 1.3.8. Show that J is definable from I, B, W and C, and the latter three are definable from the first two (i.e., J and I).

It is obvious that the combinator K is *not definable* from the combinators I, B, S, C, W, M, B′, J and Y. This suggests that it is important to make clear which combinators we assume to be available for us.

Exercise* 1.3.9. Prove informally (or by structural induction) that K is not definable from the other combinators mentioned in the previous paragraph.

A set of combinators determines a set of combinators that are either in the set or definable from it, which makes the next definition useful.

DEFINITION 1.3.7. Let $\mathfrak{B}$ be a finite nonempty set of combinatory constants. $\mathfrak{B}$ is called a *basis*.

The set of CL-terms, given a basis, is defined as in definition 1.1.1, where the constants are those in the basis. Exercise 1.3.7 asked you to show that B and C are definable by CL-terms over the basis $\{\mathsf{S},\mathsf{K}\}$, whereas exercise 1.3.8 asked you to show the *equivalence* of the combinatory bases $\{\mathsf{I},\mathsf{J}\}$ and $\{\mathsf{I},\mathsf{B},\mathsf{C},\mathsf{W}\}$. Just as expected, two combinatory bases are equivalent iff all the combinators definable from one are definable from the other. Defining the primitive combinators of the other basis (in both directions) is obviously sufficient to prove the equivalence of two bases. Clearly, not all bases are equivalent, because K is not definable from { I, J }, for instance.

The set of all possible combinators is infinite. To substantiate this claim we could simply consider all the identity combinators I^n (for all $n \in \mathbb{N}$) — there are $\aleph_0$-many of them. An interesting question to ask is whether there is a basis, preferably containing some "natural" or "not-too-complicated" combinators, that allows us to build up all the possible combinators there are.

DEFINITION 1.3.8. A combinatory basis is *combinatorially complete* iff for any function f such that $fx_1 \ldots x_n \triangleright_{\mathrm{w}} M$, where M is a CL-term built from some of the variables $x_1, \ldots, x_n$, there is a combinator Z in the basis or in the set of combinators definable from the basis such that $\mathsf{Z}x_1 \ldots x_n \triangleright_{\mathrm{w}} M$.

f may be thought of as if it were a proper combinator itself. However, combinatorial completeness could be defined more broadly to include certain types of improper combinators too. For example, occurrences of any proper combinator in M may be allowed. In fact, a fixed point combinator is definable from any basis that is combinatorially complete in the sense of the above definition. In fact, in the definition, M could be allowed to contain occurrences of f itself. The main purpose of the limitation to proper combinators in the definition is to exclude from our consideration all the weird improper combinators that introduce new variables into their reduct (i.e., into the term that results from an application of the combinator to sufficiently many variables). We will see in chapter 2 that a fixed point combinator is definable from S and K, which will underpin more firmly our thinking about combinatorial completeness as having a wider class of functions definable. (Cf. example 2.3.4, as well as the exercises 2.3.2 and 2.3.5.)

LEMMA 1.3.9. *The combinatory bases* {S, K}, {I, B, C, W, K} *and* {I, J, K} *are* combinatorially complete.

Proof: First, we outline the proof that the first basis is combinatorially complete. Then we collect the facts that show that the latter two bases are equivalent to the first.
1. The idea behind showing that S and K suffice is to define a pseudo-λ-abstraction. This is usually called *bracket abstraction* or λ^*-abstraction.[9] There are many ways to define a λ^* operation, and in chapter 4, we will look at some of those. The way the λ^* is defined determines its features; hence, it affects the properties of the transition from the λ-calculus to CL. At this point our goal is merely to show that any function, in the sense of definition 1.3.8, can be simulated by a combinator over the { S, K } basis.

We introduce for the purpose of talking about functions such as f the notation $[x].M$. The $[x]$ is a meta-language λx, and the [ ]'s around a variable motivate the

[9] Schönfinkel's work preceded the formulation of the notion of an algorithm in a precise sense. Nevertheless, the steps in the elimination of bound variables described in his paper may be seen to amount to a bracket abstraction algorithm.

name "the bracket abstraction." M may or may not contain an occurrence of x, and $[x].M$ is interpreted as the function that returns M_x^N when applied to N. (M_x^N is M itself if $x \notin \mathrm{fv}(M)$.[10])

If f is an n-ary function, and so $f(x_1,\ldots,x_n) = M$, where M is a term over the set of variables $\{x_1,\ldots,x_n\}$, then to find a suitable combinator we iteratively form $[x_1]\ldots[x_n].f(x_1,\ldots,x_n)$, which is just f, and on the other side of the equation we form $[x_1]\ldots[x_n].M$. We apply the algorithm below starting with x_n and M. The shape of the resulting combinatory term depends on the concrete M, hence, as a general notation, we denote the result by $[x_n].M$. If $n > 1$, then the next variable is x_{n-1} and the term to which the algorithm is applied is $[x_n].M$, etc. (Because of the structure of the algorithm together with the finiteness of CL-terms, the recursion is well-founded.)

(1) $[x_i].M'$ is SKK, if M' is x_i;

(2) $[x_i].M'$ is $\mathsf{K}M'$, if x_i does not occur in M' (i.e., $x_i \notin \mathrm{fv}(M')$);

(3) $[x_i].M'$ is $\mathsf{S}([x_i].N)([x_i].P)$, if M' is NP.

We think of clauses (1)–(3) as numbered and ordered. If there are two clauses that are both applicable, then the earlier one is applied. Incidentally, (1) and (2) are never both applicable to a CL-term, and the same is true of (1) and (3). However, (3) could be applied to SKK, for example, giving $\mathsf{S}([x_i].\mathsf{SK})([x_i].\mathsf{K})$. If we do not want to rely on the order of the clauses, then $x_i \in \mathrm{fv}(NP)$ may be added as a proviso to (3).

After each of $x_1,\ldots,x_n$ has been abstracted, we have $[x_1]\ldots[x_n].M$, which is the combinator for f. To show that this is really the term that can be taken for f, we show by induction that each subterm behaves as desired.

If the subterm is x_i, and this is the variable that is being abstracted, then after an application to the term N we have to get N, and $\mathsf{SKK}N \rhd_{\mathrm{w}} N$. The other base case is when M' does not contain x_i. Then the function should return M' itself for any N, and $\mathsf{K}M'N \rhd_{\mathrm{w}} M'$. If x_i occurs in N or P when the term is (NP), then by the hypothesis of the induction, we assume that $[x_i].N$ and $[x_i].P$ are the CL-terms corresponding to N's and P's x_i abstracts. Then by applying $\mathsf{S}([x_i].N)([x_i].P)$ to Q, we get $([x_i].NQ)([x_i].PQ)$, as desired.

2. First, we note that the latter two bases are equivalent due to the results in exercise 1.3.8. Supplementing exercise 1.3.7 with two definitions, namely, SKK for I and $\mathsf{S(CI)}$ for W, the equivalence of the first two bases follows. S and K can define any f, and the combinators in the two other bases suffice too. qed

Definition 1.3.6 made precise the notion of defining a combinator from other combinators. The above lemma then implies that any *proper combinator* is definable

[10]We are cheating here a bit, because we have not yet defined substitution for λ-abstraction-like expressions. However, $[x].M$ eventually will turn out to be a combinator, so this small gap is harmless. — We use the usual notation fv to denote the set of free variables of a term. See definition A.1.12 in the Appendix.

from the basis $\{\mathsf{S},\mathsf{K}\}$. The combinatorial completeness of this basis and clear informal interpretation of S and K explain why these are the best-known combinators.

You have surely discovered that B is definable as $\mathsf{S}(\mathsf{KS})\mathsf{K}$ (which is a solution to a question in exercise 1.3.7). Now we look at the above procedure to find a definition of B, where we think of this combinator as the function f. (We take x, y and z to be x_1, x_2 and x_3, that is, $\mathsf{B}x_1x_2x_3 \rhd x_1(x_2x_3)$, but we continue to write x,y,z for the sake of brevity.)

Example 1.3.10. The resulting term is $x(yz)$, and in the first round, we want to find the CL-term for $[z].x(yz)$. $z \in \mathrm{fv}(x(yz))$, so (2) is not applicable, but neither is (1), because the term is not simply z. Then by (3), we get $\mathsf{S}([z].x)([z].yz)$. However, we are not finished yet, because there are two $[z]$'s left. $[z].x$ turns into $\mathsf{K}x$, by (2), and $[z].yz$ yields $\mathsf{S}([z].y)([z].z)$. Having put the pieces together, we obtain $\mathsf{S}(\mathsf{K}x)(\mathsf{S}([z].y)([z].z))$. Now we have gotten two $[z]$'s again, but, of course, now they are prefixed to different subterms than before. The finiteness of the subterms terminates the abstraction of $[z]$, since $[z].y$ gives $\mathsf{K}y$ and $[z].z$ gives SKK, by (2) and (1), respectively. Having gathered the bits together, we can see that the meta-term $[z].x(yz)$ refers to the CL-term $\mathsf{S}(\mathsf{K}x)(\mathsf{S}(\mathsf{K}y)(\mathsf{SKK}))$.

Before we turn to the next abstraction, it may be helpful to emphasize that $\mathsf{S}(\mathsf{K}x)(\mathsf{S}(\mathsf{K}y)(\mathsf{SKK}))$ contains *no occurrences* of z or of $[z]$. Of course, $[z]$ is a metalanguage λ, and so $[z].x(yz)$ contains only bound occurrences of z — both in $[z]$ and in the term $x(yz)$. Another name for bound variables is *apparent* variables (as distinguished from *real* variables). The CL-term $\mathsf{S}(\mathsf{K}x)(\mathsf{S}(\mathsf{K}y)(\mathsf{SKK}))$ makes the "nonreal" character of z transparent.

The next round is to find $[y][z].x(yz)$, where $[z].x(yz)$ is the pure CL-term that we already have: $\mathsf{S}(\mathsf{K}x)(\mathsf{S}(\mathsf{K}y)(\mathsf{SKK}))$. That is, we want to find the CL-term that the meta-term $[y].\mathsf{S}(\mathsf{K}x)(\mathsf{S}(\mathsf{K}y)(\mathsf{SKK}))$ denotes. Now we list — without detailed justifications — the meta terms that successively result, and finally the CL-term itself.

(1) $\mathsf{S}([y].\mathsf{S}(\mathsf{K}x))([y].\mathsf{S}(\mathsf{K}y)(\mathsf{SKK}))$

(2) $\mathsf{S}(\mathsf{K}(\mathsf{S}(\mathsf{K}x)))(\mathsf{S}([y].\mathsf{S}(\mathsf{K}y))([y].\mathsf{SKK}))$

(3) $\mathsf{S}(\mathsf{K}(\mathsf{S}(\mathsf{K}x)))(\mathsf{S}(\mathsf{S}([y].\mathsf{S})([y].\mathsf{K}y))(\mathsf{K}(\mathsf{SKK})))$

(4) $\mathsf{S}(\mathsf{K}(\mathsf{S}(\mathsf{K}x)))(\mathsf{S}(\mathsf{S}(\mathsf{KS})(\mathsf{S}([y].\mathsf{K})([y].y)))(\mathsf{K}(\mathsf{SKK})))$

(5) $\mathsf{S}(\mathsf{K}(\mathsf{S}(\mathsf{K}x)))(\mathsf{S}(\mathsf{S}(\mathsf{KS})(\mathsf{S}(\mathsf{KK})(\mathsf{SKK})))(\mathsf{K}(\mathsf{SKK})))$

The CL-term in (5) is $[y][z].x(yz)$. The next round is to find $[x][y][z].x(yz)$, where $[y][z].x(yz)$ is the just-mentioned CL-term.

Exercise 1.3.10. Finish the example, that is, find the CL-term denoted by the metaterm $[x].\mathsf{S}(\mathsf{K}(\mathsf{S}(\mathsf{K}x)))(\mathsf{S}(\mathsf{S}(\mathsf{KS})(\mathsf{S}(\mathsf{KK})(\mathsf{SKK})))(\mathsf{K}(\mathsf{SKK})))$. (Hint: You should finally get a CL-term with 14 occurrences of S and 17 occurrences of K.)

It is already clear from the example, but even more from the exercise, that the term that results by the algorithm in the proof of lemma 1.3.9 is not $\mathsf{S}(\mathsf{KS})\mathsf{K}$. We could obtain that term along the lines of the algorithm, if we would take some *shortcuts*, for instance, simply replacing $[z].yz$ by y. Recall that CL-terms are interpreted as functions. $z \notin \mathrm{fv}(y)$, therefore, y as a function followed by an argument is yN, and $[z].yz$ followed by an argument is $([z].yz)N$, which we stipulated (by a meta-language β-reduction) to be $[z/N]yz$, which is yN. Having applied the above algorithm, we would get the CL-term $\mathsf{S}(\mathsf{K}y)(\mathsf{SKK})$. Once this term is applied to N, we have a redex; thus, $\mathsf{S}(\mathsf{K}y)(\mathsf{SKK})N \triangleright_w \mathsf{K}yN(\mathsf{SKK}N)$. The two new redexes yield y and N, respectively. In sum, $\mathsf{S}(\mathsf{K}y)(\mathsf{SKK})N \triangleright_w yN$ too.

With this insight, we want to find again $[x][y][z].x(yz)$. First, we get $[x][y].\mathsf{S}([z].x)([z].yz)$; then we apply (1) and our shortcut to get $[x][y].\mathsf{S}(\mathsf{K}x)y$. The next step is another shortcut step, leading to $[x].\mathsf{S}(\mathsf{K}x)$. Then we get $\mathsf{S}([x].\mathsf{S})([x].\mathsf{K}x)$, and by (2) and a shortcut, we get $\mathsf{S}(\mathsf{KS})\mathsf{K}$.

The step that we labeled "shortcut" here will be called η in the λ-calculus in chapter 4. η may be thought of as an *extensionality* step, because we do not distinguish between $[x].Mx$ and M itself, when $x \notin \mathrm{fv}(M)$.

An informal rendering of the same abstraction process, which incorporates some heuristics about taking into account the effect of the combinators S and K, goes like this. We try to abstract z, and we know that $\mathsf{S}MNz \triangleright_w Mz(Nz)$. We could prefix S to the term $x(yz)$ if x would be followed by z too, as in $xz(yz)$. But to insert a dummy variable, we can use K, because $\mathsf{K}Mz \triangleright_w M$. That is, $\mathsf{K}xz(yz) \triangleright_w x(yz)$, and $\mathsf{S}(\mathsf{K}x)yz \triangleright_w \mathsf{K}xz(yz)$. $\mathsf{S}(\mathsf{K}x)yz$ is a left-associated term, and z is in the last argument place, so as the next move we just drop it. But the same is true for $\mathsf{S}(\mathsf{K}x)y$ with respect to y — we drop that variable too.

$\mathsf{S}(\mathsf{K}x)$ is very similar to $x(yz)$, if we look at x as S, y as K and z as x. So we might take instead $\mathsf{KS}x(\mathsf{K}x)$ (since $\mathsf{KS}x(\mathsf{K}x) \triangleright_w \mathsf{S}(\mathsf{K}x)$), and then $\mathsf{S}(\mathsf{KS})\mathsf{K}x$, and finally we drop x.

The example and the exercise show that the algorithm is not guaranteed to produce the shortest possible definition for a function, but it *surely produces one*. In fact, the above algorithm typically yields quite a lengthy term. Shorter terms may be obtained by including η, as well as by expanding the combinatory basis and replacing (3) with several other more refined clauses. (See chapter 4 for details.)

In arithmetic and in elementary mathematics, there is a weaker relation (than computing the value of a function) that is commonly used, namely, *equality*. For an illustration, let us consider $2^5 \cdot 3^5$. Performing the exponentiations first, the expression computes to $32 \cdot 243$, which further yields $7,776$. However, instead of $2^5 \cdot 3^5 \mapsto 7,776$ (which is, probably, not even a standard or widely used notation), simply $2^5 \cdot 3^5 = 7,776$ is written. Sometimes, for instance, for encryption protocols to work, it is interesting that given a positive integer — such as $7,776$ — that number can be factored into primes.[11] As the numerical example suggests, we can consider the *converse* of weak reduction, possibly, combined with $\triangleright_w$ itself.

[11] There is a certain similarity between the lemma 1.3.9 and the prime factorization theorem in the sense that both yield an expression that computes to the given term or number. A dissimilarity is that there are only two primitive combinators (S and K), whereas there are infinitely many primes.

DEFINITION 1.3.11. (ONE-STEP EXPANSION) If $M \triangleright_1 N$, then $[N/M]P$, that is, the replacement of an occurrence of N in P by M is a *one-step expansion* of P. We will use the notation $P\,{}_1\!\triangleleft\,[N/M]P$ for one-step expansion.

Example 1.3.12. The CL-term xx can be one-step expanded to any of the following terms. $\mathsf{I}xx$, $x(\mathsf{I}x)$, $\mathsf{I}(xx)$, $\mathsf{M}x$, $\mathsf{K}xyx$, $\mathsf{K}(xx)y$, $\mathsf{K}xxx$, etc.

It might seem that expansion is completely arbitrary, but of course, it is not. The apparent arbitrariness is the result of the disappearance of the head of the reduced redex. $\triangleright_1$ leaves few clues about the redex that is gone; hence, the converse step allows much freedom in creating a redex.

DEFINITION 1.3.13. (WEAK EXPANSION) The reflexive and transitive closure of one-step expansion is the *weak expansion* relation on the set of CL-terms, which is denoted by ${}_{\mathrm{w}}\!\triangleleft$.

Exercise* 1.3.11. Prove that weak expansion is the converse of weak reduction. That is, as we have suggested by the choice of notation, $M \triangleright_{\mathrm{w}} N$ iff $N\,{}_{\mathrm{w}}\!\triangleleft\,M$.

Equality on numbers and other equality-like relations are *equivalence relations*, that is, they are *reflexive*, *transitive* and *symmetric*. The relation $\triangleright_{\mathrm{w}}$ is, obviously, not symmetric; neither is weak expansion symmetric.

DEFINITION 1.3.14. (WEAK EQUALITY) The transitive reflexive symmetric closure of $\triangleright_1$, the one-step reduction relation is *weak equality*, which is denoted by $=_{\mathrm{w}}$.

Weak equality may be inductively characterized by (1)–(4).

(1) If $M \triangleright_1 N$, then $M =_{\mathrm{w}} N$;

(2) if M is a CL-term, then $M =_{\mathrm{w}} M$;

(3) if $M =_{\mathrm{w}} N$, then $N =_{\mathrm{w}} M$;

(4) if $M =_{\mathrm{w}} N$ and $N =_{\mathrm{w}} P$, then $M =_{\mathrm{w}} P$.

The last clause may be put succinctly as $=_{\mathrm{w}}^{+} \subseteq =_{\mathrm{w}}$, where $^{+}$ denotes the transitive closure of a binary relation.

Exercise 1.3.12. Verify that $=_{\mathrm{w}}$ is the transitive symmetric closure of $\triangleright_{\mathrm{w}}$. (Hint: $=_{\mathrm{w}}$ above is characterized starting with $\triangleright_1$ in (1), and so is $\triangleright_{\mathrm{w}}$ in definition 1.3.4.)

Weak reduction is a stronger relation than weak equality in the sense that there are ordered pairs of terms that belong to $=_{\mathrm{w}}$ but not to $\triangleright_{\mathrm{w}}$. For example, $\langle x, \mathsf{K}xy\rangle$ is in the $=_{\mathrm{w}}$ relation (i.e., $x =_{\mathrm{w}} \mathsf{K}xy$), but not in the $\triangleright_{\mathrm{w}}$ relation (i.e., $x \not\triangleright_{\mathrm{w}} \mathsf{K}xy$). The relationship between the binary relations we have so far on the set of CL-terms is

$$\triangleright_1 \subsetneq \triangleright_{\mathrm{w}} \subsetneq =_{\mathrm{w}} .$$

We have shown that both inclusions are proper (as the symbol $\subsetneq$ indicates), but we might wonder now whether we have too many pairs included in $=_{\mathrm{w}}$. Perhaps, all CL-terms are weakly equal to each other; that is, $=_w$ is the total relation on the set of CL-terms. In fact, not all terms are weakly equal, and this is what *consistency* means for CL. We will prove this and other consistency theorems in chapter 2.

Not only is the weak equality relation not the total relation, but we can keep adding pairs of terms to it. We briefly mentioned η, which we justified by $\mathsf{S}(\mathsf{K}M)(\mathsf{SKK})N$ weakly reducing to MN. In effect, the shortcut allowed us to identify M and $\mathsf{S}(\mathsf{K}M)(\mathsf{SKK})$ by an appeal to what happens when these terms are applied to a term N. In a similar fashion, definition 1.3.6 says that a combinator Z is defined by a (compound) combinator Z' if the application of Z' to sufficiently many arguments yields the same term as the application of Z (to the same arguments) does. As yet another example, we may consider SKK and SKS. The two terms neither are the same nor weakly reduce to each other (or to any other term, for that matter) — they are in wnf. However, $\mathsf{SKK}x \rhd_w x$ and $\mathsf{SKS}x \rhd_w x$. (It is not difficult to see that any other combinator would do in the place of the second occurrence of S, without affecting the result of weak reduction.)

DEFINITION 1.3.15. (EXTENSIONAL WEAK EQUALITY, 1) $[M_1/N_1,\ldots,M_m/N_m]R$ and R are *extensionally weakly equal*, denoted by $[M_1/N_1,\ldots,M_m/N_m]R =_{w\zeta} R$, iff for each pair of terms M_i and N_i (where $1 \leq i \leq m \in \mathbb{N}$), there is an n_i (where $n_i \in \mathbb{N}$) such that (1) holds.

(1) For all terms $P_{i,1},\ldots,P_{i,n_i}$, there is a Q_i such that $M_iP_{i,1}\ldots P_{i,n_i} \rhd_w Q_i$ and $N_iP_{i,1}\ldots P_{i,n_i} \rhd_w Q_i$.

Extensionality means that the shapes of the CL-terms, or the exact reduction steps are disregarded — as long as the end result remains the same. The name for this equality is a bit lengthy, and perhaps even cumbersome; thus, we will tend to use $=_{w\zeta}$ instead. The reason to use yet another Greek letter is that, roughly speaking, η is an example of extensionality, but not a sufficiently general one. In chapter 5, we add the rule ext to $EQ_{\mathfrak{B}_1}$. (This rule is sometimes labeled by ζ instead of ext.) If we would add a rule based on η to the same calculus, then a weaker system would result.

It may be useful to note that the definition stipulates that M and N reduce to the same term whenever they are applied to the same terms $P_1,\ldots,P_n$. We already saw in the discussion of associators that, if a combinator is n-ary, then forming the term with $x_1,\ldots,x_n,\ldots,x_{n+m}$ as arguments gives a term by weak reduction in which $x_{n+1},\ldots,x_{n+m}$ remain unaffected. Thus, $\mathsf{B}P_1$ does not contain a redex headed by B. However, for $n=3$, $\mathsf{B}P_1P_2P_3 \rhd_w P_1(P_2P_3)$ and so does $\mathsf{S}(\mathsf{KS})\mathsf{K}P_1P_2P_3$. Adding further terms, $P_4,P_5,\ldots$, does not prevent the two terms from reducing to the same term. That is, the existence of some n is sufficient for the existence of *infinitely many* (larger) n's, for which the defining condition in definition 1.3.15 holds.

The P's in the definition are *arbitrary*. But in the illustrations, we looked only at what happens when two terms are applied to x or $x_1,\ldots,x_n$. However, our choice of these variables represented another sort of generality: none of them occurred in the terms like SKS or $\mathsf{S}(\mathsf{KS})\mathsf{K}$ (the definitions for I and B). In fact, the following definition yields the same concept.

DEFINITION 1.3.16. (EXTENSIONAL WEAK EQUALITY, 2) Q and $[M_1/N_1,\ldots,M_m/N_m]Q$ are *extensionally weakly equal*, denoted by $Q =_{w\zeta} [M_1/N_1,\ldots,M_m/N_m]Q$, iff for each $\langle M_i,N_i\rangle$ (where $1 \leq i \leq m \in \mathbb{N}$), there is an n_i (where $n_i \in \mathbb{N}$) such that (1) is true.

(1) For a series of distinct variables $x_{i,1},\ldots,x_{i,n_i}$ that do not occur in M_i or N_i (i.e., $x_{i,1},\ldots,x_{i,n_i} \notin \mathrm{fv}(M_i) \cup \mathrm{fv}(N_i)$), there is a P_i with $M_i x_{i,1}\ldots x_{i,n_i} \triangleright_w P_i$ and $N_i x_{i,1}\ldots x_{i,n_i} \triangleright_w P_i$.

Exercise 1.3.13.** Prove that the relations introduced in definitions 1.3.15 and 1.3.16 are the same relation on the set of CL-terms. (Hint: One direction is nearly trivial.)

In the case of $=_w$, we first introduced a *reduction* relation $\triangleright_w$ that was not symmetric (not even antisymmetric), and then we took the transitive symmetric closure of $\triangleright_w$ to get $=_w$. Both definitions 1.3.15 and 1.3.16 are, obviously, symmetric in M and N. There seems to be no reasonable way to carve out a reduction-like relation from either definition. $Mx_1\ldots x_n \triangleright_w P$ does not specify the shape of P, and if we require $Mx_1\ldots x_n \triangleright_w Px_1\ldots x_n$ for M to reduce to P in some sense, then we get a rather restricted notion.

The algorithm to define bracket abstraction provides an alternative approach. The interesting cases of $M =_{w\zeta} N$ are when M and N are combinators, and they define the same function with respect to their input–output pairs. This suggests that if we fix a particular algorithm to define bracket abstraction, then that would provide a way to *standardize* combinators.

Example 1.3.17. Let us consider the CL-term $\mathsf{SK(KK)}$. $\mathsf{SK(KK)}x \triangleright_w x$ and so does $\mathsf{I}x$, as well as $\mathsf{SKK}x$. Of course, these three combinators are not literally (letter by letter) the same, and they are not weakly equal either. However, we have $\mathsf{SK(KK)} =_{w\zeta} \mathsf{I} =_{w\zeta} \mathsf{SKK}$. For each of these three combinators, there are infinitely many other combinators that are $=_{w\zeta}$ to it, but one of all those combinators plays a distinguished role. SKK is *the* CL-term denoted by $[x].x$, when we assume the algorithm in the proof of lemma 1.3.9.

For any combinator Z, given the fixed bracket abstraction algorithm, if $\mathsf{Z}x \triangleright x$ then the standard version of Z is SKK, because Z is just the unary identity combinator — extensionally speaking.

The idea of the *standard* (or *canonical*) form for a combinator makes sense only in the extensional context. (Otherwise $\mathsf{SK(KK)}$, I and SKK are simply distinct combinators.) The possibility of selecting a distinguished element of each $=_{w\zeta}$ equivalence class allows a reasonable definition of a reduction relation that is related to $=_{w\zeta}$.

DEFINITION 1.3.18. (STRONG REDUCTION) Let us assume that a bracket abstraction algorithm has been specified.[12] The binary relation on the set of CL-terms called *strong reduction*, and denoted by $\triangleright_s$, is inductively defined by (1)–(3).

[12]For concreteness, we assume that the algorithm is the one defined in the proof of lemma 1.3.9 with η included. In the literature, sometimes the combinatory basis is expanded to $\{\mathsf{I},\mathsf{S},\mathsf{K}\}$ and SKK in (1) is replaced by I.

(1) $M \triangleright_s N$, if $M \triangleright_w N$;

(2) $[x].M \triangleright_s [x].N$, if $M \triangleright_s N$;

(3) $P \triangleright_s [M_1/N_1,\ldots,M_m/N_m]P$, if $M_1 \triangleright_s N_1, \ldots, M_m \triangleright_s N_m$.

Strong reduction is a weaker relation, in the set theoretic sense, than weak reduction is.[13] Obviously, every pair of CL-terms that is in $\triangleright_w$ is in $\triangleright_s$ by clause (1). But we also have $\mathsf{SKS} \triangleright_s \mathsf{SKK}$ and $\mathsf{I} \triangleright_s \mathsf{SKK}$, though no $\triangleright_w$ holds in place of $\triangleright_s$.

Example 1.3.19. $\mathsf{B}x =_{w\zeta} \mathsf{S(KS)K}x$, because after the addition of y and z to the terms they both reduce to $x(yz)$. $\mathsf{B}xyz \triangleright_w x(yz)$, so, by (1), $\mathsf{B}xyz \triangleright_s x(yz)$. Repeated applications of (2) give $[z].\mathsf{B}xyz \triangleright_s [z].x(yz)$, $[y][z].\mathsf{B}xyz \triangleright_s [y][z].x(yz)$ and $[x][y][z].\mathsf{B}xyz \triangleright_s [x][y][z].x(yz)$. But $[x][y][z].\mathsf{B}xyz$ is just B and $[x][y][z].x(yz)$ is $\mathsf{S(KS)K}$ (when η is included). That is, $\mathsf{B} \triangleright_s \mathsf{S(KS)K}$, and $\mathsf{S(KS)K} \triangleright_s \mathsf{S(KS)K}$. In other words, $\mathsf{S(KS)K}$ is the *standard* form of B (modulo the stipulated bracket abstraction algorithm).

Example 1.3.20. $\mathsf{SK(KK)} =_{w\zeta} \mathsf{SKS}$, because $\mathsf{SK(KK)}x \triangleright_w x$ and also $\mathsf{SKS}x \triangleright_w x$. However, $\mathsf{SK(KK)} \not\triangleright_w \mathsf{SKS}$, and $\mathsf{SKS} \not\triangleright_w \mathsf{SK(KK)}$. Since $[x].x$ is neither $\mathsf{SK(KK)}$ nor SKS; neither of these two combinators strongly reduces to the other. (Strictly speaking, we would need to scrutinize (3) too, but neither S nor K strongly reduces to anything but itself, so there is not much wiggle room for replacement.)

The relationship between the reduction-like relations is

$$\triangleright_1 \subsetneq \triangleright_w \subsetneq \triangleright_s .$$

Strong reduction is reflexive, because $\triangleright_w$ is reflexive (already). Therefore, for any term M, there is at least one term to which M strongly reduces. Weak normal forms are defined in 1.3.5, and there we used the absence of redexes in a term. However, a $\triangleright_s$ step does not always involve a redex; thus, neither does the next definition.

DEFINITION 1.3.21. (STRONG NORMAL FORM) A CL-term M is in *strong normal form*, or in *snf* for short, iff $M \triangleright_s N$ only if N is the same CL-term as M.

Example 1.3.22. The term SKK is in snf (and also in wnf). On the other hand, SKS is not in snf (though in wnf).

All the variables are in snf, and S and K are in snf due to the inclusion of η.

Exercise 1.3.14. SK and $\mathsf{K(K(SK))}$ are in wnf. However, they are not in snf. Prove this claim by finding the snf's for these terms.

It is tempting now to quickly conclude that both normal forms (the wnf and the snf) can be characterized by saying that the respective reduction relations have a

[13]The names of these relations are, probably, motivated by the use of "strength" to connote multiplicity or a large amount.

"dead end loop" on the term that is in nf. However, $M \rhd_{\mathrm{w}} N$ iff N is M does not imply that M is in wnf (at least if no restriction is placed on the combinatory basis).

We have seen that $\rhd_{\mathrm{w}}$ and $=_{\mathrm{w}}$ depend on the set of combinators chosen. However, the set of all CL-terms is also dependent on the combinatory basis in the same way. Strong reduction differs from each of $\rhd_1$, $\rhd_{\mathrm{w}}$, $=_{\mathrm{w}}$ and $=_{\mathrm{w}\zeta}$ in that it also depends on *an algorithm* that can be defined in various ways even if a combinatory basis has already been selected. $\rhd_{\mathrm{s}}$ is really a family of relations united by resemblance.

A remarkable feature of $=_{\mathrm{w}\zeta}$ is that definition 1.3.15 (or definition 1.3.16) can be replaced by a (small) *finite set* of additional equalities. This means that instead of doing one-step reductions on terms with sufficiently many new variables put into argument places, one can take $=_{\mathrm{w}}$ and replace certain subterms comprising combinators by other subterms (comprising combinators).

Depending on the combinatory basis, different sets of axioms suffice. It is not difficult (though somewhat laborious) to verify that the equalities hold; we forecast this by using $=_{\mathrm{w}\zeta}$ in them. However, to show that, if two terms are $=_{\mathrm{w}\zeta}$, then that equality may be obtained using replacements according to the given equations (and given $=_{\mathrm{w}}$), requires a suitable bracket abstraction algorithm. This explains why the axiom sets are usually presented in connection with an algorithm, which at the same time "explains" the shape of the sometimes lengthy terms in the equalities.

Let us assume that the *combinatory basis* is $\{\mathsf{S},\mathsf{K}\}$. Axioms (A1)–(A6) added to $=_{\mathrm{w}}$ yield $=_{\mathrm{w}\zeta}$.[14]

(A1) $\mathsf{K} =_{\mathrm{w}\zeta} \mathsf{S}(\mathsf{S}(\mathsf{KS})(\mathsf{S}(\mathsf{KK})\mathsf{K}))(\mathsf{K}(\mathsf{S}(\mathsf{KK})))$

(A2) $\mathsf{S} =_{\mathrm{w}\zeta} \mathsf{S}(\mathsf{S}(\mathsf{KS})(\mathsf{S}(\mathsf{K}(\mathsf{S}(\mathsf{KS})))(\mathsf{S}(\mathsf{K}(\mathsf{S}(\mathsf{KK})))\mathsf{S})))(\mathsf{K}(\mathsf{K}(\mathsf{SKK})))$

(A3) $\mathsf{S}(\mathsf{KK}) =_{\mathrm{w}\zeta} \mathsf{S}(\mathsf{S}(\mathsf{KS})(\mathsf{S}(\mathsf{KK})(\mathsf{S}(\mathsf{KS})\mathsf{K})))(\mathsf{KK})$

(A4) $\mathsf{S}(\mathsf{KS})(\mathsf{S}(\mathsf{KK})) =_{\mathrm{w}\zeta} \mathsf{S}(\mathsf{KK})(\mathsf{S}(\mathsf{S}(\mathsf{KS})(\mathsf{S}(\mathsf{KK})(\mathsf{SKK})))(\mathsf{K}(\mathsf{SKK})))$

(A5) $\mathsf{S}(\mathsf{K}(\mathsf{S}(\mathsf{KS})))(\mathsf{S}(\mathsf{KS})(\mathsf{S}(\mathsf{KS}))) =_{\mathrm{w}\zeta}$
$\mathsf{S}(\mathsf{S}(\mathsf{KS})(\mathsf{S}(\mathsf{KK})(\mathsf{S}(\mathsf{KS})(\mathsf{S}(\mathsf{K}(\mathsf{S}(\mathsf{KS})))\mathsf{S}))))(\mathsf{KS})$

(A6) $\mathsf{SKK} =_{\mathrm{w}\zeta} \mathsf{S}(\mathsf{S}(\mathsf{KS})\mathsf{K})(\mathsf{K}(\mathsf{SKK}))$

Exercise 1.3.15.** Verify that the combinators in the equations weakly reduce to the same term once sufficiently many (distinct) variables have been added in argument places.

If the *combinatory basis* also includes I, that is, if it is $\{\mathsf{I},\mathsf{S},\mathsf{K}\}$, then axioms (A1)–(A5) added to $=_{\mathrm{w}}$ give $=_{\mathrm{w}\zeta}$.

(A1) $\mathsf{S}(\mathsf{S}(\mathsf{KS})(\mathsf{S}(\mathsf{KK})(\mathsf{S}(\mathsf{KS})\mathsf{K})))(\mathsf{KK}) =_{\mathrm{w}\zeta} \mathsf{S}(\mathsf{KK})$

[14](A1) and (A2) are redundant if we allow shortcuts, that is, if η is included in the definition of the bracket abstraction algorithm.

(A2) $\mathsf{S(S(KS)K)(KI)} =_{\mathrm{w}\zeta} \mathsf{I}$

(A3) $\mathsf{S(KI)} =_{\mathrm{w}\zeta} \mathsf{I}$

(A4) $\mathsf{S(KS)(S(KK))} =_{\mathrm{w}\zeta} \mathsf{K}$

(A5) $\mathsf{S(K(S(KS)))(S(KS)(S(KS)))} =_{\mathrm{w}\zeta}$
$\mathsf{S(S(KS)(S(KK)(S(KS)(S(K(S(KS)))S))))(KS)}$

Exercise* 1.3.16. Show that (A1)–(A5) hold according to definition 1.3.16. (Hint: This exercise is similar to 1.3.15 with a different axiom system.)

Before we introduce one of the earliest axiom sets, we need to present some new combinators. S is called the *strong composition* combinator, because it is informally interpreted as taking the value of two functions for the same argument and applying the first to the second: $\mathsf{S}fgx \rhd fx(gx)$. There are other ways to compose functions and the next definition adds infinitely many combinators that are similarly motivated.

DEFINITION 1.3.23. (COMPOSITION COMBINATORS) The combinators Ψ, Φ, $\ldots$, Φ_n (for $n > 1$) are defined by (1)–(4).

(1) $\Psi xyzv \rhd x(yz)(yv)$

(2) $\Phi xyzv \rhd x(yv)(zv)$

(3) Φ_2 is Φ

(4) $\Phi_n xy_1 \ldots y_n z \rhd x(y_1 z) \ldots (y_n z)$ (where $n \geq 2$)

Example 1.3.24. If we use f, g, etc., to emphasize that we are dealing with functions, then we see that $\Psi fgxy \rhd_{\mathrm{w}} f(gx)(gy)$, whereas $\Phi fghx \rhd_{\mathrm{w}} f(gx)(hx)$.

Exercise* 1.3.17. Suppose a combinatory basis that contains at least B, K and Φ. Find a definition for Ψ.

Exercise 1.3.18. The Φ_n's are definable from the basis $\{\mathsf{B},\mathsf{S}\}$. Define Φ_3, which has as its axiom $\Phi_3 xy_1y_2y_3z \rhd x(y_1z)(y_2z)(y_3z)$.

Exercise 1.3.19. Find snf's for Φ and Ψ. (Hint: Use the algorithm in the proof of lemma 1.3.9 together with η.)

Let us assume that the *combinatory basis* is $\{\mathsf{I},\mathsf{S},\mathsf{K},\mathsf{B},\mathsf{Z}_3,\Psi,\Phi_2,\Phi_3\}$. In order to obtain $=_{\mathrm{w}\zeta}$, (A1)–(A6) are added to $=_{\mathrm{w}}$.[15]

(A1) $\Phi_2\mathsf{K} =_{\mathrm{w}\zeta} \mathsf{BK(SB(KI))}$

(A2) $\Phi_3\mathsf{S} =_{\mathrm{w}\zeta} \Psi(\Phi_2\mathsf{S})\mathsf{S}$

[15] Z_3 is the combinator from exercise 1.2.2; another notation for this combinator is B^2. (A6) is in fact redundant.

(A3) $\Psi\mathsf{SK} =_{\mathrm{w}\zeta} \mathsf{BK}$

(A4) $\mathsf{B(SB(KI))K} =_{\mathrm{w}\zeta} \mathsf{K}$

(A5) $\mathsf{Z_3(SB(KI))S} =_{\mathrm{w}\zeta} \mathsf{S}$

(A6) $\mathsf{SB(KI)} =_{\mathrm{w}\zeta} \mathsf{I}$

Exercise* 1.3.20. Show that axioms (A1)–(A6) hold according to definition 1.3.16. (Hint: This exercise is similar to exercises 1.3.15 and 1.3.16 save the set of axioms.)

If we only consider $\triangleright_{\mathrm{w}}$ on the set of CL-terms, then it is easy to see that there are terms that do not reduce to one another. None of the variables reduces to any other term, nor do K and S weakly reduce to each other. It is also nearly obvious that there is no term that can reduce to all terms. In the presence of a cancellator, some subterms (beyond heads of reduced redexes) may disappear. What we have so far does not make obvious that, for instance, there is no term that reduces to both x and y (where we suppose that those are distinct variables). Even so, it follows from the finitude of terms that no term contains infinitely many variables. Therefore, as long as the combinatory basis we start with comprises proper combinators only, there cannot exist a universal term in the set of the CL-terms.

In the early 20th century, many new paradoxes were discovered. They discredited various systems (or at least made their reliability highly questionable). It is, therefore, interesting and important to ensure that CL with $=_{\mathrm{w}}$ and $=_{\mathrm{w}\zeta}$ is *consistent*. We have already claimed consistency, and the next chapter gives a proof of that claim together with other theorems.

Chapter 2

Main theorems

We take for granted in everyday practice that the number systems we use — like the real numbers with $+$, $-$, $\cdot$ and $/$ — are reliable and stable. Of course, we sometimes make mistakes adding or dividing numbers, but we know that if $5+7=12$ today, the same is true tomorrow, and this is why we can correct the person who claims that the sum is 11. $+$, $-$, $\cdot$ and $/$ are functions, and so for any pair of numbers they give a *unique* value. That is, an application of these operations cannot result in both n and m — what would lead to $n=m$ — if n and m are distinct numbers. Moreover, it is widely believed that no such equations will emerge from calculating with numerical functions; that is, the theory of such functions is *consistent*. In the case of a formalized theory, such as Peano arithmetic (PA), this property can be formalized as "$0=1$ is not provable" (in PA).[1] A similar question may be asked concerning CL: Is there an equation (with $=_{\text{w}}$ or even $=_{\text{w}\zeta}$) between CL-terms that does not hold? In this chapter, one of the paramount results is that CL is *consistent*. We give a rigorous proof of this with $=_{\text{w}}$ — though we do not aim at formalizing the notion of consistency or the proof in CL itself.

2.1 Church–Rosser property

Given an arithmetical expression, *the order of calculating* the values of subexpressions does not matter. For example, $(5+7)\cdot(2+2)$ can be computed in various ways. We could first calculate $2+2$ to get $(5+7)\cdot 4$, or we could start with $5+7$ to get $12\cdot(2+2)$. Using the distributivity of $\cdot$ over $+$ we could even start with the multiplication to obtain $5\cdot 2+7\cdot 2+5\cdot 2+7\cdot 2$.[2] Each calculation may be continued in more than one way, as, for example, $12\cdot 4$ or $5\cdot 4+7\cdot 4$ in the first case. However, all the potential ways to compute the value of the term yield the same number, namely, 48.

The first relation from CL that is likely to come to mind based on the above example is *weak equality* ($=_{\text{w}}$). However, *weak reduction* ($\triangleright_{\text{w}}$) is more fundamental

[1] A celebrated result of 20th-century logic is that PA cannot prove its own consistency, provided PA is consistent, which is Gödel's so-called second incompleteness theorem for PA.

[2] Allowing distributivity to be used may be thought to be similar to making a replacement according to $=_{\text{w}\zeta}$.

and is easier to deal with. Also, $\triangleright_{\mathrm{w}}$ is more similar to computing the value of an arithmetical expression that $=_{\mathrm{w}}$ is.

DEFINITION 2.1.1. (CHURCH–ROSSER PROPERTY) A binary relation R (on the set X) has the *Church–Rosser property* or is *confluent* iff Rxy_1 and Rxy_2 imply that Ry_1z and Ry_2z for some z.

The property is named after Alonzo Church and Barkley Rosser, who gave proofs that this property holds for certain relations in λ-calculus and CL. Yet another name for confluence is the *diamond property*.[3]

The definition is general: X may be a set of arbitrary objects. For instance, X could be the set of algebraic expressions (perhaps, containing some functions), and R could be the relation on the set of expressions obtained via calculating the value of functions. The numerical expressions above were intended to illustrate that this R is confluent. In term rewriting systems, X is a set of abstract objects called *terms* and R is a relation called *rewriting*. We restrict our attention here to binary relations on CL-terms, but we note a general fact about the Church–Rosser property.

LEMMA 2.1.2. *If R is a symmetric relation on X, then, R has* the Church–Rosser property.

Proof: If Rxy and Rxz, then by the symmetry of R, Ryx and Rzx. qed

The proof shows that confluence is trivial for symmetric relations, because x and z in definition 2.1.1 need not be distinct. The moral of the lemma is that we need to wonder about the confluence of the reduction-like relations only.

PROPOSITION 2.1.3. (ONE-STEP REDUCTION) *The $\triangleright_1$ relation on the set of* CL-*terms* does not have the Church–Rosser property.

Proof: It is easy to prove this claim: we only need a counterexample. Here is a (relatively) simple term that suffices: $\mathsf{S}xy(\mathsf{S}xyz)$. The term contains two redexes and the two possible one-step reductions give $x(\mathsf{S}xyz)(y(\mathsf{S}xyz))$ and $\mathsf{S}xy(xz(yz))$, respectively. From the latter term, all we can get by one-step reduction is a term with no combinator in it. We cannot get such a term from the former term in one step. In sum, we have the following possible sequences of one-step reductions.

1. $\mathsf{S}xy(\mathsf{S}xyz) \triangleright_1 \mathsf{S}xy(xz(yz)) \triangleright_1 x(xz(yz))(y(xz(yz)))$
2. $\mathsf{S}xy(\mathsf{S}xyz) \triangleright_1 x(\mathsf{S}xyz)(y(\mathsf{S}xyz)) \triangleright_1 x(xz(yz))(y(\mathsf{S}xyz)) \triangleright_1 x(xz(yz))(y(xz(yz)))$
3. $\mathsf{S}xy(\mathsf{S}xyz) \triangleright_1 x(\mathsf{S}xyz)(y(\mathsf{S}xyz)) \triangleright_1 x(\mathsf{S}xyz)(y(xz(yz))) \triangleright_1 x(xz(yz))(y(xz(yz)))$

qed

[3]The three expressions "Church–Rosser property," "confluence" and "diamond property" are not always used synonymously, especially, in the literature on *abstract term rewriting systems*, but we use them interchangeably. A compendium on TRSs is Terese [148].

Notice that definition 2.1.1 does not require y_1 and y_2 to be distinct (either). Moreover, it does not require that Rxy holds for any y *at all*.[4] If a CL-term does not contain a redex, then the antecedent of the conditional (i.e., $Rxy_1 \wedge Rxy_2$) is false; hence, the term surely cannot be used to prove the lack of confluence. If a term contains just one redex, then, depending on the structure of the term, it may or may not be suitable for showing the lack of confluence for $\triangleright_1$.

Exercise 2.1.1. Choose a CL-term containing one redex, and give another proof of proposition 2.1.3 with that term.

There are many CL-terms with redexes in them, such that all one-step reductions lead to terms that further reduce to a common term in one step. For example, $\mathsf{S}(\mathsf{S}xyz)(\mathsf{S}xyz)z$ contains three redexes, and after any pair of $\triangleright_1$ steps there is a common term reachable via just *one* one-step reduction.

Exercise 2.1.2. Write out all the possible one-step reduction sequences starting with the term $\mathsf{S}(\mathsf{S}xyz)(\mathsf{S}xyz)z$.

Exercise 2.1.3. Consider the complex-looking term $\mathsf{K}(\mathsf{S}(\mathsf{KSK})(\mathsf{KKS})\mathsf{S})\mathsf{K}$. Is this term suitable for showing that $\triangleright_1$ is not confluent? How many distinct terms can be obtained from $\mathsf{K}(\mathsf{S}(\mathsf{KSK})(\mathsf{KKS})\mathsf{S})\mathsf{K}$ by successive applications of $\triangleright_1$? Do the one-step reduction sequences terminate?

A prototypical failure of the diamond property for $\triangleright_1$ may be described as follows (illustrated by the term used in the proof of proposition 2.1.3). If Z is a duplicator for its jth argument place, then we can form a term comprising a redex with Z where the jth argument contains a redex. A one-step reduction may be applied to the redex in the jth argument place or to Z. No further one-step reduction leads to a common term.

An interesting exception to this general pattern may be constructed using the combinator M, which is a duplicator for its only argument. $\mathsf{M}(\mathsf{MM})$ contains two redexes, and can be one-step reduced to both $\mathsf{MM}(\mathsf{MM})$ and to itself. However, $\mathsf{MM}(\mathsf{MM}) \triangleright_1 \mathsf{MM}(\mathsf{MM})$ and $\mathsf{M}(\mathsf{MM}) \triangleright_1 \mathsf{MM}(\mathsf{MM})$, which shows confluence. Of course, if we do not require the term to be a combinator, or we allow another combinator beyond M to be included, then we can construct terms (e.g., $\mathsf{M}(\mathsf{M}x)$ or $\mathsf{M}(\mathsf{MI})$) showing the lack of confluence.

PROPOSITION 2.1.4. *If the combinatory basis $\mathfrak{B}$ does not include a duplicator or a cancellator, then $\triangleright_1$ is* confluent *on the set of* CL-*terms with at least two redexes over $\mathfrak{B}$ (with or without a denumerable set of variables).*

Exercise* 2.1.4. Prove this proposition. (Hint: It may be useful to consider separately the case where the reduced redexes overlap and the case where they do not.)

[4]The Church–Rosser property, as defined in 2.1.1, could be formalized by the first-order sentence: $\forall x \forall y_1 \forall y_2 ((Rxy_1 \wedge Rxy_2) \Rightarrow \exists z (Ry_1z \wedge Ry_2z))$. The $\forall$'s are universal quantifiers, whereas $\exists$ is an existential quantifier. In the natural language formulation of the definition, the universal quantifiers are implicit.

Exercise 2.1.5. Suppose that we can introduce a constant for any combinator definable from the basis $\{\mathsf{S},\mathsf{K}\}$. For example, $\mathsf{S}(\mathsf{KS})\mathsf{K}xyz \triangleright_{\mathrm{w}} x(yz)$, as we saw in chapter 1. Then we may introduce the constant B (as we did earlier) with axiom $\mathsf{B}xyz \triangleright x(yz)$. Demonstrate that for any n ($n \in \mathbb{N}$) there is a CL-term M such that $M \triangleright_1 N_1$ and $M \triangleright_1 N_2$ and then more than n one-step reductions are needed to arrive at a common term from N_1 (or N_2).

The following theorem motivates our interest in the confluence of one-step reduction.

THEOREM 2.1.5. *Let R be a binary relation on the set X. If R has the diamond property then R^+, the transitive closure of R, has* the diamond property.

Proof: The transitive closure of a relation may be described informally by saying that R^+xy means that there is a finite (nonempty) path along R edges leading from x to y. (See also the discussion after definition 1.3.4 and section A.2.)

Let us assume that R^+xy_1 and R^+xy_2. Then there are positive integers m and n (i.e., $m,n \in \mathbb{Z}^+$) such that R^mxy_1 and R^nxy_2, that is, y_1 and y_2 are, respectively, m and n steps away from x.[5] If both m and n are 1, then, by the assumption of the theorem, there is a z such that both Ry_1z and Ry_2z.

If m or n is not 1, then the path may be visualized — by writing the first arguments of the R's first and conflating the expressions — as $xRw_1Rw_2\ldots w_{m-1}Ry_1$ and $xRv_1Rv_2\ldots v_{n-1}Ry_2$. Since R is confluent, there is some u_1 such that Rv_1u_1 and Rw_1u_1. Depending on whether $m > 1$ or $n > 1$, successive applications — $m \times n$-many times — of the confluence of the relation R yield $u_{m\cdot n}$ with the property that $R^ny_1u_{m\cdot n}$ and $R^my_2u_{m\cdot n}$. qed

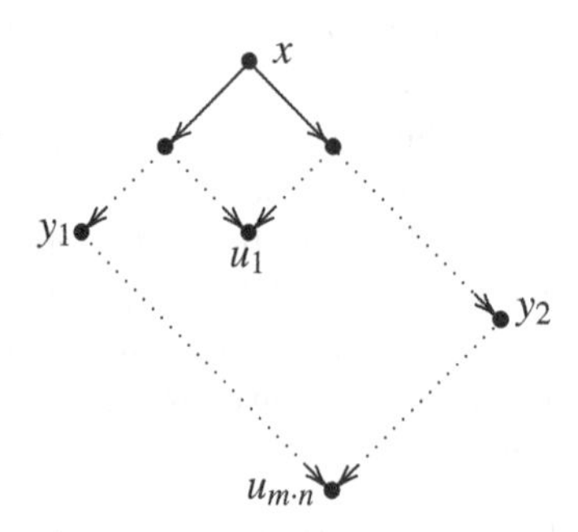

Exercise 2.1.6. Suppose that $m,n > 1$ in the preceding proof. Devise a systematic way of applying the assumption of the theorem to generate all the u's up to $u_{m\cdot n}$. (Hint: The above graph is an illustration that might be helpful.)

One-step reduction is not Church–Rosser because it is a much too strong a relation. First, exercise 2.1.5 shows that it is hopeless to place an upper bound on the

[5] By R^mxy we indicate that Rxv_1, $Rv_1v_2, \ldots, Rv_{m-1}y$ all obtain for some v's — but without naming or quantifying over these v's. This notation is used often.

length of one-step reduction sequences that might be needed to regain the diamond property. Second, $\triangleright_1$ is *not reflexive*. There are very few terms that one-step reduce to themselves. The absence of reflexivity can lead to the failure of confluence for $\triangleright_1$ in a trivial fashion. For example, $\mathsf{I}x \triangleright_1 x$, but there is no term M such that $x \triangleright_1 M$. Therefore, $\mathsf{I}x$ shows that the diamond property is false for $\triangleright_1$.

The proof of proposition 2.1.3 contains a redex with a duplicator as its head as well a redex in the argument of the duplicator that gets replicated when the duplicator is reduced. However, the copies of the duplicated redex cannot be all reduced in one step. But it is interesting to note that they do not overlap, that is, they do not have any occurrences of any subterms in common. The distinction between terms and occurrences of terms is relevant here. All the occurrences of a duplicated redex look the same. That is, they are the *same (sub)term*, but there are (at least) two *distinct occurrences* of that term; therefore, they are disjoint.

This observation led to the introduction of yet another binary relation on terms. We assume that two redex occurrences are overlapping or nonoverlapping just in case they look so. (Cf. section A.2 for a more rigorous, but also more tedious, definition.)

DEFINITION 2.1.6. (SIMULTANEOUS ONE-STEP REDUCTION) Let M be a term and $\{P_1,\ldots,P_n\}$ a (possibly empty) set of redex occurrences in M, such that for any $i \neq j$, P_i and P_j are nonoverlapping. Then $M \triangleright_{p1} M'$, that is, M' results by *a simultaneous one-step reduction* from M, when M' is $[P_1/N_1,\ldots,P_n/N_n]M$ with each $P_i \triangleright_1 N_i$.[6]

The definition is applicable both when the set of redexes is empty and when it is not. If the set of redexes is empty, then M' is the same term as M. If M is in nf, then the set of redexes is necessarily empty. However, even if M is not in nf the set of redexes still may be chosen to be empty. The upshot is that for any term M, $M \triangleright_{p1} M$. In other words, $\triangleright_{p1}$ is a *reflexive* relation.

Exercise 2.1.1 showed that $\triangleright_1$ lacks the Church–Rosser property, because terms in nf do not one-step reduce to any term whatsoever. We might entertain the idea that it is sufficient to include in $\triangleright_1$ a small amount of reflexivity (i.e., a restricted version of the identity relation), for $\triangleright_{p1}$ to become confluent. Perhaps, it is sufficient to have $M \triangleright_{p1} M$ for terms in nf. The inclusion of parallel reductions in $\triangleright_{p1}$, however, creates a new possibility for nonconfluence.

Exercise 2.1.7. Show that, if the definition of $\triangleright_{p1}$ (i.e., definition 2.1.6) is modified so that the set of redexes is required to be nonempty, and if M is in nf, then $M \triangleright_{p1} M$ is stipulated, then this modified relation is not confluent.

A source of the failure of confluence in the case of $\triangleright_1$ is that one-step reduction is not transitive. However, if a relation is transitive, for instance, if $\triangleright_{p1}$ were transitive, then a direct proof of confluence would be just as difficult as in the case of $\triangleright_w$. If

[6] We used $\triangleright_s$ as the notation for strong reduction. Now we use the subscript $_p$ on $\triangleright$ together with $_1$. The $_p$ can be thought to stand for *parallel*.

$\triangleright_{p1}$ included successive one-step reductions up to a fixed number (rather than the transitive closure of one-step reduction), then confluence would still fail.

It is helpful to rethink exercise 2.1.5, from the point of view of proving confluence for such a relation. Let us assume that we allow successive one-step reductions, for example, up to three steps, to determine a relation $\triangleright_{\leq 3}$ on CL-terms. Let us also assume that S and K are the only primitive combinators. Then the effect of $\triangleright_{\leq 3}$ is similar to the effect of allowing a new combinator to be introduced, which adds seven new instances of an argument — while retaining $\triangleright_1$.

Example 2.1.7. S duplicates its last argument. The term $\mathsf{S}xy(\mathsf{S}xy(\mathsf{S}xy(\mathsf{K}zv))) \triangleright_1 \mathsf{S}xy(\mathsf{S}xy(x(\mathsf{K}zv)(y(\mathsf{K}zv))))$. Two further $\triangleright_1$ reductions yield the term:

$$x(x(x(\mathsf{K}zv)(y(\mathsf{K}zv)))(y(x(\mathsf{K}zv)(y(\mathsf{K}zv)))))(y(x(x(\mathsf{K}zv)(y(\mathsf{K}zv)))(y(x(\mathsf{K}zv)(y(\mathsf{K}zv)))))).$$

The latter term may be obtained via one $\triangleright_{\leq 3}$ reduction. The redex $\mathsf{K}zv$ occurred once in the starting term, and occurs eight times in the resulting term.

Let Z^3 be a new combinatory constant with axiom

$$\mathsf{Z}xyz \triangleright x(x(xz(yz))(y(xz(yz))))(y(x(xz(yz))(y(xz(yz))))).$$

Then $\mathsf{Z}xy(\mathsf{K}zv)$ one-step reduces to the term displayed above.

It is a remarkable idea that the failure of the diamond property for $\triangleright_1$ can be circumvented via allowing *simultaneous* one-step reductions, rather than *successive* one-step reductions. The analysis of the counterexamples to confluence for $\triangleright_1$ completely supports this insight: sequentiality may be supplanted by the requirement that the affected redexes do not overlap.

One-step reductions may be visualized as a directed graph in which the points are terms and the edges point toward a term as $\triangleright_1$ does. The reflexive closure of $\triangleright_1$ adds a loop (an edge leading from a point to itself) to every point in the graph. $\triangleright_{p1}$ also adds certain edges — some shortcuts — to the graph. However, not all consecutive arrows are composed into one arrow. In contrast, the transitive closure of $\triangleright_1$ composes every finite sequence of arrows into one. (Sometimes, if the edge relation is known to be reflexive and transitive, then the reflexive loops and the transitive arrows are omitted from the graph to minimize clutter.)

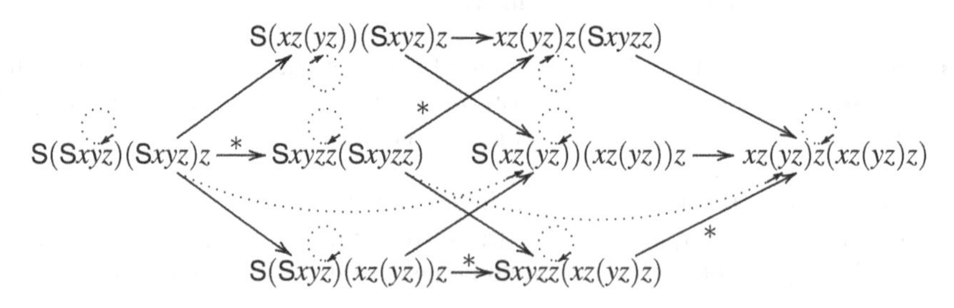

The dotted arrows indicate new edges in the $\triangleright_{p1}$ relation. The starred arrows show some places where transitivity fails.

Example 2.1.8. $\mathsf{K}MN \rhd_{p1} M$ and $\mathsf{S}MNP \rhd_{p1} MP(NP)$.

$\mathsf{K}y(\mathsf{SKK}x) \rhd_1 \mathsf{K}y(\mathsf{K}x(\mathsf{K}x))$ and $\mathsf{K}y(\mathsf{SKK}x) \rhd_1 y$. Further, $\mathsf{K}y(\mathsf{K}x(\mathsf{K}x)) \rhd_1 \mathsf{K}yx$ and $\mathsf{K}y(\mathsf{K}x(\mathsf{K}x)) \rhd_1 y$. All the $\rhd_1$'s can be replaced by $\rhd_{p1}$'s — including $\mathsf{K}y(\mathsf{SKK}x) \rhd_{p1} y$. However, no two one-step reductions compose, because there are no two disjoint redexes in any of the terms. Also, $y \rhd_{p1} y$.

It is useful to record certain formal properties of $\rhd_{p1}$ as well as its relationship to other relations on the set of CL-terms.

PROPOSITION 2.1.9. *The relation* $\rhd_{p1}$ *on* CL-*terms is* neither transitive, nor symmetric. *That is, there are terms* M, N *and* P *such that* $M \rhd_{p1} N \rhd_{p1} P$ *but not* $M \rhd_{p1} P$; *and there are terms* M *and* N *for which* $M \rhd_{p1} N$ *holds but* $N \rhd_{p1} M$ *does not. The relation* $\rhd_{p1}$ *is* reflexive. *That is, for any term* M, $M \rhd_{p1} M$.

Exercise 2.1.8. We already discussed reflexivity. Prove the rest of the proposition.

Exercise* 2.1.9. Consider the basis $\{\mathsf{S},\mathsf{K}\}$. Prove that the relation $\rhd_{p1}$ on the set of CL-terms over this basis is neither *asymmetric* nor *antisymmetric*. (Hint: The lack of both of these properties may be shown by finding two (distinct) terms M and N such that $M \rhd_{p1} N$ and $N \rhd_{p1} M$.)

LEMMA 2.1.10. *The* relationship *between the reduction-like relations on the set of* CL-*terms is*

$$\rhd_1 \subseteq \rhd_{p1} \subseteq \rhd_{\mathrm{w}} \subseteq \rhd_{\mathrm{s}}\,.$$

Proof: To prove the first inclusion, $\rhd_1 \subseteq \rhd_{p1}$, let us assume that $M \rhd_1 N$. Then there is a redex P in M such that N is $[P/P']M$ (where $P \rhd_1 P'$ is by substitution in an axiom). Certainly, $M \rhd_{p1} N$ when the set of redexes is $\{P\}$.

To prove the second inclusion, $\rhd_{1p} \subset \rhd_{\mathrm{w}}$, let us assume that $M \rhd_{p1} N$, and that there is a nonempty set of nonoverlapping redexes $\{P_1,\ldots,P_n\}$ in M such that $P_1 \rhd_1 Q_1, \ldots, P_n \rhd_1 Q_n$, and N is $[P_1/Q_1,\ldots,P_n/Q_n]M$. Then the following chain of one-step reductions links M and N: $M \rhd_1 [P_1/Q_1]M \rhd_1 [P_1/Q_1,P_2/Q_2]M \rhd_1 \ldots \rhd_1 [P_1/Q_1,\ldots,P_n/Q_n]M$. Weak reduction is defined to include the transitive closure of $\rhd_1$; therefore, $M \rhd_{\mathrm{w}} N$ is immediate.

Now let us consider $M \rhd_{p1} N$ but with the set of redexes empty. Then M is identical to N. Weak reduction includes the reflexive closure of $\rhd_1$, that is, $\rhd_{\mathrm{w}}$ is reflexive; therefore, $M \rhd_{\mathrm{w}} M$ holds.

The third inclusion, $\rhd_{\mathrm{w}} \subseteq \rhd_{\mathrm{s}}$, has been established in chapter 1 — it is in this lemma for the sake of completeness. qed

To obtain a proof that the Church–Rosser property holds for $\rhd_{\mathrm{w}}$, we will put together well-known facts about the reflexive transitive closure of a relation and what we have established about $\rhd_{p1}$ and its connection to $\rhd_1$ and $\rhd_{\mathrm{w}}$.

Recall from chapter 1 that the reflexive transitive closure of a binary relation R is denoted by R^*.

CLAIM 2.1.11. (1) $R \subseteq R^*$. (2) *If* $R_1 \subseteq R_2$ *then* $R_1^* \subseteq R_2^*$. (3) $R^* = R^{**}$.

Exercise* **2.1.10.** Prove (1)–(3) from the previous claim for an arbitrary binary relation R.

Assuming the truth of the claim, we can state the following lemma.

LEMMA 2.1.12. *The transitive reflexive closure of the simultaneous one-step reduction is* weak reduction, *that is,* $\triangleright_{p1}^{*} = \triangleright_{\mathrm{w}}$.

Proof: By definition, the transitive reflexive closure of one-step reduction is weak reduction, that is, $\triangleright_1^* = \triangleright_{\mathrm{w}}$. We have shown that $\triangleright_1 \subseteq \triangleright_{p1}$. It follows by (2), that $\triangleright_1^* \subseteq \triangleright_{p1}^*$. This is half of the identity we want to prove: $\triangleright_{\mathrm{w}} \subseteq \triangleright_{p1}$. Further, $\triangleright_{p1} \subseteq \triangleright_{\mathrm{w}}$; thus, again by (2), $\triangleright_{p1}^* \subseteq \triangleright_{\mathrm{w}}^*$. However, $\triangleright_{\mathrm{w}}$ is $\triangleright_1^*$, and $\triangleright_1^{**} = \triangleright_1^*$ by (3), that is, $\triangleright_{p1}^* \subseteq \triangleright_{\mathrm{w}}$ as desired. qed

The careful reader has surely noticed that an important piece is still missing from the proof of confluence for $\triangleright_{\mathrm{w}}$ that we are assembling: we have not yet established that $\triangleright_{p1}$ is confluent.

LEMMA 2.1.13. *The relation* $\triangleright_{p1}$ *has* the Church–Rosser property.

Proof: The proof of this lemma is by structural induction on M, which is the initial term that plays the role of x in definition 2.1.1. The starting assumption is that $M \triangleright_{p1} N_1$ and $M \triangleright_{p1} N_2$. We divide the proof into two main cases.
1. If M is an atomic term, such as a variable or a combinator, then a $\triangleright_{p1}$ step is only possible due to the reflexivity of the relation. However, $M \triangleright_{p1} M$ is sufficient to show that both N_1 and N_2, which are M, reduce to a common term.
2. Let us now assume that M is a complex term, that is, M is of the form $(M_1 M_2)$. From M being complex, it does not follow that there is a redex in M. Even if M contains one or more redexes, the $\triangleright_{p1}$ step does not compel us to suppose that N_1 or N_2 is obtained by one or more $\triangleright_1$ steps.
2.1 If N_1 and N_2 are M, then the reasoning is exactly as in case **1**.
2.2 If one of N_1 and N_2 is M, then that term, let us say for the sake of concreteness N_1, reduces to the other term, N_2, by $\triangleright_{p1}$. The starting assumption is of the form $M \triangleright_{p1} M$ and $M \triangleright_{p1} N_2$. But $M \triangleright_{p1} N_2$ and $N_2 \triangleright_{p1} N_2$.

The critical case in this proof is when both N_1 and N_2 are obtained by one-step reductions of a nonempty set of nonoverlapping redexes.
2.3 If the same set of redexes is reduced in both cases, then N_1 and N_2 are identical. Then, there is nothing special about the starting term having been complex, and the reasoning in this case may be completed as in case **1**.

Definition 2.1.6, the definition of simultaneous one-step reduction, implies that if M is a redex, then no other redex can be reduced within the same step as the whole M. This leads to two further subcases.
2.4 Let N_1 be the term obtained by reducing the whole of M, and let N_2 be the result of $\triangleright_{p1}$ in which all the reduced redexes are in one of M_1 and M_2. If M is a redex, then it is of the form $\mathsf{Z}^n P_1 \ldots P_n$, where P_n is the previous M_2. The axiom for Z^n is $\mathsf{Z} x_1 \ldots x_n \triangleright Q$; then, N_1 is $[x_1/P_1, \ldots, x_n/P_n]Q$. The term M is left-associated with respect to $\mathsf{Z}, P_1, \ldots, P_n$. Therefore, the redexes that are reduced to obtain N_2 have to

be contained in $P_1, \ldots, P_n$. Let $P'_1, \ldots, P'_n$ be the resulting terms, when those redexes are reduced in the P's standing separately. (Some of the P'_i's may be identical to their counterpart P_i's, but there is no need to be able to pinpoint those terms for the purposes of the proof.) N_2 is $\mathsf{Z}P'_1 \ldots P'_n$.

The result of performing a $\rhd_1$ step on N_2 with the redex headed by Z reduced is $[x_1/P'_1, \ldots, x_n/P'_n]Q$, by substitution in the axiom of Z. The result of reducing the same redex occurrences in the P's as were reduced to obtain N_2 is the result of the replacement $[P_1/P'_1, \ldots, P_n/P'_n]N_1$, where we assume that the particular occurrences of the P's that originated from the substitution in Q are the occurrences being replaced. Given this proviso together with the fact that N_1 looks like $[x_1/P_1, \ldots, x_n/P_n]Q$, we have $[P_1/P'_1, \ldots, P_n/P'_n][x_1/P_1, \ldots, x_n/P_n]Q$. That is, we have $[x_1/[P_1/P'_1]P_1, \ldots, x_n/[P_n/P'_n]P_n]Q$. The latter term is $[x_1/P'_1, \ldots, x_n/P'_n]Q$ showing confluence.

2.5 Lastly, it may be the case that N_1 and N_2 both result by reductions of some redexes within M_1 and M_2. Then $M_1 \rhd_{p1} N_{11}$ and $M_1 \rhd_{p1} N_{12}$, and $M_2 \rhd_{p1} N_{21}$ and $M_2 \rhd_{p1} N_{22}$. At this point we use the hypothesis of the induction, namely, that $\rhd_{p1}$ is confluent for terms that are structurally less complex than M itself. Then there are terms R_1 and R_2 such that, on one hand, $N_{11} \rhd_{p1} R_1$ and $N_{12} \rhd_{p1} R_1$, on the other hand, $N_{21} \rhd_{p1} R_2$ and $N_{22} \rhd_{p1} R_2$. However, (R_1R_2) is a term for which it is true that $N_1 \rhd_{p1} R_1R_2$ and $N_2 \rhd_{p1} R_1R_2$, establishing the diamond property in this case. This concludes the proof. qed

Now we can state and prove that $\rhd_w$ has the diamond property.

THEOREM 2.1.14. (CHURCH–ROSSER) *The weak reduction relation, $\rhd_w$, has* the Church–Rosser property.

Proof: The theorem is proved by simply piecing together what we have already established. $\rhd_{p1}$ is confluent, and the reflexive, transitive closure of $\rhd_{p1}$ is $\rhd_w$. Therefore, by theorem 2.1.5, $\rhd_w$ is confluent too. qed

The Church–Rosser theorem may be stated equivalently using $=_w$. The assumption is weakened: the two terms N_1 and N_2 (y_1 and y_2 in definition 2.1.1) are weakly equal. If $M \rhd_w N_1$ and $M \rhd_w N_2$, then $N_1 =_w N_2$. But the converse implication does not hold.

THEOREM 2.1.15. (CHURCH–ROSSER, 2) *If the terms N_1 and N_2 are weakly equal, that is, $N_1 =_w N_2$, then* there is a term P *such that $N_1 \rhd_w P$ and $N_2 \rhd_w P$.*

Proof: First, we note that $=_w$ is a reflexive transitive and symmetric closure of $\rhd_1$. Thus, there are certain obvious cases that we may quickly go through. By reflexivity, N_1 and N_2 may be the same term. Then this term itself is sufficient for the truth of the claim. N_1 and N_2 may be related by a finite sequence of $\rhd_1$'s, in which case either $N_1 \rhd_w N_2$ or $N_2 \rhd_w N_1$. The existence of a suitable P is then obvious, because N_2 (or N_1) is a term on which $\rhd_w$ converges. If both N_1 and N_2 are obtained by $\rhd_w$ from a term M, then the claim is exactly as in theorem 2.1.14 (which we have already proved).

The less obvious case is when none of the previous situations occurs, and then there are n terms $M_1,\ldots,M_n$ ($n > 1$) such that $M_1 \rhd_w N_1$, $M_1 \rhd_w Q_1$, …, $M_i \rhd_w Q_{i-1}$, $M_i \rhd_w Q_i$, …, $M_n \rhd_w Q_{n-1}$, $M_n \rhd_w N_2$. This chain of $\rhd_w$ steps is a justification for $N_1 =_w N_2$, and looks as follows.

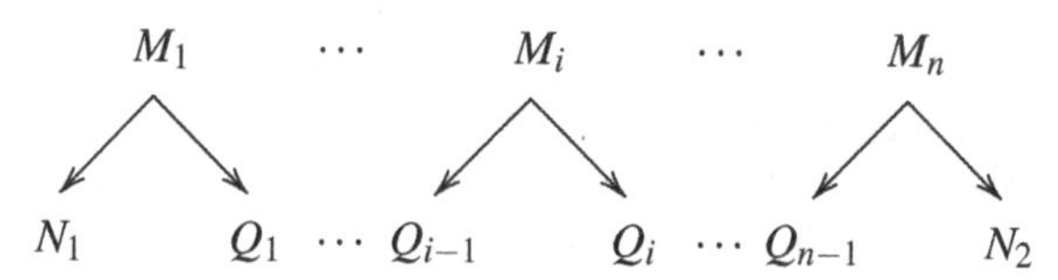

We can apply theorem 2.1.14 n times to get $P_{1,1},\ldots,P_{1,n}$, which complete the little diamonds in the picture. Notice that each Q_i ($1 \leq i \leq n-1$) is related to a pair of P's: $Q_i \rhd_w P_{1,i}$ and $Q_i \rhd_w P_{1,i+1}$. Therefore, by $n-1$ applications of theorem 2.1.14, there are terms $P_{2,1},\ldots,P_{2,n-1}$ such that $P_{1,i} \rhd_w P_{2,i}$ and $P_{1,i+1} \rhd_w P_{2,i}$. $\rhd_w$ is transitive, so $N_1 \rhd_w P_{2,1}$ and $N_2 \rhd_w P_{2,n-1}$.

At each level (indicated by the first index on the P's), the number of P's decreases by one; on the nth level there is exactly one (possibly) new term: $P_{n,n-(n-1)}$ (i.e., $P_{n,1}$). For each $P_{i,1}$, $P_{i,1} \rhd_w P_{i+1,1}$, and for each $P_{i,n-(i-1)}$, $P_{i,n-(i-1)} \rhd_w P_{i+1,n-i}$. Hence, also $N_1 \rhd_w P_{i+1,1}$ and $N_2 \rhd_w P_{i+1,n-i}$. However, then $N_1 \rhd_w P_{n,1}$ and $N_2 \rhd_w P_{n,1}$, which establishes the truth of the claim. In sum, if there are $M_1,\ldots,M_n$ to start with, then $\frac{(n+1)\cdot n}{2}$ applications of theorem 2.1.14 lead to $P_{n,1}$. This concludes the proof. qed

The above proof shows that once we have proved theorem 2.1.14, we can prove theorem 2.1.15. However, we claimed that the two theorems are equivalent, by which we mean that the theorems imply each other.

It is quite easy to see that theorem 2.1.15 implies theorem 2.1.14 as a special instance. The assumption of theorem 2.1.14 is that $M \rhd_w N_1$ and $M \rhd_w N_2$. By the definition of $=_w$, it is immediate that $N_1 =_w N_2$. Therefore, by an application of theorem 2.1.15, there is a suitable P.

2.2 Normal forms and consistency

The Church–Rosser theorem provides freedom in computing with terms. For example, we may be sure that if we devise different *strategies* for reducing terms, then this cannot lead to a "dead-end term" from which the results of other reduction sequences are not reachable.

Example 2.2.1. The term $\mathsf{K}(\mathsf{CM}x\mathsf{I})(\mathsf{W}(\mathsf{MC})(\mathsf{WI}(\mathsf{WI}x)))$ has several $\rhd_1$ reduction sequences, however, all of them are finite. It is not difficult to verify that they all lead to x, although some of the sequences seem to make a lengthy detour via reductions in subterms that are eventually omitted as an effect of the reduction of the redex with the head K.

Or let us consider $\mathsf{SKK}(\mathsf{CII}x)$. This term can be reduced in various ways; however, one-step reductions terminate in $x\mathsf{I}$, which is a term in nf.

The above examples are simple in the sense that although the terms have infinite $\triangleright_w$ reduction sequences (as any term does), their $\triangleright_w$ reduction sequences end in an infinite subsequence comprising one and the same term infinitely many times.

There are terms that have infinite $\triangleright_1$ reduction sequences comprising *infinitely many different* terms.

Example 2.2.2. The fixed point combinator Y was introduced in chapter 1 (see p. 12). The term YY has infinite $\triangleright_1$ reduction sequences only. Some of the terms in those sequences — for example, $\mathsf{YY}(\mathsf{Y}(\mathsf{YY}))$ and $\mathsf{Y}(\mathsf{Y}(\mathsf{Y}(\mathsf{Y}(\mathsf{Y}(\mathsf{YY})))))$ — appear diverging, but the Church–Rosser theorem guarantees that they can be reduced to a common term. One of the infinitely many terms to which both of these terms reduce is

$$\mathsf{Y}(\mathsf{Y}(\mathsf{Y}(\mathsf{Y}(\mathsf{YY}))))(\mathsf{Y}(\mathsf{Y}(\mathsf{Y}(\mathsf{Y}(\mathsf{Y}(\mathsf{YY})))))).$$

The term YK has exactly one $\triangleright_1$ reduction sequence and it is infinite. Each $\triangleright_1$ reduction yields a term with exactly one redex. (Incidentally, any other non-unary combinator would do in place of K.) But YK has a captivating property: after n $\triangleright_1$ reductions we get a term that can be applied to n arguments, and n further $\triangleright_1$ reductions lead back to YK. Metaphorically speaking, YK is like a black hole. For instance, $\mathsf{K}(\mathsf{K}(\mathsf{YK}))M_1M_2 \triangleright_w \mathsf{YK}$.

Definition 1.3.5 delineated a proper subclass of CL-terms: those that have a weak normal form. Clearly, not all terms are *in nf.* Moreover, some CL-terms do not reduce to a term in nf, that is, some CL-terms *do not have an nf.*

Example 2.2.3. Some of the terms mentioned above have a wnf. The normal form of $\mathsf{SKK}(\mathsf{CII}x)$ is $x\mathsf{I}$. The term $\mathsf{KM}(\mathsf{YK})$ has an nf too, which is M.

On the other hand, some of the already-mentioned terms do not have a wnf, for example, YK. Neither does MM have an nf.

The following lemma is an application of the Church–Rosser theorem.

LEMMA 2.2.4. (UNIQUENESS OF NF'S) *If a term M has an nf, that is, if there is a term N such that $M \triangleright_w N$ and N contains no redexes, then N is* unique.

Proof: Let us assume that there are two (distinct) nf's for M, let us say N_1 and N_2. $M \triangleright_w N_1$ and $M \triangleright_w N_2$; therefore, by theorem 2.1.14, $N_1 \triangleright_w P$ and $N_2 \triangleright_w P$, for some P. Since N_1 and N_2 are nf's, they do not contain a redex, which means that N_1 is P, and so is N_2. Then N_1 and N_2 are not distinct, contrary to the starting assumption. qed

We can make the following general statement.

COROLLARY 2.2.4.1. *Every CL-term has* at most one *weak normal form.*

The term KM(YK) has a normal form, but at the same time, it also has an infinite $\triangleright_1$ reduction sequence. This shows that the Church–Rosser theorem guarantees that reductions always *converge*, not that all possible reduction sequences lead to the nf of the term (if there is one). The following diagram illustrates this situation using the term KM(YK). The arrows indicate $\triangleright_1$ reductions, whereas the ...'s show that as long as only the redexes of the form YK are reduced, the $\triangleright_1$ reduction sequence does not terminate.

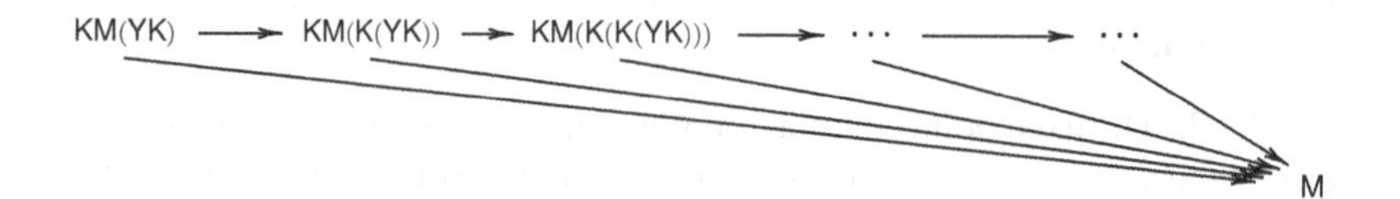

The convergence of reduction sequences ensures that we may choose a strategy for the reductions. We might miss the nf of a term, but we surely cannot get another nf. In the above example, the nf is reached only when the redex with head K is reduced. So far, we have only functions that take arguments from the right-hand side. (We will introduce other kinds of functions in chapter 7.) The redex occurrence of the combinator K is in the leftmost redex in the term.

The concept of the leftmost redex can be made precise, using the indexing introduced in section A.1. However, the following definition reflects the label well.

DEFINITION 2.2.5. If M is a CL-term and N is a redex in M, then N is *the leftmost redex* iff all redex occurrences P in M that are distinct from N are proper subterms of N, or they do not overlap with N and are to the right from N.

A CL-term, obviously, either is in weak normal form, or contains exactly one leftmost redex. Then the next definition describes a *deterministic procedure*, because at each step either there is exactly one redex to be reduced or there is none.

DEFINITION 2.2.6. Let M be a term. If N is the leftmost redex in M and reduces to N', then $[N/N']M$ is produced from M by the *one-step reduction of the leftmost redex*. This is denoted as $M \triangleright_1^{lm} [N/N']M$. The *leftmost reduction strategy*, which is abbreviated as *lmrs*, consists of performing $\triangleright_1^{lm}$ steps as long as there is a redex in the term.

<u>*Example*</u> 2.2.7. The *lmrs* applied to the term $\mathsf{S}(\mathsf{S}xy)(\mathsf{S}zv)(\mathsf{S}www)$ yields the following sequence of one-step reductions.

$\mathsf{S}(\mathsf{S}xy)(\mathsf{S}zv)(\mathsf{S}www) \triangleright_1^{lm}$
$\mathsf{S}xy(\mathsf{S}www)(\mathsf{S}zv(\mathsf{S}www)) \triangleright_1^{lm}$
$x(\mathsf{S}www)(y(\mathsf{S}www))(\mathsf{S}zv(\mathsf{S}www)) \triangleright_1^{lm}$
$x(ww(ww))(y(\mathsf{S}www))(\mathsf{S}zv(\mathsf{S}www)) \triangleright_1^{lm}$
$\ldots \qquad \triangleright_1^{lm}$

$x(ww(ww))(y(ww(ww)))(z(ww(ww))(v(ww(ww))))$.

Focusing on the leftmost redexes is motivated by the fact illustrated above, that the lmrs yields an nf if there is one. Sometimes, other reduction strategies lead to the nf too, in fact, they might comprise fewer $\triangleright_1$ reductions.

Exercise 2.2.1. Consider the starting term $\mathsf{S}(\mathsf{S}xy)(\mathsf{S}zv)(\mathsf{S}www)$ from example 2.2.7. Does this term have an infinite $\triangleright_1$ reduction sequence? What is the length of the shortest $\triangleright_1$ reduction sequence ending in the nf of the term?

LEMMA 2.2.8. *If a* CL-*term* M *has an nf, then the lmrs applied to* M yields the nf.

Proof: If a CL-term has an nf, then there is a $\triangleright_1$ reduction sequence of finite length. If the leftmost reduction sequence is a finite $\triangleright_1$ reduction sequence, then by theorem 2.1.14, the last term is the nf of the starting term.

The other possibility is that the lmrs gives us an infinite $\triangleright_1$ reduction sequence. Therefore, if some redex other than the leftmost one is reduced in another $\triangleright_1$ reduction sequence, then the leftmost redex remains in the term as long as the lmrs is not followed, i.e., the term will not reduce to its nf. qed

A normal form does not contain any redexes. Thus, if M has an nf, then either all the $\triangleright_1$ reduction sequences (starting with any of the redexes) of M are finite, or some redexes lead to infinite $\triangleright_1$ reduction sequences but they are in an omitted argument of a cancellator. Notice that in the latter case the redex, which leads to an infinite $\triangleright_1$ reduction sequence, is *inside* another *redex*. If we allowed infinitely many one-step reductions to precede the reduction of the outside redex, then we would not get to N. As an analogy, consider enumerating $\mathbb{N}\cup\{-1\}$ starting with an enumeration of $\mathbb{N}$ versus starting with -1 and continuing with an enumeration of $\mathbb{N}$. The former procedure will not succeed in enumerating the whole set, because we never get to -1.[7] The set of CL-terms that have an nf is called the set of *weakly normalizable terms*. The above lemma gives an algorithm that eventually yields the nf of a CL-term — if the term has an nf. However, we do not know how long the reductions, in accordance with the lmrs, should continue. In other words, the algorithm is not a decision procedure.

Definition 1.3.5 gives an effective description of the shape the CL-terms that *are in nf*, because it is obviously decidable if a CL-term contains a redex. We do not have though a characterization of the shape of CL-terms that *have an nf*. Of course, the above lemma does not provide anything of this sort. The best we will have is that, in chapter 9, weakly and strongly normalizable terms turn out to be CL-terms typable by certain types in a type assignment system.

Some terms *do have* nf's, and this fact has a remarkable consequence for CL. As long as only proper combinators are allowed in combinatory bases (possibly together with improper combinators that are definable from S and K), the consistency of $\triangleright_w$ and $=_w$ is quite obvious — if we have theorem 2.1.14.

[7] The analogy would be more faithful if the enumeration of -1 would somehow relieve us from enumerating $\mathbb{N}$ at all.

The consistency of CL with $\triangleright_{\text{w}}$ though may be proved without much difficulty and without any appeal to the Church–Rosser theorem. It is easy to see that no CL-term M can weakly reduce to a variable x unless x occurs in M, if we continue to assume that the combinatory basis is $\{\mathsf{S},\mathsf{K}\}$. All CL-terms are finite; therefore, there is no term M that reduces to all CL-terms.

The consistency of CL with $=_{\text{w}}$ is not so obvious. Weak reduction is symmetric, thus it could happen that somehow a chain of weak reductions and weak expansions could be construed to link two variables, for instance, x and y. Even if we restrict our consideration to the combinatory basis $\{\mathsf{S},\mathsf{K}\}$, which is combinatorially complete, there are infinitely many potential chains (each of finite length) to consider. However, we can show that CL is consistent with respect to $=_{\text{w}}$, that is, $=_{\text{w}}$ is not the total relation on CL-terms using the Church–Rosser theorem.

THEOREM 2.2.9. *There are* CL-*terms M and N such that $M =_{\text{w}} N$ does not hold, that is,* CL is consistent.

Proof: Let us consider two terms M and N that do not contain a redex. These can be two variables or two constants, for example, but they can be more complex terms too. Let us suppose — to the contrary — that $M =_{\text{w}} N$. Then by theorem 2.1.15, there is a term P such that $M \triangleright_{\text{w}} P$ and $N \triangleright_{\text{w}} P$. However, $M \triangleright_{\text{w}} M$ only, and, similarly, $N \triangleright_{\text{w}} N$ only; therefore, M and N are the same term. This is a contradiction, so $M =_{\text{w}} N$ is false. qed

The consistency of CL with respect to $=_{\text{w}}$ shows that weak equality is a reasonable relation in the sense that it partitions the set of CL-terms into more than one set. In other words, $=_{\text{w}}$ can distinguish between certain terms.

Consistency is important from several points of view. It is well known that both Cantor's informal set theory and one of Frege's formal systems were inconsistent (in the sense of negation inconsistency). Hilbert's program, which intended to place mathematics on firm foundations, included the aim to prove all mathematical theories to be consistent. Theorem 2.2.9 provides consistency for CL, though the proof itself is not formalized within CL.[8] As we will see in chapter 3, CL with the basis $\{\mathsf{S},\mathsf{K}\}$ and $=_{\text{w}}$ is a sufficiently powerful system to formalize computable functions. Therefore, analogues of Gödel's incompleteness theorems apply to CL. On the other hand, it is well known that Peano arithmetic can be (and has been — see, e.g., [71]) proved consistent. Thus there is no inconsistency or incoherence between this chapter and the next one.

The consistency result concerns *pure* combinatory logic and does not guarantee that all extensions of CL by arbitrary constants or other expressions such as formulas retain consistency. Indeed, Curry's paradox shows that a simple combination of classical sentential logic with combinatory logic is inconsistent.[9] However, CL by itself is provably consistent.

[8] The notions involved in the definitions of terms, redexes, reductions and weak equality, as well as in the proof, could be formalized in ZFC, the Zermelo–Fraenkel set theory with the axiom of choice.

[9] Two versions of Curry's paradox are detailed in section 8.1.

We mentioned that the consistency of CL with $\triangleright_w$ can be proved without appeal to powerful theorems like the Church–Rosser theorem. We state this consistency as a lemma and give an elementary proof of it.

LEMMA 2.2.10. *There are* CL-*terms* M *and* N *such that* neither $M \triangleright_w N$ nor $N \triangleright_w M$. *There is* no term P *such that* $P \triangleright_w Q$ *for all* Q.

Proof: We consider terms over the combinatory basis $\{\mathsf{S}, \mathsf{K}\}$ together with an infinite set of variables. $\mathsf{S} \triangleright_w \mathsf{K}$ and $\mathsf{K} \triangleright_w \mathsf{S}$ are both false, which proves the first part of the claim.

For the second part, we note that $\triangleright_w$ is defined from $\triangleright_1$ as its reflexive transitive closure. Let $\mathsf{S}M_1M_2M_3$ be a redex occurrence in N and let this particular redex occurrence be replaced by $M_1M_3(M_2M_3)$ to get N'. Then a variable x is free in N' only if it is free in N, because x may occur in N' outside the new subterm $M_1M_3(M_2M_3)$, or in the latter subterm. If the former, then x occurs in N outside $\mathsf{S}M_1M_2M_3$; indeed, x has the same occurrences in N and N'. If the latter, then x occurs in M_1, M_2 or M_3; hence, x also occurs in N. Considering a redex occurrence of the form $\mathsf{K}M_1M_2$ is similar.

Let us assume that a term P exists with the property that it reduces to any CL-term. If $P \triangleright_w Q$, then there is an $n \in \mathbb{Z}^+$ such that n successive one-step reductions starting with P lead to Q, or P is Q itself. In the second case, it is clear that all variables that occur in P occur in Q and vice versa, because P is Q. If $n = 1$, then — because of the first claim in the lemma — all the variables of Q are variables of P. Let a similar claim be true of a sequence of one-step reductions of length n from P to Q'. Adding yet another $\triangleright_1$ step from Q' to Q maintains the relationship between the variables of the resulting term Q and the starting term Q', and therefore, also between the variables of Q and P. That is, if x is a variable of Q, then it is a variable of P. There are infinitely many variables; however, no term can contain infinitely many of them. This completes the proof of the claim. qed

It is straightforward to ask whether considering which variables occur in some CL-terms could be used to prove the consistency of the weak equality relation.

Exercise 2.2.2. Consider $=_w$ instead of $\triangleright_w$. Why cannot the proof appealing to occurrences of variables be used to show the consistency of CL with $=_w$?

Exercise 2.2.3. Define a restricted set of CL-terms so that the proof of the above lemma can be modified to obtain a proof of consistency.

We introduced the relation $\triangleright_s$ in definition 1.3.18, and then pointed out that $\triangleright_s$ is a weaker relation than $\triangleright_w$. Therefore, the consistency of $\triangleright_w$ and $=_w$ does not imply the consistency of $\triangleright_s$. However, it can be verified that $\mathsf{S} \triangleright_s \mathsf{S}$ and $\mathsf{K} \triangleright_s \mathsf{K}$, but neither is strongly reducible to the other. That is, there are CL-terms M and N such that neither $M \triangleright_s N$ nor $N \triangleright_s M$.

LEMMA 2.2.11. *The set of* CL-*terms with* $\triangleright_s$ is consistent. *There is* no CL-term M *such that for all* N, $M \triangleright_s N$.

Exercise* 2.2.4. Prove the lemma. (Hint: $\triangleright_{\mathrm{w}}$ and $\triangleright_{\mathrm{s}}$ are defined differently; still, the proof of lemma 2.2.10 may be adapted to $\triangleright_{\mathrm{s}}$.)

Recall that $\triangleright_{\mathrm{s}}$ depends on how we define bracket abstraction, which is a pseudo-λ-abstraction in CL. We state (but do not prove) the next two theorems.

THEOREM 2.2.12. (CHURCH–ROSSER FOR $\triangleright_{\mathrm{s}}$) *If $M \triangleright_{\mathrm{s}} N_1$ and $M \triangleright_{\mathrm{s}} N_2$, then there is a term P such that $N_1 \triangleright_{\mathrm{s}} P$ and $N_2 \triangleright_{\mathrm{s}} P$.*

THEOREM 2.2.13. *There are* CL*-terms M and N such that $M =_{\mathrm{w}\zeta} N$* does not hold. *That is,* CL *is* consistent *with respect to extensional weak equality.*

2.3 Fixed points

Once the consistency of weak equality is guaranteed, it becomes interesting to look for analogues of other properties of functions. Some functions have *fixed points*, whereas some others do not.[10] For example, the "plus one" function on $\mathbb{N}$ (i.e., $f(x) = x+1$) has no fixed points, because $n+1 \neq n$ for all n. Another function, the "plus zero" on $\mathbb{N}$ (i.e., $f(x) = x+0$) has infinitely many fixed points, because $n+0 = n$, for all n. Zero is the identity element for addition, which explains why this function is rarely considered.

There are functions that have some fixed points, and it is crucial that they have a fixed point. For example, the classical consequence operator, which is often denoted by Cn, is a *closure operator* (possibly with additional properties postulated too).[11] This means that the set of consequences of the consequences of a set of formulas Δ is the set of consequences of Δ. To put it differently, the Cn operator has a fixed point given a set of formulas Δ, namely, $Cn(\Delta)$.

Closure operators, in general, have fixed points: for any x, $\mathfrak{c}(x)$ is a fixed point of the closure operator $\mathfrak{c}$. Weak reduction, $\triangleright_{\mathrm{w}}$, is defined in 1.3.4 as the reflexive transitive closure of $\triangleright_1$. The reflexive closure of $\triangleright_{\mathrm{w}}$ is $\triangleright_{\mathrm{w}}$ itself. Similarly, the transitive closure of $\triangleright_{\mathrm{w}}$ is $\triangleright_{\mathrm{w}}$ too.

A well-known *fixed point theorem* concerns monotone operations on complete lattices.[12] In chapter 6, we will encounter a fixed point theorem for complete partial orders (CPOs), which implies the fixed point theorem for complete lattices. The fixed point theorem for CPOs is important in the construction of a certain kind of a model for CL.

---

[10] Sometimes, "fixpoint" is used instead of "fixed point." I use the latter expression as a technical term.

[11] See definition A.2.7 in the appendix.

[12] This theorem is sometimes called the Knaster–Tarski or simply Tarski's fixed point theorem. Lattices are briefly introduced in section A.2; for more detailed expositions, see Birkhoff [27], Davey and Priestley [55] and Grätzer [78].

THEOREM 2.3.1. *Let* $\mathfrak{A} = \langle A, \wedge, \vee, \circledast \rangle$ *be a complete lattice with* $\circledast$ *a monotone unary operation on* A. $\circledast$ has a fixed point; *in particular,* $\circledast$ *has a* least *fixed point.*

Proof: A complete lattice has a top and a bottom element, which we denote by $\top$ and $\bot$, respectively. The top element is the greatest element, that is, for all $a \in A$, $a \leq \top$. $\leq$ is the usual partial order on a lattice, which is defined by stipulating $a \leq b$ iff $a = a \wedge b$. Then it is obvious that $\circledast\top \leq \top$, because $\circledast$ is a total function of type $A \longrightarrow A$. We associate to each element a of the lattice a set $f(a)$, which is $\{b\colon a \leq b \text{ and } \circledast b \leq b\}$. For each element $a \in A$, $f(a) \neq \emptyset$, because $\top \in f(a)$.[13]

We only consider the set $f(\bot)$, which may be defined equivalently as $f(\bot) = \{b\colon \circledast b \leq b\}$. The inequation $\circledast b \leq b$ is "half" of the fixed point equation for $\circledast$. $f(\bot)$ contains all elements from A that satisfy this inequation. We take $\bigwedge f(\bot)$, the greatest lower bound of $f(\bot)$. The set $f(\bot)$ may contain infinitely many elements. However, the lattice is complete, so $\bigwedge f(\bot)$ exists. We let c denote the element $\bigwedge f(\bot)$.

Let us assume that $b \in f(\bot)$. $c \leq b$, because c is a lower bound. $\circledast$ is monotone; thus, $\circledast c \leq \circledast b$. From $b \in f(\bot)$, we know that $\circledast b \leq b$, and, by the transitivity of $\leq$, $\circledast c \leq b$. To sum up, if $b \in f(\bot)$, then $\circledast c \leq b$. But c is the greatest lower bound of $f(\bot)$, which implies that $\circledast c \leq c$. This inequality means that $c \in f(\bot)$, by the definition of the latter.

The operation $\circledast$ is monotone. Then from $\circledast c \leq c$ we get $\circledast\circledast c \leq \circledast c$. The latter means that $\circledast c \in f(\bot)$, because $\circledast c$ satisfies the defining inequality. Then $c \leq \circledast c$, because c is a lower bound of $f(\bot)$. We have $\circledast c \leq c$ and $c \leq \circledast c$, which shows that c is a fixed point of $\circledast$.

The proof of the second part of the claim is the next exercise. qed

Exercise 2.3.1. Prove that c is *the least* fixed point of $\circledast$. (Hint: You should show that for all $d \in A$, $\circledast d = d$ implies $c \leq \circledast d$.)

In logic, mathematics and computer science many concepts are *defined inductively* or *recursively*.[14] The objects that are so defined exist because they are the least fixed points of the defining clauses.

Example 2.3.2. An infinite sequence of natural numbers is named after Fibonacci. A function F that gives the nth element of the Fibonacci sequence can be defined inductively as follows.

(1) $F(0) = 1$ and $F(1) = 1$;

(2) $F(n+2) = F(n) + F(n+1)$, where $n \in \mathbb{N}$.

[13] A series of related theorems are proved in [146]. For instance, it is shown that the set of fixed points of $\circledast$ is a complete lattice itself, and that the existence of a fixed point for a monotone operation is equipotent to the underlying lattice being complete.

[14] In section A.1, we have two sample inductive definitions and associated inductive proofs, which are representative of the kind of inductions we need here. Of course, in general, induction on all ordinals is used in mathematics, which implies the use of set theory of various strengths.

The function that is defined is the least fixed point of the defining conditions understood as a function of F.[15]

A proof of Gödel's incompleteness theorem for formalized arithmetic uses a fixed point of a function. The idea is that it is first shown that certain expressions, including formulas, can be represented by numbers, which nowadays are usually called Gödel numbers. Then a one-place predicate (i.e., a unary function) is constructed to express that the formula the Gödel number of which is the argument of the predicate is not a theorem of formalized arithmetic. From the fact that there is a fixed point for this predicate combined with the Gödel number function, it can be deduced that formalized arithmetic is incomplete if consistent. We return to the analogue of Gödel's theorem for CL in chapter 3.

These examples are to illustrate that fixed points occur in many places in logic, computer science and mathematics. In particular, we intend to show the computational power of combinatory logic, by showing that in CL all functions have fixed points.

THEOREM 2.3.3. (FIXED POINT) *Let M be a* CL*-term.* There is a CL-term N *such that* $MN =_w N$.

Proof: We may assume that we have a primitive or a defined combinator Y. (Exercise 1.2.7 justifies this assumption.) The term $\mathsf{Y}M$ can be taken in place of N, because $\mathsf{Y}M =_w M(\mathsf{Y}M)$. qed

If the fixed point combinator Y is primitive, then it is sufficient to have just *one* combinator with the axiom $\mathsf{Y}x \rhd x(\mathsf{Y}x)$. But we may have different combinatory bases and there might be many, even infinitely many, fixed point combinators. Indeed, the latter is the case as long as we do not consider $=_{w\zeta}$.

In chapter 1, we introduced a handful of combinators, including Y. Combinators that are like Y to some extent can be defined from the combinatory basis $\{\mathsf{S},\mathsf{K}\}$, but we can also take Y as a primitive constant: $\{\mathsf{S},\mathsf{K},\mathsf{Y}\}$.

Example 2.3.4. $\mathsf{BM}(\mathsf{CBM})$ is a term with the property that

$$\mathsf{BM}(\mathsf{CBM})x =_w x(\mathsf{BM}(\mathsf{CBM})x).$$

The following $\rhd_1$ reductions show that this is so. $\mathsf{BM}(\mathsf{CBM})x \rhd_1 \mathsf{M}(\mathsf{CBM}x) \rhd_1 \mathsf{CBM}x(\mathsf{CBM}x) \rhd_1 \mathsf{B}x\mathsf{M}(\mathsf{CBM}x) \rhd_1 x(\mathsf{M}(\mathsf{CBM}x))$. It is also true that $x(\mathsf{BM}(\mathsf{CBM})x) \rhd_1 x(\mathsf{M}(\mathsf{CBM}x))$, but $\mathsf{B}x\mathsf{M}(\mathsf{CBM}x)$ does not reduce to $x(\mathsf{BM}(\mathsf{CBM})x)$. To sum up, $\mathsf{BM}(\mathsf{CBM})x \rhd_w x(\mathsf{BM}(\mathsf{CBM})x)$ does not hold, unlike for Y, though the weak equality obtains.

Taking Y as an undefined combinator ensures that it behaves as desired with respect to $\rhd_w$. However, it is possible to define a combinator like Y from the basis $\{\mathsf{S},\mathsf{K}\}$.

[15]In chapter 3, it will become clear how to turn the defining conditions into a function of F, when we will define combinators such as Z_B, Z'_p and Z_{pr}. (See sections 3.2 and 3.3.)

Fixed point combinators have been defined in different ways by at least two well-known logicians: Curry and Turing. We already saw that B is definable from S and K, and W is definable too.

Curry's fixed point combinator is $\mathsf{S(BWB)(BWB)}$. This term has the fixed point property with respect to $=_\mathrm{w}$, but not with respect to $\rhd_\mathrm{w}$. On the other hand, the term is in weak nf.

The connection between proper combinators and the fixed point combinator becomes transparent if we think along the lines Turing did. $\mathsf{Y}x \rhd_\mathrm{w} x(\mathsf{Y}x)$ could be thought of as having a term MM in place of Y, that is, as $MMx \rhd_\mathrm{w} x(MMx)$. This would not make M a proper combinator, but similarly to an introduction of the existential quantifier (i.e., $\exists$) in first-order logic, we could replace *not all*, but only the last three occurrences of M, by a variable. We get the term $x(yyx)$ from $Myx \rhd_\mathrm{w} x(yyx)$, because we want to keep the newly introduced variable distinct from x. Now M is obviously a binary proper combinator, which is definable from S and K due to combinatorial completeness. Once M is defined, it only remains to apply that term to itself. The construction itself guarantees that the resulting term will have the fixed point property with respect to $\rhd_\mathrm{w}$.

Given this idea, it remains to define a combinator that has the above reduction pattern. The combinator may be defined even in standard form.

Exercise 2.3.2. Apply the algorithm from the proof of lemma 1.3.9 (but with η included) to obtain a combinator like M containing only S and K in standard form.

If we assume some of the combinators that we introduced in chapter 1, then we can define M. For example, M can be taken to be $\mathsf{BW(BB'M)}$, or replacing $\mathsf{B'}$ by its definition from B and C, M can be defined as $\mathsf{BW(B(CB)M)}$. It is also possible to replace $\mathsf{B'}$ by $\mathsf{B(CI)}$ or BT, resulting in $\mathsf{BW(B(B(CI))M)}$.[16] $\mathsf{B'}$ and $\mathsf{B(CI)}$ are not the same function (even in the sense of $=_{\mathrm{w}\zeta}$). It simply so happens that in the context of the whole term they can be exchanged.

Exercise 2.3.3. Explain why the replacement of $\mathsf{B'}$ by $\mathsf{B(CI)}$ in the above CL-term is possible, though $\mathsf{B'} =_{\mathrm{w}\zeta} \mathsf{B(CI)}$ does not hold.

Once we have a CL-term for M, *Turing's fixed point combinator* is obtained by applying that term to itself:

$$\mathsf{BW(BB'M)(BW(BB'M))}.$$

The term is not in nf. However, we can easily turn it into a term in weak normal from: $\mathsf{W(B'(M(BW(BB'M))))}$. If we take $\mathsf{BW(BB'M)}$ as a primitive, e.g., X with the axiom $\mathsf{X}xy \rhd y(xxy)$, then the combinator XX applied to x weakly reduces to $x(\mathsf{XX}x)$.

Exercise 2.3.4. Verify that both $\mathsf{BW(BB'M)(BW(BB'M))}$ and $\mathsf{W(B'(M(BW(BB'M))))}$ have the fixed point property with respect to $\rhd_\mathrm{w}$. (Hint: You can construct a single sequence of $\rhd_1$ reductions involving both CL-terms.)

[16] T is Church's notation for CI and is often used by others too; T's axiom is $\mathsf{T}xy \rhd yx$.

Let us return to Curry's fixed point combinator and analyze the idea behind it. First of all, let us look at the one-step reductions $\mathsf{BWB}xy \rhd_1 \mathsf{W}(\mathsf{B}x)y \rhd_1 \mathsf{B}xyy \rhd_1 x(yy)$. The role of the initial S in Curry's fixed point combinator is to get $\mathsf{BWB}x(\mathsf{BWB}x)$ from $\mathsf{S}(\mathsf{BWB})(\mathsf{BWB})x$. The former term reduces in three steps to $x(\mathsf{BWB}x(\mathsf{BWB}x))$. Of course, the variable x may be replaced by any term without affecting the reduction steps; thus, we can reconstruct the process of finding a fixed point combinator as follows.

We want to find a combinator like the function f in $fx \rhd_\mathrm{w} x(fx)$, and so we first will find a term M such that $Myx \rhd_\mathrm{w} x(yx)$. Again, this M is a proper combinator, and it is surely definable by S and K. Now we take $\mathsf{BWB}M(\mathsf{BWB}M)$ for y, that is, we consider $M(\mathsf{BWB}M(\mathsf{BWB}M))x$. By M's axiom, $M(\mathsf{BWB}M(\mathsf{BWB}M))x \rhd_\mathrm{w} x(\mathsf{BWB}M(\mathsf{BWB}M)x)$. By our previous observation, the latter term further reduces to $x(M(\mathsf{BWB}M(\mathsf{BWB}M))x)$. This gives $M(\mathsf{BWB}M(\mathsf{BWB}M))$ as a combinator that has the fixed point property with respect to weak reduction, that is, $M(\mathsf{BWB}M(\mathsf{BWB}M))x \rhd_\mathrm{w} x(M(\mathsf{BWB}M(\mathsf{BWB}M))x)$.

Exercise 2.3.5. Apply the algorithm from the proof of lemma 1.3.9 (with η added) to obtain a combinator like the latter M containing only S and K in standard form.

Clearly, there are a lot of CL-terms that have the fixed point property at least with respect to $=_\mathrm{w}$.

Exercise 2.3.6. Define a combinator that has the fixed point property for $=_\mathrm{w}$ using only B, C and M.

Exercise 2.3.7. Define a combinator that has the fixed point property for $=_\mathrm{w}$ using only B, I and W.

Let us assume that the basis is $\{\mathsf{S}, \mathsf{K}\}$. We have the following lemma.

LEMMA 2.3.5. *There are* infinitely many *distinct terms that have the fixed point property with respect to* $=_\mathrm{w}$.

Proof: The reduct $xz(yz)$ of the redex $\mathsf{S}xyz$ closely resembles $x(\mathsf{Y}x)$. Indeed, if we put I (i.e., SKK) into the first argument of S, then we have that $\mathsf{S}(\mathsf{SKK})yz \rhd_1 \mathsf{SKK}z(yz)$, which in turn reduces to $z(yz)$. Lastly, if we take additionally any fixed point combinator as the second argument of S (with SKK in place of the first argument), then we have a fixed point combinator. Thus if we have a Y defined in terms of S and K, then $\mathsf{S}(\mathsf{SKK})\mathsf{Y}$, $\mathsf{S}(\mathsf{SKK})(\mathsf{S}(\mathsf{SKK})\mathsf{Y})$, $\mathsf{S}(\mathsf{SKK})(\mathsf{S}(\mathsf{SKK})(\mathsf{S}(\mathsf{SKK})\mathsf{Y}))$, ..., $\mathsf{S}(\mathsf{SKK})(\ldots(\mathsf{S}(\mathsf{SKK})(\mathsf{S}(\mathsf{SKK})\mathsf{Y}))\ldots)$, ... are all combinators that have the fixed point property with respect to $=_\mathrm{w}$. qed

We can think about the multiplicity of CL-terms with the fixed point property along different lines.

LEMMA 2.3.6. *There are* infinitely many CL-*terms, each of which has the fixed point property with respect to* $\rhd_\mathrm{w}$.

Exercise 2.3.8. Prove that there are infinitely many distinct terms Y such that $\mathsf{Y}x \rhd_{\mathrm{w}} x(\mathsf{Y}x)$. (Hint: It is straightforward to generalize Turing's idea to arbitrary n.)

Exercise 2.3.9.** Prove that there are infinitely many distinct terms Y with the fixed point property for $\rhd_{\mathrm{w}}$ via a suitable generalization of Curry's idea. (Hint: Generalize Curry's idea to arbitrary n.)

We mentioned that fixed points for unary functions are quite common in many areas, and so the idea of having a fixed point for any function might not seem very surprising by now. To return to some of the remarks we made in chapter 1 about combinatorial completeness, we state *the general fixed point principle.*

THEOREM 2.3.7. (GENERAL FIXED POINT PRINCIPLE) *Let $\mathfrak{B}$ be a combinatorially complete basis in the sense of definition 1.3.8. If f is an n-ary function such that $fx_1 \ldots x_n \rhd_{\mathrm{w}} M$, where M is a term that is built from (some of) the variables $x_1, \ldots, x_n$, proper combinators and f itself, then* there is a combinator Z *definable from $\mathfrak{B}$ such that $\mathsf{Z}x_1 \ldots x_n \rhd_{\mathrm{w}} M$, where the occurrences of f in M (if there are any) are replaced by Z.*

Proof: We divide the proof into three steps, depending on the kind of atomic subterms in M.

1. If the set of atomic subterms of M is a subset of $\{x_1, \ldots, x_n\}$, then the claim coincides with lemma 1.3.9; hence, the proof of that lemma can be used now with a minor addition. If $\mathfrak{B}$ is combinatorially complete, then, by definition 1.3.8, S and K are definable from $\mathfrak{B}$ or they are in $\mathfrak{B}$. The defining CL-terms can be used instead of S and K in the proof. In chapter 4, we will see that, if there are more combinators in a basis, then sometimes shorter CL-terms result from a bracket abstraction. However, using the CL-terms in place of S and K always works — independently of the concrete combinatory basis.

2. If the set of atomic subterms of M contains proper combinators in addition to some (or none) of the variables $x_1, \ldots, x_n$, then we can apply lemma 1.3.9, and obtain definitions of the proper combinators that occur in M. Then we re-apply the bracket abstraction algorithm, which is guaranteed to produce a CL-term for Z. Notice that the presence of combinators in M is unproblematic, because they do not contain variables. In fact, combinators might occur in the meta-terms that contain pseudo-λ-operators. The only difference now is that we *start* with a term that contains some combinators and possibly variables.

3. If the set of atomic subterms of M contains f itself, then we can follow Turing's idea to reduce the problem to that in **1** or **2**. First, we replace all the occurrences of f in M by yy, where y is a new variable with respect to $x_1, \ldots, x_n$. We also replace f on the left of the $\rhd_{\mathrm{w}}$ by gy, where g is a new function symbol. g is definable and gg is a CL-term suitable as a definition of f. qed

Exercise* 2.3.10. Generalize Curry's idea to give a proof of theorem 2.3.7.

So far we have focused on fixed points of single functions. Recursive definitions of numerical functions can be turned into single equations when we think of the two equations in the definition as filling in parts of an "if $n = 0$, then ..., else ..." statement. (We will see in chapter 3 how this can be done with CL-terms representing natural numbers and functions.) However, sometimes *simultaneous induction* is used to define several functions or sets at once. Thus it is an interesting question whether fixed points for two functions can be found.

DEFINITION 2.3.8. Let f_1 and f_2 be binary functions. f_1 and f_2 have *mutual fixed points* when there are x and y such that both $f_1xy \triangleright_w x$ and $f_2xy \triangleright_w y$.

The analogous definition in CL is as follows.

DEFINITION 2.3.9. Let M_1 and M_2 be CL-terms. M_1 and M_2 have *mutual fixed points* just in case there are terms N_1 and N_2 such that both $M_1N_1N_2 =_w N_1$ and $M_2N_1N_2 =_w N_2$.

It is useful to compare the above definitions to $fx = x$, that is the single fixed-point equation, and to $MN =_w N$, that is the single fixed-point weak equality. The fixed point combinator Y ensures that for any function M, $\mathsf{Y}M$ can be taken for the term N; that is, $\mathsf{Y}M$ is the fixed point of M.

In the case of two functions, we are looking for a combinator (call it Y_2) that, for any f_1 and f_2, can serve as the head both of an N_1 (such that $N_1 \triangleright_w f_1N_1N_2$) and of an N_2 (such that $N_2 \triangleright_w f_2N_1N_2$). Since N_1 and N_2 must contain f_1 and f_2, and since both $f_1N_1N_2$ and $f_2N_1N_2$ contain N_1 and N_2, one of Y_2's argument places must contain f_1 in both N_1 and N_2, and another must contain f_2 in both N_1 and N_2. A third argument place must contain f_1 in N_1 and f_2 in N_2, because $f_1N_1N_2$ starts with f_1 and $f_2N_1N_2$ starts with f_2. In sum, Y_2 must be at least ternary. If the arguments x, f_1 and f_2 are in that order, then we get $\mathsf{Y}_2f_1f_1f_2$ and $\mathsf{Y}_2f_2f_1f_2$, respectively, when we substitute f_1 or f_2 for x. The two terms, $\mathsf{Y}_2f_1f_1f_2$ and $\mathsf{Y}_2f_2f_1f_2$ are the mutual fixed points of the functions f_1 and f_2. Therefore, Y_2 should have the axiom

$$\mathsf{Y}_2x_1x_2x_3 \triangleright x_1(\mathsf{Y}_2x_2x_2x_3)(\mathsf{Y}_2x_3x_2x_3).$$

We can use Turing's idea and find a combinator Z such that $\mathsf{Z}vxyz \triangleright x(vvyyz)(vvzyz)$. By combinatorial completeness, this function can be expressed as a combinator built from S and K only. Then Y_2 is ZZ, which justifies the next claim.

LEMMA 2.3.10. *Any pair of* CL*-terms* M_1 *and* M_2 *has* mutual fixed points.

Exercise 2.3.11. Define the combinator Z using the combinators introduced so far.

Exercise* 2.3.12. Define Z using only the combinators S and K. (Hint: You may apply the abstraction algorithm from the proof of lemma 1.3.9, perhaps together with η, to obtain such a combinator.)

Exercise* 2.3.13. Adapt Curry's idea for defining Y, and find another Y_2.

Exercise* 2.3.14. Express Y_2 from the previous exercise using S and K only.

Finding mutual fixed points for several functions may be generalized easily for any $n \in \mathbb{N}$.

DEFINITION 2.3.11. Let $M_1, \ldots, M_n$ be CL-terms. The *mutual fixed points* of these functions are CL-terms $N_1, \ldots, N_n$ such that

$$M_1 N_1 \ldots N_n =_w N_1$$
$$\vdots$$
$$M_n N_1 \ldots N_n =_w N_n.$$

LEMMA 2.3.12. *If $M_1, \ldots, M_n$ are CL-terms, then they have* mutual fixed points.

Proof: The search for a suitable combinator Y_n can proceed as before. Let us assume that there is such a combinator. Then Y_n may be an $n+1$-ary function that yields the fixed point of the function in its first argument place — when that function is identical to one of the functions in the remaining n argument places. More explicitly, we want to have the following weak equality hold.

$$\mathsf{Y}_n x y_1 \ldots y_n =_w x(\mathsf{Y}_n y_1 y_1 \ldots y_n) \ldots (\mathsf{Y}_n y_n y_1 \ldots y_n).$$

By theorem 2.3.7, Y_n is definable from any combinatorially complete basis. The mutual fixed points are $\mathsf{Y}_n M_i M_1 \ldots M_n$, where $1 \leq i \leq n$. qed

Exercise* 2.3.15. Find a combinator that gives the mutual fixed point for three CL-terms M_1, M_2 and M_3. (Hint: You may find it helpful to use some of the combinators (Ψ, $\Phi, \ldots, \Phi_n$) that were introduced in section 1.3 or to define a combinator $\mathsf{Z}xyzv \rhd xy(xz)(xv)yzv$ separately.)

Exercise* 2.3.16. Describe a general procedure for finding a suitable Y_n for an arbitrary n — based on Curry's method of finding Y. (Hint: See example 2.3.4 and the subsequent description of discovering Y according to Curry's approach.)

The fixed point theorems may be used to show the existence of CL-terms that satisfy weak equalities of some prescribed shape. We list some exercises that are solvable by applying the ideas used in the proofs of the fixed point theorems in this section.

Exercise 2.3.17. Prove that there is a CL-term M such that for any CL-term N, $MN =_w M$. (Hint: This weak equality closely resembles $MN =_w N$ from the first fixed point theorem. However, the equations are not exactly the same, and the task now is to find one particular M for all N.)

Exercise 2.3.18. Prove that there is a CL-term M such that $M \rhd_w MM$. In other words, there is a CL-term that weakly reduces to its fixed point. (Hint: $MM \rhd_w M$ can be easily seen to hold with I in place of M, but $\mathsf{I} \rhd_w \mathsf{II}$ is certainly false.)

Exercise* 2.3.19. Prove that, for any $n \in \mathbb{N}$, there is a CL-term M such that $M \rhd_w \underbrace{M \ldots M}_{n}$. (Hint: Generalize the solution of the previous exercise from 2 to n.)

Exercise 2.3.20. For any combinator Z^n, where $n > 1$, there are CL-terms M and N such that $\mathsf{Z}M =_{\mathrm{w}} N$, where N is distinct from $\mathsf{Z}M$. Show that this is true in two different ways: using some fixed point ideas and not using them. (Hint: Recall that a combinator is a function such that when supplied with fewer arguments than its arity, in general, does not weakly reduce to any CL-term.)

Exercise 2.3.21. Prove that there are CL-terms M_1 and M_2 such that $M_1 =_{\mathrm{w}} M_1M_2$ and $M_1 =_{\mathrm{w}} M_1M_2M_2$. (Hint: There is a certain similarity between this exercise and exercise 2.3.18.)

Exercise 2.3.22. Prove that there are distinct CL-terms M_1 and M_2 such that $M_1 \triangleright_{\mathrm{w}} M_2M_1$ and $M_2 \triangleright_{\mathrm{w}} M_1(M_2M_1)$. (Hint: Recall that $\triangleright_{\mathrm{w}}$ is transitive.)

2.4 Second fixed point theorem and undecidability

As a brief concluding section, we mention two theorems here, because of their importance for CL. The proofs of these theorems will follow in chapter 3.

CL is sufficiently powerful to allow encodings of recursive functions; therefore, the language of CL can be arithmetized. In particular, it is possible to encode by CL-terms the "Gödel number" function, which is denoted by G. Thus $\mathsf{G}(M)$ gives the CL-term that encodes the Gödel number of the CL-term M.

THEOREM 2.4.1. (SECOND FIXED POINT THEOREM) *For any* CL-*term* M, there is a CL-term N *such that* $M(\mathsf{G}N) =_{\mathrm{w}} N$.

The motivation for calling this theorem a fixed point theorem is obviously the shape of the weak equality. The proof that we will give does not involve an application of Y. However, the theorem is equivalent to the existence of a certain fixed point combinator, the existence of which follows from theorem 2.3.7 (by step **3** in the proof). Once G is defined, we are looking for a CL-term for a unary function f that is characterized by $fx = x(\mathsf{G}(fx))$. (Actually, G will occur in f itself in the concrete CL-term that we construct in chapter 3.)

The next theorem is due to Scott, and it shows once again that CL is a highly nontrivial theory.

THEOREM 2.4.2. (SCOTT'S THEOREM) *Let X and Y be nonempty sets of* CL-*terms closed under* $=_{\mathrm{w}}$.[17] *Then X and Y are* not recursively separable.

This claim means that it is not possible to find a set X' that is a superset of X, that is decidable (i.e., recursive), yet that does not overlap with Y.

COROLLARY 2.4.2.1. *The set of* CL-*terms that have a weak normal form (wnf) is* not decidable.

[17] "Closure" is applied here in the usual sense, that is, if $M =_{\mathrm{w}} N$, then if either is an element of a set, so is the other.

Chapter 3

Recursive functions and arithmetic

Combinatory logic is a *powerful formalism.* This is demonstrated by its ability to formalize the notion of a computable function. Some other formal systems that can model the same class of functions are Turing machines, Post calculi, recursive functions, λ-calculi, Markov algorithms and register machines. (There are further formalisms that are just as expressive as those mentioned.) Some of these formal systems have several versions. For instance, variants of Turing machines can have a one-way or a two-way infinite tape, which does not alter the class of functions computable by Turing machines; bounding the size of the available tape by a linear function restricts, whereas adding an oracle expands, the class of computable functions.[1]

Some of the formalisms, let us say Markov algorithms, might *appear* to be more powerful than others (e.g., register machines), because, for instance, the input and output of a Markov algorithm may be a word that can contain letters from an alphabet, that is, the inputs and the output are not limited to natural numbers. This impression is misleading. However, a concrete notion of computable functions might be easier to represent in another model because of accidental features of the functions such as the permitted kinds of inputs and outputs.

Computable functions — informally speaking — are functions that can be computed effectively, perhaps, as in Turing's conception of computability: using paper and pencil. Effectiveness incorporates certain idealizations such as an unbounded supply of time (and, possibly, of other resources that are part of the concrete notion of computation). Thus the lack of sufficient time for a computation to be carried out — for instance, due to the limited life span of the solar system — does not show that a function is not computable. It shows only that there is no hope of actually computing that particular value given the function and the input. With the advancement of computer science and the ubiquity of computers, questions of practical computability, which go beyond theoretical computability, became paramount. But here we are concerned only with the theoretical aspects of computability, that is, with "computability in principle."

In order to demonstrate the expressive power of CL, we have to choose a notion of computability. *Recursive functions* (which subsequently we will call *partial recursive* functions) are functions on the set of natural numbers. They can be represented

[1]For introductions to computability, see Harel [80], Boolos and Jeffrey [28] and Sipser [135]. Other classic treatments and papers include Rogers [125], Davis [57] and Davis [56].

in both CL and λ-calculus. *Representing* recursive functions in CL means that we have to select a suitable set of CL-terms to stand for the natural numbers, and we have to show that, for any recursive function, there is a CL-term that behaves with respect to the terms that stand for the natural numbers like that recursive function behaves with respect to the natural numbers.

In this chapter, we make precise the claim about the power of CL, and afterward we look at some consequences along the lines of Gödel's incompleteness theorem for arithmetic.

3.1 Primitive and partial recursive functions

There are different but equivalent ways to define the classes of primitive and partial recursive functions.[2] *Primitive recursive functions* are generated from a class of functions by two operations. The functions in the base set are thought to be simple or easily calculable, in some sense.

First of all, *constant functions* are not difficult to calculate, since they take the same value for all inputs. There is a *constant zero function*, which takes the value 0 whenever the argument is a natural number. Let us denote this unary function by $\mathfrak{z}$. Then we have that $\mathfrak{z}(0) = 0$, $\mathfrak{z}(1) = 0$, $\mathfrak{z}(2) = 0$, etc.

The natural numbers may be defined inductively from 0 by "+1," that is, by the operation of counting, which is also conceived as calculable. We denote this operation, *the successor function*, by $\mathfrak{s}$. This function is unary too, and its values are $\mathfrak{s}(0) = 1$, $\mathfrak{s}(1) = 2$, $\mathfrak{s}(2) = 3$, etc.

We may consider functions that return as their value one of their inputs. These functions are called *projection functions* — because of an analogy with geometry — and denoted by π's. The superscript on π indicates the arity of the function, whereas the subscript shows which argument becomes the value of the function. Of course, for any π^n_m, the subscript m must fall within the range from 1 to n. These functions are evidently similar to certain combinators. For instance, π^1_1 is a unary function such that $\pi^1_1(x) = x$, that is, π^1_1 is like the unary identity combinator I. The function π^2_1 is binary, and always returns its first argument: $\pi^2_1(x,y) = x$, that is, this function behaves like K. Similarly, π^2_2 is like KI. All the projection functions of arity greater than or equal to 2 resemble cancellators in CL, more precisely, those cancellators that drop all but one of their arguments.

DEFINITION 3.1.1. The *base set* of primitive recursive functions comprises $\mathfrak{z}$, $\mathfrak{s}$ and π^n_m, where m and n are positive integers (i.e., $n,m \in \mathbb{Z}^+$) and $m \leq n$.

The base set has *infinitely many* functions in it. Nevertheless, there are a lot of functions not in this set that are informally thought to be calculable. A frequently

[2] We give here an often-used definition — see, e.g., Boolos and Jeffrey [28] and Mendelson [105].

exploited idea in building up a set of objects is to find operations yielding new objects that preserve some desirable property of the starting objects. *Composition* and *primitive recursion* are the operations that generate more functions, and they are thought to preserve computability.

Given the unary functions f and g, their composition is $f \circ g$, meaning that $f \circ g$ is a unary function such that the value of $f \circ g$ for x is $f(g(x))$.[3] The case of unary functions straightforwardly generalizes as follows. Let us assume that $f_1^n, \ldots, f_m^n$ are all n-ary functions, and g^m is an m-ary function. Then we can put all these functions together to obtain a compound function that takes n arguments. The λ notation makes precise the way the composition operation creates an n-ary function.

(fc) $$\lambda x_1 \ldots \lambda x_n . g(f_1(x_1, \ldots, x_n), \ldots, f_m(x_1, \ldots, x_n)).$$

The domain (dom) of g is $\underbrace{\mathbb{N} \times \ldots \times \mathbb{N}}_{n}$, and $\mathrm{cod}(f_i) = \mathbb{N}$, for all i, $1 \leq i \leq n$.

The composition of two unary functions is often denoted by $\circ$ (as above), and we adopt this symbol (in a prefix position) for the general composition operation. Thus the above function can be written (omitting the λ's) as

$$\circ(g, f_1^n, \ldots, f_m^n).$$

The subscript on the last f indicates the arity of g, whereas the superscripts on the f's, which are all the same, show the arity of the function that results via composition.

Example 3.1.2. Let us consider $+$, the addition function, as well as two unary functions, taking the square and the cube of a number x: x^2 and x^3. We can compose the three functions into a unary function: $\lambda x. x^2 + x^3$. (In more customary notation, the new function is written as $f(x) = x^2 + x^3$.) If we denote $+$ by f_1, $\lambda x. x^2$ by f_2 and $\lambda x. x^3$ by f_3, then f is $\circ(f_1, f_2, f_3)$. Having defined f, we can calculate the values of f and we can write $f(0) = 0$, $f(1) = 2$, $f(2) = 12$, etc.

There are combinators that function as composition operations. For example, B can be seen to compose two unary functions, let us say f and g. $\mathsf{B}fgx \triangleright_1 f(gx)$, which, according to the convention concerning parentheses in CL-terms, means that g is applied to x and then f is applied to $g(x)$. B' is also a *composition combinator* like B except that the order of the two functions is reversed — just as in some notational variants of function composition. (Occasionally, $(f \circ g)(x)$ is spelled out as $g(f(x))$.) The combinator S is called the *strong composition combinator*. If we consider the first two arguments in $\mathsf{S}fgx$ as functions (as the notation, f and g, intends to suggest) then by one-step reduction we get the term $fx(gx)$. The latter term can be thought of as f and g each being applied to x separately, and then the former

[3]For function composition, in general, it is often assumed — either as part of the notation, or explicitly — that the value of $g(x)$ is in the domain of f. The codomain (cod) of arithmetic functions is always $\mathbb{N}$, thus unary arithmetic functions always can be composed.

result being applied to the latter one. S' is related to S as B' is to B: the axiom for S' is $\mathsf{S}'fgx \rhd_1 gx(fx)$.

Function composition may appear to be one operation in virtue of the functional expression in (fc) above. The combinators that we considered, however, make clear that for any pair $\langle m,n \rangle$, there is a function composition operation. (If we permit permutations of functions like those reflected by B' and S', then for any $\langle m,n \rangle$, where $m=n$ or $m \geq 2$, there is more than one function composition operation.) The perspective taken in CL makes clear some of the assumptions hidden in the informal notation used in $\circ(g, f_1^n, \ldots, f_m^n)$. An assumption that CL does not reveal, but rather drops completely, is that the domains and codomains of the functions composed are appropriately related.

The second operation generating further primitive recursive functions is *primitive recursion*. The structure of the set of the natural numbers suggests that calculations could proceed according to that structure. To calculate the value of a function for a given n, one could start at 0; once $f(0)$ is obtained, then $f(1)$ could be calculated according to some function applied to the value $f(0)$ and, perhaps, to 1. This is the idea behind recursion. The term "recursion" is derived from the calculations recursing (or "running back") to calculations of values for smaller arguments.

A binary function $f(x,y)$ is defined by primitive recursion when, first, $f(x,0)$ is defined as the value of a unary function, let us say of g, at x. Then $f(x,y)$, for $y>0$, is the value of a ternary function h with x, $y-1$ and $f(x,y-1)$ as its arguments. Recursive definitions are often written in the form of two equations, and the general schema of primitive recursion, defining an n-ary function can also be formulated in this way. We will first consider the case where the function defined is binary. Let g and h be a unary and a ternary function, respectively. Then the binary function f is defined by (1) and (2).

(1) $f(x,0) = g(x)$,

(2) $f(x,\mathfrak{s}(y)) = h(x,y,f(x,y))$.

In order to avoid the need to postulate a side condition, we used $\mathfrak{s}$, the successor function in (2). Since 0 is not a successor of a natural number, $\mathfrak{s}(y)$ is a natural number strictly greater than 0.

Example 3.1.3. Probably the simplest example of a primitive recursive function that is defined via primitive recursion is addition. A more descriptive name for this function would perhaps be "addition with recursion on the right." We use $+$ in a prefix position for this function.

(1) $+(x,0) = x$,

(2) $+(x,\mathfrak{s}(y)) = \mathfrak{s}(+(x,y))$.

These equations do not conform exactly to the pattern prescribed by the general schema. First of all, x on the right-hand side of the equation in (1) has to be replaced by a unary function of x that gives the value x for each x. It is easy to replace the first equation by $+(x,0) = \pi_1^1(x)$.

The second equation also has to be modified in order to have the required shape. On the right-hand side of the equation sign, we have to have a ternary function, and π_3^3 seems to be an obvious candidate. However, $+(x,\mathfrak{s}(y)) = \pi_3^3(x,y,+(x,y))$ does not work, because the two equations together define the constant x function. Composition can also be understood as a sort of substitution. The equation $+(x,\mathfrak{s}(y)) = \mathfrak{s}(+(x,y))$ and the equation that did not work (i.e., $+(x,\mathfrak{s}(y)) = \pi_3^3(x,y,+(x,y))$) together suggest that $\pi_3^3(x,y,+(x,y))$ needs to be substituted into $\mathfrak{s}$, yielding $\mathfrak{s}(\pi_3^3(x,y, +(x,y)))$. Using the $\circ$ notation, $\circ(\mathfrak{s},\pi_3^3)$ is the desired ternary function h. Thus the second equation becomes $+(x,\mathfrak{s}(y)) = \circ(\mathfrak{s},\pi_3^3)(x,y,+(x,y))$.

We can look at definitions by primitive recursion as "If ..., then ..., else ..." statements. The first equation for $+$ fills in the first two empty spots: "If $y=0$, then $+(x,y) = \pi_1^1(x)$, ..." The second equation could be written as "If $y \neq 0$, then $+(x,y) = \mathfrak{s}(+(x,y-1))$. However, in this rendering the "-1" or predecessor function appears. Noting that $y \neq 0$ is true just when $y = \mathfrak{s}(n)$ for some n, we can get another formulation of the second equation as "If $y = \mathfrak{s}(n)$, then $+(x,\mathfrak{s}(n)) = \mathfrak{s}(+(x,n))$." It is always true that $y=0$ or $y = \mathfrak{s}(n)$, but not both. Therefore, the two clauses can be straightforwardly merged into "If $y \neq \mathfrak{s}(n)$, then $+(x,y) = \pi_1^1(x)$, else $+(x,y) = \mathfrak{s}(+(x,n))$."

There is a similarity — but there is also a difference — between equations involving CL-terms and definitions by primitive recursion. CL-terms are viewed as functions. Thus if a combinator occurs as the head of some redexes on both sides of an equation, then the equation resembles the second equation in the definition of a function by primitive recursion. $\mathsf{Y}x =_{\mathsf{w}} x(\mathsf{Y}x)$ is a prototypical example of an equation (with $=_{\mathsf{w}}$ as the equality relation) in which a combinator is the head of redexes on both sides of $=_{\mathsf{w}}$. A difference between this equation and an equation like $+(x,\mathfrak{s}(y)) = \mathfrak{s}(+(x,y))$ is that the equation containing Y has no restrictions or accompanying equations for the "base case." To put it differently, although $+$ occurs on both sides of the equation sign, the arguments of $+$ are (more precisely, the second argument of the $+$ function is) different, whereas Y's argument on each side is x.

A definition by primitive recursion can be thought of as a pair of conditions the least fixed point of which is the function that is defined. This allows the use of the fixed point combinator in modeling primitive recursion in CL, because the fixed point combinator yields a fixed point for any function — primitive recursive or not.

DEFINITION 3.1.4. The set of *primitive recursive functions* is defined inductively by (1)–(3).[4]

(1) The functions $\mathfrak{z}$, $\mathfrak{s}$ and π_m^n (where $n,m \in \mathbb{Z}^+$, $m \leq n$) are primitive recursive.

(2) If f^n and $g_1^m,\ldots,g_n^m$ are primitive recursive functions, then $\circ(f,g_1,\ldots,g_n)$ is an m-ary primitive recursive function.

[4]As usual, we leave the closure clause tacit in the definition. The notation $\vec{x}$ indicates a finite sequence of arguments.

(3) If f^n and g^{n+2} are primitive recursive functions, then so is h^{n+1} defined by the following equations.[5]

1. $h(\vec{x},0) = f(\vec{x})$,

2. $h(\vec{x},\mathfrak{s}(y)) = g(\vec{x},y,h(\vec{x},y))$.

Exercise 3.1.1. Once we have addition, multiplication is definable as a primitive recursive function. Show that $\times(x,y)$ is a primitive recursive function. (Hint: First, give an "informal" definition as in example 3.1.3, and then adjust the form of the definition to conform to clause (3) in the above definition.)

Exercise 3.1.2. The factorial of a positive natural number n (denoted by $n!$) is the product of all positive integers that are less than or equal to n. The factorial of 0 is defined to be equal to 1. Show that $f(x) = x!$ is a primitive recursive function. (Hint: Start by putting the given definition into a form where $n!$ is calculated from the factorial of a number less than n.)

There are many other functions commonly used in arithmetic that are primitive recursive. Our goal here has been only to look at some functions as examples and to recognize them as primitive recursive functions.[6]

We have two generating operations so far: composition and primitive recursion. It is thought, however, that there is another operation, *minimization*, that can be used to define a function that is computable in an informal sense. The range of the functions under consideration is $\mathbb{N}$, the set of natural numbers, which has a least element, 0. We can ask if a function takes 0 as its value on any argument, and, if it does, what is the least argument for which this happens.

DEFINITION 3.1.5. Let f be a function of $n+1$ arguments. We use $\mathfrak{y}_f^{\vec{x}}$ as an abbreviation for $\inf\{y\colon f(x_1,\ldots,x_n,y) = 0\}$. The n-ary function g is defined by *minimization* as follows.

$$g(x_1,\ldots,x_n) = \begin{cases} \mathfrak{y}_f^{\vec{x}} & \text{if } \mathfrak{y}_f^{\vec{x}} \text{ exists (i.e., } \mathfrak{y}_f^{\vec{x}} \in \mathbb{N}), \\ \star & \text{otherwise,} \end{cases}$$

where $\star$ means that $g(x_1,\ldots,x_n)$ is undefined. The minimization operation that yields g is denoted by μ, that is, μf is g.

Sometimes, the notation $\mu x_{n+1} f$ is used instead together with the convention that minimization affects the last argument of f. Having a variable after μ may seem like a good idea, because x_{n+1} is not an argument of μf. However, one of the insights CL provides for us is that μ can be considered to be an operation on a function and need not be viewed as a variable binding operator like $\forall$ or λ.

---

[5]If $n = 0$, then $f(\vec{x})$ is a constant.

[6]We have included a few more examples in section A.3.

The abbreviation inf stands for infimum (greatest lower bound), and we assume that $x_1,\ldots,x_n,x_{n+1}$ and y are natural numbers, which come with the usual $\leq$ ordering on $\mathbb{N}$. If $g(x_1,\ldots,x_n)=y$, that is, the value of g is obtained from the first case in the definition of g, then this implies that $f(x_1,\ldots,x_n,x_{n+1})=0$ for at least one x_{n+1}. If there is no such x_{n+1}, then $\{x_{n+1}\colon f(x_1,\ldots,x_n,x_{n+1})=0\}=\emptyset$. However, inf $\emptyset$ does not exist in $\mathbb{N}$, because there is no largest natural number.

Minimization allows us to define an n-ary function from another function that has arity $n+1$. The minimization happens with respect to the last argument of f.[7] The first line in the definition gives the value of g for $\langle x_1,\ldots,x_n\rangle$ (i.e., y) if the function f takes the value 0 for $\langle x_1,\ldots,x_n,y\rangle$ and y is the least such number in the $n+1$ st argument place of f with $\langle x_1,\ldots,x_n\rangle$ fixed. The second line in the definition indicates that the function g is undefined for $\langle x_1,\ldots,x_n\rangle$ if f never takes the value 0 with $\langle x_1,\ldots,x_n\rangle$ as its first n arguments.

The definition may seem to be "wildly non-computable." But inf can be read here as "the least natural number such that ...." Then it is clear that it is not necessary to find all the members of $\{x_{n+1}\colon f(x_1,\ldots,x_n,x_{n+1})=0\}$ before returning the least element of the set as the value of $g(x_1,\ldots,x_n)$. There is another question concerning computability to which we will return a bit later.

Example 3.1.6. Let us consider $\mathfrak{s}$ for a moment. $\mu\mathfrak{s}$ is constantly undefined, because $\mathfrak{s}(x)\neq 0$ for any x. Thus $\mu\circ(\mathfrak{s},+)$ is a unary function that is totally undefined. To see what is going on, let us write $\circ(\mathfrak{s},+)$ (informally) as $\mathfrak{s}(+(x,y))$, where x and y are the arguments, in that order. The composition with $\mathfrak{s}$ results in the successor of the sum of x and y. Since $\mathfrak{s}$ itself never takes the value 0, it does not matter what is the sum of x and y; $\mathfrak{s}(+(x,y))$ is never 0 for any x. $\mu\mathfrak{s}(+(x,y))$ is a function of one variable, namely, of x, and $\mu\circ(\mathfrak{s},+)(x)=\star$, for all x.

The example shows that minimization may result in a function that is *not everywhere defined* (i.e., *not total*), even if the function to which μ has been applied is itself total. In terms of computation — both in the informal sense and in the sense of a computing device — if the value of a function is undefined for some input, then the computation does not terminate; it is divergent. One of the popular models of computation is the Turing machines. If a Turing machine does not stop in a standard configuration (for instance, because it enters a loop), then the value of the function being computed cannot be determined; hence, it is undefined. This computational viewpoint hints at a potential difficulty with respect to the minimization operation, if it is applied repeatedly to a primitive recursive function. If $\star$ appears as a value of f for $\langle x_1,\ldots,x_n,y_1\rangle$ before a y_2 with $f(x_1,\ldots,x_n,y_2)=0$ is found, then it is not clear how to compute $(\mu f)(x_1,\ldots,x_n)$. In other words, the difficulty is that $\star$ is not a genuine value for a function; it is merely a symbol that we use to indicate the lack of a value.

Let us consider a concrete function: $\dot{-}$, a sort of subtraction, which is primitive recursive and it is undefined for any pair $\langle x,y\rangle$, when $y>x$ but otherwise equals x

[7] There is nothing special about the last argument of f; it is simply notationally convenient to select this argument "to disappear."

minus y (i.e., the result of subtraction in the usual sense). Given $\dot{-}$, we can define a function, which we denote by $\overset{\mathfrak{s}}{-}$, such that $\overset{\mathfrak{s}}{-}$ is undefined for $\langle 0,y\rangle$ when $y>0$ but also for $\langle 0,0\rangle$ (unlike $\dot{-}$, which gives 0). For all $x>0$, there is a y such that $\langle x,y\rangle$ yields a genuine value (i.e., a value other that $\star$). To round out the description of $\overset{\mathfrak{s}}{-}$, we make $\overset{\mathfrak{s}}{-}$ a binary function that takes values as follows.[8]

$$x \overset{\mathfrak{s}}{-} y = \begin{cases} x-\mathfrak{s}(y) & \text{if } \ x>y, \\ \star & \text{if } \ x=y, \\ \star & \text{if } \ x<y. \end{cases}$$

We will use the successor function in the definition of $\overset{\mathfrak{s}}{-}$ (this is what "justifies" the notation). Let us compose $\dot{-}$ in its second argument with $\mathfrak{s}$, that is, let $x \overset{\mathfrak{s}}{-} y = x \dot{-} \mathfrak{s}(y)$. The smallest number that can be put in the place of both x and y is 0. However, $0 \overset{\mathfrak{s}}{-} 0$, $0 \overset{\mathfrak{s}}{-} 1$, $0 \overset{\mathfrak{s}}{-} 2$, $0 \overset{\mathfrak{s}}{-} 3$, etc., are all undefined, that is, $0 \overset{\mathfrak{s}}{-} 0=\star$, $0 \overset{\mathfrak{s}}{-} 1=\star$, $0 \overset{\mathfrak{s}}{-} 2=\star,\ldots$.

If we apply minimization to the function $\overset{\mathfrak{s}}{-}$, then we get a function, $\mu\overset{\mathfrak{s}}{-}$, which is undefined for 0 only. Of course, we could think of this unary function in set-theoretic terms as a set of ordered pairs over $\mathbb{N}\cup\{\star\}$, namely, $\{\langle 0,\star\rangle,\langle 1,0\rangle,\langle 2,1\rangle,\langle 3,2\rangle,\ldots\}$, but $\star\notin\mathbb{N}$. Thinking computationally about minimization, we could assume that the least y is found by going through all the possible values starting with 0. But $0 \overset{\mathfrak{s}}{-} 0$ is undefined, $0 \overset{\mathfrak{s}}{-} 1$ is undefined, and so on. It is important to note that we are not given the "bird's-eye" view of this function that we provided in the above definition by cases. We do not know at the point when $0 \overset{\mathfrak{s}}{-} y$ is considered that, if the function is undefined for this y, then so it is for any inputs greater than this particular y. Thus infinitely many possible inputs have to be considered to determine that $\mu\overset{\mathfrak{s}}{-}(0)=\star$. The undefined value is inherited through function composition, and if an undefined value is like a diverging computation, then that is not a problem with respect to calculating the value of the composed function. However, if the 0 of a function follows undefined values, then the computation does not reach the 0. A diverging computation or a loop in a computation is itself similar to stepping through an inifinity of steps.

The problem of stepping through infinitely many numbers before getting to another stage is similar to a problem that can emerge if a naive approach is taken toward showing that there are $\aleph_0$ many rational numbers.[9] Rational numbers may be defined to be fractions of positive integers signed with $+$ or $-$ and with 0 thrown into the set. (Alternatively, $\mathbb{Q}=\{\frac{m}{n}: m,n\in\mathbb{Z}\wedge n\neq 0\}$, where $\mathbb{Z}$ is the set of all integers.) The positive rational numbers may be listed as shown in the following table.

[8] The previous sentence did not give a full description of the function. Our intention is to get an easily definable function that can still serve as an illustration.

[9] The analogy with a naive attempt at an enumeration is quite appropriate here. $\overset{\mathfrak{s}}{-}$ will not be excluded from the set of functions to which minimization can be applied, because no initial segment of $\mathbb{N}$ that yields $\star$ everywhere is followed by a natural number n such that $x \overset{\mathfrak{s}}{-} n=0$. For all $n\in\mathbb{N}$, $0 \overset{\mathfrak{s}}{-} n=\star$.

$\mathbb{Q}^+$	1	2	3	4	5	6	7	...
1	1/1	2/1	3/1	4/1	5/1	6/1	7/1	...
2	1/2	2/2	3/2	4/2	5/2	6/2	7/2	...
3	1/3	2/3	3/3	4/3	5/3	6/3	7/3	...
4	1/4	2/4	3/4	4/4	5/4	6/4	7/4	...
5	1/5	2/5	3/5	4/5	5/5	6/5	7/5	...
6	1/6	2/6	3/6	4/6	5/6	6/6	7/6	...
7	1/7	2/7	3/7	4/7	5/7	6/7	7/7	...
⋮	...	...	...	...	...	...	...	⋱

This arrangement is obviously redundant. For example, all n/n's are notational variants of $1/1$, that is, 1. However, if the listing with repetitions can be shown to have cardinality $\leq \aleph_0$, then $\mathrm{card}(\mathbb{Q}^+) = \aleph_0$, because $\mathbb{Z}^+ \subseteq \mathbb{Q}^+$.

If $\mathbb{Q}^+$ is enumerated line by line, then we get an infinite sequence of length $\aleph_0$ from the first line alone. Of course, the point of arranging $\mathbb{Q}^+$ into the infinite matrix above is to reveal the short diagonals, which can be used to enumerate $\mathbb{Q}^+$ as one denumerable sequence. Such an enumeration will have each positive rational repeated infinitely many times — just as in the matrix.

The operation of minimization could not be seen to yield a computable function if for some x infinitely many steps would precede reaching a y yielding the value 0 in the calculation. From this computational angle, an undefined value may be thought of as a loop in a Turing machine. This means that if we incorporate minimization into the definition of recursive functions, we have to restrict the class of functions to which minimization may be applied.

DEFINITION 3.1.7. (MINIMIZATION) Let f^{n+1} be a function that has the property in (1). (All $x_1,\ldots,x_n,y,z$ and v range over elements of $\mathbb{N}$. As previously, we abbreviate $\inf\{y\colon f(x_1,\ldots,x_n,y)=0\}$ by $\mathfrak{y}_f^{\vec{x}}$.)

(1) $\forall y.(\exists v.\, v = \mathfrak{y}_f^{\vec{x}} \wedge y < v) \Rightarrow \exists z.\, f(x_1,\ldots,x_n,y) = z$

The *minimization* on f gives the n-ary function g that is defined as

$$g(x_1,\ldots,x_n) = \begin{cases} \mathfrak{y}_f^{\vec{x}} & \text{if } \mathfrak{y}_f^{\vec{x}} \text{ exists,} \\ \star & \text{otherwise.} \end{cases}$$

Condition (1) on f says that starting with 0 for y, and counting upward, as long as $f(x_1,\ldots,x_n,y) \neq 0$, $f(x_1,\ldots,x_n,y)$ gives a genuine value z (i.e., a z such that $z \in \mathbb{N}$).

Definitions 3.1.5 and 3.1.7 are really the same except that the latter definition restricts the type of functions to which the operation can be applied. If a function f is total, that is, it returns a natural number as a value for every $n+1$-tuple of natural numbers, then obviously μf is the same function no matter which definition is used. Example 3.1.6 showed that the minimization operation can lead to a function g that is not everywhere defined even though f is a total function. $\overset{\mathfrak{s}}{-}$ satisfies condition (1) for f in definition 3.1.7, and, therefore, the two definitions give the same result. In general, if both definitions are applicable, then they yield the same function. However, with functions that are not everywhere defined, care has to be taken to make sure that the resulting function is computable, which is ensured when the function is defined "below its least zero." Thus we take the definition 3.1.7 as the definition of minimization.

DEFINITION 3.1.8. A function is *partial recursive* iff it is

1. one of the *basic primitive recursive functions* $\mathfrak{z}$, $\mathfrak{s}$ and π^n_m ($m,n \in \mathbb{N}$ and $m \leq n$), or

2. obtained from partial recursive functions by *composition* and *primitive recursion*, or

3. obtained from a partial recursive function by *minimization*.

The name "partial recursive" is sometimes replaced by its permuted variant "recursive partial." The latter expression, perhaps, is less likely to imply that the functions in question are not everywhere defined. The intended meaning of "partial recursive functions" is that the set of these functions contains not only total functions but also functions that are partial. Each function is (definitely) either partial or total. Thus another way to explicate the meaning of "partial" in "partial recursive" is to say that it is *not assumed* that a function belonging to this set is total. Totality contrasts primitive recursive functions with partial recursive functions, which are not all total. Sometimes, "partial recursive" is used only for those functions from definition 3.1.8 that are indeed partial or that are not primitive recursive, and then "recursive" refers to all or to the rest of the functions.[10]

The set of partial recursive functions is identical to the set of *general recursive functions*; however, the latter class of functions is defined by sets of equations over primitive recursive functions rather than via a minimization operation.[11]

LEMMA 3.1.9. *All primitive recursive functions are* total, *that is,* everywhere defined.

Proof: The set of primitive recursive functions is defined inductively, and so the proof that we sketch is by induction. The basic functions can be seen to be total at once.

The inductive step is split into two, since there are two operations in the definition of primitive recursive functions that generate primitive recursive functions. Let us assume, as the hypothesis of induction, that $g^n_1, \ldots, g^n_m$ are total functions, and so is

[10] As we mentioned already, we use "recursive" as a short name for partial recursive functions, and we include "primitive" when it is important that a function belongs to that smaller class of functions.

[11] We do not introduce general recursive functions in detail here. However, see [136], [125] and [105].

h^m. Then, given $x_1,\ldots,x_n$, $g_i(x_1,\ldots,x_n) \in \mathbb{N}$ for each i (where $1 \leq i \leq m$), which then implies that $\circ(h,g_1,\ldots,g_m)$ has a value — a natural number — for $\langle x_1,\ldots,x_n \rangle$. The proof is completed by the step in exercise 3.1.3. qed

Exercise 3.1.3. The definition of a primitive recursive function may involve primitive recursion (see definition 3.1.4). Finish the proof of the preceding lemma by proving that a function defined by primitive recursion is total, if the functions used in the definition are total.

Obviously, not all recursive functions are total — as we have already demonstrated by examples. The term "partial recursive" indicates only that totality is not assumed, but one might inquire "How partial are the partial recursive functions?" If we look at functions as ordered tuples, then there are only a few distinctions that can be made. A function may have no (genuine) value: for no inputs, for one or more but finitely many inputs, for infinitely many inputs or for all inputs. However, in terms of generating a function we may correlate (potential) partiality with minimization.

THEOREM 3.1.10. (KLEENE NORMAL FORM) *If f is a partial recursive function, then f may be defined (from the basic functions using composition and primitive recursion) with at most one application of minimization.*[12]

We included this theorem here merely to emphasize the difference between the roles that the three generating operations play in the definition of partial recursive functions. Of course, minimization cannot be completely eliminated, because then we would be left with primitive recursive functions only.

The informal notion of computable functions is wider than the set of primitive recursive functions. A well-known example is the Ackermann function. The Ackermann function is defined by a set of equations, which, however, do not adhere to the primitive recursion schema. Ackermann's original function and its definition have been simplified and modified by others without the characteristic property of the function being lost. The modified functions are often labeled as Ackermann functions too.

DEFINITION 3.1.11. Let the binary function g be defined by (i)–(iii).

(i) $g(x,0) = 0$,

(ii) $g(x,1) = 1$,

(iii) $g(x,\mathfrak{s}(\mathfrak{s}(y))) = x$.

The Ackermann function $\mathfrak{a}$ is ternary and is defined by (1)–(3).

(1) $\mathfrak{a}(x,y,0) = x+y$,

(2) $\mathfrak{a}(x,0,\mathfrak{s}(z)) = g(x,z)$,

[12] The claim is more formally stated (and also proved) in Rogers [125, §1.10]. A strengthening is given by Smullyan [136, §14, theorem 9].

(3) $\mathfrak{a}(x, \mathfrak{s}(y), \mathfrak{s}(z)) = \mathfrak{a}(x, \mathfrak{a}(x, y, \mathfrak{s}(z)), z)$.

The function g is merely an auxiliary function in the definition, and it is not difficult to see that g is primitive recursive. It may be overlooked that (1)–(3) are not of the form of the primitive recursive schema.[13] The discrepancy is in (3), where on the left hand side of the equation both y and z are taken with the successor function, that is, a recursion happens on two variables.[14]

Exercise* 3.1.4. Show that the function g in the above definition is indeed primitive recursive.

Ackermann proved that $\mathfrak{a}$ is *not primitive recursive*. Ackermann's function and similar functions have been garnering more attention lately. The characteristic property of these functions is that they *dominate* all primitive recursive functions. A function f *dominates* g iff there is an $n \in \mathbb{N}$ such that for all $n' > n$, $f(n') > g(n')$. We have seen that addition and multiplication are primitive recursive, but so is exponentiation. Domination of all primitive recursive functions implies that the function $\mathfrak{a}$ after a certain point grows faster than exponentiation, which is considered to be growing unfeasibly fast from the point of view of complexity theory and actual computations on computers.

We only wish to reiterate that $\mathfrak{a}$ is *computable*; therefore, the primitive recursive functions cannot be taken to capture the informal notion of computable functions.

Exercise 3.1.5. Ackermann's function is sometimes called "generalized exponentiation." Which value for which argument of $\mathfrak{a}$ does give the "real" exponentiation function?

Exercise* 3.1.6. Calculate the value of $\mathfrak{a}(2,2,4)$. (Hint: Write out the steps of the calculation of values of $\mathfrak{a}$ with smaller arguments that are used in calculating $\mathfrak{a}(2,2,4)$, which is an "enormous" number.)

3.2 First modeling of partial recursive functions in CL

A modeling of partial recursive functions requires us to show that for any such function there is a CL-term that computes as expected on the set of CL-terms, which are taken to represent the natural numbers.

Natural numbers and functions on them may be represented in various ways using CL-terms. These numbers are extremely familiar to all of us, and we might think at first that we really know very well what natural numbers are. However, when natural

[13]The function + is used in the usual infix notation, but + is the same addition function as above, and hence primitive recursive.

[14]See also Gödel [77, §9].

numbers are taken to be mathematical objects, then — unsurprisingly — it turns out that there are different approaches to defining or characterizing them. For instance, natural numbers can be represented by various sets. 3 may be the set of all sets that have three elements, or 3 may be the set with elements 0, 1 and 2, where the latter are also certain sets. Having more than one way to describe natural numbers and partial recursive functions within the set of CL-terms is completely unproblematic and does not diminish the significance of a representation.[15]

Church in [44] gave a representation of natural numbers by λ-terms, and defined the notion of λ-computable functions on them. λ-calculi and CL are closely related — as we will see in more detail in chapter 4 — and so are Church's representation and the following one.

DEFINITION 3.2.1. (NATURAL NUMBERS, 1) The CL-terms that stand for natural numbers are as follows.

(1) 0 is KI;

(2) if M is x, then $\mathfrak{s}(x)$ is $\mathsf{SB}(M)$.

In other words, *zero* is KI and *the successor function* is SB.

Partial recursive functions are defined inductively, and this means that it is sufficient to find terms that represent the basic functions and the generating operations to have an inductively defined set of CL-terms that represent partial recursive functions.

The *successor* function, $\mathfrak{s}$ is already given; and the *constant zero* function $\mathfrak{z}$ can be taken to be $\mathsf{K}(\mathsf{KI})$. We need to have that $\mathfrak{z}(x) = 0$ for *any* argument x, and $\mathsf{K}(\mathsf{KI})x \triangleright_{\mathrm{w}} \mathsf{KI}$, which is 0.

We can use $\mathfrak{z}$ to highlight a feature of our enterprise that may seem slightly unusual at first. When we defined recursive functions, we tacitly assumed that numbers and functions are different sorts of objects. We would not even consider $\mathfrak{s}(\mathfrak{z})$, for instance, to be a well-formed expression. This is in sharp contrast with what happens when CL-terms stand in for numbers and functions. $\mathsf{SB}(\mathsf{K}(\mathsf{KI}))$ is a CL-term, and this term is in weak nf just as $\mathsf{SB}(\mathsf{KI})$ is. The dissimilarity is even greater in the case of $\mathsf{K}(\mathsf{KI})(\mathsf{SB})$. $\mathfrak{z}(\mathfrak{s})$ is not a well-formed expression in the context of recursive functions, because $\mathfrak{s} \notin \mathrm{dom}(\mathfrak{z})$ (i.e., $\mathfrak{s} \notin \mathbb{N}$), and so $\mathfrak{z}(\mathfrak{s}) = 0$ is senseless. On the other hand, $\mathsf{K}(\mathsf{KI})(\mathsf{SB})$ is not in nf, and $\mathsf{K}(\mathsf{KI})(\mathsf{SB}) \triangleright_{\mathrm{w}} \mathsf{KI}$ just as $\mathsf{K}(\mathsf{KI})M \triangleright_{\mathrm{w}} \mathsf{KI}$ for terms M that are natural numbers.

On second thought, we can see that there is nothing strange about natural numbers and recursive functions both being CL-terms. The lack of apparent distinctions between terms representing functions and terms representing numbers is not unusual.

[15]In a certain sense, it is not clear how to characterize the natural numbers, because the first-order axiomatization of arithmetic (Peano arithmetic or PA) is not categorical — as demonstrated by the existence of nonstandard models. We do not go into a detailed discussion of these questions no matter how interesting or important they are. We only need here a more or less common-sense understanding of the natural numbers and of the fact that there are alternative set theoretical modelings of them.

Some terms can be interpreted as a "legitimate" application of a function to a number, whereas other terms cannot. This is completely in parallel with representations of numbers and functions in set theory. One might think that functions obviously differ from numbers, because they are ordered pairs (or tuples). However, ordered pairs are just sets, which have a particular structure, perhaps like $\{\{a\},\{a,b\}\}$, which is $\langle a,b\rangle$. Unless we know which encoding of ordered pairs into sets is used, we will see only $\{\{\{a\},0\},\{\{b\}\}\}$ or $\{\{a,0\},\{b,1\}\}$ without knowing which (if either) is $\langle a,b\rangle$.[16]

All the *projection functions* except π_1^1 omit at least one of their arguments. The corresponding combinators are *cancellators*. The projection functions return one of their arguments. This implies that the corresponding combinators are proper. One of these combinators is K, which is often taken to be an element of the combinatory basis. By lemma 1.3.9, for any function f with the rule of computation $f^n(x_1,\ldots,x_n)=x_i$ (where $1\leq i\leq n$), there is a combinator Z that is definable from $\{\mathsf{S},\mathsf{K}\}$, or any other combinatorially complete basis, such that $\mathsf{Z}x_1\ldots x_n \rhd_{\mathrm{w}} x_i$.

Exercise 3.2.1. Define combinators that can represent the projection functions π_1^3, π_3^4 and π_4^4. (Hint: In order to get shorter CL-terms, you might wish to use some further combinators beyond S and K.)

Next we have to show that *composition* is expressible by combinators. We assume — as our inductive hypothesis — that there are combinators that represent the functions that are being composed. Let M_g and $M_{f1},\ldots,M_{fm}$ be the combinators for g and $f_1,\ldots,f_m$, respectively. This means that if $N_1,\ldots,N_n$ are CL-terms for some natural numbers $l_1,\ldots,l_n$, then $M_{fi}N_1\ldots N_n \rhd_{\mathrm{w}} N_j$ just in case $f_i(l_1,\ldots,l_n)=l_j$, where $1\leq i\leq m$. Then the representation of composition simplifies to the issue of finding a CL-term Z_c^{m+n+1} such that $\mathsf{Z}_cM_gM_{f1}\ldots M_{fm}N_1\ldots N_n$ weakly reduces to $M_g(M_{f1}N_1\ldots N_n)\ldots(M_{fm}N_1\ldots N_n)$. Slightly more abstractly, we have to have a combinator Z with arity $m+n+1$ such that

$$\mathsf{Z}_cx_1\ldots x_{n+m+1} \rhd x_1(x_2x_{m+2}\ldots x_{m+n+1})\ldots(x_{m+1}x_{m+2}\ldots x_{m+n+1}).$$

Without the preceding context, it could seem puzzling why we are interested in this particular combinator. The indices, of course, capture the intended structure of the resulting term, but that might seem arbitrary or at least unmotivated without the idea of composition. The desired axiom for Z_c makes it clear that Z_c is an easily definable combinator. There are no variables on the right-hand side of $\rhd$ that do not occur on the left-hand side, and there are no constants either. Moreover, Z_c itself does not occur on the right-hand side. In other words, composition is represented by proper combinators, for all m and n. The reduction pattern displayed above also reminds us that we need a combinator for each pair $\langle m,n\rangle$ (where $1\leq m,n$ and $m,n\in\mathbb{N}$), because the operation that is called composition of functions in the theory of recursive functions is more complicated than we first might have thought. From

[16]In fact, some models of CL rely on the possibility of encoding functions as natural numbers. See section 6.3.

the axiom for Z_c, we can see that if $m = 1$, then Z_c only needs to be an *associator*, whereas if $m > 1$, then Z_c must be also a duplicator for each of its arguments from x_{m+2} to x_{m+n+1}.

Example 3.2.2. The case of $\langle 1,1 \rangle$ is quite simple, indeed, this is what is most often denoted by $\circ$. $f \circ g$ is the function that takes one argument and $(f \circ g)(x) = f(g(x))$, for any x. If the CL-terms for f and g are M_f and M_g, and N is the term representing the argument, then B does the job. Indeed, $\mathsf{B}M_fM_gN \rhd_1 M_f(M_gN)$, which is the desired CL-term.

Example 3.2.3. Let us assume that we would like to find a combinator for $\langle 2,1 \rangle$ composition. That is, g_1^1 and g_2^1 are to be composed with h^2. The superscript on the g's tells us that the resulting function is unary, thus we want to have a combinator, let us say $\mathsf{Z}_{2,1}^4$, such that $\mathsf{Z}_{2,1}^4 x_1x_2x_3x_4 \rhd x_1(x_2x_4)(x_3x_4)$. We might notice the similarity between the term on the right and the term $x_2x_4(x_3x_4)$ that results from the reduction of $\mathsf{S}x_2x_3x_4$. In fact, $\mathsf{Z}_{1,2}^4$ is one of the Φ's, which we introduced in definition 1.3.23 — the Φ without subscripts. Using this insight, we can take instead of $x_1(x_2x_4)(x_3x_4)$ the term $\mathsf{B}x_1x_2x_4(x_3x_4)$. Then $\mathsf{S}(\mathsf{B}x_1x_2)x_3x_4$ would work, except that x_1 and x_2 are grouped together with B. Of course, we can add another B (or two) to "ungroup" $\mathsf{B}x_1x_2$, first to get $\mathsf{BS}(\mathsf{B}x_1)x_2x_3x_4$, and then to arrive at $\mathsf{B}(\mathsf{BS})\mathsf{B}x_1x_2x_3x_4$. B is definable in terms of S and K, and so we can replace all the occurrences of B with $\mathsf{S}(\mathsf{KS})\mathsf{K}$, if we want to.

This example shows that with some ingenuity and "backward thinking," that is, utilizing *expansions* rather than reductions, we can find a suitable proper combinator for $\langle 2,1 \rangle$ composition. But we certainly cannot give an infinite list of combinators if there is no formally describable systematic pattern in their structure. In order to guarantee that there is a proper combinator for each composition $\langle m,n \rangle$, we again appeal to the lemma 1.3.9 stating *combinatorial completeness*.

Exercise 3.2.2. Find a composition combinator for $\langle 2,2 \rangle$ and $\langle 3,4 \rangle$. You may use either heuristics together with the various combinators introduced in chapter 1, or the bracket abstraction algorithm from the proof of lemma 1.3.9.

The next operation that we have to show to be representable by CL-terms is *primitive recursion*. An application of the two equations in clause 3 of definition 3.1.4, may be thought of as checking whether y is 0, and then performing different calculations in accordance with the answer. This is how we would calculate the value of a function defined by primitive recursion — at least until we have some other algorithmic descriptions of the function. For instance, the addition algorithm taught in school is obviously different from the algorithm obtainable from example 3.1.3. The former relies on a small addition table (which is memorized) together with some features of the decimal notation. If $y = 0$ is true, then the value of $h(\vec{x}, 0)$ is calculated as $f(\vec{x})$ (where $\vec{x}$ are all the other arguments of h). If $y \neq 0$ — therefore, $y > 0$, that is, $y = \mathfrak{s}(z)$ for some z — then $g(\vec{x}, y, h(\vec{x}, y))$ is calculated.

Then clearly, it would be useful to have a predicate that is only true of 0, an "if ..., then ..., else ..." construction and a function that is the inverse of $\mathfrak{s}$ (for $x > 0$)

— together with a couple of other components. We do not yet have combinators for these concepts, thus we consider them one by one.

DEFINITION 3.2.4. (TRUTH AND FALSITY) The combinators K and KI represent the two truth values, *truth* and *falsity*, respectively.

Two-valued logic (*TV*) has an interpretation based on these truth values. In algebraic presentations of classical sentential logic, the truth values are often replaced by 0 and 1. Then the connectives are interpreted as numerical functions, such as $\times$ and $+$ (mod 2). Falsity is the same combinator as the combinator for 0, but truth is not the same combinator as 1. The choice of K and KI for the truth values is motivated by their behavior when they are applied to a pair of terms. $\mathsf{KI}xy \rhd_{\mathrm{w}} \mathsf{I}y \rhd_{\mathrm{w}} y$, whereas $\mathsf{K}xy \rhd_{\mathrm{w}} x$ — one of the combinators retains the second argument, the other one keeps the first. This leads to a straightforward definition of *branching* conditioned on a true or false expression.

Having a *predicate that is true of* 0 (i.e., of KI), and false of all numbers greater than 0 (i.e., of all terms of the form SB(...(SB(KI))...)) means finding a $\mathsf{Z}_{[0]}$ such that given an argument it reduces to K or KI. This task may seem not doable in CL, because a remarkable feature of combinators is that there is no restriction on what their arguments are. The arguments of a combinator may be — but cannot be required to be — the same or to be variables. The arguments cannot even be required to be different from the combinator that is applied to them. The trick is to position K or KI in the reduct of the combinator $\mathsf{Z}_{[0]}$, and to take into account the reduction patterns of the CL-terms representing natural numbers.

Example 3.2.5. Let us assume that x and y are CL-terms to which a CL-term standing for a natural number is applied. Natural number terms are listed on the top line, and the result of weak reductions are on the bottom line.

KI	SB(KI)	SB(SB(KI))	SB(SB(SB(KI)))	...
y	xy	$x(xy)$	$x(x(xy))$	...

The pattern may be discovered quite easily: as many x's are in front of a y as there are SB's in the combinator. That is, if the CL-term stands for n, then there are n x's. Furthermore, the resulting term with the x's and y's (which are atomic) is right-associated.

The term $\mathsf{KI}z\mathsf{K}$ reduces to K (i.e., to truth) no matter what z is. On the other hand, $\mathsf{K}(\mathsf{KI})z$ reduces to KI (i.e., to falsity) no matter what z is. It remains to integrate all these observations into one term defining the zero predicate. We pick K(KI) for x and K for y from the above example. That is, the combinator $\mathsf{Z}_{[0]}$ should have as its axiom $\mathsf{Z}_{[0]}x \rhd x(\mathsf{K}(\mathsf{KI}))\mathsf{K}$. Such a combinator is, certainly, definable from any combinatorially complete basis. For example, C(B(CIK))(K(KI)) is a suitable CL-term for the combinator $\mathsf{Z}_{[0]}$.

Exercise 3.2.3. Verify that $\mathsf{Z}_{[0]}$ applied to the combinatory terms, which stand for 0, 1, 2 and 7, gives the desired result.

Once we have $\mathsf{Z}_{[0]}$, we do not need the "if ..., then ..., else ..." construction in its full generality for primitive recursion. Rather we only need "if K, then M, else N," because we can use the zero predicate to test if a number is zero.

Let us recall that truth and falsity are both *binary* combinators, moreover, they are both *cancellators*. We may think about their two arguments as an *ordered pair*, since the order of the arguments, in general, is not arbitrary. Thus we are looking for a ternary combinator such that if its first argument reduces to K, then the result is the second argument, and if the first argument reduces to KI, then the result is the third argument. Despite the importance of the "if ..., then ..., else ..." constructor in programming languages, usually, there is no corresponding primitive connective in formulations of two-valued logic. This may cause us to anticipate to find a complicated combinator. Then it may come as a surprise that the combinator we are looking for is the *ternary identity* combinator I^3. This combinator is definable from the primitives we already have, for instance, as the CL-term $\mathsf{B}(\mathsf{BI})$, where I is the unary identity combinator that was introduced in chapter 1.

If the first argument of I^3 is not (or does not reduce to) truth or falsity, then the resulting term may not represent a number. Take, for example, CI (i.e., T) in place of K as the first argument. $\mathsf{I}^3(\mathsf{CI})MN \rhd_{\mathrm{w}} NM$, and this term may compute the application of the number represented by N to the number represented by M, if those terms are numbers. (Of course, application is not an operation on numbers, but on terms, thus we might prefer to think of NM simply as two juxtaposed numbers, though NM is not in nf when N stands for a number.)

We encountered the need for "counting down" (or having an $x-1$ function) when we introduced primitive recursion. Accordingly, we need a combinator for the function that is the inverse of the successor function, more precisely, the inverse of the successor function with the ordered pair $\langle 0,0\rangle$ added. This function is usually called the *predecessor* function; we denote this function by p. Informally, $p(x)=x-1$ whenever $x>0$, and 0 otherwise (i.e., when $x=0$).

The predecessor function is crucial, because it is often involved in definitions by primitive recursion not only nominally, but also in the actual calculation of the value of a function. For instance, the predecessor function is needed if we wish to define $\dot{-}$, a *subtraction* function. $\dot{-}$ in its full generality is not definable on the set of natural numbers, because $\mathbb{N}$ is not closed under this operation. However, just as p is a version of $x-1$ amended with $0-1=0$, a version of subtraction enhanced with $x-y=0$ (when $x<y$) is a definable and useful function.

DEFINITION 3.2.6. The predecessor function p is not a basic recursive functions, however, it is easily definable via primitive recursion as follows.

1. $p(0)=0$;
2. $p(\mathfrak{s}(x))=\pi_1^2(x,p(x))$.

π_1^2 in the second clause means that we take $p(\mathfrak{s}(x))$ to be x, and disregard the second argument, namely, $p(x)$. We might note that this definition suggests that given an alternative representation of natural numbers by CL-terms as pairs, the predecessor function might become definable without emulating primitive recursion or computing. (We set aside this observation for now, but we will return to it in the next section, where such an alternative representation is given.)

The above definition by primitive recursion is quite simple, hence we will use this definition as an example for the definition of a combinator that represents primitive recursion. To justify the idea that it is reasonable to look for one combinator for each potential arity for h, we point out that each definition by primitive recursion is completely characterized by *two functions*.[17] Those functions are the function that is specific to a definition's first clause and the function that is specific to the second clause. In case of p, these are the nullary function 0 (as f) and π_1^2 (as g). Given these two functions the whole definition can be reconstructed, because the *format* of primitive recursion does not change from function to function. The "if ..., then ..., else ..." construction provides the bare-bone skeleton for primitive recursion.

However, the *modeling* of primitive recursion is not yet solved, because the second equation includes the function that is being defined by the equations. Following the two clauses in definition 3.1.4, we want to have a ternary combinator Z such that $\mathsf{Z}XYZ$ reduces to X if Z is KI (i.e., zero), and $\mathsf{Z}XYZ$ reduces to $YZ^-(\mathsf{Z}XYZ^-)$, where we used Z^- for the term the successor of which is Z.[18] We already saw that conditional branching on zero can be defined by combinators — by $\mathsf{Z}_{[0]}$ and I^3.

Again a way to define Z is to exploit the functional properties of the numbers rather than to assume that a combinator has different reductions depending on the shape of the terms in its arguments. We already saw, in the previous chapters, that it is possible to define certain improper combinators, in particular, fixed point combinators from combinatorially complete bases. Thus it is reasonable to expect that a combinator such as Z above is definable.

First we define a combinator D that is the specialized version of "if ..., then ..., else ..." that we mentioned above. $\mathsf{D}xyz$ can be read as "if $z = 0$, then x, else y." The third argument is expected to be a number, and if it is zero, which is the CL-term KI, then x should be selected. If the third argument is of the form SB(...(SB(KI))...), then y should result. This can be accomplished if the axiom for this combinator is $\mathsf{D}xyz \triangleright z(\mathsf{K}y)x$. A term that defines D using well-known proper combinators is C(BC(B(CI)K)).

Exercise 3.2.4. Take as the first two arguments for D the terms M and N. Let the third argument be a term representing a number. For instance, let KI and then let SB(SB(SB(KI))) be z. Verify by writing out the one-step reductions that D behaves as "if $z = 0$, then x, else (i.e., if $z > 0$ then) y," that is, $\mathsf{D}MN(\mathsf{KI}) \triangleright_{\mathrm{w}} M$ and $\mathsf{D}MN(\mathsf{SB}(\mathsf{SB}(\mathsf{SB}(\mathsf{KI})))) \triangleright_{\mathrm{w}} N$.

Now, using D we define a *binary* combinator Q that can be thought to mimic the recursive part of primitive recursion. The idea here is to calculate from 0 up to n

[17] Each of the functions that we denoted by f and g in definition 3.1.4 determines *the arity* of h.

[18] It is convenient from the point of view of CL-terms to have the term corresponding to x as the last one. This is how the definition is usually constructed; we only note that the combinator CC would yield our desired order of arguments from the order suggested by the natural language expression "if ..., then ..., else ...."

via repeating the calculations in the recursive clause in the definition of the primitive recursive function. In order not to be overly specific, the function that is used in the second clause of a primitive recursive definition is abstracted out (as x), that is, it is turned into an argument of Q. The axiom for Q is

$$\mathsf{Q}xy \rhd \mathsf{D}(\mathsf{SB}(y(\mathsf{KI})))(x(y(\mathsf{KI}))(y(\mathsf{SB}(\mathsf{KI})))).$$

Exercise* 3.2.5. Q is, obviously, not a proper combinator, however, there is no principal obstacle that would prevent us from defining it. Find a CL-term that defines Q. (Notice that some of the subterms are terms for 0, 1 and the successor function, which might suggest that some temporary abbreviations might be useful in order to find a definition for Q quickly.)

If Q is applied to the terms Y and $\mathsf{D}ZX$ (where Z stands for a number), then $\mathsf{Q}Y(\mathsf{D}ZX) \rhd_{\mathrm{w}} \mathsf{D}(\mathsf{SB}Z)(YZX)$.

Exercise 3.2.6. Go through the reduction steps leading to $\mathsf{D}(\mathsf{SB}Z)(YZX)$.

If Z is KI, then $\mathsf{D}(\mathsf{SB}(\mathsf{KI}))(Y(\mathsf{KI})X)$ — viewed informally — is the ordered pair $\langle 1,(Y0X)\rangle$. A selection of the first or of the second element can be made by supplying a number as the third argument of D. So far we have that

$$\mathsf{Q}Y(\mathsf{D}(\mathsf{KI})X) \rhd_{\mathrm{w}} \mathsf{D}(\mathsf{SB}(\mathsf{KI}))(Y(\mathsf{KI})X).$$

From $\mathsf{Q}Y(\mathsf{D}(\mathsf{KI})X)$ we could move on to another term, in which this term is the second argument of Q with Y being the first. Then we get the following reductions.

$$\begin{aligned}&\mathsf{Q}Y(\mathsf{Q}Y(\mathsf{D}(\mathsf{KI})X)) \rhd_{\mathrm{w}}\\ &\qquad \mathsf{Q}Y(\mathsf{D}(\mathsf{SB}(\mathsf{KI}))(Y(\mathsf{KI})X)) \rhd_{\mathrm{w}}\\ &\qquad\qquad \mathsf{D}(\mathsf{SB}(\mathsf{SB}(\mathsf{KI})))(Y(\mathsf{SB}(\mathsf{KI}))(Y(\mathsf{KI})X)).\end{aligned}$$

The second argument of D in the resulting term can be seen to accumulate subterms that can be turned into *calculations* of the value of Y on $0,1,\ldots$. The first argument of D functions as a *counter*, which is incremented by each iteration of Q. The numbers themselves are of the form $f(\ldots f(f(g))\ldots)$, and so the iteration of $\mathsf{Q}Y$ is itself similar to counting.

Looking at the terms $(Y(\mathsf{KI})X)$, $(Y(\mathsf{SB}(\mathsf{KI}))(Y(\mathsf{KI})X))$, ..., we may notice that the number immediately following the first occurrence of Y is characteristic for the whole term. That is, the rest of the term can be restored once this number is known (together with Y and X, of course), because the term is of the form

$$Y(\underbrace{\mathsf{SB}\ldots(\mathsf{SB}}_{n}(\mathsf{KI}))\ldots)(\ldots(Y(\mathsf{SB}(\mathsf{KI}))(Y(\mathsf{KI})X))\ldots).$$

We denote these terms by W_0, W_1, ... Clearly, for each number these terms can be written out in full, however, in order to be able to refer to these terms with an arbitrary number, we have to have some sort of abbreviation, which hides their internal structure.

The whole reduct is

$$\underbrace{\mathsf{Q}Y(\ldots(\mathsf{Q}Y}_{n}(\mathsf{D}(\mathsf{KI})X))\ldots) \rhd_{\mathrm{w}}$$

$$\mathsf{D}(\underbrace{\mathsf{SB}(\ldots(\mathsf{SB}}_{n}(\mathsf{KI}))\ldots))(Y(\underbrace{\mathsf{SB}\ldots(\mathsf{SB}}_{n-1}(\mathsf{KI}))\ldots)(\ldots(Y(\mathsf{KI})X)\ldots)),$$

or more compactly and somewhat informally (with natural numbers inserted into the CL-terms):

$$(\mathsf{Q}Y)^n(\mathsf{D}(\mathsf{KI})X) \rhd_{\mathrm{w}} \mathsf{D}n(Y(n-1)W_{n-1}).$$

W_n can be obtained as the reduct of $\mathsf{D}NW_n(\mathsf{SB}(\mathsf{KI}))$, where N is the CL-term for n; this is immediate from the axiom of D. $\mathsf{D}NW_n(\mathsf{SB}(\mathsf{KI}))$ is the term we obtained from filling out the arguments of Q by Y and $\mathsf{D}(\mathsf{KI})X$, but further applied to the CL-term representing 1. An effect of applying $\mathsf{Q}Y$ to $\mathsf{D}NX$ was incrementing N by one. Thus $\mathsf{D}NW_n(\mathsf{SB}(\mathsf{KI}))$ is the reduct of

$$\underbrace{\mathsf{Q}Y(\ldots(\mathsf{Q}Y}_{n}(\mathsf{D}(\mathsf{KI})X))\ldots)(\mathsf{SB}(\mathsf{KI})).$$

The numbers are of the form of function application and they yield a function applied to an argument, hence, by yet another expansion, we get that W_n can be obtained from $N(\mathsf{Q}Y)(\mathsf{D}(\mathsf{KI})X)(\mathsf{SB}(\mathsf{KI}))$, where N is the CL-term for the number n.

The last term contains three terms N, X and Y, which are, respectively, a natural number and the two functions from the primitive recursion schema. Thus the combinator that we need to emulate primitive recursion has the following axiom.[19]

$$\mathsf{Z}_B x_1 x_2 x_3 \rhd x_3(\mathsf{Q}x_2)(\mathsf{D}(\mathsf{KI})x_1)(\mathsf{SB}(\mathsf{KI})).$$

Z_B is yet another combinator which is not proper. However, there is no difficulty to find a definition for Z_B, since both Q and D, as well as $\mathsf{I},\mathsf{B},\mathsf{S}$ and K are either proper combinators or they are definable from proper combinators.

Exercise 3.2.7. Find a definition of Z_B using $\mathsf{Q},\mathsf{D},\mathsf{I},\mathsf{B},\mathsf{S},\mathsf{K}$, and, possibly, other proper combinators that were introduced in chapter 1.

Let us now see how we can find a combinator to stand for p, the predecessor function given that we have the combinator Z_B. The definition of p using primitive recursion is in definition 3.2.6. The two functions that are used with the schema of primitive recursion are 0 and π_1^2. The corresponding CL-terms are KI and K, that is, the combinator representing p is $\mathsf{Z}_B(\mathsf{KI})\mathsf{K}$. First, $p(0) = 0$ should be true. $\mathsf{Z}_B(\mathsf{KI})\mathsf{K}(\mathsf{KI})$ reduces to $\mathsf{KI}(\mathsf{QK})(\mathsf{D}(\mathsf{KI})(\mathsf{KI}))(\mathsf{SB}(\mathsf{KI}))$. By two one-step reductions, we get $\mathsf{D}(\mathsf{KI})(\mathsf{KI})(\mathsf{SB}(\mathsf{KI}))$. The third argument of D stands for 1, which means that the result is KI, that is, 0. (It is incidental that the first argument of D would work as the final result in this case too.)

[19]The idea of this combinator is due to Paul Bernays; hence the subscript on the combinator.

The predecessor of 3 is 2, that is, $p(3) = 2$. The following term and some reduction steps illustrate the structure of the calculation.

$\mathsf{Z}_B(\mathsf{KI})\mathsf{K}(\mathsf{SB}(\mathsf{SB}(\mathsf{SB}(\mathsf{KI})))) \triangleright_{\mathrm{w}}$
$\mathsf{SB}(\mathsf{SB}(\mathsf{SB}(\mathsf{KI})))(\mathsf{QK})(\mathsf{D}(\mathsf{KI})(\mathsf{KI}))(\mathsf{SB}(\mathsf{KI})) \triangleright_{\mathrm{w}}$
$\mathsf{QK}(\mathsf{QK}(\mathsf{QK}(\mathsf{D}(\mathsf{KI})(\mathsf{KI}))))(\mathsf{SB}(\mathsf{KI})) \triangleright_{\mathrm{w}}$
$\mathsf{QK}(\mathsf{QK}(\mathsf{D}(\mathsf{SB}(\mathsf{KI}))(\mathsf{K}(\mathsf{KI})(\mathsf{KI}))))(\mathsf{SB}(\mathsf{KI})) \triangleright_{\mathrm{w}}$
$\mathsf{QK}(\mathsf{D}(\mathsf{SB}(\mathsf{SB}(\mathsf{KI})))(\mathsf{K}(\mathsf{SB}(\mathsf{KI}))(\mathsf{KI})))(\mathsf{SB}(\mathsf{KI})) \triangleright_{\mathrm{w}}$
$\mathsf{D}(\mathsf{SB}(\mathsf{SB}(\mathsf{SB}(\mathsf{KI}))))(\mathsf{K}(\mathsf{SB}(\mathsf{SB}(\mathsf{KI})))(\mathsf{SB}(\mathsf{KI})))(\mathsf{SB}(\mathsf{KI})) \triangleright_{\mathrm{w}}$ *
$\mathsf{K}(\mathsf{SB}(\mathsf{SB}(\mathsf{KI})))(\mathsf{SB}(\mathsf{KI})) \triangleright_{\mathrm{w}}$
$\mathsf{SB}(\mathsf{SB}(\mathsf{KI}))$

The last term, indeed, represents 2, as desired.

Exercise 3.2.8. The above terms do not show all the one-step reductions. Write out the missing one-step reductions (taking for granted the axioms for Q and D).

The term on the line that we flagged with * may be rewritten as $\langle 3, \pi_1^2\langle 2, 1\rangle\rangle$ and 1, where the latter indicates that the second element of the preceding pair will be chosen. If we would not have shortened the term by performing some reductions with the head K, then the second element of the pair would be more lengthy. The term, which more fully encodes the computation of the predecessor of 3 is $\langle 3, \pi_1^2\langle 2, \pi_1^2\langle 1, \pi_1^2\langle 0, 0\rangle\rangle\rangle\rangle$ together with 1, which selects the second element of the whole complex pair.

Exercise 3.2.9. Translate this complex ordered pair into a CL-term.

Incidentally, notice that we gave an informal rewriting of a CL-term in which we turned into ordered pairs two sorts of terms: one preceded with K, which represents π_1^2, and another preceded with D, which itself is ambiguous (or, perhaps, undecided) between π_1^2 and π_2^2 with the selection depending on the value of its third argument. As the axiom and the definition of D, that we gave above, makes clear, D does not include KI, that is, π_2^2. The selective effect that D can have partly stems from the concrete representation of the natural numbers combined with some permutation. The rendering of terms with different heads (such as K, D or KI) as ordered pairs does not cause a problem. There are no angle brackets, $\langle\ \rangle$, in CL, and in all those cases the *first* and *second* argument of a combinator are viewed, respectively, as the first and second element of *an ordered pair*.

Exercise 3.2.10. Use the combinator Z_B to model addition defined in example 3.1.3. Calculate the value of the addition function with 0 for x and 4 for y.

The crucial idea behind Z_B is the *accumulation* of CL-terms that represent the *calculation* of a function up to n. This general idea could be implemented in other ways. A different idea is to use the expressive power of CL, namely, the definability of fixed point combinators.

We could think about the predecessor function as the fixed point of the equation described in words as: $p(x) = (\text{If } x = 0, \text{ then } 0, \text{ else } \pi_1^2(p(x), p(p(x))))$. Of course, the application of π_1^2 simplifies to $p(x)$. Then we are trying to find a CL-term defining the combinator Z_p from its defining equation, where $\mathsf{Z}_p x = \mathsf{I}^3(\mathsf{Z}_{[0]}x)(\mathsf{KI})(\mathsf{Z}_p x)$. The two combinators I^3 and $\mathsf{Z}_{[0]}$ are as above, in particular, I^3 is the ternary identity combinator and $\mathsf{Z}_{[0]}$ is the zero test. Thus the equation further simplifies to $\mathsf{Z}_p x = \mathsf{Z}_{[0]}x(\mathsf{KI})(\mathsf{Z}_p x)$.

We can find a term for Z_p, for instance, using Turing's method of finding a fixed point combinator. The equation $\mathsf{Z}_{pT}yx = \mathsf{Z}_{[0]}x(\mathsf{KI})(yyx)$ can be viewed as the axiom of an improper combinator, when $=$ is replaced by $\triangleright$: $\mathsf{Z}_{pT}yx \triangleright \mathsf{Z}_{[0]}x(\mathsf{KI})(yyx)$. We already have the combinators $\mathsf{Z}_{[0]}$ and KI, thus Z_{pT} is definable. For instance, the term $\mathsf{W}(\mathsf{B}(\mathsf{B}(\mathsf{C}(\mathsf{CZ}_{[0]}(\mathsf{KI}))))\mathsf{M})$ can be taken for Z_{pT}, and then $\mathsf{Z}_{pT}\mathsf{Z}_{pT}$ is Z_p.

Exercise 3.2.11. Verify that Z_p applied to the CL-term for 0 yields the desired result.

Unfortunately, the above combinator does not produce a CL-term standing for a number when applied to a CL-term standing for a number — except for KI. Of course, this could have been expected in a sense, because the defining equation contains $\mathsf{Z}_p x$ on both sides of the equation. $\mathsf{Z}_p(\mathsf{KI})$ reduces to KI, only because the occurrence of $\mathsf{Z}_p(\mathsf{KI})$ is dropped by the K that results from $\mathsf{Z}_p(\mathsf{KI})$. One could, perhaps, think that the problem is with our rendering of the definition of the definition of p via the primitive recursive schema into CL-terms. But, of course, $\mathsf{Z}_p x \triangleright \mathsf{I}^3(\mathsf{Z}_{[0]}x)(\mathsf{KI})(\mathsf{SB}(\mathsf{Z}_p(\mathsf{Z}_p x)))$ suffers from the same problem within the third argument of I^3.

First, this attempted definition of Z_p using the idea behind the definition of a fixed point combinator illustrates that this construction should be used carefully, because we typically want not simply a solution to an equation, but a CL-term that *calculates* on a class of CL-terms as a recursive function computes on natural numbers. We could have used the fixed point combinator itself to obtain a solution. There is a CL-term corresponding to $[x].\mathsf{Z}_{[0]}x(\mathsf{KI})(\mathsf{Z}_p x)$, and further this equals to $([\mathsf{Z}_p]\,[x].\mathsf{Z}_{[0]}x(\mathsf{KI})(\mathsf{Z}_p x))\mathsf{Z}_p$, which is Z_p. Then Z_p is $\mathsf{Y}([\mathsf{Z}_p]\,[x].\mathsf{Z}_{[0]}x(\mathsf{KI})(\mathsf{Z}_p x))$.

Second, the equation $\mathsf{Z}_p x = \mathsf{I}^3(\mathsf{Z}_{[0]}x)(\mathsf{KI})(\mathsf{Z}_p x)$ that we obtained from the definition of p by the primitive recursion schema elucidates the role of p in the transcription of instances of this schema into CL-terms. It so happens, that in the schema $\mathfrak{s}$ is explicit, however, when a CL-term is taken to be an argument of a CL-term that stands for a function the presence or absence of SB at the left of the term cannot be used by the function explicitly.

In order to calculate the predecessor function, what we would *really like to do* is to take a look at the term, if it is KI, then we return the term itself, but if it is of the form $\mathsf{SB}(M)$, then we return M. We simply want to get rid of one occurrence of SB. Because of the structure of numbers as CL-terms using the Church notation, it does not matter *which* occurrence of SB is deleted.

The natural numbers in the Church notation are all binary functions, and it is one of the nice properties of this representation of the natural numbers in CL. Thus we can take advantage of the structure of the numbers. Counting up to a preset number can always be achieved computationally by as many symbols or states as the number is. The number we have to count off is *one*: the predecessor of a number $\mathsf{SB}(\ldots(\mathsf{SB}(\mathsf{KI}))\ldots)$ is $\mathsf{SB}(\ldots(\mathsf{KI})\ldots)$. If we apply a number to x and y, then we get a term the general structure of which matches the structure of the number, with x's in place of SB's and a y in place of KI. (Cf. example 3.2.5)

We can easily construct a term with a zero test to ensure that we get $p(0) = 0$. Thus we have to worry only about the case when the input is greater than 0. We need to find a term for x and y with the following properties.

$$xw \triangleright_{\mathrm{w}} \begin{cases} \mathsf{KI} & \text{if } \ w \text{ is } y, \\ \mathsf{SB}(\mathsf{KI}) & \text{if } \ w \text{ is } \mathsf{KI}, \\ \mathsf{SB}w & \text{otherwise.} \end{cases}$$

Our idea is to utilize the functional character of the numbers as CL-terms. xy (which is to be a term distinct from $\mathsf{SB}(\mathsf{KI})$) will reduce to KI, thereby accomplishing the elimination of exactly one occurrence of SB. All the other occurrences of x (if there are any), will revert back to SB to restore the term to one that stands for a natural number.

We gave a definition for $\mathsf{Z}_{[0]}$ as $\mathsf{C}(\mathsf{B}(\mathsf{CIK}))(\mathsf{K}(\mathsf{KI}))$. Of course, originally, our intention was only to be able to get K if the input was KI and to get KI if the input was another number, that begins with SB. However, there are many other CL-terms beyond those that stand for numbers. If we test BBB for zero, then we obtain a term that takes two arguments, cancels those arguments, and returns KI. BBB is the term that we want to count minus one with.

If M stands for a number greater than 0, then to ensure that $\mathsf{Z}_{[0]}M \triangleright_{\mathrm{w}} \mathsf{SB}(M)$, we can take the term $\mathsf{Z}_{[0]}MN(\mathsf{SB}(M))$ (where N is still to be determined). However, in case BBB has already triggered the change to KI, we also want to have $\mathsf{Z}_{[0]}(\mathsf{KI})$ to yield $\mathsf{SB}(\mathsf{KI})$, that is, N should be $\mathsf{SB}(M)$ or $\mathsf{SB}(\mathsf{KI})$. To put all these bits and pieces together, we define Z'_p as follows.

$$\mathsf{Z}'_p x \triangleright \mathsf{Z}_{[0]}xx(x(\mathsf{S}(\mathsf{CZ}_{[0]}(\mathsf{SB}(\mathsf{KI})))(\mathsf{SB}))(\mathsf{BBB}))$$

The combinator Z'_p is obviously definable, because it involves combinators that we already defined or we might suppose to have in the basis. The axiom Z'_p does not contain occurrences of Z'_p on the right-hand side at all.

The first occurrence of $\mathsf{Z}_{[0]}$ caters for the case when x is KI, that is, 0. Otherwise, the term starting with the third occurrence of x is evaluated. x stands for a number, let us say n, which is a binary function. x applies its first argument n times to its second argument, which is BBB. An application of the first argument to y may be calculated in the following steps.

$$\begin{aligned} &\mathsf{S}(\mathsf{CZ}_{[0]}(\mathsf{SB}(\mathsf{KI})))(\mathsf{SB})y \triangleright_{\mathrm{w}} \\ &\mathsf{CZ}_{[0]}(\mathsf{SB}(\mathsf{KI}))y(\mathsf{SB}y) \triangleright_{\mathrm{w}} \\ &\mathsf{Z}_{[0]}y(\mathsf{SB}(\mathsf{KI}))(\mathsf{SB}y) \end{aligned}$$

The meaning of the last CL-term is what we described above; the role of S and C is merely to abstract out y.

Exercise 3.2.12. Use the new Z'_p to calculate the predecessor of KI and $\mathsf{SB(KI)}$.

Exercise 3.2.13. Verify that $\mathsf{Z}'_p(\mathsf{SB(SB(SB(SB(KI)))))}$ is $\mathsf{SB(SB(SB(KI)))}$.

Exercise 3.2.14. Find a combinator for Z'_p. (You may assume that $\mathsf{Z}_{[0]}$ is given, that is, it already has been defined.)

The above combinator Z'_p is "hand-crafted," but *elegant*. An attractive feature of this combinator is that it is much more straightforward than generating and storing terms for successive calculations. In a sense, our Z'_p parallels the predecessor function that we will encounter in the next section: it utilizes the specific structure of the CL-terms that represent the natural numbers. However, we have retained *the Church numerals*, that are in certain ways *superior* to the representation given in the next section. In addition, our Z'_p is motivated by insights garnered from models of computation investigated in theoretical computer science.

Once we have a definition for p, the idea behind the fixed point combinator can be used, in general, to find a combinator to represent primitive recursion. The schema of primitive recursion takes two functions, let us say x and y, and a number, let us say z. We assume that we have a CL-term that represents p, and we denote that by Z_p — it may be either $\mathsf{Z}_B(\mathsf{KI})\mathsf{K}$ or our Z'_p. Z_{pr} applied to x, y and z yields x when $z = 0$, and otherwise gives $y(\mathsf{Z}_p z)(\mathsf{Z}_{pr}xy(\mathsf{Z}_p z))$. Using D we can turn the "if ..., then ..., else ..." statement into a term, and place that on the right-hand side of the equation, as in

$$\mathsf{Z}_{pr}xyz =_{\mathrm{w}} \mathsf{D}x(y(\mathsf{Z}_p z)(\mathsf{Z}_{pr}xy(\mathsf{Z}_p z)))z.$$

By combinatorial completeness, in effect, using a bracket abstraction, x, y and z may be abstracted out from the term on the right-hand side, leaving $\mathsf{Z}_{pr} = [x,y,z].\,\mathsf{D}x\,(y(\mathsf{Z}_p z)(\mathsf{Z}_{pr}xy(\mathsf{Z}_p z)))z$. Of course, the right-hand side of the equation is just another combinator that takes three arguments and yields the CL-term that was on the right-hand side of the equation above. Recall that if we have an equation $x = fx$, then $\mathsf{Y}f$ is a solution to the equation, that is, $\mathsf{Y}f$ is the term that we can put into the place of x. This implies that we have to transform $[x,y,z].\,\mathsf{D}x(y(\mathsf{Z}_p z)(\mathsf{Z}_{pr}xy(\mathsf{Z}_p z)))z$ into the form where some function f is applied to Z_{pr}. This can be achieved via yet another application of bracket abstraction: $([v]\,[x,y,z].\,\mathsf{D}x(y(\mathsf{Z}_p z)(vxy(\mathsf{Z}_p z)))z)\mathsf{Z}_{pr}$. (We gave v its own brackets to emphasize the importance of this abstraction.) Then, "as if by magic," Z_{pr} is defined as $\mathsf{Y}[v]\,[x,y,z].\,\mathsf{D}x(y(\mathsf{Z}_p z)(vxy(\mathsf{Z}_p z)))z$.

Exercise* 3.2.15. Given $\mathsf{D}x(y(\mathsf{Z}_p z)(\mathsf{Z}_{pr}xy(\mathsf{Z}_p z)))z$, find a CL-term for the bracket abstract $[x,y,z].\,\mathsf{D}x(y(\mathsf{Z}_p z)(\mathsf{Z}_{pr}xy(\mathsf{Z}_p z)))z$. (Hint: Using various combinators from chapter 1 you may find a shorter term than utilizing the abstraction algorithm with S and K only.)

Exercise* 3.2.16. Abstract out Z_{pr} from the term $[x,y,z].\,\mathsf{D}x(y(\mathsf{Z}_p z)(\mathsf{Z}_{pr}xy(\mathsf{Z}_p z)))z$. (Hint: You need the CL-term that is the result of the previous exercise.)

Exercise 3.2.17. Calculate the value of $\mathsf{Z}_{pr}(\mathsf{KI})\mathsf{K}(\mathsf{KI})$ and $\mathsf{Z}_{pr}(\mathsf{KI})\mathsf{K}(\mathsf{SB(SB(KI))})$. (Hint: You may assume that Z_{pr} behaves like in the displayed weak equality above.)

Now we have not just one, but two combinators that can represent primitive recursion, which amply ensures — together with the representation of the basic functions and composition — the truth of the following claim.

THEOREM 3.2.7. (REPRESENTATION OF PRIMITIVE RECURSIVE FUNCTIONS) *All primitive recursive functions are* representable *by* CL*-terms.*

The definability of a fixed point combinator is a remarkable aspect of CL. However, a fixed point combinator or a similar combinator that is not proper may not have an nf. This is an obvious consequence of the axiom of the fixed point combinator Y, which contains that combinator itself in a redex position within its reduct. And we have seen an example in the case of the predecessor function, where a fixed point type construction did not lead to a term that represented the function. Z_{pr} certainly has an *infinite reduction sequence*.

Exercise* 3.2.18. Explain how the combinator Z_{pr} (with an occurrence of Z_{pr} in its reduct) can be used to define the exponential function in CL, so that the resulting function takes CL-terms standing for natural numbers into CL-terms standing for natural numbers. (Hint: Each number is represented by a CL-term, which is in nf.)

The term Z_B, and the definitions of p as $\mathsf{Z}_B(\mathsf{KI})\mathsf{K}$ or Z'_p, on the other hand, *do not rely* on the definability of a fixed point combinator. Rather these terms produce or perform — via reductions — a series of well-founded computational steps. $\mathsf{Z}_B(\mathsf{KI})\mathsf{K}$ accumulates a series of predecessors, whereas our predecessor function Z'_p exploits the specific functional character of the Church numerals together with our capability of counting up to one by selecting a CL-term that does not stand for a natural number. As a result, each of these terms have an nf.

Exercise 3.2.19. Define $+$ using Z_B and Z_{pr}. Compare the resulting CL-terms.

Exercise 3.2.20. Find CL-terms for the factorial function $!$ using Z_B and Z_{pr}. (Hint: Take a look at exercise 3.1.2 first.)

Partial recursive functions may be generated by an application of μ (in addition to composition and primitive recursion). Let us now turn to *minimization*. Here again, we may choose between constructions. A simpler construction involves an application of the fixed point combinator, and the resulting CL-term does not have an nf. Alternatively, we may opt for a more complicated construction, which, however, yields a CL-term with an nf. We start with the latter.

The computation of y for which the function f takes the value 0 can be viewed as finding a combinator Z_i such that

$$\mathsf{Z}_i XY \rhd_{\mathrm{w}} \begin{cases} Y & \text{if } XY =_{\mathrm{w}} \mathsf{KI}, \\ \mathsf{Z}_i X(\mathsf{SB}Y) & \text{otherwise.} \end{cases}$$

Here Y is the CL-term for y, X is the CL-term for f and $XY =_{\mathrm{w}} \mathsf{KI}$ stands for $f(y) = 0$. It is straightforward to define a combinator that satisfies the first case.

We could consider $\mathsf{Z}_{[0]}(XY)Y$ for the "if ..., then ..." part of the statement, which returns Y if XY is zero, and false (i.e., KI) otherwise. For instance, $\mathsf{BW}(\mathsf{BZ}_{[0]})$ could be the CL-term standing for the first case. However, the second case repeats an application of Z_i, thus we will define Z_i using another combinator Z_o.

Recall that D is a combinator that selects the first or the second element of a pair that comprises the first two arguments of D. On the other hand, KI (i.e., 0) selects the second one from its two arguments. Thus, we would like to find a unary combinator Z_o such that its argument ends up inside a term that is in the second argument of D. The first argument of D will be KI, and we place XY into the third argument of D. In sum, $\mathsf{Z}_o x \rhd \mathsf{D}(\mathsf{KI})M$, where the shape of M is still to be determined, but we already know that x occurs in M. To determine the shape of M, we should consider what happens when XY is 0.

$\mathsf{Z}_o X(XY)$ yields $\mathsf{D}(\mathsf{KI})M(XY)$ which reduces to KI when $XY =_{\mathrm{w}} \mathsf{KI}$. In order to get Y as the output, we need to have a term like $\mathsf{Z}_o X(XY)NY$, where the shape of N is to be determined. If XY is not 0, then we get $\mathsf{D}(\mathsf{KI})M(\mathsf{SB}\ldots(\mathsf{KI})\ldots)NY$ and then MNY. The latter term should reduce to $\mathsf{Z}_o X(X(\mathsf{SB}Y))N(\mathsf{SB}Y)$. If we take N to be $\mathsf{Z}_o X$, then we can determine what shape M should have. $M(\mathsf{Z}_o X)Y \rhd_{\mathrm{w}} \mathsf{Z}_o X(X(\mathsf{SB}Y))(\mathsf{Z}_o X)(\mathsf{SB}Y)$. Thus, M is $[v,y].\,v(X(\mathsf{SB}y))v(\mathsf{SB}y)$. Then $\mathsf{Z}_o x \rhd_{\mathrm{w}} \mathsf{D}(\mathsf{KI})[v,y].\,v(x(\mathsf{SB}y))v(\mathsf{SB}y)$, hence Z_o is $[x].\,\mathsf{D}(\mathsf{KI})[v,y].\,v(x(\mathsf{SB}y))v(\mathsf{SB}y)$.

Exercise 3.2.21. Find a CL-term for Z_o using D and any of the well-known combinators, which we introduced in chapter 1.

We still have to find Z_i. We have seen that $\mathsf{Z}_o X(XY)(\mathsf{Z}_o X)Y \rhd_{\mathrm{w}} Y$ if $XY =_{\mathrm{w}} \mathsf{KI}$, but $\mathsf{Z}_o X(XY)(\mathsf{Z}_o X)Y \rhd_{\mathrm{w}} \mathsf{Z}_o X(X(\mathsf{SB}Y))(\mathsf{Z}_o X)(\mathsf{SB}Y)$ if $XY =_{\mathrm{w}} \mathsf{SB}(\ldots(\mathsf{KI})\ldots)$. Then Z_i is $[x,y].\,\mathsf{Z}_o x(xy)(\mathsf{Z}_o x)y$.

Exercise 3.2.22. Suppose that Z_o has been defined (e.g., as in the previous exercise). Find a CL-term for Z_i.

In definition 3.1.7, we did not assume that the function f is unary. Thus the last step is to separate the $n+1$st argument from the others (which should be left-associated with the CL-term for f). Let us assume that Z_f is a combinator of arity $n+1$, and it represents the function f. Then μf is represented by a CL-term $\mathsf{Z}_{\mu f}$, such that

$$\mathsf{Z}_{\mu f} x_1 \ldots x_n \rhd \mathsf{Z}_i(\mathsf{Z}_f x_1 \ldots x_n)(\mathsf{KI}).$$

Here the term $\mathsf{Z}_f x_1 \ldots x_n$ plays the role of the previous X. The variables $x_1,\ldots,x_n$ can be abstracted out of the right-hand side term in a usual way.

Exercise 3.2.23. Find a CL-term for $[x_1,x_2].\,\mathsf{Z}_i(\mathsf{Z}_f x_1 x_2)(\mathsf{KI})$.

Exercise* 3.2.24. Find CL-terms (i.e., describe the shape of suitable CL-terms) for all n, that represent $[x_1,\ldots,x_n].\,\mathsf{Z}_i(\mathsf{Z}_f x_1 \ldots x_n)(\mathsf{KI})$.

Now we can take a closer look at the function μf, that is, $\mathsf{Z}_{\mu f}$ when it is applied to n arguments, let us say to $X_1,\ldots,X_n$. We assume that $s(0)$ is the least y for which $f(x_1,\ldots,x_n,y) = 0$ — so as not to make the calculation overly lengthy.

1. $\mathsf{Z}_{\mu f} X_1 \ldots X_n \rhd_{\mathrm{w}}$

2. $\mathsf{Z}_i(\mathsf{Z}_f X_1 \ldots X_n)(\mathsf{KI}) \rhd_{\mathrm{w}}$
3. $\mathsf{Z}_o(\mathsf{Z}_f X_1 \ldots X_n)(\mathsf{Z}_f X_1 \ldots X_n(\mathsf{KI}))(\mathsf{Z}_o(\mathsf{Z}_f X_1 \ldots X_n))(\mathsf{KI}) \rhd_{\mathrm{w}}$
4. $\mathsf{D}(\mathsf{KI})([v,y].\,v(\mathsf{Z}_f X_1 \ldots X_n(\mathsf{SB}(\mathsf{KI})))v(\mathsf{SB}(\mathsf{KI})))(\mathsf{Z}_f X_1 \ldots X_n)$
   $(\mathsf{Z}_o(\mathsf{Z}_f X_1 \ldots X_n))(\mathsf{KI}) \rhd_{\mathrm{w}}$
5. $\mathsf{Z}_o(\mathsf{Z}_f X_1 \ldots X_n)(\mathsf{Z}_f X_1 \ldots X_n(\mathsf{SB}(\mathsf{KI})))(\mathsf{Z}_o(\mathsf{Z}_f X_1 \ldots X_n))(\mathsf{SB}(\mathsf{KI})) \rhd_{\mathrm{w}}$
6. $\mathsf{KI}(\mathsf{Z}_o(\mathsf{Z}_f X_1 \ldots X_n))(\mathsf{SB}(\mathsf{KI})) \rhd_{\mathrm{w}}$
7. $\mathsf{SB}(\mathsf{KI})$

The step from 1 to 2 is justified by the definition of $\mathsf{Z}_{\mu f}$. The definition of Z_i gives 3, where $\mathsf{Z}_f X_1 \ldots X_n Y$ is not KI — due to our (ad hoc) assumption about the function f. The next term is obtained using the fact that Z_o includes D. But D here selects its second argument, hence we get the CL-term on line 5. Since now $\mathsf{Z}_f X_1 \ldots X_n(\mathsf{SB}(\mathsf{KI})) =_{\mathrm{w}} \mathsf{KI}$ (by our stipulation), D selects its first argument, which is KI. Two further one-step reductions of the CL-term on line 6 yield the CL-term on line 7, which represents the successor of 0.

We used a hypothetical function f and the corresponding CL-term Z_f to create the CL-term $\mathsf{Z}_{\mu f}$. Once we have $[x_1,\ldots,x_n].\,\mathsf{Z}_i(\mathsf{Z}_f x_1 \ldots x_n)(\mathsf{KI})$, then we can view $\mathsf{Z}_{\mu f}$ as $\mathsf{Z}_\mu \mathsf{Z}_f$ with an axiom for Z_μ as

$$\mathsf{Z}_\mu y \rhd [x_1,\ldots,x_n]\,\mathsf{Z}_i(y x_1 \ldots x_n)(\mathsf{KI}).$$

Z_μ is obviously not a proper combinator, however, once again it is easily seen to be a definable combinator.

Exercise 3.2.25. Find a CL-term for Z_μ. (Hint: Exercise 3.2.24 asked you to find a CL-term for $[x_1,\ldots,x_n].\mathsf{Z}_i(\mathsf{Z}_f x_1,\ldots,x_n)(\mathsf{KI})$, that is, for $\mathsf{Z}_{\mu f}$.)

The definitions of Z_o and Z_i are independent of the arity of the function f to which μ is applied. However, $\mathsf{Z}_{\mu f}$ is an n-ary combinator, and so Z_μ is $n+1$-ary. This shows another case when the usual notation used in recursion theory masks the multiplicity of similar operations. In other words, for each n, we have to define an $n+1$-ary *minimization combinator* Z_μ.

Another way to define minimization is to describe it as "if …, then …, else …" statement, then to turn that into an equation and to solve it. We reuse the notation Z_f from above for the CL-term that represents the function f to which μ is applied.

$\mathsf{Z}_m Y X_1 \ldots X_n$ is defined as "if $\mathsf{Z}_f X_1 \ldots X_n Y =_{\mathrm{w}} \mathsf{KI}$, then Y, else $\mathsf{Z}_m(\mathsf{SB}Y)X_1 \ldots X_n$." The branching depends on 0, thus we can use, for example, D again to form the CL-term: $\mathsf{D}Y(\mathsf{Z}_m(\mathsf{SB}Y)X_1 \ldots X_n)(\mathsf{Z}_f X_1 \ldots X_n Y)$. If $\mathsf{Z}_f X_1 \ldots X_n Y$ reduces to KI, then the term reduces to Y, as needed. Otherwise, the reduct is $\mathsf{Z}_m(\mathsf{SB}Y)X_1 \ldots X_n$, which in turn reduces to

$$\mathsf{D}Y(\mathsf{Z}_m(\mathsf{SB}(\mathsf{SB}Y))X_1 \ldots X_n)(\mathsf{Z}_f X_1 \ldots X_n(\mathsf{SB}Y)).$$

This gives the following equation.

$$\mathsf{Z}_m Y X_1 \ldots X_n =_{\mathrm{w}} \mathsf{D}Y(\mathsf{Z}_m(\mathsf{SB}Y)X_1 \ldots X_n)(\mathsf{Z}_f X_1 \ldots X_n Y)$$

Z_m is not a proper combinator, but we can find a CL-term for it without difficulty.

Exercise 3.2.26. Carry out the bracket abstractions to get a CL-term for the pseudo-λ-abstract $[y,x_1,\ldots,x_n].\,\mathsf{D}y(\mathsf{Z}_m(\mathsf{SB}y)x_1\ldots x_n)(\mathsf{Z}_f x_1\ldots x_n y)$.

Now we define $\mathsf{Z}_{\mu f}$, which is the CL-term that represents μf as $\mathsf{Z}_m(\mathsf{KI})$. Next, we want to define a CL-term for Z_μ itself, which can take any n-ary functions as its argument. This requires the replacement of the occurrence of Z_f in the definition of Z_m with a variable, and then abstracting that variable out.

Exercise 3.2.27. Find a CL-term for the combinator Z_μ. (Hint: You need the term from the previous exercise.)

This completes the second definition of the minimization operation. To sum up what we have shown — in more than one way — in this section, we state the following theorem.

THEOREM 3.2.8. (REPRESENTATION OF PARTIAL RECURSIVE FUNCTIONS) *There is a set of* CL*-terms that represent the* natural numbers *and the* partial recursive functions.

Truth functions. Classical sentential logic (TV) is very well understood nowadays. Indeed, the two-valued (or true–false) interpretation of Boole's logic was already developed in quite a detail together with the truth tables by Charles S. Peirce. All the connectives of TV are interpreted by functions of types $\{F,T\}^n \longrightarrow \{F,T\}$, where n is the arity of the connective. The truth values may be replaced by 0 and 1, and then the connectives are interpreted by functions of types $\{0,1\}^n \longrightarrow \{0,1\}$, where n is as before. Usually, 0 is taken to stand for falsity and 1 for truth — though it is possible to take the dual view.

The CL-term KI stands both for falsity and for 0, however, the representation of truth differs from that of 1. T is K, whereas 1 is $\mathsf{SB}(\mathsf{KI})$. The following are truth matrices for some often used two-valued connectives.

	$\sim$
T	F
F	T

$\wedge$	T	F
T	T	F
F	F	F

$\vee$	T	F
T	T	T
F	T	F

$\supset$	T	F
T	T	F
F	T	T

$\dagger$	T	F
T	F	F
F	F	T

The meaning of the connectives may be rendered in English as follows. $\sim$ is "it is not the case that ...," $\wedge$ is "... and ...," $\vee$ is "... or ...," $\supset$ is "if ... then ...," and $\dagger$ is "neither ..., nor ...." Sometimes other symbols are used for these connectives, and other names are given to them than what we list here: $\sim$ is *negation*, $\wedge$ is *conjunction*, $\vee$ is *disjunction*, $\supset$ is *conditional*, and $\dagger$ is *nor*.

Exercise 3.2.28. Find CL-terms for the listed truth-functional connectives assuming that the functions operate with truth values. (Hint: For instance, $\mathsf{Z}_\wedge\mathsf{KK}$ should reduce to K, but if either K is replaced by KI, then the reduct must be KI.)

Taking 0 and 1 instead of F and T, the truth functions can be viewed as *numerical functions* in addition to their meaning as the interpretation of connectives.

Then the connectives may be viewed as the following functions. $\sim(x) = 1 - x$, $\wedge(x,y) = \min(x,y)$, $\vee(x,y) = \max(x,y)$, $\supset(x,y) = \max(1-x,y)$, and $\dagger(x,y) = \min(1-x, 1-y)$.

Exercise 3.2.29. Find CL-terms for the functions listed. Suppose that x and y range over $\{0,1\}$. (Hint: For instance, $\mathsf{Z}_{\max}(\mathsf{KI})(\mathsf{KI})$ should reduce to KI. However, if either occurrence of KI is replaced by $\mathsf{SB}(\mathsf{KI})$, then the reduct must be $\mathsf{SB}(\mathsf{KI})$.)

3.3 Second modeling of partial recursive functions in CL

Another *representation of natural numbers* within CL selects a different set of CL-terms to stand for them, and then proceeds to the construction of terms for recursive functions accordingly. We saw above that in the CL representation of primitive recursive functions, in particular, in modeling the primitive recursion schema, the *predecessor function* becomes crucial — just as in an actual calculation of the value of a function defined by primitive recursion. The previous representation of the natural numbers is similar to the successor function's application to a number, or as a generalization of the unary notation for natural numbers (in which 7 is |||||||). The new representation is similar to viewing natural numbers as *sets of lesser numbers*, and it allows a simple definition of the predecessor function.

First of all, the starting point is I, that is, the *unary identity* combinator represents 0. The successor function is different than before. Recall that a combinator V with axiom $\mathsf{V}xyz \rhd zxy$ was introduced in exercise 1.2.4 and V is definable from $\{\mathsf{B},\mathsf{C},\mathsf{I}\}$, for instance. The *successor function* $\mathfrak{s}$ is $\mathsf{V}(\mathsf{KI})$. The CL-terms representing the first four natural numbers are listed below together with their weak reduct when they are applied to x.

I	$\mathsf{V}(\mathsf{KI})\mathsf{I}$	$\mathsf{V}(\mathsf{KI})(\mathsf{V}(\mathsf{KI})\mathsf{I})$	$\mathsf{V}(\mathsf{KI})(\mathsf{V}(\mathsf{KI})(\mathsf{V}(\mathsf{KI})\mathsf{I}))$	$\cdots$
x	$x(\mathsf{KI})\mathsf{I}$	$x(\mathsf{KI})(\mathsf{V}(\mathsf{KI})\mathsf{I})$	$x(\mathsf{KI})(\mathsf{V}(\mathsf{KI})(\mathsf{V}(\mathsf{KI})\mathsf{I}))$	$\cdots$

The shape of these reducts obviously differs from the reducts we obtained in the previous representation (after applying the CL-terms representing natural numbers to x and y). Now each reduct — save that of $\mathsf{I}x$ — is of the form x followed by *falsity* (which is unchangeably KI), and then followed by the predecessor of the number that was applied to x to start with. The structure of this CL-term has a certain similarity to the shape of the second clause in definition 3.2.6, which defines the predecessor function by primitive recursion.

Ordered pairs can be represented in various ways by sets, and we already saw that certain CL-terms can be viewed as ordered pairs. If we think of the term $[x].xMN$ as the *ordered pair* $\langle M,N\rangle$, then 1 is $\langle F,0\rangle$. The function $[x].xMN$ is $\mathsf{V}MN$, which helps us to decipher the other numbers. 2 is $\langle F,\langle F,0\rangle\rangle$, 3 is $\langle F,\langle F,\langle F,0\rangle\rangle\rangle$ or $\langle F,2\rangle$, etc. Thus each number contains its predecessor as the second element of the ordered pair.

If x is I, then all numbers weakly reduce to I, that is, 0. This gives an easy definition of $\mathfrak{z}$, the *constant zero* function as CII, because $\mathsf{CII}x \rhd_w \mathsf{I}x\mathsf{I}$, hence further to $x\mathsf{I}$. Of course, we can disregard the shape of the CL-terms that stand for the natural numbers and we can simply take the CL-term KI to be $\mathfrak{z}$. KI applied to any CL-term M reduces to I — including those CL-terms that represent numbers. We take CII for $\mathfrak{z}$.

We continue to assume that K stands for *truth*, just as we already assumed above that KI stands for falsity. If we apply a number to K, then we get K for 0, and KI for all the other numbers. Then a CL-term, which has a similar structure to CII (i.e., to the constant zero function $\mathfrak{z}$) can play the role of the *zero predicate*. $\mathsf{CIK}x \rhd_w \mathsf{I}x\mathsf{K} \rhd_w x\mathsf{K}$. That is, the combinator $\mathsf{Z}_{[0]}$ is now defined by the CL-term CIK.

Exercise 3.3.1. Verify by going through the one-step reductions that for 0, 3 and 11 the zero predicate $\mathsf{Z}_{[0]}$ returns the term that is expected.

The *projection functions* can be defined as previously — they are not dependent on the concrete representation of the numbers. This completes the representation of the basic primitive recursive functions.

We may view ordered n-tuples as $\langle x_1, \langle x_2, \ldots, \langle x_{n-1}, x_n \rangle \ldots \rangle \rangle$. Given that we took V to form an ordered pair, the CL-term standing for the natural numbers shows this pattern: each positive natural number n is an ordered $n+1$-tuple. Instead of the iterative construction of n-tuples as embedded ordered pairs, we may think of n-tuples as "flat" constructions, for instance, as $\langle x_1, x_2, x_3 \rangle$, $\langle x_1, x_2, x_3, x_4 \rangle$, etc.

Exercise 3.3.2. Define a combinator $\mathsf{Z}_{\langle 3 \rangle}$ such that it takes 3 arguments x_1, x_2 and x_3, and returns $\langle x_1, \langle x_2, x_3 \rangle \rangle$.

Exercise* 3.3.3. Define combinators like $\mathsf{Z}_{\langle 3 \rangle}$ for each n (where $n \geq 3$).

Once we have the notion of n-tuples derived from pairs, we might want to have *selection functions* that are like the projection functions to some extent, however, they return the ith element from an *n-tuple*.

Exercise* 3.3.4. Let M be an n-tuple, and let $1 \leq i \leq n$. Describe CL-terms for each i that behave as selection functions.

Returning to the observation about the reducts resulting from a number being applied to x, we may notice that if x is KI, then for every number greater that 0, we get the predecessor of that number. That is, $\mathsf{CI(KI)}$ is the inverse of the successor function. This is not quite the predecessor function, because $\mathsf{CI(KI)I} \rhd_w \mathsf{I(KI)}$, which is not one of the CL-terms that represent a number. We denote this *partial predecessor function* by p'.

Exercise 3.3.5. Define the *total predecessor function* p for this representation of natural numbers. p' is undefined for 0, but $p(0) = 0$, that is, $\mathsf{Z}_p\mathsf{I}$ should reduce to I. (Hint: Notice that all numbers save 0, contain the permutator V, which causes an asymmetry in the reducts of numbers applied to x.)

It is good to have the usual predecessor function p. However, in the primitive recursion schema we can get by p'. We do not need a CL-term for p to model primitive recursion, because if the input to the function is 0, then the value is calculated according to the first equation in which the predecessor function typically does not occur. To be more precise, the predecessor of 0 could occur in the first equation if f is given as a complex function. However, all such occurrences of $p(0)$ can be replaced by the constant zero function $\mathfrak{z}$ or simply by 0, since $p(0) = \mathfrak{z}(0) = 0$.

Let us suppose that the primitive recursive functions g and h are represented by Z_g and Z_h, respectively. Then the primitive recursive schema defining the function f takes the form "if $\mathsf{CIK}x$, then Z_g, else $\mathsf{Z}_h x(\mathsf{Z}_f(\mathsf{Z}_p x))$," where Z_f is the term that represents the function that is being defined. If we suppose that the only variable of f is x, that is, f is unary, then we can abstract that out, which gives us the equation

$$\mathsf{Z}_f =_{\mathrm{w}} [x].\,\mathsf{CIK}x\mathsf{Z}_g(\mathsf{Z}_h x(\mathsf{Z}_f(\mathsf{Z}_p x))).$$

Next Z_f can be abstracted out similarly, giving

$$\mathsf{Z}_f =_{\mathrm{w}} [y]\,[x].\,\mathsf{CIK}x\mathsf{Z}_g(\mathsf{Z}_h x(y(\mathsf{Z}_p x)))\mathsf{Z}_f.$$

Then using the fixed point combinator Y,

$$\mathsf{Z}_f =_{\mathrm{w}} \mathsf{Y}([y]\,[x].\,\mathsf{CIK}x\mathsf{Z}_g(\mathsf{Z}_h x(y(\mathsf{Z}_p x)))).$$

Exercise 3.3.6. Find CL-terms that could be used in place of the two CL-terms with bracket abstraction, $[x].\,\mathsf{CIK}x\mathsf{Z}_g(\mathsf{Z}_h x(\mathsf{Z}_f(\mathsf{Z}_p x)))$ and $[y]\,[x].\,\mathsf{CIK}x\mathsf{Z}_g(\mathsf{Z}_h x(y(\mathsf{Z}_p x)))\mathsf{Z}_f$. (Hint: You may use the bracket abstraction algorithm, possibly with η, or you may rely on heuristics and any primitive combinators introduced so far.)

Exercise** 3.3.7. Find a CL-term without any use of Y that can stand for primitive recursion given the current representation of the natural numbers. (Hint: Recall that Z_B was defined based on the idea of iterated calculations.)

To conclude showing that primitive recursive functions may be represented with the numbers and functions chosen as in this section, we note that *function composition* can be carried out as before. Composition is an operation that is independent of the concrete representation of the natural numbers, because the application of a function to its arguments remains the application operation on CL-terms.

The last operation we have to show to be representable is *minimization*. We can take a look at minimization again in the form of an "if ..., then ..., else ..." statement. If we are to minimize on f, then we calculate $f(0)$ and check if the result is 0; if it is, then we return 0 and stop. Otherwise, we calculate the value of f for the successor of the previous input as long as the value is not 0.

Let Z_m be defined as the function that is characterized by the equation derived from the "if ..., then ..., else ..." statement describing minimization.

$$\mathsf{Z}_m =_{\mathrm{w}} [y,x_1,\ldots,x_n].\,\mathsf{CIK}(\mathsf{Z}_f x_1\ldots x_n y)y(\mathsf{Z}_m(\mathsf{V}(\mathsf{KI})y)x_1\ldots x_n)$$

Then Z_m is the fixed point of the function

$$[v]\,[y,x_1,\ldots,x_n].\,\mathsf{CIK}(\mathsf{Z}_f x_1 \ldots x_n y)y(v(\mathsf{V}(\mathsf{KI})y)x_1 \ldots x_n).$$

To obtain a CL-term for the function μf, we apply Z_m to 0 (i.e., to I): $\mathsf{Z}_{\mu f} =_{\mathrm{w}} \mathsf{Z}_m\mathsf{I}$.

Lastly, we define the *minimization combinator* for n-ary functions. This means that we abstract out a variable that is put in place of Z_f. In other words,

$$\mathsf{Z}_\mu =_{\mathrm{w}} [w].\,\mathsf{Y}[v]\,[y,x_1,\ldots,x_n].\,\mathsf{CIK}(wx_1 \ldots x_n y)(v(\mathsf{V}(\mathsf{KI})y)x_1 \ldots x_n).$$

Exercise 3.3.8. Find CL-terms for the combinators Z_m and Z_μ.

Exercise 3.3.9. Are there any similarities or differences between the combinators that represent minimization here and in the previous section?

General number systems. We have looked, in some detail, at two representations of the natural numbers and the recursive functions on them. We have seen that some of the definitions in the two representations do not differ at all. The requirements for a set of CL-terms to serve as a modeling of partial recursive functions can be generalized.

DEFINITION 3.3.1. A set of CL-terms $\mathfrak{R}$ is a *representation* of the natural numbers and partial recursive functions iff (1)–(5) obtain.

(1) There is a term $M_{\mathfrak{s}} \in \mathfrak{R}$ that is called the *successor function*.

(2) There is a set of CL-terms $\mathfrak{N}$, which represents the *natural numbers*, $\mathfrak{N} \subseteq \mathfrak{R}$ such that the terms are pairwise not weakly equal. Further, there is an $M_0 \in \mathfrak{N}$ (which stands for 0), and for all $n > 0$, there is an $M_n \in \mathfrak{N}$ that is obtained from M_0 by n applications of $\mathfrak{s}$.[20]

(3) There is a term $\mathsf{Z}_p \in \mathfrak{R}$ with the property that for any $n > 0$, $\mathsf{Z}_p M_n \triangleright_{\mathrm{w}} M_{n-1}$ and $\mathsf{Z}_p M_0 \triangleright_{\mathrm{w}} M_0$. Z_p stands for the *predecessor function*.

(4) There is a term $\mathsf{Z}_{[0]} \in \mathfrak{R}$ such that $\mathsf{Z}_{[0]} M_0 \triangleright_{\mathrm{w}} \mathsf{K}$, whereas for all other elements M_n of $\mathfrak{N}$, $\mathsf{Z}_{[0]} M_n \triangleright_{\mathrm{w}} \mathsf{KI}$. $\mathsf{Z}_{[0]}$ is the *zero predicate*.

(5) $\mathfrak{R}$ contains the *constant zero*, the *projection functions*, and is closed under *composition*, *primitive recursion* and *minimization*. There are CL-terms that represent the latter *three operations*.

The point of the definition is to underscore that a numerical system is specific in some aspects such as the terms that represent the numbers, and some of the related functions. However, some components may be used in all representations. Of course, all the recursive functions are related to the numbers, however, the *successor*, the *constant zero*, the *predecessor* functions and the *zero test* predicate appeal

[20] We may require that each element of $\mathfrak{N}$ is one of the M_n's, that is, $\mathfrak{N}$ is the least set generated by $\mathfrak{s}$ from M_0 — in order to avoid bogus terms to be included into the representation.

to the concrete shape and structure of the numbers.[21] On the other hand, the *projection* functions and *function composition* are invariant. Primitive recursion and minimization may be defined in a representation independent way, if Z_p, $\mathsf{Z}_{[0]}$, etc., are assumed to have been defined concretely.

Exercise* 3.3.10.** Take into consideration what a representation of natural numbers and partial recursive functions requires. Find a set of CL-terms that is a representation (and is not identical to either of the representations in this chapter). (Hint: This is an open-ended exercise; you may find a completely new representation that is elegant and useful.)

3.4 Undecidability of weak equality

We stated theorem 2.4.2 (in chapter 2) without providing a proof. Now we have accumulated results that allow us to sketch the proof of undecidability. The *undecidability of first-order classical logic* is based on *Gödel's incompleteness theorem.* The undecidability of CL will be based on a "Gödel-type" theorem.[22]

We have seen that CL-terms can stand for natural numbers and functions on natural numbers. At the same time, CL-terms are finite objects, which can be regarded as (finite) strings. The arithmetization of the syntax means that CL-terms are *encoded* as natural numbers. The encoding does not need to be thought of as a complicated process or as encryption. Rather, the simpler is the encoding, the better is it. But simplicity may not hinder certain features of the encoding, which ensure that the numbers faithfully represent the CL-terms.

The number that is assigned to a CL-term M is called *the Gödel number* of that expression and denoted by $\mathfrak{g}(M)$.

We require that no two strings have the same Gödel number, that is, $\mathfrak{g}$ must be *injective.* The function $\mathfrak{g}$ itself should be *computable.* Furthermore, $\mathfrak{g}$ must be "reversible" in the sense that given a natural number n, there should be an effectively computable function that allows us to reconstitute the term M if and only if $n = \mathfrak{g}(M)$. As a matter of fact, all strings get a number assigned to them, however, we are interested in the Gödel numbers of CL-terms, in particular, that of combinators.

Strings are formed by the operation of *concatenation.* We rarely think about the decimal notation in terms of concatenation, because the places where the digits appear carry significance. For instance, 13 (in decimal notation) is not 1 and 3 concatenated per se; rather 13 is the superposition of 10 and 3. That is, $13 = 1 \cdot 10^1 + 3 \cdot 10^0$. However, just because we usually do not think about numbers in terms of concatenation, that does not mean that we cannot think about them that way. Gödel originally placed the numbers that stand for symbols into powers of

---

[21] $\mathfrak{z}$ depends on what 0 is, but $\mathsf{K}M_0$ is always suitable as $\mathfrak{z}$.

[22] In section A.3, we list some related theorems for Peano arithmetic.

prime numbers to indicate the place of each symbol within a string. His numbers even for relatively short expressions are enormous.[23]

We start with assigning numbers to the *individual symbols*. We know that the combinators S and K constitute a combinatorially complete basis. This means that we can encode all the combinators (over this basis) using only four numbers. For instance, 1, 2, 3 and 4 can be taken for S, K, (and), respectively. The operation of function application was denoted by juxtaposition, hence we do not need to allocate a number to that operation. We intend to consider the decidability of $=_{\rm w}$, hence we assign 5 to the weak equality symbol. This will allow us to assign a number to certain *statements* about pairs of CL-terms.[24]

Of course, CL-terms are *structured* strings, as it is apparent from definition 1.1.1. However, we can consider the parentheses simply to be symbols, and then CL-terms are strings. Each symbol has been given a Gödel number, therefore, the string of symbols can be represented by *a string of numbers*.[25]

LEMMA 3.4.1. *The concatenation of strings of digits as natural numbers in decimal notation is* a partial recursive function, *that is, concatenation is* computable.

Proof: We give only a sketch of the proof here. Let us consider the string $a_1 \ldots a_n$ and $b_1 \ldots b_m$, where each a and b is from the set $\{1,2,3,4,5\}$. We consider the string in which the a's are followed by the b's: $a_1 \ldots a_n b_1 \ldots b_m$. The numerical interpretation of the strings we start with are $a_1 \cdot 10^{n-1} + \ldots + a_n \cdot 10^0$ and $b_1 \cdot 10^{m-1} + \ldots + b_m \cdot 10^0$. The numerical interpretation of the resulting string is $a_1 \cdot 10^{n+m-1} + \ldots + a_n \cdot 10^m + b_1 \cdot 10^{m-1} + \ldots + b_m \cdot 10^0$. The same resulting string may be thought of as the sum (or superposition) of $a_1 \ldots a_n \cdot 10^m$ and $b_1 \ldots b_m$.

In either case, we need to be able to calculate m from $b_1 \ldots b_m$. The problem may seem to be a trifle until we recall that we will want to compute with a combinator on the CL-terms that represent the numbers $a_1 \ldots a_n$ and $b_1 \ldots b_m$. CL-terms do not come in decimal notation, which perhaps, would allow us to compute the length of the string. But for now, we assume that there is a function ℓ, which is like the function in definition A.1.5, that gives *the length of a string* of digits. The actual definition of this function in CL has to take into account that we are apparently not given a natural number in decimal notation, when a CL-term representing a natural number is given.

Exercises 3.1.1 and 3.1.2 asked you to define *multiplication* and *factorial* as primitive recursive functions. *Exponentiation* is primitive recursive too — see exercise

[23]Gödel numberings based on concatenation have been defined for various languages — including CL. See Boolos and Jeffrey [28], as well as Smullyan [137], the presentation that we more or less follow here, and Smullyan [138].

[24]Notice that we have completely omitted the variables, because we do not need them at this point. They could be included though. For example, we could define the Gödel numbering so that 6 stands for x, which is followed by as many 7's as the number in the subscript of x.

[25]The Roman numerals hint at what a numerical system based on concatenation rather than decimal places could look like. (E.g., 74 is *LXXIV*, that is, $50+10+10+(-1+5)$, where the minus is concatenation on the left.)

3.4.1 below. If so, then there is a combinator, let us say, Z_{exp} that represents the exponentiation function.

We wish to calculate $a_1 \ldots a_n \cdot 10^{\ell(b_1 \ldots b_m)} + b_1 \ldots b_m$. We already have addition and multiplication, and now, together with exponentiation and the length function the number resulting from concatenation is partial recursive. Then, again, there is a combinator Z_{con}, that takes two CL-terms that stand for natural numbers, such that $\mathsf{Z}_{\text{con}}MN$ reduces to P, where P is the CL-term standing for the natural number that results by concatenating in decimal notation the numbers represented by the CL-terms M and N. qed

Exercise 3.4.1. Define exponentiation as a primitive recursive function. (Hint: Similarly to exercises 3.1.1 and 3.1.2, you may find it easier first to define exponentiation by two equations, and then adjust their form to match precisely the primitive recursion schema.)

Exercise* 3.4.2. Based on the solution to the previous exercise, find a CL-term that represents the exponentiation function. (Hint: There is more that one way to represent natural numbers by CL-terms, and there is more that one way to turn an instance of the primitive recursive schema into a CL-term.)

In the proof we emphasized at each step the existence of suitable combinators. For our purpose, what we need is that Z_{con} exists.

Example 3.4.2. Let us assume that the term is (SK). The string of numbers that stands for this term is 3124. To obtain this number, we can apply the concatenation function to $\mathfrak{g}(()$ and $\mathfrak{g}(\mathsf{S})$. Similarly, we can take the CL-terms that stand for $\mathfrak{g}(\mathsf{K})$ and $\mathfrak{g}())$, and apply Z_{con} to them.

$$\mathsf{Z}_{\text{con}}(\mathsf{SB}(\mathsf{SB}(\mathsf{KI})))(\mathsf{SB}(\mathsf{SB}(\mathsf{SB}(\mathsf{SB}(\mathsf{KI})))))$$

yields $\{\mathsf{SB}\}^{24}(\mathsf{KI})$. We introduce the ad hoc notation $\{\ \}$ with a superscript to abbreviate the CL-term that is a natural number, because the Gödel numbers of CL-terms quickly lead to lengthy CL-terms even in the approach based on concatenation. Finally, the concatenation of the two strings of numbers (i.e., 31 and 24) yields 3124 represented by $\{\mathsf{SB}\}^{3124}(\mathsf{KI})$.

The example illustrates that the length of a string (in the decimal notation of the number) is not immediate from the CL-term that represents that number. A little thought should convince you that the other representation of the natural numbers (i.e., the one in section 3.3) does not produce CL-terms that more closely resemble lengthwise the decimal notation either.

The length of a number in decimal notation (or in general, base-n notation) may be determined by checking if the number falls into a *particular range*. For instance, a number requires two digits (i.e., a string of length two) in decimal notation when it is strictly greater than 9 and strictly less than 100.

Relations may or may not be functions. The *less than* relation is clearly not a function on any set of natural numbers with at least two elements. However, for each relation R, there is a *characteristic function* that returns "true" when the relation obtains between its arguments, and "false" otherwise. (This is a usual trick that turns relations into functions.)

We used the "if ..., then ..., else ..." construction previously, and it also helps when we define $<$, "strictly less than."

$$m < n = \begin{cases} \text{false} & \text{if } n = 0, \\ \text{true} & \text{if } n \neq 0 \text{ and } m = 0, \\ \text{true} & \text{if } n \neq 0 \text{ and } m-1 < n-1, \\ \text{false} & \text{otherwise.} \end{cases}$$

The conditional clauses reflect the idea of extracting the answer recursively from the comparison of the predecessors of m and n — together with two base cases. The first two conditionals ensure that the recursion is well-founded, which terminates as soon as one of the two numbers becomes 0.

Exercise 3.4.3. Find a combinator for $<$. (Hint: Recall that K stands for "true" and KI is "false." We have already established that there are combinators for the components of the definition, such as the zero predicate combinator.)

The length of a number in decimal notation may be obtained by comparing the number to 10, 100, etc. However, there are *arbitrarily large* natural numbers, whereas CL-terms are *finite*. This means that we cannot explicitly build into the combinator ℓ the numbers with which the argument is to be compared. We can turn around the reasoning though. The numbers 10, 100, 1000, etc., are all whole positive powers of 10; moreover, we know that exponentiation is computable. Thus given m, we are trying to find the number n that satisfies both $10^{n-1} \leq m$ and $m < 10^n$. The first part of the condition could be omitted if we could limit n in another way than by requiring that m is not less than 10^{n-1}. *Minimization* is exactly the limitation we are looking for, because the n we wish to find is the least n for which $m < 10^n$ is true. Notice that once we have found n, we have the value of $\ell(m)$.

Exercise 3.4.4. Define ℓ as a partial recursive function. (Hint: The preceding paragraph describes informally a way to define ℓ.)

Exercise 3.4.5. Based on the solution of the previous exercise, give a combinator that can be used to stand for ℓ. (Hint: We have defined earlier a combinator for minimization.)

The combinator for ℓ is the last piece that we needed in the proof of lemma 3.4.1.

Concatenating the numbers that correspond to the symbols in a string gives a number that corresponds to that string of symbols. (S) is not a CL-term, but it has a number, namely, 314. For our purposes, this number is useless, because it corresponds to a string that is not a CL-term.

The Gödel number of the CL-term representing 0, $\mathfrak{g}((\mathsf{K}((\mathsf{SK})\mathsf{K})))$ is 3233124244. The CL-terms for the rest of the numbers are obtained by prefixing SB sufficiently many times to $(\mathsf{K}((\mathsf{SK})\mathsf{K}))$. B is, of course, definable as $((\mathsf{S}(\mathsf{KS}))\mathsf{K})$. Each time, when SB is prefixed, a pair of parentheses is added too. Purely syntactically, the pattern looks like filling in the space inside $(\mathsf{SB}\ \ )$ or replacing M in the CL-term $(\mathsf{SB}M)$.

Example 3.4.3. Let us assume that M is the CL-term for m. $\mathfrak{g}(m)$ is the Gödel number of m. The successor of m is represented by $((\mathsf{S}((\mathsf{S}(\mathsf{KS}))\mathsf{K}))M)$. The Gödel number of the latter term is obtained by concatenating 33133132144244 and $\mathfrak{g}(M)$ and 4.

LEMMA 3.4.4. *The Gödel number function for* CL*-terms that stand for natural numbers is* partial recursive, *therefore,* computable.

Proof: We can easily define a function that takes the value 3233124244 for 0, and for $\mathfrak{s}(x)$, the successor of x yields the number that is the concatenation of 33133132144244 with the Gödel number of x and then with 4. All the functions involved are partial recursive. qed

Exercise 3.4.6. Show that $\mathfrak{g}(M)$ is partial recursive — according to the definition of partial recursive functions in section 3.1 — when M is a CL-term for a natural number. (Hint: In the proof we gave a somewhat informal description of $\mathfrak{g}$. Some abbreviations, for instance, for large numbers may be helpful.)

Exercise* 3.4.7. The Gödel number function (as restricted above) is partial recursive; hence, there is a CL-term that represents it. Find a suitable combinator.

The calculation of the Gödel numbers for CL-terms standing for numbers can be extended to a calculation of the Gödel numbers of arbitrary CL-terms. There are two atomic terms — if we ignore the variables. The Gödel numbers of these terms are 1 and 2, respectively. The Gödel number of any other term, which must be of the form (MN), can be calculated by concatenating: 3, $\mathfrak{g}(M)$, $\mathfrak{g}(N)$ and 4.

Exercise 3.4.8.** Define a combinator G that is applicable to any CL-term M, and yields the CL-term encoding the Gödel number of M. (Hint: Reformulate the definition of the function with two "if ..., then ..." constructions.)

We denote the combinator that represents $\mathfrak{g}$ by G (as in the previous exercise).

The *second fixed point theorem* (see theorem 2.4.1) includes a CL-term followed by a CL-term representing the Gödel number of a CL-term (i.e., $(M(\mathsf{G}N))$). We want to be able to construct a CL-term with a somewhat similar structure: $(M(\mathsf{G}M))$.

Assuming that M stands for a number, we need a combinator Z_{gn} such that $\mathsf{Z}_{\mathrm{gn}}M$ yields the CL-term that represents the number obtained by concatenating 3, the number denoted by M with the Gödel number of M, and with 4. In other words, $\mathsf{Z}_{\mathrm{gn}}M$ yields

$$\mathsf{Z}_{\mathrm{con}}(\mathsf{Z}_{\mathrm{con}}(\{\mathsf{SB}\}^3(\mathsf{KI}))M)(\mathsf{Z}_{\mathrm{con}}(\mathsf{G}M)(\{\mathsf{SB}\}^4(\mathsf{KI}))).$$

Example 3.4.5. Let M be $(\mathsf{SB}(\mathsf{KI}))$, which is the CL-term representing 1, where B and I are abbreviations. The Gödel number of $((\mathsf{S}((\mathsf{S}(\mathsf{KS}))\mathsf{K}))(\mathsf{K}((\mathsf{SK})\mathsf{K})))$ is

331331321442443233124244 4. The concatenation of 1 and the previous number is 1331331321442443233124244 4, which we denote by n. The combinator $\mathsf{Z_{gn}}$ applied to the CL-term representing 1 yields the CL-term $\{\mathsf{SB}\}^n(\mathsf{KI})$.

Exercise 3.4.9. Find a combinator $\mathsf{Z_{gn}}$. (Hint: We have gotten all the "pieces" that can be put together to obtain $\mathsf{Z_{gn}}$.)

The combinator $\mathsf{Z_{gn}}$ has the remarkable property that if it is applied to a CL-term that is the Gödel number of an expression, then it yields a CL-term that is the Gödel number too — despite the fact that in general $\mathsf{Z_{gn}}$ does not result in a CL-term that stands for a Gödel number of a CL-term.

LEMMA 3.4.6. *If M denotes the Gödel number of a* CL*-term N, then $\mathsf{Z_{gn}}M$ is the* CL*-term that represents the* Gödel number of $(N(\mathsf{G}N))$.

Proof: The Gödel number of N is $\mathfrak{g}(N)$. Then M is $\{\mathsf{SB}\}^{\mathfrak{g}(N)}(\mathsf{KI})$. $\mathsf{Z_{gn}}M$ yields $\{\mathsf{SB}\}^{3\mathfrak{g}(N)\mathfrak{g}(M)4}(\mathsf{KI})$. However, $3\mathfrak{g}(N)\mathfrak{g}(M)4$ is $\mathfrak{g}((NM))$, which means that the resulting term is $\{\mathsf{SB}\}^{\mathfrak{g}((NM))}(\mathsf{KI})$. We started with the assumption that M stands for $\mathfrak{g}(N)$, hence we have $\{\mathsf{SB}\}^{\mathfrak{g}((N(\mathsf{G}N)))}(\mathsf{KI})$. qed

Now we restate theorem 2.4.1.

THEOREM 3.4.7. (SECOND FIXED POINT THEOREM) *For any* CL*-term M*, there is a CL-term N, *such that* $M(\mathsf{G}N) =_{\mathrm{w}} N$.

Proof: Let M be a CL-term. Then

$$M(\mathsf{G}(\mathsf{B}M\mathsf{Z_{gn}}(\mathsf{G}(\mathsf{B}M\mathsf{Z_{gn}})))) =_{\mathrm{w}} \mathsf{B}M\mathsf{Z_{gn}}(\mathsf{G}(\mathsf{B}M\mathsf{Z_{gn}})).$$

In the proof of the fixed point theorem (see theorem 2.3.3), we used $\mathsf{Y}M$ as the solution for the weak equality $MN =_{\mathrm{w}} N$. Here we take $\mathsf{B}M\mathsf{Z_{gn}}(\mathsf{G}(\mathsf{B}M\mathsf{Z_{gn}}))$ in a similar role.

We start with the right-hand side term, $\mathsf{B}M\mathsf{Z_{gn}}(\mathsf{G}(\mathsf{B}M\mathsf{Z_{gn}}))$. According to B's axiom, $M(\mathsf{Z_{gn}}(\mathsf{G}(\mathsf{B}M\mathsf{Z_{gn}})))$ results by $\triangleright_1$. Now the first occurrence of $\mathsf{Z_{gn}}$ is the head of a redex, which further gives the term $M(\mathsf{G}(\mathsf{B}M\mathsf{Z_{gn}})(\mathsf{G}(\mathsf{G}(\mathsf{B}M\mathsf{Z_{gn}}))))$. By lemma 3.4.6, the latter term is the same term as $M(\mathsf{G}(\mathsf{B}M\mathsf{Z_{gn}}(\mathsf{G}(\mathsf{B}M\mathsf{Z_{gn}}))))$, which in turn, is the same term as the one on the left-hand side of the above weak equality. This concludes the proof that the displayed weak equality holds. qed

Gödel's incompleteness theorem for PA, can be seen to imply the statement that the set of true sentences of PA is *not recursive*. For CL, the similar result is *Scott's theorem* that we stated without proof as theorem 2.4.2.

A set is recursive when the set and its complement are recursively enumerable.

Example 3.4.8. The set of odd natural numbers, $\mathbb{N}_O$ is recursive. $-\mathbb{N}_O$ is the set of even natural numbers, $\mathbb{N}_E$. Either set can be enumerated by stepping through $\mathbb{N}$ and dividing each number n by 2. (The remainder function of $2\,|$ is partial recursive.) If the remainder is 1, then $n \in \mathbb{N}_O$; if the remainder is 0, then $n \in \mathbb{N}_E$.

If a set of numbers is recursive then there is a combinator that represents the function f, which is the characteristic function of the set. That is, for any natural number n, $f(n)$ is true if n is an element, and $f(n)$ is false if n is not an element of the set. We will focus on Gödel numbers, which are just some of the natural numbers, perhaps, somewhat "scarcely spread out" within $\mathbb{N}$.

DEFINITION 3.4.9. If X is a set of natural numbers, then *the truth set of* X, denoted by X^t, is $X^t = \{n\colon n52 \in X\}$.

Truth is represented by K (with Gödel number 2), and $=_\mathrm{w}$ has Gödel number 5. Thus X^t is the set of numbers that are Gödel numbers of true sentences according to X. It is not difficult to see that X^t is recursive provided X is.

Exercise 3.4.10. Assume that Z_X is the combinator that represents the characteristic function of X. That is, Z_X takes CL-terms representing numbers as its argument and returns K or KI, depending if the number does or does not belong to X. Define Z_{X^t} that does the same for X^t.

Exercise 3.4.11. Show that, if Z_{X^t} exists, then there is a combinator representing the characteristic function of X; that is, if Z_{X^t} exists, so does Z_X. (Hint: This is the converse of the claim in the previous exercise.)

The combinator Z_{X^t} is a CL-term, hence it can be the CL-term M in the second fixed point theorem. Then $\mathsf{BZ}_{X^t}\mathsf{Z}_{\mathrm{gn}}(\mathsf{G}(\mathsf{BZ}_{X^t}\mathsf{Z}_{\mathrm{gn}}))$ is a fixed point (in the sense of theorem 3.4.7) of Z_{X^t}. For the sake of brevity, we denote the term $\mathsf{BZ}_{X^t}\mathsf{Z}_{\mathrm{gn}}(\mathsf{G}(\mathsf{BZ}_{X^t}\mathsf{Z}_{\mathrm{gn}}))$ as N, and we assume that its Gödel number is n. $N =_\mathrm{w} \mathsf{K}$ is a sentence, and its Gödel number is $n52$.

If $N =_\mathrm{w} \mathsf{K}$ is true, then by an instance of the weak equality in the second fixed point theorem and the transitivity of $=_\mathrm{w}$, $\mathsf{Z}_{X^t}(\mathsf{G}(\mathsf{BZ}_{X^t}\mathsf{Z}_{\mathrm{gn}}(\mathsf{G}(\mathsf{BZ}_{X^t}\mathsf{Z}_{\mathrm{gn}})))) =_\mathrm{w} \mathsf{K}$ is true. This weak equality simply says that Z_{X^t} applied to $\mathsf{G}N$ equals true. Since Z_{X^t} computes X^t, $n \in X^t$. By the definition of X^t, $n52 \in X$, which is the Gödel number of $N =_\mathrm{w} \mathsf{K}$.

To show the converse of what we have just shown, namely, that the truth of $N =_\mathrm{w} \mathsf{K}$ implies $n52 \in X$, let us assume $N =_\mathrm{w} \mathsf{K}$. By the definition of X^t, $n \in X^t$. Then $\mathsf{Z}_{X^t}(\mathsf{G}N) =_\mathrm{w} \mathsf{K}$ is true, because Z_{X^t} computes X^t. However, $\mathsf{Z}_{X^t}(\mathsf{G}N) =_\mathrm{w} N$ is by an application of the second fixed point theorem. Therefore, $N =_\mathrm{w} \mathsf{K}$, by the transitivity of $=_\mathrm{w}$. This proves the following lemma.

LEMMA 3.4.10. *Let* $n = \mathfrak{g}(\mathsf{BZ}_{X^t}\mathsf{Z}_{gn}(\mathsf{G}(\mathsf{BZ}_{X^t}\mathsf{Z}_{gn})))$, *and let* N *be the* CL-*term representing* n. *Then* $n52 \in X$ if and only if $N =_\mathrm{w} \mathsf{K}$ *is true.*

Exercise 3.2.28 asked you to find combinators for truth functions. One of the truth functions in classical logic is negation, denoted by $\sim$. If X is recursive, then so is its complement, which is the set that contains a number just in case X *does not* contain that number.

Exercise 3.4.12. Show that, if the combinator Z_X represents the characteristic function of X, then there is a combinator that represents the characteristic function of $-X$, the complement of X.

We reformulate Scott's theorem (i.e., theorem 2.4.2) from chapter 2.

THEOREM 3.4.11. (SCOTT'S THEOREM) *The weak equality relation on* CL*-terms is* not decidable.

Proof: Let us consider the set of weak equalities that are true in CL. If this set is recursive, then the weak equality relation between CL-terms is *decidable*. A set is not recursive if either the set or its complement is not recursively enumerable. In order to show that the set of all true weak equalities is not recursive, it is sufficient to show that the complement of the set of the Gödel numbers of all true weak equalities is not recursively enumerable.

Let $n \in X$ if and only if $n = \mathsf{G}M$ and M is a true weak equality. Then $-X$ contains the Gödel numbers of all false weak equalities. We would like to have $J =_{\mathrm{w}} \mathsf{K}$ true, where J is like N, but for $-X$ rather than X. Then $J =_{\mathrm{w}} \mathsf{K}$ is true iff $j52 \in -X$, where j is the Gödel number of J. However, if $J =_{\mathrm{w}} \mathsf{K}$ were true, then $j52$ cannot be in $-X$, because $-X$ does not contain Gödel numbers of true equalities. If $J =_{\mathrm{w}} \mathsf{K}$ were false, then $j52$ would have to be in X, because the complement of $-X$ is X and the construction of $J =_{\mathrm{w}} \mathsf{K}$ is characteristic for $-X$. However, the Gödel numbers of false weak equalities are in $-X$, hence so is $j52$. We have not assumed anything special about $J =_{\mathrm{w}} \mathsf{K}$ except that it characterizes X in the sense of lemma 3.4.10, thus we may conclude that no suitable J exists.

Therefore, $-X$ is not recursive, and X is not recursive either. qed

Chapter 4

Connections to λ-calculi

Combinatory logic is sufficiently expressive to allow a definition of a *function abstraction operation* — as we already saw in chapter 1. The bracket abstraction operation is not part of the language of CL, rather it is a meta-notation for certain CL-terms (or certain classes of CL-terms, depending on the definition of the bracket abstraction). It is one of the great intellectual discoveries of the 20th century that a mathematically rigorous theory of functions does not need to include a *variable binding abstraction operator.*

Thinking about functions as *incomplete* expressions that are void of their arguments has a long tradition, although the empty places had been often filled with variables as "placeholders." The first successful theory that made a precise distinction between functions per se and functions applied to some arguments (even if the arguments were variables) was invented by Alonzo Church in the 1930s.[1] The λ in the λ-calculi is a variable binding operator, and in this respect the λ is similar to the quantifiers $\exists$ and $\forall$. The notion of *λ-definable functions* is equivalent to the notion of *computable functions*, which underscores that there should be a connection between CL and the λ-calculus.

The theory of the λ-calculus, more precisely, of the *λ-calculi* (or Λ, for short), is very extensive by itself. We do not provide an exhaustive exposition of Λ here, rather we look for some similarities and differences between Λ and CL. As the expressive equipotence of Λ and CL lead us to expect, there are ways to move back and forth (i.e., to translate) between CL and Λ, and we discuss some of these.

4.1 λ-calculi: Λ

We often talk about CL, however, we have already seen that we can choose different combinatory bases, and not all of those are equivalent. Further, there is more than one interesting relation on the set of CL-terms. That is, strictly speaking, CL is a whole family of logics.

The situation is similar in the case of Λ. Certain restrictions on the abstraction operation yield the $\lambda\mathsf{I}$- and the linear λ-calculi from the $\lambda\mathsf{K}$-calculus. The restrictions

[1]Church [44] is a classic, but it is still a very readable presentation of some of the λ-calculi.

are similar to selecting a combinatory basis. In addition, we can consider various relations on Λ-terms too. This is the reason why we use Λ — just as we use CL — to refer to a family of closely related theories.

The *language* of Λ contains denumerably many *variables*, a binary operation *application* and the *abstraction operator*, λ. The auxiliary symbols include (and). Often a *dot*, that is, . is also permitted as a delimiter, and we use it to separate the operator prefix. The application operation is usually denoted by *juxtaposition* — similarly, as in the language of CL.

DEFINITION 4.1.1. (Λ-TERMS) The set of Λ-*terms* is inductively defined by clauses (1)–(3).

(1) If x is a variable, then x is a Λ-term;

(2) if M and N are Λ-terms, so is (MN);

(3) if M is a Λ-term and x is a variable, then $(\lambda x.M)$ is a Λ-term.

This definition parallels definition 1.1.1. The language of Λ may include constants, and then (1) in the above definition is modified so that the base set includes those constants too. In the λI-*calculus*, a λ-abstraction $(\lambda x.M)$ is well-formed only if x occurs in M. Accordingly, for λI-terms, clause (3) is modified to reflect this restriction.

It is usual to omit the outside parentheses, as well as to abbreviate a series of λ's by writing $\lambda x_1 \ldots x_n.M$ instead of $(\lambda x_1.\ldots(\lambda x_n.M)\ldots)$. The presence of a single λ followed by a series of variables *does not* mean that one λ can bind more than one variable, or that the abstraction with respect to the variables $x_1,\ldots,x_n$ happens at once.[2] In the *scope* of a λ operator, that is, within the term M, the convention of omitting parentheses from left-associated terms is permitted — exactly as in CL-terms.

Exercise 4.1.1. Simplify the following Λ-terms — without introducing ambiguity. (a) $(\lambda y.(\lambda x.(xy)))$, (b) $(\lambda x.(((yy)y)(\lambda y.(x(\lambda x.(yx))))))$, (c) $((\lambda x.(\lambda y.(xx)))(yx))$ and (d) $((\lambda y.(\lambda x.(x(yz))))((\lambda x.(\lambda y.(x(yz))))(\lambda z.(x(yz)))))$.

Exercise 4.1.2. Restore all the λ's and all the parentheses in the following Λ-terms. (a) $\lambda xyzv.xy(xvz)$, (b) $\lambda y.(\lambda x.x)y$, (c) $\lambda x.\lambda yz.yz(xz)$ and (d) $(\lambda x.xy)(\lambda xy.x)$.

Every λ-operator *binds* at least one variable. A λ binds, first of all, the variable that immediately follows the λ, but more interestingly, it binds all the free occurrences of the same variable in the scope of the λ. A difference between definitions 1.1.1 and 4.1.1 is that the latter — specifically, clause (3) — depends on a *proper subset* of Λ-terms, namely, the variables. An occurrence of a λ in a Λ-term is *always* followed by a variable, and then by an arbitrary Λ-term.

[2]There are λ-calculi with so-called *one-sweep* abstraction where no equivalent piecemeal abstraction is stipulated. We do not consider those systems here.

Every *occurrence* of every variable is either free or bound but not both. An occurrence of a variable x is *free* if it is not in the scope of λx; otherwise, it is *bound*. The *set of free variables* of a Λ-term M contains those variables that have a free occurrence in the term M, and the *set of bound variables* of M contains those variables that have a bound occurrence in M. The latter two sets for the term M are often denoted by $\mathrm{fv}(M)$ and $\mathrm{bv}(M)$. These two sets do not need to be disjoint.[3]

Exercise 4.1.3. Determine for each occurrence of each variable in the following terms if that is free or bound, then list the elements of fv and bv. (a) $\lambda xy.z(xy)$, (b) $\lambda x.(\lambda x.x)y(\lambda xyz.xz)$, (c) $x\lambda x.xy$ and (d) $(\lambda x.y)(\lambda y.xy)$. (Hint: It may be useful to first restore all the λ's and parentheses in order to delineate the scope of each λ.)

Exercise 4.1.4. Create a Λ-term in which x has only free occurrences, y has only bound occurrences, and z has both.

The complications that emerge from variable binding can be illustrated by the following two terms.

$$\lambda x.y(\lambda x.x) \qquad\qquad \lambda xyz.xz(yz)$$

The term on the left contains two λ's and they both bind x. Beyond the two occurrences of x immediately after the λ-operators, there is yet another occurrence of x in the term. That occurrence is not bound by the first λ, but it is bound by the second one. This example shows that the *scope* of the λ's have to be kept track of in order to determine *binding relationships* between λ's and variable occurrences. Incidentally, this term also illustrates that λx may be prefixed to a Λ-term that does not contain any free occurrences of x.

The term on the right is a *closed Λ-term*; that is, it does not contain free occurrences of any variable — unlike the term on the left. The Λ-term $xz(yz)$ is in the scope of each λ — there are three λ's in this term, not only the one visible λ. This implies that if we would substitute a term M for x such that M contains free occurrences of y or z, then those occurrences would become bound. Substitution is a way to combine two Λ-terms, and it is undesirable for this operation to *introduce* variable bindings, which are not already present in the Λ-terms that are being combined. The problem of substitution in Λ is very similar to the problem of substitution in other formalisms that contain variable binding operators. For example, first-order logic includes the quantifiers $\forall$ and $\exists$, which bind a variable. The problem of substitution in first-order logic is what motivated the invention of combinatory logic.

Exercise 4.1.5. Determine which λ operators bind which variables in the following terms. (a) $\lambda xy.(\lambda x.x)(xyy)$, (b) $\lambda x.x\lambda xy.x$, (c) $\lambda xyz.(\lambda xy.x)(\lambda xy.x)\lambda xyz.xz(yz)$.

Exercise 4.1.6. What are the sets $\mathrm{fv}(M)$ and $\mathrm{bv}(M)$ for each of the following terms? (a) $(\lambda xy.(x(yy)))$, (b) $(\lambda xyz.((xy)(zy)))(\lambda x.(xy))$, (c) $((\lambda x.((\lambda y.((\lambda z.(((xx)(yy))(zz)))y))x))z)$.

[3] Inductive definitions of these sets may be found in section A.4 — see definitions A.4.1 and A.4.2. Definition A.1.12 is similar, but concerns CL-terms.

A way to deal with the problem of accidentally capturing free variables by λ-operators, that is, with what is sometimes called "variable clashes," is to allow the *renaming* of some *bound variables* in the substitution itself. Another way to avoid the problem is to stipulate a separate rule that is applicable to all Λ-terms, which allows the renaming of bound variables (independently of substitutions).[4] Let us assume that M is of the form $\lambda x.N$, and y is a variable that has no occurrences in M. Then the term $\lambda y.[x/y]N$ *α-reduces to M*. The term $[x/y]N$ is the result of substituting y for all the free occurrences of x in N; we define this operation below. The requirement that $y \notin \mathrm{fv}(M)$ could be somewhat relaxed, and then the resulting notion would lead to a larger class of Λ-terms.

The transitive, symmetric and reflexive closure of α-reduction is *α-equality*, which is denoted by $=_\alpha$. α-equality is an *equivalence* relation, and Λ-terms that fall into the same equivalence class can be viewed as *notational variants* of each other. Such terms also turn out to denote the same function. If $M =_\alpha N$ (but they are not the same term), then we will consider N to be M with some *bound variables renamed*, and vice versa.

Exercise 4.1.7. Rename all the bound variables in the following terms. Notice that the renaming of bound variables in the case of some terms may increase or decrease the number of bound variables. (a) $\lambda zyx.y$, (b) $\lambda x.x(\lambda y.yyx)(\lambda y.yyx)$, (c) $\lambda xy.(\lambda xy.x)(\lambda xy.y)xyy$. (Hint: Try to use as many and then as few bound variables as you can.)

The λ-operator forms a function, and every λ-term of the form $\lambda x.M$ is *a unary function*. Then $(\lambda x.M)N$ is a Λ-term, which is the application of a function to the argument N. The application of a function to an argument is the main step of manipulating terms in Λ, but before we can define that step that will yield the result of function application, we describe the notion of substitution.

DEFINITION 4.1.2. (SUBSTITUTION) The *substitution* of the term N for the variable x in the term M is defined inductively by (1)–(5) below. (We use the notation $[x/N]M$ as well as M_x^N for the term that results.)

(1) If M is x, then M_x^N is N;

(2) if M is y, where y is a variable distinct from x, then M_x^N is y;

(3) if M is (PQ), then M_x^N is $(P_x^N Q_x^N)$;

(4) if M is $\lambda x.P$, then M_x^N is $\lambda x.P$;

(5) if M is $\lambda y.P$, where y is a variable distinct from x, then M_x^N is $\lambda y.Q_x^N$, where Q is α-equal to P, and $\mathrm{bv}(Q) \cap \mathrm{fv}(N) = \emptyset$ (i.e., the bound variables of P has been renamed, if necessary, so that no bound variables are free in N).

[4]Church introduced a rule like this. Nowadays this rule is often called α-*conversion*. It is also possible to incorporate the renaming of bound variables into the definition of β-reduction steps.

This definition of substitution makes substitution a *total* operation on terms in virtue of the possibility to rename bound variables during the process of substitution. Incidentally, clause (5) includes more renaming of variables that may be absolutely necessary to perform. The only bound variables that *have to* be renamed are those that are free in the term that is substituted, and at the same time there is a free occurrence of the variable for which a term is substituted within the scope of a λ, which binds a variable that is free in the substituted Λ-term. Clause (5) is easier to state — and perhaps, easier to understand — than an alternative clause in which the modifications in the Λ-term M are limited to the absolute minimum.

Example 4.1.3. Consider the terms $((\lambda xyz.\,yz)x)$ and (yz). Definition 4.1.2 applied to $((\lambda xyz.\,yz)x)_x^{yz}$ results in the term $((\lambda xvw.\,vw)(yz))$. However, no variable capture would occur without the renaming either, because $((\lambda xyz.\,yz)(yz))$ is α-equal to the former term. In other words, x has no free occurrences in yz that are in the scope of a λy or λz — simply because x has no free occurrences in yz at all.

Exercise 4.1.8. Give a formulation of clause (5) so that renaming of bound variables is limited to the absolutely necessary ones. (Hint: It may be easier first to introduce a notion to describe the relation between the occurrences of a variable and the substituted term.)

Bound variables can always be renamed, that is, for any term there are sufficiently many α-equal terms and α-equality is well-defined. *First* of all, Λ-terms are *finite*, hence given any Λ-term, there are infinitely many variables that have neither free nor bound occurrences in that Λ-term. Λ-terms may be arbitrarily long, and so the need for "fresh" variables is what impels us to postulate that there are $\aleph_0$-many variables in the language. Sometimes the notion of a "fresh" variable is made more unambiguous by taking the first variable from a fixed sequence of all variables such that that variable has no occurrence in the given Λ-term.[5] We assume simply that we can choose a suitable variable.

Second, any Λ-term has subterms that do not contain further occurrences of the λ operator, because $\lambda x.M$ is not an atomic Λ-term. Thus, the renaming of bound variables can be ordered — starting with a λ that has no λ's in its scope — so that either clause (5) is not used at all, or it is used only with P and Q being identical. In other words, the bound variable renaming in the definition of substitution does not lead to unending sequences of variable renaming steps.

The language of Λ may contain *constants* too. For example, we could add combinators or numbers. If there are any primitive constants in the language of Λ, then definition 4.1.2 has to be extended by the following clause.

(6) If M is c, a constant, then M_x^N is c.

This clause is similar to (2) and (4) to the extent that substitution does not change the Λ-term in any way. Constants always fall into the class of *closed* Λ*-terms*, because they never contain occurrences of variables.

[5]In implementations of Λ, for instance, in versions of LISP and other functional programming languages, it may be necessary to pick the first variable in order to have a deterministic program.

Exercise 4.1.9. Carry out step by step the substitutions in the following terms — according to definition 4.1.2, but applied simultaneously.

(a) $[x/\lambda xy.x, y/\lambda yx.y, z/x]\,(\lambda xyz.xz(yz))$

(b) $[v/(xy)y, w/zv]\,(\lambda xyz.(\lambda x.xzvw)(\lambda x.xyvw)(\lambda z.z(wv)))$

(c) $[x/yv, y/\lambda x.x, z/y]\,((\lambda x.xy)(\lambda yz.xz(yz))(z(x(yy))))$

(Hint: The slash notation is the same as the superscript–subscript notation, but more convenient when the substituted terms are longer.)

Having scrutinized substitution in Λ, we can proceed to the application of a λ-abstract to its argument.

DEFINITION 4.1.4. (β-REDEX) A term of the form $(\lambda x.M)N$ is a β-*redex*, with $(\lambda x.M)$ as the *head*.

The notion of a β-redex is simpler than the notion of a redex in CL, because any λ-abstract followed by a Λ-term forms a β-redex — independently of the concrete form of M.

DEFINITION 4.1.5. (ONE-STEP β-REDUCTION) Let the term P contain an occurrence of the β-redex $(\lambda x.M)N$. The term P *one-step* β-*reduces* to the term $[(\lambda x.M)N/M_x^N]P$, which is denoted by $P \rhd_{1\beta} [(\lambda x.M)N/M_x^N]P$, where the latter term is the result of the replacement of $(\lambda x.M)N$ by M_x^N.

One-step β-reduction, $\rhd_{1\beta}$ has a similar place in the theory of Λ as one-step reduction, $\rhd_1$ has in the theory of CL: $\rhd_{1\beta}$ is the fundamental "reduction-type" concept in the theory of Λ. α-reduction and α-equality, which have no analogues in CL, can be viewed as supplementary concepts that make the theory of Λ smoother.

DEFINITION 4.1.6. (β-REDUCTION) A Λ-term M β-*reduces* to the Λ-term N iff (1) or (2) obtains.

(1) M is α-equal to N;

(2) there is a nonempty finite sequence of one-step β-reductions leading from M to N', where N' is α-equal to N.

In other words, β-reduction, which is denoted by $\rhd_\beta$, is the *reflexive transitive closure* of one-step β-reduction together with α-equality.

We incorporated an α-equality step into the notion of substitution; therefore, it is sufficient to allow one extra α-equality step in β-reductions.

λ-abstraction is a way to form a Λ-term from a Λ-term (plus a variable together with a λ). Abstraction — broadly understood — as the intellectual operation of isolating certain aspects of a situation or extracting properties from objects is like the converse of β-reductions.

DEFINITION 4.1.7. (β-EXPANSION) The converse of the one-step β-reduction is *one-step β-expansion*. That is, if $M \triangleright_{1\beta} N$, then $N\ {}_{1\beta}\!\triangleleft\ M$.

The converse of β-reduction is *β-expansion*. That is, $M \triangleright_\beta N$, then $N\ {}_\beta\!\triangleleft\ N$.

Example 4.1.8. Consider the Λ-term $\lambda y.xxy(xxy)$. Each of the following Λ-terms one-step β-reduces to this term. $(\lambda zy.xxy(xxy))x$, $(\lambda zy.zxy(zxy))x$, $(\lambda zy.xzy(xzy))x$, $(\lambda zy.zxy(zzy))x$, $(\lambda zy.zzy(zzy))x$.

The above list is not exhaustive. The choice of the number of z occurrences that are inserted into x occurrences in $\lambda y.xxy(xxy)$ is accidental, so to speak. We also have $(\lambda zy.zy(zy))(xx) \triangleright_{1\beta} \lambda y.xxy(xxy)$, and $(\lambda zy.zy(xxy))(xx) \triangleright_{1\beta} \lambda y.xxy(xxy)$ too. Furthermore, $(\lambda z.z)(\lambda y.xxy(xxy)) \triangleright_{1\beta} \lambda y.xxy(xxy)$.

We alluded to the informal or broadly understood abstraction operation that is significant in sciences and other disciplines. However, the examples show that typically there is a range of possible terms to choose from, each of which β-reduces to a given Λ-term. Given a Λ-term in which a particular β-redex has been selected, $\triangleright_{1\beta}$ is an operation that yields a unique Λ-term (up to α-equality). We do not have a notion that would be the dual of a β-redex, and which would allow us to obtain a unique term by one-step β-expansion. Roughly speaking, given a Λ-term M, we would need to select a pair $\langle N, \langle P_{i\in I}\rangle\rangle$ such that the P_i's are subterm occurrences of a Λ-term in N, and N is a subterm occurrence in M. Every occurrence of every variable that is free in any of the P_i's must be a free occurrence in N too. Then abstracting out the P_i's with respect to N, and replacing N by the obtained abstraction in M is a one-step β-expansion, which gives a unique Λ-term modulo α-equality.

Example 4.1.9. Given the Λ-term $\lambda y.xxy(xxy)$, and the pair $\langle (xxy)_1, \langle x_1, x_2\rangle\rangle$, we get the Λ-term $\lambda y.(\lambda z.zzy)x(xxy)$.[6]

From the same Λ-term, we could obtain $\lambda y.((\lambda z.zy(zy))(xx))$ by ${}_{1\beta}\!\triangleleft$ with respect to $\langle xxy(xxy), \langle (xx)_1, (xx)_2\rangle\rangle$.

Exercise 4.1.10. Give an example that shows that if the proviso about free variable occurrences is violated, then the P_i's cannot be abstracted out.

Exercise* 4.1.11. The procedure of selecting subterm occurrences of a Λ-term for abstraction has an inherent limitation. Find an example of a ${}_{1\beta}\!\triangleleft$ that cannot be conceived in this way. (Hint: For the ${}_{1\beta}\!\triangleleft$ steps that we left out, uniqueness cannot be guaranteed.)

β-reduction and β-expansion may be combined, and the resulting relation is an equivalence relation.

DEFINITION 4.1.10. (β-EQUALITY) A Λ-term M is *β-equal* to N, denoted by $=_\beta$ iff there are Λ-terms $P_0, \ldots, P_n$ (where $n \in \mathbb{N}$) such that P_0 is M, P_n is N and for each i (where $0 \leq i \leq n$), $P_i \triangleright_\beta P_{i+1}$ or $P_{i+1} \triangleright_\beta P_i$.

[6] We have not defined indexing for subterm occurrences for Λ-terms that would parallel definition A.1.7 for CL-terms. The indexes here are used informally, counting from left to right.

The notions of β-reduction and β-equality may seem to make too many distinctions between Λ-terms that denote, or perhaps, *should* denote the same function. Recall that a λ-abstract is a function, and it is not required that the variable x, which is "abstracted out," occurs in the term M that follows the λx. If it so happens that $x \notin \text{fv}(M)$, then $\lambda x.Mx$ behaves extremely similarly to M, what may be seen as follows. Let us assume that M is applied to the Λ-term N; the result is MN. Now let us also consider what ensues, when $\lambda x.Mx$ is applied to N. We get $(\lambda x.Mx)N$, which is not identical to MN, but β-reduces to MN. Since the Λ-terms M and $\lambda x.Mx$ yield the same result when they are applied to a Λ-term, one might think that they should be, or at least in a suitable sense *could* be equal.

DEFINITION 4.1.11. (η-REDEX AND $\triangleright_{1\eta}$) If $x \notin \text{fv}(M)$, then $\lambda x.Mx$ is an η-*redex*. The Λ-term P *one-step* η-*reduces* to $[\lambda x.Mx/M]P$, which is denoted by $P \triangleright_{1\eta} [\lambda x.Mx/M]P$.

The converse of one-step η-reduction is *one-step* η-*expansion*. P one-step η-expands to $[M/\lambda x.Mx]P$, when $x \notin \text{fv}(M)$; this is denoted by $P\,{}_{1\eta}\!\triangleleft\, [M/\lambda x.Mx]P$.

η-reduction captures the *extensionality* of functions. Extensionality means that functions that compute to the same term when they are applied to the same term are not distinguished. Extensionality in this sense is widely used. For example, conjunction is called an extensional connective in logic, because it cannot distinguish between component formulas that differ in their meaning but not in their truth value. In contrast, necessity and possibility are called intensional connectives. Two formulas may be both true, but one of them not be necessarily true. Similarly, two formulas may be both false, still one may express a statement that is possible, that is, its possibility is true.

The *intensionality* of functions is a well-known phenomenon. Indeed, until the set theoretic understanding of functions became common sometime in the 20th century, the intensional understanding of functions was prevalent. Functions were often thought to be "recipes," so to speak, about how to calculate values. The two ways to understand functions have their respective uses and importance. In mathematics, a function is often thought of as a set of ordered pairs, whereas in computer science, a function may be primarily conceived as an algorithm.[7]

DEFINITION 4.1.12. ($\beta\eta$-REDUCTION) A Λ-term M $\beta\eta$-*reduces* to N, denoted by $\triangleright_{\beta\eta}$ iff (1) or (2) holds.

(1) M is α-equal to N;

(2) there is a finite sequence of one-step β and one-step η-reductions leading from M to N', and N' is α-equal to N.

[7] In category theory functions are treated as primitive objects, and mathematical structures are characterized by properties of functions. This view of functions does not quite fit into the extensional versus intentional distinction, though more in vein of the latter.

In other words, $\beta\eta$-reduction, which is denoted by $\triangleright_{\beta\eta}$, is the reflexive transitive closure of the union of one-step β- and one-step η-reductions together with α-equality.

One-step η-reduction does not require substitution at all, which means that allowing one α-equal step is sufficient for obtaining all the Λ-terms that can be obtained by intertwining β-reductions and one-step η-reductions.

Example 4.1.13. The Λ-term $\lambda x.(\lambda x.yx)x$ one-step η-reduces to $\lambda x.yx$. The latter term further reduces to y by an η step. Thus we have $\lambda x.(\lambda x.yx)x \triangleright_{\beta\eta} \lambda z.yz$, as well as $\lambda x.(\lambda x.yx)x \triangleright_{\beta\eta} y$. This shows that a $\beta\eta$-reduction sequence need not contain any one-step β-reductions.

The Λ-term $(\lambda x.(\lambda y.y(yy))x)(xx)(\lambda z.zz)$ contains both η- and β-redexes. The term one-step η-reduces to $(\lambda y.y(yy))(xx)(\lambda z.zz)$, when the η-redex $\lambda x.(\lambda y.y(yy))x$ is reduced. If we were to reduce the β-redex $(\lambda x.(\lambda y.y(yy))x)(xx)$, then we would get the same term. On the other hand, if we start the reductions with the β-redex $(\lambda y.y(yy))x$, then the resulting term is $(\lambda x.x(xx))(xx)(\lambda z.zz)$. From either term, we can obtain via another β-reduction $xx(xx(xx))(\lambda z.zz)$. Combining some of these steps with an α-step, we have that $(\lambda x.(\lambda y.y(yy))x)(xx)(\lambda z.zz) \triangleright_{\beta\eta} xx(xx(xx))\lambda y.yy$.

We can consider the converse of the $\beta\eta$-reduction relation too.

DEFINITION 4.1.14. ($\beta\eta$-EXPANSION) The converse of $\triangleright_{\beta\eta}$ is called $\beta\eta$-*expansion* and it is denoted by ${}_{\beta\eta}\triangleleft$.

The combination of $\beta\eta$-reduction and $\beta\eta$-expansion is the next relation on the set of Λ-terms that we define.

DEFINITION 4.1.15. ($\beta\eta$-EQUALITY) A Λ-term M is $\beta\eta$-*equal* to N, denoted by $=_{\beta\eta}$ iff there are Λ-terms $P_0,\ldots,P_n$ (where $n \in \mathbb{N}$) such that P_0 is M, P_n is N, and for each i (where $0 \leq i < n$), $P_i \triangleright_{\beta\eta} P_{i+1}$ or $P_{i+1} \triangleright_{\beta\eta} P_i$.

To simplify our presentation, from now on, we disregard all the α-equality steps.

The weak reduction relation in CL has the Church–Rosser property, which provides further pleasant consequences such as the consistency of CL with $\triangleright_w$ as well as with $=_w$. We have similar results for Λ.

THEOREM 4.1.16. (CHURCH–ROSSER THEOREM FOR $\triangleright_\beta$) *For any* Λ*-term* M *such that* $M \triangleright_\beta N_1$ *and* $M \triangleright_\beta N_2$, there is a Λ-term P *such that* $N_1 \triangleright_\beta$ *and* $N_2 \triangleright_\beta P$.

Proof: The proof of theorem 2.1.14 uses the relation $\triangleright_{1p}$ defined in 2.1.6, which is sandwiched between $\triangleright_1$ and $\triangleright_w$ and at the same time has the diamond property.

The suitable notion in this case is the *composite* β*-reduction* of a set of β-redexes (defined below). Once we have shown that that relation is confluent and its transitive closure is $\triangleright_\beta$, the proof is finished by an application of theorem 2.1.5. qed

DEFINITION 4.1.17. (COMPOSITE β-REDUCTION) The Λ-term M *composite* β*-reduces* to N, which is denoted by $\triangleright_{c\beta}$, iff one of (1)–(4) obtains.

(1) M is identical to N;

(2) M is $\lambda x.P$ and N is $\lambda x.P'$ where $P \rhd_{c\beta} P'$;

(3) M is (PQ) and N is $(P'Q')$ where $P \rhd_{c\beta} P'$ and $Q \rhd_{c\beta} Q'$;

(4) M is (PQ) and N is $[x/Q']R'$ where P is $\lambda x.R$ and $R \rhd_{c\beta} R'$, as well as $Q \rhd_{c\beta} Q'$. (I.e., M is $(\lambda x.R)Q$ and N is $[x/Q']R'$.)

This notion is more complicated than the simultaneous one-step reduction of a set of nonoverlapping redexes in CL. (See definition 2.1.6.) To start with, it is clear that the relation $M \rhd_{c\beta} N$ is a *reflexive* relation. Then it is obvious that $\rhd_{1\beta}$ is a subset of the relation $\rhd_{c\beta}$, because if we have $(\lambda x.M)N$, then $(\lambda x.M)N \rhd_{c\beta} [x/N]M$ by (4), provided that $M \rhd_{c\beta} M$ and $N \rhd_{c\beta} N$, which obtain by reflexivity.

<u>Example</u> 4.1.18. $(\lambda x.x)y((\lambda z.zz)v) \rhd_{c\beta} y(vv)$, by clause (3). $(\lambda x.x)y \rhd_{c\beta} y$ and $(\lambda z.zz)(vv) \rhd_{c\beta} vv$, by clause (4). Both clauses (3) and (4) contain $\rhd_{c\beta}$ itself, which makes possible parallel one-step β-reductions.

Certain sequences of one-step β-reductions can also become instances of $\rhd_{c\beta}$. $\lambda x.x((\lambda y.y((\lambda z.z)v))w)$ contains overlapping β-redexes, namely, the redex headed by λy and the one headed by λz. We can obtain the Λ-term $\lambda x.xwv$ from $\lambda x.x((\lambda y.y((\lambda z.z)v))w)$, by clause (2) in the definition of $\rhd_{c\beta}$.

It might seem that $\rhd_{c\beta}$ is a way too powerful notion of reduction for the purposes of a proof of confluence. We might wonder if there are any $\rhd_{\beta}$ reductions that are *not* $\rhd_{c\beta}$ reductions too.

<u>Example</u> 4.1.19. Consider the Λ-term $\lambda x.xy(\lambda z.z)$. This term contains only one β-redex, and it one-step β-reduces to $(\lambda z.z)y$. This term contains a *new* β-redex, hence the term can be further reduced to y. Notice that clause (4) does not allow $c\beta$-steps to be performed after a one-step β-reduction unless the β-redexes were inside of the substituted term or inside the scope of the λ-abstract to start with.

We have illustrated some of the properties of $\rhd_{c\beta}$, and we can state the next lemma.

LEMMA 4.1.20. *The transitive closure of* $\rhd_{c\beta}$ is $\rhd_{\beta}$.

Proof: $\rhd_{c\beta}$ is reflexive, and a one-step β-reduction is a $\rhd_{c\beta}$ step. This means that $\rhd_{c\beta} \subseteq \rhd_{\beta}$. However, by claim 2.1.11, this implies that $\rhd^*_{c\beta} \subseteq \rhd^*_{\beta}$; but the latter is $\rhd_{\beta}$ itself.

To prove the other direction of the inclusion, we note that $\rhd_{1\beta} \subseteq \rhd_{c\beta}$. Therefore, by the monotony of the * operation, $\rhd^*_{1\beta} \subseteq \rhd^*_{c\beta}$. $\rhd_{\beta} = \rhd^*_{1\beta}$, hence, $\rhd^*_{c\beta} = \rhd_{\beta}$, as we intended to prove. qed

The next piece in the proof establishing the confluence of $\rhd_{\beta}$ is to show that $\rhd_{c\beta}$ has the diamond property.

LEMMA 4.1.21. *For any Λ-term M, if $M \rhd_{c\beta} N_1$ and $M \rhd_{c\beta} N_2$, then* there is a term P *such that $N_1 \rhd_{c\beta} P$ and $N_2 \rhd_{c\beta} P$.*

Proof: The proof is by structural induction on the Λ-term M. The proof can be presented along the lines of the proof of lemma 2.1.13. However, both in that proof and in the present proof, we may bunch together some cases that are alike. We consider two easy cases and leave the rest of the proof to the reader.
1. If N_1 and N_2 are the same Λ-term as M, then M is suitable as the term P. If M is atomic, or it does not contain any β-redexes, then N_1 and N_2 are necessarily M. However, even if M contains β-redexes, $M \rhd_{c\beta} M$ holds; therefore, this possibility must be considered as a subcase.
2. If either of N_1 and N_2 is not the same term as M, but the other one is, then the term that is not M suffices as P. We have $M \rhd_{c\beta} N$ and $M \rhd_{c\beta} M$ as the first pair of steps, and then $N \rhd_{c\beta} N$ and $M \rhd_{c\beta} N$ as the second pair of steps. qed

Exercise 4.1.12. Finish the proof of lemma 4.1.21. (Hint: In the remaining cases, M has to be complex and contain at least one β-redex, which may occur in various places within M.)

The other reduction-type relation that we considered is $\rhd_{\beta\eta}$.

THEOREM 4.1.22. (CHURCH–ROSSER THEOREM FOR $\rhd_{\beta\eta}$) *For any Λ-term M, if $M \rhd_{\beta\eta} N_1$ and $M \rhd_{\beta\eta} N_2$, then* there is a Λ-term P *such that both $N_1 \rhd_{\beta\eta} P$ and $N_2 \rhd_{\beta\eta} P$.*

We leave the proof of this theorem to the reader as a series of exercises.

Exercise 4.1.13. Consider the notion of composite β-reduction from definition 4.1.17. Modify that notion — if necessary — to make it usable for the proof of theorem 4.1.22.

Exercise 4.1.14. Prove theorem 4.1.22. (Hint: A proof may be constructed along the lines of proofs of theorems 4.1.16 and 2.1.14. Then analogues of lemmas 4.1.21 and 4.1.20 have to be established.)

We can reformulate the two Church–Rosser theorems in this chapter so that they parallel theorem 2.1.15.

THEOREM 4.1.23. (CHURCH–ROSSER THEOREM WITH $=_\beta$) *For any Λ-terms M and N, if $M =_\beta N$ then* there is a Λ-term P *such that $M \rhd_\beta P$ as well as $N \rhd_\beta P$.*

Exercise 4.1.15. Prove that theorems 4.1.16 and 4.1.23 are equivalent. (Hint: One direction is trivial, whereas the other direction may be proved like theorem 2.1.15.)

THEOREM 4.1.24. (CHURCH–ROSSER THEOREM WITH $=_{\beta\eta}$) *For any Λ-terms M and N, if $M =_{\beta\eta} N$ then* there is a Λ-term P *such that $M \rhd_{\beta\eta} P$ and $N \rhd_{\beta\eta} P$.*

Exercise 4.1.16. Prove the theorem. (Hint: To prove the more difficult direction, theorem 4.1.22 has to be applied possibly repeatedly.)

The diamond property ensures us that the order in which reductions are carried out in a term is not important from the point of view of the final result. The Church–Rosser theorems have other desirable consequences. To be able to articulate them, we first introduce the notion of normal forms for Λ-terms.

DEFINITION 4.1.25. (β NORMAL FORMS) A Λ-term M is in β *normal form* iff M has no occurrences of β-redexes.

Example 4.1.26. $\lambda x.x$ is in β nf, just as $\lambda x.y(\lambda y.x)$ is. We could modify the latter term by introducing a pair of parentheses as in $(\lambda x.y)(\lambda y.x)$. The latter Λ-term is *not* a notational variant of the former, and it is no longer in β normal form. Indeed, $(\lambda x.y)(\lambda y.x) \triangleright_{1\beta} y$.

The β normal forms are somewhat similar to the weak normal forms of CL-terms. However, in Λ, a functional expression (i.e., λ-abstract) followed by a term always creates a β-redex, whereas in CL, a combinator followed by a term forms a redex only when the combinator is unary.

DEFINITION 4.1.27. ($\beta\eta$ NORMAL FORMS) A Λ-term M is in $\beta\eta$ *normal form* iff M has no occurrences of β-redexes or η-redexes.

Example 4.1.28. Consider the Λ-term $\lambda x.yx$, which obviously does not contain a β-redex. This term is in β normal form, but not in $\beta\eta$ normal form.

This term might suggest that finding Λ-terms that are in $\beta\eta$ normal form is not at all easy. But all variables are — trivially — in $\beta\eta$ normal form. Quite a few closed Λ-terms ($\aleph_0$ many) are also in $\beta\eta$ normal form. For example, a slight modification of the above term is $\lambda xy.yx$ and it is in $\beta\eta$ normal form.

Λ-terms form redexes of one or another kind quite often. Nonetheless, Λ are consistent with the reductions and equalities introduced. We first prove two lemmas.

LEMMA 4.1.29. *If M is a Λ-term that does not contain β- or η-redexes, then* no Λ-term M' *such that M and M' are α-equal contains a β- or η-redex.*

Proof: The term M may not contain any λ's, in which case, M' is identical to M. Otherwise, M contains a λ in a subterm of the form $\lambda x.N$, but then this subterm is not followed by a term. $\lambda x.N$ may be part of a subterm of the form $x(\lambda x.N)$, for example. A Λ-term that is α-equal to $\lambda x.N$ will be of the form $\lambda y.N$ for a suitable y. If N is of the form $N'z$, where $z \notin \mathrm{fv}(N')$, then due to the definition of α-equality, z and y are distinct. The place of the subterm remains the same within the whole term; hence, no new β- or η-redex is created. qed

The lemma and its proof make very explicit that β- and η-redexes exist due to the *structure* of the term, which is not altered by α-equality. To put it from another perspective, the renaming of bound variables is sufficiently constrained so that it cannot create new redexes. The role of α-equality is limited to providing some freedom to carry out β-reductions, when a variable clash would happen.

The Church–Rosser theorem implies that normal forms, if exist, are unique.

LEMMA 4.1.30. (UNIQUENESS OF NORMAL FORMS) *If a Λ-term M has a β normal form (or a $\beta\eta$ normal form), then that is* unique *up to α-equality.*

Proof: By theorem 4.1.16 reduction sequences converge. If a Λ-term M has a β normal form M', then M' does not contain a β-redex, and if M and M' are distinct, then M' may be obtained from M by one-step β-reductions followed by an α-equality step. M' may be in β normal form, nonetheless, can contain λ's. A Λ-term that contains a bound variable has infinitely many "alphabetic variants," which are Λ-terms but with some or all bound variables renamed. Although "absolute" uniqueness cannot be guaranteed, uniqueness up to α-equality can, and this is the analogue of lemma 2.2.4 for CL-terms.

The proof for $\beta\eta$ is similar, but relies on theorem 4.1.22. We only note that a Λ-term may be in $\beta\eta$ normal form and still may contain λ's. If a Λ-term has a β normal form (or a $\beta\eta$ normal form) without occurrences of λ, then the normal form is unique, because it has no distinct alphabetic variants. qed

THEOREM 4.1.31. (CONSISTENCY OF β-REDUCTION) *The set of Λ-terms is* consistent with $\triangleright_\beta$.

Proof: To prove this claim, it is sufficient to find two terms such that neither of them β-reduces to the other. A pair of distinct variables, let us say, x and y prove the claim. A variable does not contain a β-redex; it β-reduces to itself, but to no other term. Obviously, there are infinitely many other Λ-terms — including nonatomic Λ-terms — that are in β normal form and could be used in this proof. qed

THEOREM 4.1.32. (CONSISTENCY OF $\beta\eta$-REDUCTION) *The set of Λ-terms is* consistent with $\triangleright_{\beta\eta}$.

Exercise 4.1.17. Prove the theorem. (Hint: The proof is straightforward and can be constructed similarly to the proof of theorem 4.1.31.)

THEOREM 4.1.33. (CONSISTENCY OF $\beta\eta$-EQUALITY) *The set of Λ-terms is* consistent with $=_{\beta\eta}$.

Proof: The proof is quite like the proof of theorem 4.1.31, except that we need two Λ-terms that are not $\beta\eta$-equal. In the proof of theorem 2.2.9, we appealed to the Church–Rosser theorem for CL. Now we use the Church–Rosser theorem for Λ, more precisely, theorem 4.1.24. Let us consider the terms $(\lambda yx.\,yx)\lambda xy.\,x$ and $(\lambda xy.\,yx)\lambda xy.\,y$. We have the following $\beta\eta$-reduction sequences.

$$(\lambda yx.\,yx)\lambda xy.\,x \triangleright_{1\beta} \lambda x.\,(\lambda xy.\,x)x \triangleright_{1\eta} \lambda xy.\,x$$
$$(\lambda xy.\,yx)\lambda xy.\,y \triangleright_{1\eta} (\lambda y.\,y)\lambda xy.\,y \triangleright_{1\beta} \lambda xy.\,y$$

The Λ-terms on the right-hand side are not the same, though they are both closed Λ-terms. Also, both terms are in $\beta\eta$ normal form without being α-equal. Hence, by the Church–Rosser theorem for $=_{\beta\eta}$, the Λ-terms we started with are not $\beta\eta$-equal. Again, there are infinitely many other Λ-terms that we could have used instead. We added a bit of complexity to the Λ-terms to demonstrate and emphasize that it may not be obvious that two Λ-terms are not $\beta(\eta)$-equal. Some Λ-terms are not $\beta(\eta)$-equal, even though they are not atomic. qed

The β-equality relation is a *proper subset* of the $\beta\eta$-equality relation. Therefore, if $=_{\beta\eta}$ is *not* the total relation on the set of Λ-terms, let alone is $=_\beta$ the total relation. We phrase this conclusion as a corollary.

COROLLARY 4.1.33.1. *The set of* Λ*-terms is* consistent with $=_\beta$.

4.2 Combinators in Λ

We are interested in the *connections* between CL and Λ, rather than in Λ per se. It so happens that the λK-calculus, that is, Λ without any restriction on the formation of Λ-terms contains terms that behave more or less like the combinators in CL. The difference between the weak normal forms and β normal forms means that the Λ-terms that are like CL combinators form redexes more often that their CL counterparts.

DEFINITION 4.2.1. (COMBINATORS IN Λ) If $\mathrm{fv}(M) = \emptyset$ then M is a *combinator*, that is, a closed λ-term is a combinator in Λ.

Example 4.2.2. The Λ-term $\lambda x.x$ is closed, and this term quite closely resembles I. $(\lambda x.x)M \triangleright_{1\beta} M$ just as $\mathsf{I}M \triangleright_1 M$.

Another closed Λ-term is, for instance, $\lambda xy.y((\lambda zvw.z(vw))(\lambda u.u)x)$. This could be likened to the CL-term $\mathsf{B}(\mathsf{CI})(\mathsf{BI})$, which reduces to $y(\mathsf{BI}x)$ when applied to x and y. We have that $(\lambda xy.y((\lambda zvw.z(vw))(\lambda u.u)x))xy \triangleright_\beta y((\lambda zvw.z(vw))(\lambda u.u)x)$. $\mathsf{B}(\mathsf{CI})(\mathsf{BI})$ is obviously not a proper combinator, but it is in weak nf. The reduct of $\mathsf{B}(\mathsf{CI})(\mathsf{BI})xy$, $y(\mathsf{BI}x)$ is also in weak nf. In contrast, $\lambda xy.y((\lambda zvw.z(vw))(\lambda u.u)x)$ is not in β normal form, and neither is $y((\lambda zvw.z(vw))(\lambda u.u)x)$ in β nf. Their β nf's are, respectively, $\lambda xy.y(\lambda z.xz)$ and $y(\lambda z.xz)$.

The definition might appear a bit strange at first glance. However, if all atomic terms in the language are variables, then closed terms are one of two possible forms $\lambda x.M$ or $(\lambda x.M)(\lambda y.N)$, where M and N may be complex Λ-terms. Of course, in both cases the set of free variables of the term must be empty, which implies that $\mathrm{fv}(M) \subseteq \{x\}$, and $\mathrm{fv}(N) \subseteq \{y\}$.

The Λ-term $\lambda x.M$ may or may not be in β nf — depending on the structure of M. The other Λ-term, $(\lambda x.M)(\lambda y.N)$, however, is a β-redex itself. That is, $(\lambda x.M)(\lambda y.N) \triangleright_{1\beta} [x/\lambda y.N]M$. The resulting term is closed, because $\lambda y.N$ is closed and M may contain only x as its free variable, for which $\lambda y.N$ is substituted. Once again, depending on the shape of M, the Λ-term $[x/\lambda y.N]M$ may or may not be in β normal form.

Exercise 4.2.1. Construct a Λ-term that is both closed and does not have a β normal form.

In order to be able to find Λ-terms that correspond to proper combinators of CL, it is sufficient to restrict our attention to closed terms that are in β nf. Let Z be a proper combinator in CL with axiom $\mathsf{Z}x_1 \ldots x_n \rhd M$. The assumption that Z is *proper* implies that M does not contain any other atomic subterms except some of $x_1, \ldots, x_n$ (possibly, with multiple occurrences). We can obtain a closed Λ-term by replacing Z with a λ and by inserting . in place of $\rhd$. That is, the transformation resulting in the Λ-term that corresponds to a combinator in CL is

$$\mathsf{Z}x_1 \ldots x_n \rhd M \qquad \rightsquigarrow \qquad \lambda x_1 \ldots x_n . M.$$

The following is immediate from the description of the transformation when the definition of proper combinators is taken into account.

LEMMA 4.2.3. *For any proper combinator* Z *in* CL, *there is* a combinator N in Λ *such that* (1) *and* (2) *are true.*

(1) *N is in β normal form;*

(2) *if $\mathsf{Z}^n x_1 \ldots x_n \rhd M$ then $Nx_1 \ldots x_n \rhd_\beta M$.*

In section 1.2, we gave a list of often-used combinators together with their axioms. Using the above recipe to turn a proper combinator in CL into a combinator in Λ, we get the following list of Λ-terms matching the list of well-known combinators in CL.

$\lambda x.x$	$\lambda xyz.x(yz)$	$\lambda xyz.xz(yz)$
$\lambda xy.x$	$\lambda xyz.xzy$	$\lambda xy.xyy$
$\lambda x.xx$	$\lambda xyz.y(xz)$	$\lambda xyzv.xy(xvz)$

Having selected closed Λ-terms to stand for CL combinators, we can consider the similarities and differences more meticulously.

In CL, the list of combinators was a list of *axioms*. Indeed, without the axioms, it would be fully indeterminate what function the combinator represents.[8] The situation in Λ is quite different: the closed Λ-terms are not supplemented by any axioms. These terms are not "special" from the point of view of function application. Rather it just so happens that they are closed.

There are infinitely many unary combinators in CL. If Z is one of them, then $\mathsf{Z}M$ is a redex in CL. The corresponding Λ-term is of the form $\lambda x.N$, where $\mathrm{fv}(N) \subseteq \{x\}$. The Λ-term $(\lambda x.N)M$ is certainly a β-redex.

On the other hand, if a combinator in CL has arity > 1, then there is a difference between the combinators in Λ and those in CL with respect to forming redexes of the appropriate kind. For example, W is a combinator of arity 2 and the corresponding Λ-term is $\lambda xy.xyy$. The term $\mathsf{W}M$ is not a redex in CL, but $(\lambda xy.xyy)M$ is a β-redex in Λ. Given a bracket abstraction in CL, it is possible to define a notion of reduction that would allow us to consider $\mathsf{W}M$ to be a redex. The issue is how to specify the result of a reduction step if $\mathsf{W}M$ is a redex. In the case of W, we could perhaps simply consider the term $\mathsf{W}M$ itself to be the result, because W is a regular combinator.[9] Essentially, the same problem arises if we consider the term $\mathsf{S}MN$.

[8]Originally, Schönfinkel introduced letters that were intended to be mnemonic. For instance, I was chosen for identity. However, the assignment of particular letters to certain combinators is nothing more than good notational convention.

[9]We introduced regular combinators in section 1.2, though we have not placed special emphasis on them.

$(\lambda xyz.xz(yz))MN$ reduces to $\lambda z.Mz(Nz)$, which is the Λ-term that we would need to *emulate* in CL. If we use the bracket abstraction with η, then the result is just $\mathsf{S}MN$.

The equivalence between unary and n-ary functions is a deep insight. Nevertheless, it seems to be worthwhile to emphasize that Λ with β-reduction reflects the view that functions are *unary*, whereas CL with weak reduction reflects the view that functions have a *fixed arity*, and some of them are *not* unary.

Lastly, we may note that the Λ-terms listed above lack "special status" in yet another sense: we could have chosen any α-equal Λ-term instead.

Example 4.2.4. The above procedure does not work if the combinator is improper. Let us consider CII. This is a unary combinator and its axiom is $\mathsf{CII}x \rhd x\mathsf{I}$. The replacements would give $\lambda x.x\mathsf{I}$, however, I is not in the language of Λ. Of course, I itself is a proper combinator, and we could attempt to replace I with $\lambda y.y$ to obtain $\lambda x.x\lambda y.y$.

We could arrive at the same result if we would have considered first CII as C with two arguments filled in — both with I. Above we gave the λ equivalent of C as $\lambda xyz.xzy$ and of I as $\lambda x.x$. We can form the λ-term $(\lambda xyz.xzy)(\lambda x.x)(\lambda x.x)$, which β-reduces, in two steps, to $\lambda z.(\lambda x.x)z\lambda x.x$. Perhaps, the term would be more "transparent" if the two I's would have been λ-terms with disjoint sets of bound variables, for instance, $\lambda x.x$ and then $\lambda y.y$. But $\lambda z.(\lambda x.x)z\lambda y.y$ is, of course, α-equal to the previous term. The latter Λ-term further β-reduces to $\lambda z.z\lambda y.y$, because the subterm $(\lambda x.x)z$ is a β-redex. In sum, we obtained the term $\lambda z.z\lambda y.y$ as before — modulo α-equality.

The method we have just illustrated does not work for the *fixed point combinator* Y, because its axiom is $\mathsf{Y}x \rhd x(\mathsf{Y}x)$. However, we have shown that Y may be defined from proper combinators. Y is certainly definable from $\{\mathsf{S},\mathsf{K}\}$, but it is also definable from combinatory bases that are not functionally complete, for example, from $\{\mathsf{B},\mathsf{I},\mathsf{W},\mathsf{C}\}$. Thus given a definition of Y, the translation of the proper combinators composed appropriately is a closed Λ-term, which can be taken to be a fixed point combinator in Λ.

Finding definitions for combinators in CL that behave like Y can proceed more directly — along the lines of ideas that we associated with Curry's and Turing's names. The steps in those constructions can be replicated in Λ.

Let us start with *Turing's idea.* We want to find a Λ-term M such that $Mx \rhd_\beta x(Mx)$. First, we replace M by NN, which gives us $NNx \rhd_\beta x(NNx)$, and then we replace all but the first occurrence of N by y in the terms on the left and on the right of $\rhd_\beta$. That is, we have $Nyx \rhd_\beta x(yyx)$; accordingly, we are looking for a closed Λ-term that β-reduces to $x(yyx)$ when applied to y and then to x. N is just like a proper combinator, which we can emulate in Λ, by lemma 4.2.3. Indeed, we can take the closed Λ-term $\lambda yx.x(yyx)$ or $\lambda xy.y(xxy)$.

Of course, we should not forget that N is not the Λ-term we wanted to find. M is NN, hence, M is $(\lambda yx.x(yyx))(\lambda yx.x(yyx))$. This term is not in β normal form. A one-step β-reduction gives $\lambda x.x((\lambda yx.x(yyx))(\lambda yx.x(yyx))x)$. The initial λx has a "wide scope," that is, the rest of the term is in its scope, hence, the *whole term* is not a β-redex. However, the $\triangleright_{1\beta}$ step created a new β-redex, which is exactly of the form NN, the term we started with. It is obvious then that $(\lambda yx.x(yyx))(\lambda yx.x(yyx))$ does not have a β normal form — unlike Y or, for instance, the CL-term $\mathsf{BW(BB'M)}$, which have weak normal forms.

We have to verify that the Λ-term behaves as we claimed it does. For the sake of transparency, let us consider the α-equal term

$$\lambda x.\,x((\lambda yz.\,z(yyz))(\lambda yz.\,z(yyz))x).$$

We call this Λ-term *Turing's fixed point combinator* in Λ, and we denote this term (or any of its alphabetic variants) by Y_T. If Y_T is applied to a Λ-term, for instance to P, then a one-step β-reduction yields $P(\lambda yz.\,z(yyz)(\lambda yz.\,z(yyz))P)$. The latter is, certainly, $P(Y_TP)$, as we intended to show.

To implement *Curry's idea*, we start with $Mx =_\beta x(Mx)$. We replace M — in all occurrences — with y: $yx =_\beta x(yx)$. Of course, the β-equality is *false* after the replacement. However, $Nyx =_\beta x(yx)$ holds for some N. Again, N is like a proper combinator, hence by lemma 4.2.3, we can easily find a suitable Λ-term. $\lambda yx.x(yx)$ applied to y and x β-reduces to $x(yx)$. Returning to $Nyx =_\beta x(yx)$, we cannot substitute either N or NN for y to arrive at a β-equality that has the form $Mx =_\beta x(Mx)$.

We need another Λ-term P, such that $Pxy \triangleright_\beta x(yy)$. P can be taken to be $\lambda xy.x(yy)$. Now we combine N and P with x into $N(PN(PN))x$. The β-equality for N gives $x(PN(PN)x)$, and the β-equality for P — with N in place of x and PN in place of y — gives $x(N(PN(PN))x)$. In two steps we got from $N(PN(PN))x$ to $x(N(PN(PN))x)$, which shows that the complex term $N(PN(PN))$ is like a fixed point combinator.

So far we were using N and P in Λ as if they were combinators in CL. Let us make the construction more precise by using the actual Λ-terms. We use α-equal terms in some places for clarity.

$$
\begin{aligned}
&(\lambda xy.\,y(xy))((\lambda xy.\,x(yy))(\lambda xy.\,y(xy))((\lambda xy.\,x(yy))(\lambda xy.\,y(xy)))) \triangleright_{1\beta}\\
&\lambda y.\,y(((\lambda xy.\,x(yy))(\lambda xy.\,y(xy))((\lambda xy.\,x(yy))(\lambda xy.\,y(xy))))y) \triangleright^2_{1\beta}\\
&\lambda y.\,y(((\lambda x.\,(\lambda zv.\,v(zv))(xx))(\lambda x.\,(\lambda zv.\,v(zv))(xx)))y) \triangleright^2_{1\beta}\\
&\lambda y.\,y(((\lambda xv.\,v(xxv))(\lambda xv.\,v(xxv)))y) \triangleright_{1\beta}\\
&\lambda y.\,y((\lambda v.\,v((\lambda xz.\,z(xxz))(\lambda xz.\,z(xxz))v))y) \triangleright_{1\beta}\\
&\lambda y.\,y(y((\lambda xv.\,v(xxv))(\lambda xv.\,v(xxv))y))
\end{aligned}
$$

The above steps show how one-step β-reductions "simplify" the term. (Two lines show the result of two one-step β-reductions on a pair of alike β-redexes.) It is easy to see that the structure of the second term is $\lambda y.\,y(Qy)$, which is like the right-hand side term in the axiom of the fixed point combinator Y. However, the first term is not $\lambda y.\,Qy$, and so to make it clear that Q really behaves like a fixed point combinator in further reductions, we performed further one-step β-reductions. We chose the names for the variables in the fourth and sixth Λ-terms so that it is obvious that the slightly simplified Q, let us say, Q' takes y as its argument and produces $y(Q'y)$. On the last line, we have a term of the form $\lambda y. y(y(Q'y))$.

Exercise 4.2.2. Scrutinize the steps from $(\lambda xy. x(yy))(\lambda xy. y(xy))$ to $(\lambda xv. v(xxv))$ and from $((\lambda xv. v(xxv))(\lambda xv. v(xxv)))$ to $(\lambda v. v((\lambda xz. z(xxz))(\lambda xz. z(xxz))v))$. (Hint: Carefully write out the renaming of the variables and the one-step reductions.)

Of course, the Λ-term that we denoted as Q' above *does not have* a β normal form. The Λ-terms allow β-reduction steps where the corresponding CL-terms are in weak normal form, which may be thought to be an advantage of Λ. If we take a look at the last term in the displayed sequence of Λ-terms, then we can see that a fixed point combinator could be constructed from two combinators slightly differently than we did. One of the combinators would be N, however, the other, let us say O, would have to have an axiom $Oxy \rhd y(xxy)$, or as a Λ-term, $\lambda xy. y(xxy)$. The last displayed term then is of the form $N(OO)$. $N(OO)x \rhd x(OOx)$, and (OOx) further reduces to $x(OOx)$. Of course, this O is just the N from the definition of a fixed point combinator along Turing's ideas. Thus using Λ-terms we even discovered a similarity between the two approaches.

Y in CL is *in nf*, and someone also might find it more easy to follow the definition of the fixed point combinator using combinators, rather than Λ-terms. The latter quickly become lengthy and perhaps, somewhat difficult to parse.[10] Incidentally, the fixed point combinators in Λ also demonstrate that the renaming of the bound variables is a *necessity* rather than a convenience. Even if — cautiously — we start with $(\lambda xy. y(xxy))(\lambda zv. v(zzv))$ instead of $(\lambda zv. v(zzv))(\lambda zv. v(zzv))$, after just a single one-step β-reduction, we end up having a subterm of the latter form, which then forces some renaming of bound variables.

4.3 Back and forth between CL and Λ

We drew some, more or less informal, parallels between CL and Λ already. Now we would like to establish a *precise relationship* between CL-terms and λ-terms. Then we can also compare the relations on the sets of terms in a rigorous fashion.

We need a translation from CL-terms into Λ-terms, and a translation taking us into the opposite direction from Λ-terms to CL-terms. The first one is the easier; in fact, we have already described such a translation for *proper* combinators. In order to extend that translation to CL-terms, which might contain some improper combinators too, we start with delineating a subclass of improper combinators.

DEFINITION 4.3.1. (FAIR COMBINATORS) Let Z be a combinator such that its axiom is one of the following two forms. Then Z is called a *fair combinator*.

[10] This piffling grumbling shows that I like combinatory logic more than the λ-calculi. Well, this is a book about *combinatory logic*, not about the λ-calculi!

(1) $\mathsf{Z}x_1 \ldots x_n \rhd M$, where M contains some of the variables $x_1, \ldots, x_n$, but no other variables or constants;

(2) $\mathsf{Z}x_1 \ldots x_n \rhd M$, where M contains some of the variables $x_1, \ldots, x_n$ and proper combinators.

Clause (1) specifies proper combinators, which were introduced in section 1.2 already. The second clause would suffice by itself if the occurrence of proper combinators in M would be made optional. However, proper combinators are undoubtedly an extremely useful class of combinators, which are sufficient themselves when combinatorially complete bases are considered.

Improper combinators, in general, are unruly, which is why we do not want to consider all of them. Incidentally, from the point of view of a translation, a combinator Z_{mad} with axiom $\mathsf{Z}_{\text{mad}}x_1, \ldots, x_n \rhd y$ is unproblematic, though it would *not result in* a combinator in Λ. Z_{mad} is just an absolutely useless combinator. On the other hand, Y, which is a useful improper combinator does not fit into the translation that we give below. We have seen that Y may be defined from proper combinators, thus we exclude Y as a primitive combinator.

We gave a translation of proper combinators in the previous section. Now we add improper but fair combinators, as well as variables.

DEFINITION 4.3.2. (FROM CL TO Λ) Let $\mathfrak{B}$ be a combinatory basis that contains only fair combinators, and let M be a term from the set of terms generated by $\mathfrak{B}$ and a denumerable set of variables. The *translation* of M into Λ, which is denoted by M_λ, is defined by (1)–(4).

(1) If M is a variable x, then M_λ is x;

(2) if M is a proper combinator with axiom $Mx_1 \ldots x_n \rhd N$, then M_λ is $\lambda x_1 \ldots x_n.N_\lambda$;

(3) if M is a fair combinator with axiom $Mx_1 \ldots x_n \rhd N$, then M_λ is $\lambda x_1 \ldots x_n.N_\lambda$;

(4) if M is (NP), then M_λ is $(N_\lambda P_\lambda)$.

Notice that this $_\lambda$ translation gives the same Λ-terms for proper combinators that we listed above.

<u>Example</u> 4.3.3. The combinator S' has as its axiom $\mathsf{S}'xyz \rhd yz(xz)$. By clause (2), S'_λ is $\lambda xyz.(yz(xz))_\lambda$. Clause (4) allows us to distribute the $_\lambda$ onto the atomic subterms starting with $(yz(xz))_\lambda$, then $(yz)_\lambda(xz)_\lambda$, etc., to $y_\lambda z_\lambda(x_\lambda z_\lambda)$. According to clause (1), each variable is its own translation; more accurately, we use alike symbols for variables in both languages. Then it is immediate that we can simply drop $_\lambda$ from the very first term we got, that is, S'_λ is $\lambda xyz.(yz(xz))$.

Of course, S' or any other proper combinator may occur in a CL-term as a proper subterm. $_\lambda$ provides a translation for all such CL-terms. A fair but not proper combinator has a reduct that is composed of variables and proper combinators, that is, looks exactly like these CL-terms.

Example 4.3.4. Let us consider the CL-term $y(\mathsf{S}(\mathsf{KS})\mathsf{K}x)xy$. The following series of terms shows the steps that yield the translation of this CL-term. We use $\rightsquigarrow$ to indicate two parallel moves (after the initial step).

$$\begin{array}{llll} (y(\mathsf{S}(\mathsf{KS})\mathsf{K}x)xy)_\lambda & \rightsquigarrow & (y(\mathsf{S}(\mathsf{KS})\mathsf{K}x)x)_\lambda y_\lambda & \rightsquigarrow \\ (y(\mathsf{S}(\mathsf{KS})\mathsf{K}x))_\lambda x_\lambda y & \rightsquigarrow & y_\lambda(\mathsf{S}(\mathsf{KS})\mathsf{K}x)_\lambda xy & \rightsquigarrow \\ y((\mathsf{S}(\mathsf{KS})\mathsf{K})_\lambda x_\lambda)xy & \rightsquigarrow & y((\mathsf{S}(\mathsf{KS}))_\lambda \mathsf{K}_\lambda x)xy & \rightsquigarrow \\ y(\mathsf{S}_\lambda(\mathsf{KS})_\lambda(\lambda xy.x)x)xy & \rightsquigarrow & & \\ y((\lambda xyz.xz(yz))(\mathsf{K}_\lambda \mathsf{S}_\lambda)(\lambda xy.x)x)xy & \rightsquigarrow & & \\ y((\lambda xyz.xz(yz))((\lambda xy.x)(\lambda xyz.xz(yz)))(\lambda xy.x)x)xy & & & \end{array}$$

Now let us assume that some Z has as its axiom $\mathsf{Z}xyz \rhd y(\mathsf{S}(\mathsf{KS})\mathsf{K}x)xy$. Then by clause (3) from definition 4.3.2, Z_λ is $\lambda xyz.(y(\mathsf{S}(\mathsf{KS})\mathsf{K}x)xy)_\lambda$. Given the above steps, Z_λ is $\lambda xyz.y((\lambda xyz.xz(yz))((\lambda xy.x)(\lambda xyz.xz(yz)))(\lambda xy.x)x)xy$.

As a more "realistic" example, we may consider the combinator that has been denoted by $\mathsf{1}$.[11] The axiom for $\mathsf{1}$ is $\mathsf{1}xy \rhd y(\mathsf{B}'\mathsf{I}x)$. $\mathsf{1}$ is not a proper combinator, however, it is *fair*. The translation of $\mathsf{1}$ goes as follows. $\lambda xy.(y(\mathsf{B}'\mathsf{I}x))_\lambda$, and then by four applications of clause (4), we get $\lambda xy.y_\lambda(\mathsf{B}'_\lambda \mathsf{I}_\lambda x_\lambda)$. We know — based on the axioms of B' and I — that the translations of these proper combinators are $\lambda xyz.y(xz)$ and $\lambda x.x$, respectively. We can piece together the Λ-term that is $\mathsf{1}_\lambda$ as $\lambda xy.y((\lambda xyz.y(xz))(\lambda x.x)x)$. The latter Λ-term contains a β-redex, and in two steps reduces to $\lambda xy.y(\lambda z.x((\lambda x.x)z))$. Lastly, we get $\lambda xy.y(\lambda z.xz)$. The structure of this Λ-term still somewhat resembles the CL-term $y(\mathsf{B}'\mathsf{I}x)$, but not very closely.

LEMMA 4.3.5. *For any* CL-*term* M *over a combinatory basis comprising proper and fair combinators together with denumerably many variables,* $_\lambda$ *is* defined *and* $_\lambda$ *is a* function. *That is,* $_\lambda$ *is* a total function.

Exercise 4.3.1. Prove the lemma. (Hint: Use structural induction on CL-terms.)

We clearly *do not have* that, if a CL-term is in wnf, then its translation into Λ is in β nf. However, it might be expected, for instance, from the constructions of fixed point combinators in Λ in the previous section, that if a CL-term is *not in weak nf*, then its translation into Λ is *not in* β *nf* either.

LEMMA 4.3.6. *If* M *is not in weak nf, then* M_λ *is* not in β nf.

Proof: Let us assume that M is not in weak normal form. Then there is an occurrence of a redex in M. Let us denote the subterm, which is the redex in M by N. (If there are several redexes, we randomly pick one of them.) Let us assume that Z^n is the head of the redex, and $N_1,\ldots,N_n$ are its arguments. If the axiom for this combinator is $\mathsf{Z}x_1\ldots x_n \rhd P$, then its translation is $\lambda x_1\ldots x_n.P_\lambda$, either by (2) or by (3) from definition 4.3.2. The N_i's themselves are translated into some Λ-terms; hence, $(\mathsf{Z}N_1\ldots N_n)_\lambda$ is $(\lambda x_1\ldots x_n.P_\lambda)N_{1\lambda}\ldots N_{n\lambda}$, which a β-redex whenever $n \geq 1$. qed

[11]See [21]. The type of this combinator is the wff called *restricted assertion* — cf. section A.9.

The translation from CL into Λ is quite straightforward. We can go a bit further if we wish. Given a combinator Z^n with axiom $\mathsf{Z}x_1 \ldots x_n \rhd M$, where $M \in (\{x_i \colon i \in \mathbb{N}\} \cup \{\mathsf{Z}\})^{\oplus}$, we can find a Λ-term for Z. The ideas that we attributed to Turing and to Curry, respectively, in obtaining a suitable Λ-term for a fixed point combinator are not specific to fixed point combinators. These procedures use only that the combinator comes with a reduct in which the combinator itself occurs together with some of the variables to which the combinator is applied. Allowing some proper combinators in the reduct would be acceptable too.

DEFINITION 4.3.7. If $\mathsf{Z}x_1 \ldots x_n \rhd M$, where $M \in (\{x_i \colon i \in \mathbb{N}\} \cup \{\mathsf{Z}\})^{\oplus}$, then Z is a *repeating combinator.*

Clearly, all the fixed point combinators are repeating. None of the repeating combinators are fair, let alone are they proper.

Example 4.3.8. Let us assume that Z^2 is a repeating combinator with axiom $\mathsf{Z}xy \rhd yx\mathsf{Z}(yx)$. First, we would like to find a combinator Q such that $\mathsf{Q}xyz \rhd zy(xx)(zy)$. For example, $\mathsf{BC}(\mathsf{BB}(\mathsf{BW}(\mathsf{W}(\mathsf{BT}))))$ is a CL-term that can be taken for Q. Then Z is QQ.

Exercise 4.3.2. Translate the Z from the preceding example into Λ. Simplify the term by performing some β-reductions.

Exercise 4.3.3. Describe a procedure based on Turing's approach that generates a Λ-term for an arbitrary repeating combinator.

Exercise*4.3.4. Use Curry's approach to find a definition (by proper combinators) for Z from the last example. Then translate the CL-term into Λ.

Exercise*4.3.5. Describe a general procedure along the lines of Curry's ideas that is applicable to any repeating combinators.

We know by theorem 2.3.7 that all repeating combinators are definable from a combinatorially complete base, for instance, from $\{\mathsf{S},\mathsf{K}\}$. We also know, however, that each definable combinator has $\aleph_0$ many definitions. We have seen a tiny slice of this multiplicity in the case of the fixed point combinator in chapter 2. But, the abundance of suitable CL-terms is not an obstacle; we can simply choose a CL-term that we like. Since we can suppose that only S and K occur in the selected CL-term, we can define the term according to definition 4.3.2 — without using clause (3). We can declare (if we wish) the Λ-term that we obtained in this way to be *the* translation of our repeating combinator.

DEFINITION 4.3.9. Let $\mathfrak{B}$ be a combinatory basis that comprises proper, fair and repeating combinators. Further, for each repeating combinator Z, let Z_λ be the Λ-term that is the translation of Z. Then for any term M over $\mathfrak{B}$ together with a denumerable set of variables, the *translation* of M into Λ is M_λ, which is obtained by (1)–(4) from definition 4.3.2 with (5) added.

(5) if M is a repeating combinator, then M_λ is the Λ-term, which has been chosen as its translation.

We could consider combinators that combine features of fair and repeating combinators. Repeating combinators include the fixed point combinators, but we have not introduced so far a combinator that would be both repeating and fair.

Exercise 4.3.6. Try to find combinators that are well-motivated and combine features of fair and repeating combinators.

Exercise 4.3.7. Amend the definition of the translation from CL into Λ to accommodate the combinators from the previous exercise.

Now we turn to translations *from* Λ *into* CL. Of course, some closed Λ-terms may be just what we would get from a translation of a combinator in CL. But there is a certain *asymmetry* between CL and Λ here: in CL we have a set of constants with axioms, whereas in Λ we simply have an infinite set of Λ-terms, among them infinitely many closed terms. This discrepancy is retained even if we were to restrict the set of allowable Λ-terms, for instance, to the $\lambda\mathsf{I}$-terms.

It is useful to think about the Λ-terms as *all-purpose notation* for functions. Then, we want to translate these functions into CL just like we found a combinator for an arbitrary (or at least, almost arbitrary) function. The translation from the set of Λ-terms into the set of CL-terms is a *bracket abstraction.*[12]

The translation $_\lambda$ had the property that M_λ was *uniquely determined by* M. The atomic terms were dealt with in (1)–(3) and (5) from definition 4.3.2 and 4.3.9. Neither of these clauses can be applied in place of another. And (4) was applicable only to complex terms.

It is customary and convenient to take a slightly different view of the defining clauses in the case of bracket abstractions. We assume that the listing of the clauses constitutes a linear *ordering* of the clauses, and it is always the *first applicable* clause that is applied. This idea is not unusual; a set of instructions or commands is often taken to be ordered. For instance, many programming languages exploit the sequentiality of the executions of the commands line by line. An abstract model of computation that incorporates the ordering of commands is *Markov algorithms.*[13] We always assume that the ordering is *from top to bottom.*

We have suggested in various exercises the use of further combinators beyond S and K, and we hinted at the possibility of getting shorter CL-terms. Thus, it should be unsurprising that there have been several algorithms proposed with various undefined combinators. To distinguish the bracket abstractions below, we will place a superscript on the bracket. On the other hand, to show some of the commonalities of the algorithms, we give the same label to clauses that are identical, possibly, except the superscript on $[\,]$. We stressed in chapter 1 that $[x].M$ is a *meta-notation* for a CL-term. We specify what the actual CL-term looks like by giving an equation.

[12] We outlined one particular bracket abstraction in the proof of lemma 1.3.9.

[13] See Curry's presentation of these algorithms in [52].

However, the equation is not $=_w$ or any of the other relations we have defined on CL-terms; $=$ simply means that $[x].M$ *denotes* a particular CL-term or meta-term given certain side conditions. Some of the side conditions will be given concisely by the *shape* of M, for instance, as NP.

The *first bracket abstraction algorithm* is due to Schönfinkel — though the concept of an algorithm or the term "bracket abstraction" did not exist at the time.

(K) $[x]^1.\, M = \mathsf{K}M$ if $x \notin \mathrm{fv}(M)$;

(I) $[x]^1.\, x = \mathsf{I}$ (x is a variable);

(η) $[x]^1.\, Mx = M$ if $x \notin \mathrm{fv}(M)$;

(B) $[x]^1.\, NP = \mathsf{B}N([x]^1.\, P)$ if $x \notin \mathrm{fv}(N)$;

(C) $[x]^1.\, NP = \mathsf{C}([x]^1.\, N)P$ if $x \notin \mathrm{fv}(P)$;

(S) $[x]^1.\, NP = \mathsf{S}([x]^1.\, N)([x]^1.\, P)$.

If $\mathfrak{B}$ is a combinatorially complete basis, then all the combinators mentioned in the above clauses are definable. However, we will assume that *for each algorithm*, we have a combinatory basis that contains at least the combinators that figure into the clauses. For instance, for the above algorithm, we assume that the basis is $\mathfrak{B} \supseteq \{\mathsf{S}, \mathsf{K}, \mathsf{I}, \mathsf{B}, \mathsf{C}\}$.

Having all the combinators that are used in the clauses as primitives gives a different perspective on the combination of $_\lambda$ and $[\,]$ translations than what we would get with $\mathfrak{B} = \{\mathsf{S}, \mathsf{K}\}$.

The difference between Schönfinkel's bracket abstraction algorithm and the one we gave in chapter 1 is that it includes (B) and (C), and the abstraction of a(n identical) variable is with I (not SKK). The rationale behind including (B) and (C) is that, if a term is complex and we know whether the abstracted variable does or does not occur in each immediate subterm, then we can obtain a shorter CL-term by using B or C (instead of using S, and then K too). This allows us to limit further abstraction steps to the subterm that actually has an occurrence of the variable, which is abstracted out.

The next algorithm is almost — but not quite — the same that we gave in chapter 1.

(K) $[x]^2.\, M = \mathsf{K}M$ if $x \notin \mathrm{fv}(M)$;

(I) $[x]^2.\, x = \mathsf{I}$;

(S) $[x]^2.\, NP = \mathsf{S}([x]^2.\, N)([x]^2.\, P)$.

For the sake of comparison, we give as the third and fourth algorithms the ones from chapter 1.

(K) $[x]^3.\, M = \mathsf{K}M$ if $x \notin \mathrm{fv}(M)$;

(I) $[x]^3.\, x = \mathsf{SKK}$ (x is a variable);

(S) $[x]^3. NP = \mathsf{S}([x]^3. N)([x]^3. P)$.

We now buildin the "shortcut," that is, η-reduction.

(K) $[x]^4. M = \mathsf{K}M \quad \text{if } x \notin \text{fv}(M)$;

(I) $[x]^4. x = \mathsf{SKK}$;

(η) $[x]^4. Mx = M \quad \text{if } x \notin \text{fv}(M)$;

(S) $[x]^4. NP = \mathsf{S}([x]^4. N)([x]^4. P)$.

The next algorithm differs from the second one only in the ordering of the clauses.

(S) $[x]^5. NP = \mathsf{S}([x]^5. N)([x]^5. P)$;

(K) $[x]^5. M = \mathsf{K}M \quad \text{if } x \notin \text{fv}(M)$;

(I) $[x]^5. x = \mathsf{I}$.

Notice that (S) has no side conditions except that the term NP is complex. Thus it follows that (K) is applicable in the last algorithm only if M is *not* complex, that is, M is either one of S, K and I, or it is a variable distinct from x.

Although the fifth algorithm may seem unmotivated or even ill-conceived, we can give a straightforward justification why such an algorithm is reasonable. Let us consider how we could implement this algorithm or what *information* we need at each stage to determine which clause to apply. (S) is applicable as long as the term M is complex; we only need the information that the length of M is greater than 1. Independently of the concrete shape of the atomic terms, this information is usually easy to extract from a term. Once (S) is not applicable, M is atomic; we only need to check whether the variable that is being abstracted is identical to M. Again, this question is computationally easy to decide.

Exercise 4.3.8. Consider $[\,]^2$ and $[\,]^5$. Find CL-terms such that $[x]^2. M$ yields a longer CL-term than $[x]^5. M$ does, and vice versa. Are there terms for which the algorithms yield CL-terms of the same length?

Exercise 4.3.9. Find general characterizations of classes of terms, for which $[\,]^2$ and $[\,]^5$ result in CL-terms with relationships mentioned in the previous exercise.

We could create two more algorithms from $[\,]^1$ by omitting either (B) or (C), but not both. Another variation on $[\,]^1$ is the omission of (η), which is absent from the second, the third and the fifth algorithms. The inclusion of (η) into $[\,]^3$ gives $[\,]^4$; the inclusion of (η) into $[\,]^5$, however, would go against the ease of its computation that we laid out as a motivation. The inclusion of (η) into $[\,]^2$ though would yield an algorithm comparable to $[\,]^4$.

Exercise 4.3.10. Make some of the modifications outlined in the previous paragraph, and determine how the changes affect the algorithm.

Exercise* 4.3.11. For each algorithm, take the corresponding (minimal) combinatory basis, and translate each combinator in the basis into Λ. Replace the λ's by brackets, and find the CL-term that that particular algorithm gives.

Exercise 4.3.12. Consider the Λ-term $\lambda xyz.\, yz(xz)$. What is the CL-term that corresponds to $[x]\,[y]\,[z].\, yz(xz)$ according to each of the five bracket abstractions above?

Exercise 4.3.13. Consider the meta-term $[x_1 x_2 x_3]^1.\, x_1 x_3([y]^1.\, x_2 y)$. Which combinator does this meta-term denote?

Exercise 4.3.14. Consider the meta-term $vy\,[xyzv]^1.\, xy(xvz)$. Which CL-term is denoted by this meta-term?

Some of the algorithms appealed to properties of subterms. Obviously, we could have more combinators and more refined information about the subterms — accompanied with the expectation that the resulting CL-terms would become shorter.

We first introduce two new combinators as primitives.

$$\mathsf{C}^* xyzv \rhd x(yv)z \qquad\qquad \mathsf{S}^* xyzv \rhd x(yv)(zv)$$

These combinators have been considered not only with the aim of defining an abstraction algorithm. The notation is intended to suggest that C^* (S^*) has a certain similarity to C (S). They have the same effect once the first argument is removed. C^* and S^* are both definable from C and S, respectively, with B that provides the "distance" effect. $\mathsf{B(BC)B}$ can be taken for C^*, and $\mathsf{B(BS)B}$ can be taken for S^*.

The last algorithm uses the two new combinators together with D from among the associators mentioned in section 1.2.

(K) $[x]^6.\, M = \mathsf{K}M$ if $x \notin \mathrm{fv}(M)$;

(I) $[x]^6.\, x = \mathsf{I}$ (x is a variable);

(η) $[x]^6.\, Mx = M$ if $x \notin \mathrm{fv}(M)$;

(W) $[x]^6.\, Mx = \mathsf{W}([x]^6.\, M)$;

(D) $[x]^6.\, NPQ = \mathsf{D}NP([x]^6.\, Q)$ if $x \notin \mathrm{fv}(NP)$;

(C^*) $[x]^6.\, NPQ = \mathsf{C}^*N([x]^6.\, P)Q$ if $x \notin \mathrm{fv}(NQ)$;

(S^*) $[x]^6.\, NPQ = \mathsf{S}^*N([x]^6.\, P)([x]^6.\, Q)$ if $x \notin \mathrm{fv}(N)$;

(B) $[x]^6.\, NP = \mathsf{B}N([x]^6.\, P)$ if $x \notin \mathrm{fv}(N)$;

(C) $[x]^6.\, NP = \mathsf{C}([x]^6.\, N)P$ if $x \notin \mathrm{fv}(P)$;

(S) $[x]^6.\, NP = \mathsf{S}([x]^6.\, N)([x]^6.\, P)$.

The algorithm is essentially Schönfinkel's algorithm with four new clauses inserted in the middle. (W) pairs with (η): if the latter is not applicable because $x \in \mathrm{fv}(M)$, then (W) can be applied. The next three clauses yield shorter CL-terms, when the structure of M is at least three terms associated to the left. In each clause, x is not free in N, P or Q.

Exercise 4.3.15. Consider the $_\lambda$ translation of the combinators W, D, C^* and S^*. Replace the λ's by $[\,]$'s and determine what CL-terms result according to the latest algorithm.

Exercise 4.3.16. Find the CL-terms that correspond to $[xyz].yz([v].x)$ — according to the fifth and the sixth algorithms.

We have not mentioned all the algorithms that have been introduced, let alone did we present all the possible algorithms.

However, we have accumulated a sufficiently diverse collection of algorithms to turn to the definition of translations from Λ into CL.

DEFINITION 4.3.10. (FROM Λ TO CL) The *translation* of a Λ-term M into CL, which is denoted by M_c, is defined by (1)–(3), when a bracket abstraction algorithm has been selected.

(1) If M is a variable x, then M_c is x;

(2) if M is $\lambda x.N$, then M_c is $[x].N_c$;

(3) if M is (NP), then M_c is $(N_c P_c)$.

This definition is quite simple. But the situation is unbalanced: we have *one* translation from CL into Λ, but at least *six* translations from Λ into CL. It is trivial to state that the six algorithms do not always yield the same CL-terms, that is, for each pair of algorithms there is a Λ-term such that two (i.e., distinct) CL-terms result by the application of the algorithms.

We have also seen already in chapter 1 that $[x]\,[y]\,[z].xz(yz)$ need not be S. We might wonder whether either of the algorithms would give use back the CL-terms that we start with, that is, if $(M_\lambda)_c$ is M. (We will add the superscript of the bracket abstraction algorithm to a term for clarity, whenever we add the subscript $_c$.) We assume from now on that $\mathfrak{B}$ is as small as possible for a particular algorithm.

LEMMA 4.3.11. *For all* CL*-terms M, M is* identical to $(M_\lambda)_c^1$, $(M_\lambda)_c^4$ *and* $(M_\lambda)_c^6$.

Proof: We have just stated that the algorithms give different results, which may seem to contradict to the claim. But recall that we assumed that the algorithms determine the minimal necessary combinatory basis, and thereby, also the set of CL-terms. (That is, if we consider $[\,]^1$, then M cannot contain W, and for $[\,]^4$, M cannot even contain C or B.)

The core of the proof is the verification that, for each combinator Z in the corresponding combinatory basis, $(\mathsf{Z}_\lambda)_c$ is Z. Exercise 4.3.11 covers $[\,]^1$ and $[\,]^4$. For $[\,]^6$, see the next exercise. qed

Exercise 4.3.17. Consider the combinatory basis for $[\,]^6$. Verify that the five combinators that appeared in some of the other algorithms are their own back-and-forth translations via λ, and then ^6_c. (That is, show that $(\mathsf{K}_\lambda)^6_c = \mathsf{K}$, $(\mathsf{I}_\lambda)^6_c = \mathsf{I}$, $(\mathsf{B}_\lambda)^6_c = \mathsf{B}$, $(\mathsf{C}_\lambda)^6_c = \mathsf{C}$ and $(\mathsf{S}_\lambda)^6_c = \mathsf{S}$.) Then finish the proof of the preceding lemma.

Exercise 4.3.18. Exercise 4.3.10 asked you to look at some modifications in the algorithms. Specifically, consider $[\,]^1$ with (B) or with (C) omitted. Is the lemma above true for the resulting algorithms? (Hint: Either amend the proof of lemma 4.3.11 or find a counterexample.)

We might consider the other direction too, that is, we can start with a Λ-term. In this case we are interested in the preservation of weaker relations than identity of terms.

The following lemma states that bracket abstractions emulate λ-abstraction up to one-step β- or one-step $\beta\eta$-reductions.

LEMMA 4.3.12. *Let N be a variable that does not occur in M. If $[\,]^n$ includes (η), then $([x]^n.\, M^n_c)_\lambda N$ is $\beta\eta$*-equal to $[x/N](M^n_c)_\lambda$.
If $[\,]^n$ does not include (η), then $([x]^n.\, M^n_c)_\lambda N$ is β-equal to $[x/N](M^n_c)_\lambda$.

Exercise 4.3.19. Prove the first claim in the lemma for $[\,]^1$.

Exercise 4.3.20. Prove the second claim in the lemma for $[\,]^5$.

LEMMA 4.3.13. *For all Λ-terms M, $M =_{\beta\eta} (M^1_c)_\lambda$, $M =_{\beta\eta} (M^4_c)_\lambda$ and $M =_{\beta\eta} (M^6_c)_\lambda$.*
For all Λ-terms M, $M =_\beta (M^2_c)_\lambda$ and $M =_\beta (M^5_c)_\lambda$.

Proof: The lemma contains two claims and each claim refers to multiple algorithms. (In total, there are five "subclaims" in the lemma.) We divide the proof into several steps, the first two of which is common to each subclaim that we indicate by using ^{n} instead of a concrete number in the superscript.

1. Let us assume that M is atomic, that is, M is a variable. M is identical to M^n_c, as well as to $(M^n_c)_\lambda$.

2. If M is of the form of (NP), then $(M^n_c)_\lambda$ is $(N^n_c P^n_c)_\lambda$, and further $((N^n_c)_\lambda (P^n_c)_\lambda)$. By hypothesis, $P =_{\beta(\eta)} (P^n_c)_\lambda$ and $N =_{\beta(\eta)} (N^n_c)_\lambda$, but then also $(M^n_c)_\lambda =_{\beta(\eta)} M$.

3. Let us now assume that M is $\lambda x.N$. According to definition 4.3.10, M^n_c is $[x]^n.N^n_c$. Depending on the particular n (i.e., on the bracket abstraction algorithm) and on the side conditions on N, $[x]^n.N^n_c$ may yield different CL-terms. We take the fifth subclaim as an example.

There are three subcases: N is compound, N is x or N is some other atomic term. If N is (PQ), then $[x]^5.PQ$ is $\mathsf{S}([x]^5.P^5_c)([x]^5.Q^5_c)$. The $_\lambda$ translation of M^5_c is $(\mathsf{S}([x]^5.P^5_c)([x]^5.Q^5_c))_\lambda$, that is, $\mathsf{S}_\lambda([x]^5.P^5_c)_\lambda([x]^5.Q^5_c)_\lambda$. S_λ is $\lambda xyz.xz(yz)$, thus we get $\lambda z.([x]^5.P^5_c)_\lambda z(([x]^5.Q^5_c)_\lambda z)$. We have $\lambda z.([x/z](P^5_c)_\lambda)([x/z](Q^5_c)_\lambda)$, by lemma 4.3.12. P and Q are shorter terms than M; hence, we have $\lambda z.([x/z]P)[x/z]Q$ which α-equals to $\lambda x.PQ$.

If N is x, then M^5_c is I. The $_\lambda$ translation of the latter is $\lambda x.x$, which is M.

If N is an atom distinct from x, then M_c^5 is $\mathsf{K}N_c^5$. Then $(M_c^5)_\lambda$ is $(\mathsf{K}N_c^5)_\lambda$, hence, $\mathsf{K}_\lambda(N_c^5)_\lambda$. By hypothesis, $(N_c^5)_\lambda$ is $\beta\eta$-equal to N. K_λ is $\lambda xy.x$. Then we get $(\lambda xy.x)N$, and further, $\lambda y.N$. qed

Exercise* 4.3.21. Prove the other four subclaims from the lemma.

Chapter 5

(In)equational combinatory logic

We have so far described *relations on* CL-*terms* by, first, specifying operations on terms, and then by applying a closure to that operation. For example, we started with $\triangleright_1$, which is defined as the replacement of a redex by its reduct in a term, and then we defined $\triangleright_w$ as the reflexive transitive closure of the $\triangleright_1$ relation. Such an approach is quite satisfactory, moreover, it turned out to be useful in proving various theorems, in defining combinators for special purposes or in comparing CL-terms to Λ-terms.

Another approach, which is often used in logic and mathematics (and has certain advantages) is to select some "self-evident" principles and to add rules that allow one to derive further statements. This is called the *axiomatic approach*, and in this chapter we consider its applicability to CL.

The relations of weak reducibility and weak equality are binary relations on CL-terms. $\triangleright_w$ is *transitive* and *reflexive*, whereas $=_w$ is additionally *symmetric*. The properties of these relations suggest right away that these relations on terms, perhaps, can be formally captured by extensions of *inequational* and *equational* logics.[1]

5.1 Inequational calculi

Weak reduction is a *preorder* on the set of CL-terms. Our inequational logic comprises axioms and rules that characterize the main predicate in the logic as a *preorder*. We denote this predicate by $\geq$. This symbol is often used for weak total orders, or for a relation that is at least a weak partial order. Our use of this symbol does not imply that the relation is a partial order, let alone that it is a total order. The choice is simply motivated by the similarity of the shapes of $\triangleright$ and $\geq$.

In order to be able to extend this logic with CL-terms, we have to determine whether the components that are specific to CL are compatible with inequational logic. In particular, we have an operation — function application — that we visually suppressed by not using a symbol for it. The definition of weak reduction implies

[1]For detailed discussions of equational and inequational logics, see, for example, Dunn and Hardegree [65]. A brief summary of equational and inequational calculi in general may be found in section A.5.

that function application is a *monotone* operation with respect to the preorder that is $\triangleright_{\mathrm{w}}$.

Example 5.1.1. Let us take the axiom for C, and let M be $\mathsf{S}yyy$. We can form the CL-term MN, where N is $\mathsf{C}xyz$, that is, $\mathsf{S}yyy(\mathsf{C}xyz)$. $\mathsf{C}xyz \triangleright_{\mathrm{w}} xzy$, and $\mathsf{S}yyy \triangleright_{\mathrm{w}} yy(yy)$. If we think of $\geq$ as $\triangleright_{\mathrm{w}}$, then we should have $\mathsf{S}yyy(\mathsf{C}xyz) \geq \mathsf{S}yyy(xzy)$ and $\mathsf{S}yyy(\mathsf{C}xyz) \geq yy(yy)(\mathsf{C}xyz)$ to show that function application is monotone in both of its argument places. Those pairs of CL-terms are, indeed, in $\triangleright_{\mathrm{w}}$.

We have called pairs of CL-terms *axioms*, where the CL-terms together show the effect of a combinator, and we continue to supply combinators with axioms. We have to choose a combinatory basis, and we will take all the well-known combinators listed in section 1.2, which is more than sufficient.

We suppose definition 1.1.1 of CL-terms as before. The additional symbol $\geq$ will be used as a *binary predicate* in infix notation. *Formulas* are of the form $M \geq N$, where M and N are CL-terms. The informal understanding of a formula such as $M \geq N$ is that M weakly reduces to N.

The *axioms* are formulas, whereas the *rules* comprise one or more formulas as premises and one formula as the conclusion. The premises and the conclusion are separated by a horizontal rule; if there is more than one premise, then they are spaced apart.

The combinatory basis is $\mathfrak{B}_1 = \{\mathsf{S}, \mathsf{K}, \mathsf{I}, \mathsf{B}, \mathsf{C}, \mathsf{W}, \mathsf{M}, \mathsf{B}', \mathsf{J}\}$, and we will denote by $IQ_{\mathfrak{B}_1}$ the calculus that incorporates all these combinators. The axioms and the rules are as follows.

$$x \geq x \ \ \mathrm{id}$$

$$\mathsf{I}x \geq x \ \ \mathsf{I} \qquad \mathsf{B}xyz \geq x(yz) \ \ \mathsf{B} \qquad \mathsf{S}xyz \geq xz(yz) \ \ \mathsf{S}$$

$$\mathsf{K}xy \geq x \ \ \mathsf{K} \qquad \mathsf{C}xyz \geq xzy \ \ \mathsf{C} \qquad \mathsf{W}xy \geq xyy \ \ \mathsf{W}$$

$$\mathsf{M}x \geq xx \ \ \mathsf{M} \qquad \mathsf{B}'xyz \geq y(xz) \ \ \mathsf{B}' \qquad \mathsf{J}xyzv \geq xy(xvz) \ \ \mathsf{J}$$

$$\frac{M \geq N}{[x_1/P_1,\ldots,x_n/P_n]M \geq [x_1/P_1,\ldots,x_n/P_n]N} \ \mathrm{sub}$$

$$\frac{M \geq P \qquad P \geq N}{M \geq N} \ \mathrm{tr}$$

$$\frac{M \geq N}{MP \geq NP} \ \mathrm{m}_l \qquad \frac{M \geq N}{PM \geq PN} \ \mathrm{m}_r$$

A *derivation* is a tree, in which all nodes are formula occurrences, and all nodes except the leaves are justified by applications of rules. A derivation is a *proof* whenever all the leaves are instances of axioms. The root of a proof tree is the formula that is proved, and it is called a *theorem*.

The first axiom is *identity* for the variable x. Together with the rule of substitution, the effect of this axiom is that $M \geq M$ is provable for all CL-terms M. We call this axiom identity due to the analogy with sequent calculi. Neither $\vdash$ in the sequent calculi, nor $\geq$ here is an equivalence relation; they are more like $\rightarrow$. But $A \rightarrow A$ is called self-identity of $\rightarrow$. An alternative motivation for the label is that a binary relation is reflexive when it includes the identity relation over its domain. The *nine* other *axioms* are for combinators — one for each combinator in the combinatory basis $\mathfrak{B}_1$.

The rule of *substitution* is similar to the substitution that is built into the notion of one-step reduction. (See definition 1.3.2.) A difference is that no combinator need to be involved in this substitution rule.

The *transitivity* rule is exactly what its label suggest: $\geq$ is transitive due to the inclusion of this rule.

The *monotonicity* rules express the properties that we illustrated in the previous example.

Example 5.1.2. B' is just like B, but with x and y, the first two arguments permuted. C permutes its second and third arguments. Thus CB defines B' and CB' defines B in the sense of definition 1.3.6. In $IQ_{\mathfrak{B}_1}$, the CL-term CB applied to any M,N and P emulates B' as the following proof shows.

$$\dfrac{\dfrac{\dfrac{\mathsf{C}xyz \geq xzy}{\mathsf{CB}MN \geq \mathsf{B}NM}}{\mathsf{CB}MNP \geq \mathsf{B}NMP} \qquad \dfrac{\mathsf{B}xyz \geq x(yz)}{\mathsf{B}NMP \geq N(MP)}}{\mathsf{CB}MNP \geq N(MP)}$$

The example suggests that $IQ_{\mathfrak{B}_1}$ is sufficiently powerful, but we might wonder whether it is, perhaps *too* powerful. Ideally, we would like the axiomatic calculus to reproduce exactly the $\triangleright_w$ relation. For the next claim, we assume that the set of CL-terms is generated by the same combinatory basis $\mathfrak{B}_1$ together with a denumerable set of variables.

THEOREM 5.1.3. *The formula* $M \geq N$ *is* provable in $IQ_{\mathfrak{B}_1}$ *if and only if* $M \triangleright_w N$.

Proof: **1.** We start from left to right. If $M \geq N$ is provable, then it has a proof. We use induction on the height of the proof, where the latter is the height of the proof tree. The shortest proofs are formulas that are instances of an axiom. If the axiom is id, then it is a special instance of $M \triangleright_w M$ expressing the reflexivity of $\triangleright_w$. If the axiom is for a combinator, then we can first verify that the terms in the axioms in $IQ_{\mathfrak{B}_1}$ contain the same pair of CL-terms as the axioms we gave for the respective combinators in section 1.2. Then by definition 1.3.2, each pair of terms in an axiom in $IQ_{\mathfrak{B}_1}$ is in the $\triangleright_1$ relation, hence in the $\triangleright_w$ relation.

2. If $M \geq N$ is by a rule, then there are four cases to consider. (We provide some details of two cases and leave the other two, as exercise 5.1.1, to the reader.)

2.1 Let $M \geq N$ be the result of an application of sub. We introduced substitution for CL-terms in definition 1.2.2. However, both single and simultaneous substitution are operations on single CL-terms (rather than on pairs of CL-terms). The rule sub presupposes the substitution operation, however, the latter does not define or give rise to the rule sub by itself. We built a limited amount of substitution into $\triangleright_1$ to allow for the application of combinators to other CL-terms — beyond the variables that figure into their axioms. This way of handling substitution yielded a tightly circumscribed notion of $\triangleright_1$, and then of $\triangleright_w$, which facilitated proofs involving those relations. However, now we need that substitution is an "admissible rule," which is the content of lemma 5.1.4, below. Given the lemma, it is immediate that sub preserves $\triangleright_w$.

2.2 Suppose that $M \geq N$ is by rule m_l, that is, the formula is of the form $M'P \geq N'P$. By assumption, $M' \triangleright_w N'$. If M' is N', then $M'P \triangleright_w N'P$ is obvious. If $M' \triangleright_1 N'$, then $M'P \triangleright_1 N'P$ is by the definition of one-step reduction. If $M' \triangleright_w N'$, but neither of thc prcvious two cases hold, then $M' \triangleright_w Q$ and $Q \triangleright_1 N'$, for some Q. Then we may assume $M'P \triangleright_w QP$, by the hypothesis of the induction. $QP \triangleright_1 N'P$ by the definition of one-step reduction, and since $\triangleright_w$ is transitive, $M'P \triangleright_w N'P$, as we need.

3. For the other direction, let us assume that $M \triangleright_w N$. There are three cases to consider: when M is N, and $M \triangleright_w N$ is by reflexive closure, when $M \triangleright_1 N$, and when $M \triangleright_w P$ and $P \triangleright_w N$.

3.1 M may or may not be a variable, in particular, it may or may not be x. We construct the following proof in $IQ_{\mathfrak{B}_1}$.

$$\frac{x \vdash x}{[x/M]x \vdash [x/M]x}$$

The first line is the id axiom, the next line is by sub.

3.2 Let us assume that $M \triangleright_1 N$. By the definition of $\triangleright_1$, M contains a redex which is replaced by its reduct. There are nine combinators in $\mathfrak{B}_1$ and each combinator is treated similarly; for the sake of concreteness, we choose S. Let us assume that the CL-terms that are the arguments of the selected redex occurrence of S are P_1, P_2 and P_3. $\mathsf{S}P_1P_2P_3$ is a subterm of M, and in the formation tree of M, the subtree with root $\mathsf{S}P_1P_2P_3$ is replaced by the formation tree of $P_1P_3(P_2P_3)$. M may contain other subterms, that is, the redex need not be the whole term M. We construct the following proof to obtain the effect of $\triangleright_1$.

$$\cfrac{\cfrac{\cfrac{\mathsf{S}xyz \geq xz(yz)}{\mathsf{S}P_1P_2P_3 \geq P_1P_3(P_2P_3)}}{\mathsf{S}P_1P_2P_3Q_1 \geq P_1P_3(P_2P_3)Q_1}}{Q_2(\mathsf{S}P_1P_2P_3Q_1) \geq Q_2(P_1P_3(P_2P_3)Q_1)}$$

$$\vdots$$

Q_1 and Q_2 illustrate that, if the redex is a proper subterm of M, then finitely many applications of m_l and m_r suffice to embed the redex and its reduct to obtain $M \geq N$.

We leave finishing cases **2** and **3** for the next two exercises. qed

Exercise 5.1.1. Give a proof of the remaining subcases (**2.3** and **2.4**) in the above proof. (Hint: One of these subcases is nearly obvious given **2.2**.)

Exercise 5.1.2. Finish step **3** in the proof of the above theorem. (Hint: Only one step, namely, **3.3** remains to be proved.)

Recall that we appealed to a substitution rule for $\triangleright_{\mathrm{w}}$ in the proof of the last theorem.

LEMMA 5.1.4. *If* $M \triangleright_{\mathrm{w}} N$*, then* $[x/P]M \triangleright_{\mathrm{w}} [x/P]N$*. That is,* $\triangleright_{\mathrm{w}}$ *is* closed under uniform substitution *of* CL*-terms for a variable.*

Proof: If $M \triangleright_{\mathrm{w}} N$, then there are three possibilities according to the definition of $\triangleright_{\mathrm{w}}$. Namely, M and N may be the same CL-term. Then $[x/P]M$ is also the same CL-term as $[x/P]N$, because $[x/P]$ is an operation.

Another case is when $M \triangleright_{\mathrm{w}} N$ by $\triangleright_1$. Then N is $[Q_1/Q_2]M$, where Q_2 is a redex occurrence, and Q_1 is its reduct. We have nine combinators in $\mathfrak{B}_1$, but let us consider Z^n for the sake of generality. Given $\mathsf{Z}x_1 \ldots x_n \triangleright R$ (where $R \in \{x_1,\ldots,x_n\}^{\oplus}$), Q_2 is $\mathsf{Z}Q'_1 \ldots Q'_n$, and Q_1 is $[x_1/Q'_1,\ldots,x_n/Q'_n]R$. x may have occurrences in Q_2 and in the rest of M. N is $[[x_1/Q'_1,\ldots,x_n/Q'_n]R/\mathsf{Z}Q'_1 \ldots Q'_n]M$, which makes clear that if x occurs in Q_2, then x must occur in at least one of the Q'_i's. $\mathsf{Z}Q'_1 \ldots [x/P]Q'_i \ldots Q'_n$ is a redex, and its reduct is $[x_1/Q'_1,\ldots,x_i/[x/P]Q'_i,\ldots, x_n/Q'_n]R$.

If x occurs in M outside of the redex affected by the one-step reduction, then by the definition of substitution, P is substituted in those subterms of M and N in the same way. That is, $([\mathsf{Z}Q'_1 \ldots Q'_n]M)^P_x$ is M^P_x with the redex occurrence made explicit. Then we further have $[\mathsf{Z}(Q'_1)^P_x \ldots (Q'_n)^P_x](M^P_x)$. The substitution for N is $[[x_1/(Q'_1)^P_x,\ldots,x_n/(Q'_n)^P_x]R/\mathsf{Z}Q'_1 \ldots Q'_n](M^P_x)$. $R \in \{x_1,\ldots,x_n\}^{\oplus}$, which means that N^P_x results by one-step reduction from M^P_x.

Lastly, $M \triangleright_{\mathrm{w}} N$ may obtain because $M \triangleright_{\mathrm{w}} Q$ and $Q \triangleright_{\mathrm{w}} N$; therefore, by transitive closure, $M \triangleright_{\mathrm{w}} N$. If the previous two cases do not hold, then we may assume that the length of one-step reductions in $M \triangleright_{\mathrm{w}} Q$ and in $Q \triangleright_{\mathrm{w}} N$ is less than in $M \triangleright_{\mathrm{w}} N$. Then, by hypothesis, both $[x/P]M \triangleright_{\mathrm{w}} [x/P]Q$ and $[x/P]Q \triangleright_{\mathrm{w}} [x/P]N$. $[x/P]$ is an operation, which means that the two occurrences of $[x/P]Q$ are the same CL-term. Then $[x/P]M \triangleright_{\mathrm{w}} [x/P]N$ is immediate by the transitivity of $\triangleright_{\mathrm{w}}$. qed

The lemma suggests that we can formulate a calculus with the same set of provable formulas without a rule of substitution. More precisely, substitution may be left implicit, because we can use *meta-terms* not only in the rules but also in the axioms. We denote this calculus by $IQ'_{\mathfrak{B}_1}$.

$$M \geq M \quad \text{id}$$

$$\mathsf{I}M \geq M \quad \mathsf{I} \qquad \mathsf{B}MNP \geq M(NP) \quad \mathsf{B} \qquad \mathsf{S}MNP \geq MP(NP) \quad \mathsf{S}$$

$$\mathsf{K}MN \geq M \quad \mathsf{K} \qquad \mathsf{C}MNP \geq MPN \quad \mathsf{C} \qquad \mathsf{W}MN \geq MNN \quad \mathsf{W}$$

$$\mathsf{M}M \geq MM \ \mathsf{M} \qquad \mathsf{B}'MNP \geq N(MP) \ \mathsf{B}' \qquad \mathsf{J}MNPQ \geq MN(MQP) \ \mathsf{J}$$

$$\frac{M \geq P \qquad P \geq N}{M \geq N} \ \text{tr}$$

$$\frac{M \geq N}{MP \geq NP} \ \text{m}_l \qquad \frac{M \geq N}{PM \geq PN} \ \text{m}_r$$

We can use the same notions of *derivation* and *proof* as before. However, we can also make "official" the practice of using meta-terms in a proof. In example 5.1.2, we proved $\mathsf{CB}MNP \geq N(MP)$, which is a meta-term itself. We could have proved the formulas $\mathsf{CB}xyz \geq y(xz)$ or $\mathsf{CB}zxy \geq x(zy)$, by essentially the same proof, that is, by constructing a proof comprising the same axioms and applications of the same rules. Indeed, we can take any CL-terms for M, N and P in $\mathsf{CB}MNP \geq N(MP)$, and the resulting formula will be provable. This is yet another way to restate that the rule of substitution is *admissible* (in $IQ_{\mathfrak{B}_1}$ too).

LEMMA 5.1.5. *The two calculi $IQ_{\mathfrak{B}_1}$ and $IQ'_{\mathfrak{B}_1}$ are* equivalent.

Exercise 5.1.3. Prove the lemma. (Hint: Consider $M \geq N$ where M and N are concrete CL-terms.)

Example 5.1.2 suggests that we might consider extending either calculus by a rule that would allow an inequation to be derived when the applications of CL-terms on the left and on the right would yield a provable inequation.

Regular combinators keep their first argument in place.[2] A dual property — keeping the last argument in place — characterizes the class of final regular combinators.

DEFINITION 5.1.6. A combinator Z^n is *final regular* whenever its axiom is $\mathsf{Z}x_1 \ldots x_n \rhd Mx_n$.

None of the well-known combinators that we included into the basis $\mathfrak{B}_1$ is final regular and it is not difficult to see why. The combinators in $\mathfrak{B}_1$, perhaps except J, cause relatively simple changes in the redex CL-term that they are the head of. If we consider weak reduction, then the need for more arguments than those that are affected in some way seems to be a hindrance rather than an advantage. However, when we consider CL-terms that are combinators (but may not be in $\mathfrak{B}_1$), then we soon find some that are clearly final regular.

Example 5.1.7. Let us consider $\mathsf{BB}(\mathsf{KI})$. To form redexes with both B's, we need four arguments. (BB is itself a combinator and we denoted it by D in section 1.2; D is quaternary.) $\mathsf{BB}(\mathsf{KI})xyz \rhd_{\mathrm{w}} yz$, and it appears that we could omit z. However, $\mathsf{BB}(\mathsf{KI})xy$ does not weakly reduce to y, because $\mathsf{B}(\mathsf{KI}x)y$ does not contain a redex with B as its head.

[2]Regular combinators are defined in section 1.2.

We can disregard the fact that $\mathsf{B}\mathsf{I}y$ is in nf, and expand the notion of weak reduction to include $\mathsf{B}\mathsf{I}y \rhd' y$. To have this effect in the inequational calculi, we add a rule to $IQ_{\mathfrak{B}_1}$. The expanded calculus is denoted by $IQ^{\eta}_{\mathfrak{B}_1}$.

$$\frac{Mx \geq Nx \qquad x \notin \mathrm{fv}(MN)}{M \geq N}\ \mathrm{ext}$$

The requirement that x is not a free variable of MN excludes spurious cases such as $\mathsf{M}x \geq xx$ and then $\mathsf{M} \geq x$, which obviously would be undesirable. If we would consider $=$, the tansitive symmetric closure of $\geq$ instead of $\geq$ itself, then $\mathsf{M} = x$ (together with $\mathsf{M} = y$, by an alike step) would immediately lead to inconsistency.

Exercise 5.1.4. Investigate what the consequences of omitting $x \notin \mathrm{fv}(MN)$ in the rule ext are. (Hint: A paramount question is if the calculus remains consistent.)

In chapter 1, we gave definitions of extensional weak equality and strong reduction. (See definitions 1.3.15, 1.3.16 and 1.3.18.) The relation between CL-terms that is determined by provable inequalities in $IQ^{\eta}_{\mathfrak{B}_1}$ is not an equivalence relation. On the other hand, we have already hinted at the possibility that there are provable inequations in $IQ^{\eta}_{\mathfrak{B}_1}$ that are not provable in $IQ_{\mathfrak{B}_1}$. The next series of exercises address the relationships between these relations on CL-terms.

Exercise 5.1.5. Prove that $IQ^{\eta}_{\mathfrak{B}_1}$ is consistent. (Hint: Recall that by consistency we mean that the relation $\geq$ is not the total relation on the set of CL-terms.)

Exercise 5.1.6. Prove that there is no CL-term M such that for all N, $M \geq N$ is provable in $IQ^{\eta}_{\mathfrak{B}_1}$. (Hint: Although x is arbitrary in the ext rule, this does not change the relationship between $\mathrm{fv}(M)$ and $\mathrm{fv}(N)$ in a provable inequality.)

Exercise 5.1.7. Prove that the binary relation on the set of CL-terms determined by $\geq$ is not a symmetric relation; hence, it is not an equivalence relation either. (Hint: Consider the axiom of K, for example.)

Exercise 5.1.8. What is the relationship between $\rhd_s$ and the relation emerging from provable inequations in $IQ^{\eta}_{\mathfrak{B}_1}$?

The inequational calculi somewhat resemble *sequent calculi*.[3] For instance, in chapter 7, we introduce structurally free logics, which constitute a family of sequent calculi. (There is a clash between the usual terminology used in connection to sequent calculi and the inequational calculi here. In the structurally free logics, combinators are formulas, whereas here they are terms. We continue to use the usual terminology for both calculi and hope that no confusion will result.)

Many sequent calculi do not allow the right-hand side of a sequent to contain more that one formula. It is tempting to try to consider an inequation as a sequent, in which there is exactly one CL-term on each side of $\geq$. This view creates a sense

[3] See section A.7 for a brief exposition of sequent calculi.

of similarity between the two types of calculi, but there are also differences that are worth noting.

The *structural connective* (or connectives) are essential to sequent calculi. In an inequational calculus, the structural manipulations are limited, or they may be thought to be absent altogether. There is one rule in $IQ^{\eta}_{\mathfrak{B}_1}$ that allows altering the structure of a derivation, namely, the tr rule. (We do not consider the side condition in the ext rule as a node in the proof tree; hence, this rule is not alike to tr.)

The tr rule is quite similar to the cut rule, which is originally not even called a structural rule in sequent calculi. Well-formulated sequent calculi do not include the cut rule, rather they have it as an *admissible rule*. We might ponder if the transitivity rule is admissible in $IQ_{\mathfrak{B}_1}$ (provided tr is excluded to start with).

Exercise 5.1.9. Determine whether tr is admissible in $IQ_{\mathfrak{B}_1}$ (or $IQ^{\eta}_{\mathfrak{B}_1}$), and prove your claim. (Hint: If the rule is admissible, then a proof could use insights from proofs of the cut theorem. Otherwise, a counter example suffices.)

5.2 Equational calculi

Weak reduction is a fundamental relation, but *weak equality* is, of course, also important. $=_{\text{w}}$ is an equivalence relation, what should immediately suggest that we look for an *equational calculus* if we intend to capture this relation.

To make comparisons between the inequational and equational calculi easy, we use the same combinatory basis $\mathfrak{B}_1$ that we used in the previous section.

An *equation* is a pair of CL-terms with $=$ in between.[4] The *equational calculus* $EQ_{\mathfrak{B}_1}$ contains the equational calculus for a binary operation together with *axioms* for combinators.

$$x = x \ \text{id}$$

$$\mathsf{I}x = x \ \mathsf{I} \qquad \mathsf{B}xyz = x(yz) \ \mathsf{B} \qquad \mathsf{S}xyz = xz(yz) \ \mathsf{S}$$

$$\mathsf{K}xy = x \ \mathsf{K} \qquad \mathsf{C}xyz = xzy \ \mathsf{C} \qquad \mathsf{W}xy = xyy \ \mathsf{W}$$

$$\mathsf{M}x = xx \ \mathsf{M} \qquad \mathsf{B}'xyz = y(xz) \ \mathsf{B}' \qquad \mathsf{J}xyzv = xy(xvz) \ \mathsf{J}$$

$$\frac{M = N}{[x_1/P_1,\ldots,x_n/P_n]M = [x_1/P_1,\ldots,x_n/P_n]N} \ \text{sub}$$

[4]The symbols $=_{\text{w}}$ and $=$ are very similar, and we intend to axiomatize weak equality. Nonetheless, as symbols, $=$ and $=_{\text{w}}$ are completely distinct.

$$\frac{M = P \qquad P = N}{M = N}\ \text{tr} \qquad \frac{M = N}{N = M}\ \text{sym}$$

$$\frac{M = N}{MP = NP}\ \mathrm{m}_l \qquad \frac{M = N}{PM = PN}\ \mathrm{m}_r$$

The axioms and the rules may be viewed as straightforward modifications of the axioms and rules in $IQ_{\mathfrak{B}_1}$: the $\geq$ symbol has been replaced by $=$. The only exception is the rule sym, which is completely new here. This rule ensures that the relation induced by provable equations on the set of CL-terms is *symmetric*.

The notion of a *derivation* and of a *proof* is exactly as before. The two calculi differ only in the symbol that separates CL-terms and in the latter calculus having a rule sym, which motivates the next statement.

LEMMA 5.2.1. *For any $M \geq N$ that is provable in $IQ_{\mathfrak{B}_1}$, $M = N$ and $N = M$ are* provable *in $EQ_{\mathfrak{B}_1}$.*

The truth of this claim may be established by induction, which is quite easy due to the similarities of the calculi.

Exercise 5.2.1. Prove the previous lemma.

The equational calculus also can be formulated with meta-terms. We denote this modification of the equational calculus by $EQ'_{\mathfrak{B}_1}$.

$$M = M\ \text{id}$$

$$\mathsf{I}M = M\ \mathsf{I} \qquad \mathsf{B}MNP = M(NP)\ \mathsf{B} \qquad \mathsf{S}MNP = MP(NP)\ \mathsf{S}$$

$$\mathsf{K}MN = M\ \mathsf{K} \qquad \mathsf{C}MNP = MPN\ \mathsf{C} \qquad \mathsf{W}MN = MNN\ \mathsf{W}$$

$$\mathsf{M}M = MM\ \mathsf{M} \qquad \mathsf{B}'MNP = N(MP)\ \mathsf{B}' \qquad \mathsf{J}MNPQ = MN(MQP)\ \mathsf{J}$$

$$\frac{M = P \qquad P = N}{M = N}\ \text{tr} \qquad \frac{M = N}{N = M}\ \text{sym}$$

$$\frac{M = N}{MP = NP}\ \mathrm{m}_l \qquad \frac{M = N}{PM = PN}\ \mathrm{m}_r$$

Having the two equational calculi plus the notion of $=_{\mathrm{w}}$ leads us to the question if the relations determined by provable equations in the calculi and $=_{\mathrm{w}}$ coincide on the set of CL-terms.

LEMMA 5.2.2. *An equation $M = N$ is* provable *in $EQ_{\mathfrak{B}_1}$ if and only if it is* provable *in $EQ'_{\mathfrak{B}_1}$.*

Exercise 5.2.2. Give a proof of the lemma using induction on the height of proofs in the two calculi.

The second version of the equational calculus allows for proofs of formulas that contain two meta-terms in addition to proofs of formulas that contain two CL-terms.

THEOREM 5.2.3. *The equation $M = N$ is* provable *in $EQ'_{\mathfrak{B}_1}$ if and only if $M =_{\mathrm{w}} N$.*

Proof: We divide the proof into two parts. First, we show that the conditional holds in the direction moving from $EQ'_{\mathfrak{B}_1}$ to $=_{\mathrm{w}}$.

1. Let us assume that $M = N$ is by an axiom. If it is by id, then N is M. $=_{\mathrm{w}}$ is reflexive by definition, which means that $M =_{\mathrm{w}} M$. If $M = N$ is by a combinatory axiom, then we first of all note that the axioms in $EQ'_{\mathfrak{B}_1}$ match the axioms for the combinators from section 1.2. Second, the definition of $\triangleright_1$ provides for substitution in a redex, and $\triangleright_1 \subseteq \triangleright_{\mathrm{w}}$ as well as $\triangleright_{\mathrm{w}} \subseteq =_{\mathrm{w}}$.

Weak equality is transitive and symmetric by definition. Thus if $M = N$ and $N = P$ in $EQ'_{\mathfrak{B}_1}$ (hence, by hypothesis, $M =_{\mathrm{w}} N$ and $N =_{\mathrm{w}} P$), then it is immediate that $M =_{\mathrm{w}} P$. If $M = N$, that is, by hypothesis $M =_{\mathrm{w}} N$, then certainly, $N =_{\mathrm{w}} M$, as needed.

Lastly, the definition of $\triangleright_1$ implies monotonicity of function application both on the left and on the right; $\triangleright_1$ is included in $=_{\mathrm{w}}$, which takes care of the rules m_l and m_r.

2. To prove the converse, we consider the definition of $=_{\mathrm{w}}$ and then we show that a proof may be constructed in $EQ'_{\mathfrak{B}_1}$. We can establish this direction by using lemma 5.2.1 and theorem 5.1.3. If $M =_{\mathrm{w}} N$ is due to $M \triangleright_{\mathrm{w}} N$, then $M \geq N$ in $IQ'_{\mathfrak{B}_1}$, hence, also $M = N$ in $EQ'_{\mathfrak{B}_1}$. Otherwise, $M =_{\mathrm{w}} N$ is by transitive symmetric closure of $\triangleright_{\mathrm{w}}$. If $M \triangleright_{\mathrm{w}} N$, then by lemma 5.2.1, $M = N$ and $N = M$ in $EQ_{\mathfrak{B}_1}$, and by lemma 5.2.2, $M = N$ and $N = M$ are provable in $EQ'_{\mathfrak{B}_1}$. Given $M_1 =_{\mathrm{w}} M_2, \ldots, M_{n-1} =_{\mathrm{w}} M_n$, we have $M_1 = M_2, \ldots, M_{n-1} = M_n$, and by $n-2$ applications of the tr rule, $M_1 = M_n$. qed

Exercise 5.2.3. Construct step **2** in the proof using the definition of $\triangleright_{\mathrm{w}}$ and then $=_{\mathrm{w}}$ directly (i.e., without relying on lemma 5.2.1).

Notice that, by using the second formulation $EQ'_{\mathfrak{B}_1}$, we avoided the question about emulating steps by the rule of substitution in pairs of CL-terms that are weakly equal. However, we have the following lemma that parallels lemma 5.1.4.

LEMMA 5.2.4. *The weak equality relation is* closed under substitution *of* CL-*terms for variables in* CL-*terms.*

Exercise 5.2.4. Prove the lemma. (Hint: The lemma is similar to lemma 5.1.4.)

Exercise 5.2.5. Given the above lemma, prove theorem 5.2.3 for $EQ_{\mathfrak{B}_1}$ (instead of $EQ'_{\mathfrak{B}_1}$).

We added the rule ext to $IQ_{\mathfrak{B}_1}$, and we add its equational variant to $EQ_{\mathfrak{B}_1}$; the resulting calculus is denoted by $EQ^{\eta}_{\mathfrak{B}_1}$.

$$\frac{Mx = Nx \qquad x \notin \mathrm{fv}(MN)}{M = N}\ \mathrm{ext}$$

We saw that adding the extensionality rule yielded new provable inequations. Similarly, we get new provable equations with ext added to $EQ_{\mathfrak{B}_1}$.

Example 5.2.5. $\mathsf{SII}x \rhd_w \mathsf{I}x(\mathsf{I}x) \rhd_w xx$. This shows that M is definable as SII in the sense of definition 1.3.6. In $EQ^{\eta}_{\mathfrak{B}_1}$, we can prove $\mathsf{SII} = \mathsf{M}$ as:

$$\dfrac{\dfrac{\dfrac{\mathsf{S}xyz = xz(yz)}{\mathsf{SI I}x = \mathsf{I}x(\mathsf{I}x)} \qquad \dfrac{\dfrac{\mathsf{I}x = x}{\mathsf{I}x(\mathsf{I}x) = x(\mathsf{I}x)} \qquad \dfrac{\mathsf{I}x = x}{x(\mathsf{I}x) = xx}}{\mathsf{I}x(\mathsf{I}x) = xx}}{\mathsf{SII}x = xx} \qquad \dfrac{\mathsf{M}x = xx}{xx = \mathsf{M}x}}{\dfrac{\mathsf{SII}x = \mathsf{M}x}{\mathsf{SII} = \mathsf{M}}}.$$

The next proposition shows that, in the context of $EQ^{\eta}_{\mathfrak{B}_1}$, the name "extensionality" for the ext rule is fully justified. That is, if this rule is present, then the relation on CL-terms that is induced by provable equations is extensional.

LEMMA 5.2.6. *The equivalence relation generated by equations provable in* $EQ^{\eta}_{\mathfrak{B}_1}$ *coincides with extensional weak equality.*

Proof: Extensional weak equality is specified in definitions 1.3.15 and 1.3.16. It is more convenient here to rely on the second definition, because of the use of variables.

From right to left, let us assume that M and N — when applied to $x_1,\ldots,x_n$ — yield the term P via weak reduction. That is, $Mx_1\ldots x_n \rhd_w P$ and $Nx_1\ldots x_n \rhd_w P$. By lemma 5.2.1, we know that $Mx_1\ldots x_n = P$ is provable in $EQ^{\eta}_{\mathfrak{B}_1}$, and so is $Nx_1\ldots x_n = P$. The rules sym and tr give $Mx_1\ldots x_n = Nx_1\ldots x_n$ in two steps. An assumption in definition 1.3.16 is that $x_n \notin \mathrm{fv}(MNx_1\ldots x_{n-1})$, which makes ext applicable to $Mx_1\ldots x_n = Nx_1\ldots x_n$. We get $Mx_1\ldots x_{n-1} = Nx_1\ldots x_{n-1}$, and using the similar assumptions for $x_{n-1},\ldots,x_1$ successively, we obtain $M = N$, by $n-1$ further applications of the rule ext.

To prove that the inclusion between the two relations holds in the other direction, let us assume that $M = N$. If the proof of $M = N$ does not involve ext, then we know that $M =_w N$, by theorem 5.2.3. Then, by the Church–Rosser theorem, there is a CL-term P such that with no x's supplied $M \rhd_w P$ as well as $N \rhd_w P$. If $M = N$ is by an application of ext, then $Mx = Nx$ with x not free in either M or N is provable in $EQ^{\eta}_{\mathfrak{B}_1}$. Then, by the hypothesis of the induction, $Mxy_1\ldots y_n \rhd_w P$ for some P such that $Nxy_1\ldots y_n \rhd_w P$ too. However, the two latter weak reductions are sufficient to show that there are $x_1\ldots x_n$ (for some n) and there is a P such that $Mx_1\ldots x_n \rhd_w P$ as well as $Nx_1\ldots x_n \rhd_w P$. qed

We return to inequational and equational calculi in sections 7.1.1 and 7.2.1, where we consider calculi for dual and symmetric CL.

Chapter 6

Models

In the previous chapters, we looked at properties of CL as a formal system. The results that we mentioned can be obtained about CL without assigning a "meaning" to the CL-terms. Indeed, we have not given a *precise interpretation* for CL-terms, though we often alluded to the informal idea that they all stand for functions.

An informal interpretation of the combinators and the application operation is useful, and hardly dispensable. However, rigorous interpretations are always desirable, and they are often preferable to informal interpretations. The precision that a rigorous interpretation provides can yield *new insights* about the formal system itself. Indeed, the first functional models confirmed that the notion of function application can reasonably include the self-application of functions.

The first models for combinatory logic were constructed from the *language* of CL itself; hence, they are called *term models*. These models are closely related to the algebra of CL, and can be viewed as the canonical algebras in a class of algebras that are representations for CL. The *algebras* that can serve as models of CL are sometimes called operational models.

The most commonly used functions are numerical. At first, it might seem unlikely that functions on the set of natural numbers could interpret CL-terms. However, using a suitable encoding of functions (on natural numbers) by natural numbers, a *functional model* may be defined. Natural numbers have a lot of structure to them, not all of which is exploited in this model. The features that are needed in the construction of a model is distilled into the concept of *domains*. A certain class of functions provides another model in which all CL-terms are interpreted as functions. Notably, neither sort of functional models precludes self-application.

Implicational formulas can interpret CL-terms, which yields a concrete operational model. Implicational formulas, in turn, can be viewed as simple types. The connection between formulas and CL-terms is further developed in interpretations of typed CL. The type constructor $\rightarrow$ is an implication but not in the sense of classical logic. In the first instance, simple types are linked to the implicational fragment of intuitionistic logic, and to implicational fragments of various relevance logics. Nonclassical logics are often given a relational (or "possible worlds") semantics. CL is a nonclassical logic itself, in which function application is a binary operation similar to multiplication or fusion. The paradigm of relational semantics is applicable to CL — even though CL does not contain conjunction- and disjunction-like connectives.

6.1 Term models

Models for a formal system may be built from entities of any sort, which does not mean that all the models are equally "pleasing," "useful" or "informative." Some models are also easier to come up with.

The components of the language in which a system is formulated are readily available; thus it is natural to ask the question whether it is possible to build a model from the bits and pieces of the language itself. The objects in CL are terms, hence these kinds of models are called *term models*.

Of course, it is always possible to interpret a CL-term by itself. However, such an interpretation would be quite uninteresting. The other extreme would be to interpret all the CL-terms as one particular term, or as the set of all terms. This would be a rather uninteresting model too. An idea, that is more fruitful, is to interpret the terms so that the interpretation is neither obvious nor trivial, and some *information* about the behavior of the terms or about the relationship between them is captured.

Between the syntactic identity relation and the total relation, there are many possible relations on the set of CL-terms. Weak reduction and weak equality are relations that immediately come to mind. An advantage is that they are not some arbitrary relations in the sense that CL characterizes them; that is, we do not need to look for some independent and new description of the relation.

Weak equality is an *equivalence relation*, whereas weak reduction is only a *preorder* (when the formulation of antisymmetry is thought to contain syntactic identity rather than $=_{\mathrm{w}}$).[1] Equivalence relations facilitate the creation of classes of objects that are similar from the point of view of that relation.

DEFINITION 6.1.1. The *equivalence class* generated by the CL-term M is denoted by $[M]$. The set $[M]$ is defined as follows.

(1) $[M] = \{N\colon M =_{\mathrm{w}} N\}$.

The binary *application operation* on equivalence classes, denoted by juxtaposition, is defined as

(2) $[M][N] = \{PQ\colon P \in [M] \text{ and } Q \in [N]\}$.

It is obvious that each CL-term generates an equivalence class with respect to $=_{\mathrm{w}}$, to which the term itself belongs, because of the reflexivity of $=_{\mathrm{w}}$. The consistency of CL means that there is *more than one* equivalence class. That is, for any CL-term M, there is an N, such that $[M] \neq [N]$. We also recall that $N \in [M]$ is *not decidable*, because $=_{\mathrm{w}}$ is not decidable.

In CL, there are no bound variables, hence the interpretation may be defined without separating the interpretation of variables from that of constants.

[1] See section A.6 for definitions.

DEFINITION 6.1.2. (INTERPRETATION OF CL-TERMS) The *interpretation function I* is defined inductively according to the structure of CL-terms.

(1) $I(x) = [x]$, when x is a variable;

(2) $I(\mathsf{S}) = [\mathsf{S}]$ and $I(\mathsf{K}) = \mathsf{K}$, when $\{\mathsf{S},\mathsf{K}\}$ is the combinatory basis;

(3) $I(MN) = [M]\,[N]$, when M and N are CL-terms.

The interpretation takes an atomic CL-term to the equivalence class that is generated by that term, whereas the application operation on terms is interpreted by the application operation on equivalence classes. We have to show that the latter is, indeed, an operation on equivalence classes. The latter operation is not affected by the choice of a particular term from an equivalence class.

LEMMA 6.1.3. *If M and N are* CL*-terms, then $[M]\,[N]$ is an* equivalence class *of* CL*-terms.*

Proof: Let $P \in [M]$ and $Q \in [N]$, that is, $P =_{\mathrm{w}} M$ and $Q =_{\mathrm{w}} N$. Then $PQ =_{\mathrm{w}} MN$, or to state it differently, $[M]\,[N] = [MN]$. qed

The construction using equivalence classes and an operation that is defined from an operation on the objects in the equivalence classes is quite usual. Sometimes it is called the algebraization of a logic, and the resulting structure is often called the Lindenbaum algebra of the logic. In the case of classical logic, the equivalence classes emerge from the relation of mutual provability between formulas. For instance, A and $(A \wedge (A \vee A))$ are provably equivalent, thus they generate the same equivalence class: $[A] = [(A \wedge (A \vee A))]$. Further, $[A] \wedge [B] = [A \wedge B]$, where $\wedge$ on the equivalence classes is defined by type-lifting $\wedge$ from formulas.

Algebraically speaking, the set of CL-terms is an algebra with a binary operation, and $=_{\mathrm{w}}$ is a *congruence* relation in this algebra. A congruence is an equivalence relation, which is in harmony with the operations on the elements of the equivalence classes: the result of the operation is independent of the concrete representatives chosen from the equivalence classes. To put it in yet another way, replacement holds with respect to the equivalence relation.

The result of an application of I to the set of CL-terms is an algebra, and it behaves — somewhat unsurprisingly — as an interpretation is expected to behave.

LEMMA 6.1.4. (ADEQUACY) *If $M =_{\mathrm{w}} N$ in* CL*, then $I(M) = I(N)$. Conversely, if $I(M) = I(N)$, then $M =_{\mathrm{w}} N$.*

Proof: We will use the equational calculus with S and K from chapter 5.[2] First, we prove the first claim and then the second one.

1. The subcases dealing with id, tr and sym are practically obvious. (If you do not find them so, then you should write out the details.) Weak equality is a congruence

[2]Here, too, the combinatory basis could be varied, and it only has to be ensured that clauses that concern the combinatory constants in definition 6.1.2 match the elements of the combinatory basis.

due to the m_l and m_r rules. Let us assume that $M =_w N$ and P is a CL-term. Then, by the definition of I (i.e., clause (3) in definition 6.1.2), $I(MP) = [M][P]$, which is the same set of terms as $[MP]$. However, $NP =_w MP$, and so $NP \in [MP]$, therefore, $[NP] = [MP]$. The next series of equalities shows that $I(MP) = I(NP)$: $I(MP) = [M][P] = [MP] = [NP] = [N][P] = I(NP)$. Showing that the claim is true for the combinatory axioms is similar, and we leave the details as an exercise.
2. Let us assume that $M \neq_w N$. Then $[M] \neq [N]$, by the definition of equivalence classes. It is easy to verify that for all M, $M \in I(M)$, indeed, $I(M) = [M]$. However, $M \notin [N]$ and $N \notin [M]$, therefore, $I(M) \neq I(N)$ as we intended to show. qed

Exercise 6.1.1. Finish step **1** from the above proof.

This construction and the resulting algebra is pivotal in thinking about the meaning of CL-terms. As long as we intend to model CL with $=_w$, we can take this algebra and interpret it. There is a difference in the type level between the elements of the algebra comprising CL-terms and the latter algebra comprising equivalence classes of CL-terms. However, composing an interpretation of the algebra of CL with the map $M \mapsto [M]$ always recovers the corresponding interpretation for the terms themselves.

The other advantage gained by isolating the algebra of CL is that this algebra is *prototypical* for all the algebras that can be taken as interpretations of CL, which is the topic of the next section.

6.2 Operational models

The algebra of CL with $=_w$ suggests that we consider a class of algebras that are characterized by the equations that are true in the algebra of CL. We will call algebras in this variety *combinatory algebras*.

DEFINITION 6.2.1. (COMBINATORY ALGEBRAS) Let $\mathfrak{A} = \langle A; \cdot, \mathsf{S}, \mathsf{K} \rangle$ be an algebra of similarity type $\langle 2, 0, 0 \rangle$ in which (1)–(2) hold. (We use $a, b, c, \ldots$ for arbitrary elements of the algebra, and we omit parentheses from left-associated algebraic terms.)

(1) $\mathsf{S} \cdot a \cdot b \cdot c = a \cdot c \cdot (b \cdot c)$;

(2) $\mathsf{K} \cdot a \cdot b = a$.

The simplicity of this definition (compared to the equational calculus) is due to the tacit assumptions that the notion of an algebra implies. In particular, $=$ is supposed to be reflexive, transitive and symmetric without especially being mentioned. The replacement property is also assumed with respect to each operation (of positive arity) in the algebra. Lastly, the variables in the equations are presumed to be universally quantified; that is, the equations hold for arbitrary elements a, b and c, of the algebra.

We used S and K in the definition; however, the class of algebras in this variety includes further algebras beyond the algebra of CL. Thus S and K are *not* the well-known combinators, rather two distinguished elements of the algebra that have a certain similarity to the combinators S and K in CL (what we emphasized by using the same notation).

DEFINITION 6.2.2. An *interpretation* of CL into a combinatory algebra $\mathfrak{A}$ is a function I that satisfies conditions (1)–(3).

(1) $I(\mathsf{S}) = \mathsf{S}$ and $I(\mathsf{K}) = \mathsf{K}$;

(2) $I(x) \in A$, when x is a variable;

(3) $I(MN) = I(M) \cdot I(N)$.

The interpretation function *fixes* the meaning of the combinatory constants as the distinguished elements of the algebra. Otherwise, the application operation of CL is turned into the binary operation of the combinatory algebra; the variables are interpreted as (arbitrarily chosen) elements of the algebra.

The identity function satisfies the above conditions, and is an interpretation of CL into a combinatory algebra. This interpretation may seem uninteresting at first, but it is very useful.

LEMMA 6.2.3. *If* $M =_{\mathrm{w}} N$ *in* CL, *then* $I(M) = I(N)$ *in* every *combinatory algebra* $\mathfrak{A}$ *with any interpretation* I.

Proof: The definition of a variety, as we already pointed out, incorporates parts of the equational calculus for CL, such as the properties of $=$ and its interaction with the binary operation $\cdot$. Then, our proof shortens to verifying that the claim holds for the combinatory axioms. However, that is obvious, when the basis is $\{\mathsf{S}, \mathsf{K}\}$. qed

Combinatory algebras from the class in definition 6.2.1 look simple; at least their definition is quite simple. Nonetheless, they have remarkable properties beyond including the algebra of CL.

Recall that an algebra $\mathfrak{A}$ is *trivial* if and only if its carrier set A contains exactly one element. The triviality of the algebra of a logic corresponds to the absolute inconsistency of that logic.

LEMMA 6.2.4. *If a combinatory algebra* $\mathfrak{A}$ *is not trivial, then it is* not Abelian. *That is, if the* $\cdot$ *operation of* $\mathfrak{A}$ *is commutative, then* A *is a singleton set.*

Proof: We will prove the second formulation of the claim of the lemma. Let us assume that $\cdot$ is commutative. We start with the term $\mathsf{K} \cdot \mathsf{S} \cdot a \cdot \mathsf{K} \cdot (\mathsf{K} \cdot (\mathsf{K} \cdot b)) \cdot c$, which equals to $\mathsf{S} \cdot \mathsf{K} \cdot (\mathsf{K} \cdot (\mathsf{K} \cdot b)) \cdot c$. The latter term is simply an (overly complicated) alternative of the definition of the identity combinator (when we squint and view S and K as combinators) applied to c. Indeed, $\mathsf{S} \cdot \mathsf{K} \cdot (\mathsf{K} \cdot (\mathsf{K} \cdot b)) \cdot c = c$.

Let us now start with the same term $\mathsf{K} \cdot \mathsf{S} \cdot a \cdot \mathsf{K} \cdot (\mathsf{K} \cdot (\mathsf{K} \cdot b)) \cdot c$, and use the commutativity of $\cdot$ for the fourth occurrence of $\cdot$. This gives us the term $\mathsf{K} \cdot (\mathsf{K} \cdot b) \cdot (\mathsf{K} \cdot \mathsf{S} \cdot a \cdot$

$\mathsf{K}) \cdot c$. Having applied the characteristic equation for K to this term, we get $\mathsf{K} \cdot b \cdot c$, which further equals to b.

In sum, $c = b$, though b and c are arbitrary elements of the algebra, which means that $A = \{c\}$. qed

Of course, the lemma *does not* mean that combinators that are permutators are forbidden. S is a permutator itself — in addition to being a duplicator and an associator.

Exercise 6.2.1. Suppose — for the sake of simplicity — that T is a constant in a combinatory algebra (like the combinator T).[3] That is, $\mathsf{T} \cdot a \cdot b = b \cdot a$. Check that the steps in the proof cannot be repeated with $\mathsf{T} \cdot (\mathsf{K} \cdot \mathsf{S} \cdot a \cdot \mathsf{K}) \cdot (\mathsf{K} \cdot (\mathsf{K} \cdot b)) \cdot c$ or $\mathsf{T} \cdot \mathsf{K} \cdot \mathsf{S} \cdot a \cdot \mathsf{K} \cdot (\mathsf{K} \cdot (\mathsf{K} \cdot b)) \cdot c$ when the commutativity of $\cdot$ is not assumed.

Exercise 6.2.2. Find another term of a combinatory algebra (preferably, a shorter and simpler term than the one in the proof) to prove the same claim that Abelian combinatory algebras are trivial.

LEMMA 6.2.5. *If a combinatory algebra $\mathfrak{A}$ is not trivial, then its operation $\cdot$ is* not associative.

Exercise 6.2.3. Prove the lemma. (Hint: Recall that a binary operation is associative, when for all elements a, b and c, $(a \cdot b) \cdot c = a \cdot (b \cdot c)$.)

LEMMA 6.2.6. *If a combinatory algebra $\mathfrak{A}$ is not trivial, then its carrier set A is* not finite.

Proof: Let us assume that a combinatory algebra $\mathfrak{A}$ has a carrier set with elements $a_1, \ldots, a_n$, where $n > 1$ and $n \in \mathbb{N}$. We show that if so, then there are a_i and a_j such that $i \neq j$, but $a_i = a_j$ in $\mathfrak{A}$. First, we show that $n = 2$ is impossible.

1.1 Let us consider the elements a_1 and a_2. We may choose a_1 to be K; then $\mathsf{S} = a_1$ or $\mathsf{S} = a_2$. In the first case, we consider the term $\mathsf{S} \cdot (\mathsf{K} \cdot \mathsf{K}) \cdot \mathsf{K} \cdot a_1 \cdot a_2 \cdot a_2$. The assumption about $\mathsf{S} = a_1 = \mathsf{K}$ means that this term equals $\mathsf{K} \cdot (\mathsf{K} \cdot \mathsf{K}) \cdot \mathsf{K} \cdot a_1 \cdot a_2 \cdot a_2$. Using repeatedly the equations governing S and K, we arrive at the equalities

$$a_1 = \mathsf{S} \cdot (\mathsf{K} \cdot \mathsf{K}) \cdot \mathsf{K} \cdot a_1 \cdot a_2 \cdot a_2 = \mathsf{K} \cdot (\mathsf{K} \cdot \mathsf{K}) \cdot \mathsf{K} \cdot a_1 \cdot a_2 \cdot a_2 = a_2,$$

which mean that $a_1 = a_2$.

1.2 If $\mathsf{S} = a_2$, then we choose other terms with the intention to show that $\mathsf{K} = \mathsf{S}$, that is, $a_1 = a_2$. First, let us consider $\mathsf{K} \cdot \mathsf{K}$, which is a term in the language of the algebra, hence it denotes an element. Let us assume that $\mathsf{K} \cdot \mathsf{K} = a_1$. Then we start with the term $\mathsf{S} \cdot (\mathsf{K} \cdot \mathsf{K}) \cdot (\mathsf{K} \cdot \mathsf{K}) \cdot a_1 \cdot a_2 \cdot a_2 \cdot a_1$. Since $\mathsf{K} \cdot \mathsf{K} = \mathsf{K}$, the term equals to $\mathsf{S} \cdot \mathsf{K} \cdot \mathsf{K} \cdot a_1 \cdot a_2 \cdot a_2 \cdot a_1$. However, if $\mathsf{K} \cdot \mathsf{K} = \mathsf{K} = a_1$, then $a_1 = \mathsf{K} \cdot (\mathsf{K} \cdot (\mathsf{K} \cdot \mathsf{K}))$. Applying the equations for S and K, we get that the former term equals a_2, whereas

[3]There is a term in every combinatory algebra comprising occurrences of S, K and $\cdot$, which defines this constant. For example, $\mathsf{S} \cdot (\mathsf{K} \cdot (\mathsf{S} \cdot (\mathsf{S} \cdot \mathsf{K} \cdot \mathsf{K}))) \cdot (\mathsf{S} \cdot (\mathsf{K} \cdot \mathsf{K}) \cdot (\mathsf{S} \cdot \mathsf{K} \cdot \mathsf{K}))$ is such a term.

the latter equals a_1. The other possibility is that $\mathsf{K}\cdot\mathsf{K}=a_2$, that is, S. We start with the term $\mathsf{K}\cdot\mathsf{K}\cdot(\mathsf{K}\cdot\mathsf{K})\cdot\mathsf{K}\cdot(\mathsf{K}\cdot\mathsf{K})\cdot a_2\cdot a_2\cdot a_2$ and we replace the first two occurrences of $\mathsf{K}\cdot\mathsf{K}$ with S. Then the latter term equals a_2, that is, S, whereas the former equals K (i.e., a_1). In both subcases, we have shown that $a_1=a_2$, which shows that $n=2$ implies $n=1$.

2. Now we show that if $\mathfrak{A}$ has (no more than) $n+1$ elements, then $\mathfrak{A}$ has (no more than) n elements. Let $\mathfrak{A}$ have $n+1$ elements, which we label by $a_1,\ldots,a_{n+1}$. $\mathsf{S},\mathsf{K}\in A$; hence, without loss of generality, we may assume that $a_1=\mathsf{K}$. If $\mathsf{S}=a_1$ too, then we can proceed as in case **1.1**. Otherwise, we stipulate that $\mathsf{S}=a_2$. Again, this choice does not affect the generality of the proof. We note that there are terms of arbitrarily large finite length such as, for example, $\mathsf{K}\cdot\mathsf{K}$, $\mathsf{K}\cdot(\mathsf{K}\cdot\mathsf{K})$, etc. These terms exist in the language of $\mathfrak{A}$ independently of the size of the carrier set of $\mathfrak{A}$. Let $\mathsf{K}^{(n)}$ be defined inductively as follows.

(1) $\mathsf{K}^{(1)}$ is K;

(2) $\mathsf{K}^{(n+1)}$ is $(\mathsf{K}\cdot\mathsf{K}^{(n)})$, when $n\in\mathbb{N}^+$.[4]

In the algebra of CL, the corresponding combinators $\mathsf{K}^{(n)}$ have arity $n+1$, and $\mathsf{K}^{(n)}x_1\ldots x_nx_{n+1}\rhd_{\mathrm{w}}x_n$.

To prove that, for some elements a_i and a_j (where $i\neq j$), we have $a_i=a_j$, it is sufficient to prove that $\mathsf{K}^{(i)}=\mathsf{K}^{(j)}$ implies $a_i=a_j$. Let us assume that $\mathsf{K}^{(i)}=\mathsf{K}^{(j)}$. Again, we may arbitrarily choose whether $i>j$ or $j>i$; let us suppose the former. Then we take i-many copies of the term $\mathsf{S}\cdot\mathsf{K}\cdot\mathsf{K}$ for the first i arguments of $\mathsf{K}^{(i)}$ and a_i for its $i+1$st argument. The term equals to $\mathsf{S}\cdot\mathsf{K}\cdot\mathsf{K}$, by applying i times the equation for K. Replacing $\mathsf{K}^{(i)}$ by $\mathsf{K}^{(j)}$, we get that the term equals to one or more copies of $\mathsf{S}\cdot\mathsf{K}\cdot\mathsf{K}$ followed by a_i, hence, to a_i. Starting with a similar term but with a_j in place of a_i, we get a sequence of equations: $a_i=\mathsf{S}\cdot\mathsf{K}\cdot\mathsf{K}=a_j$. qed

Exercise 6.2.4. Prove the claims we made in the proof about the $\mathsf{K}^{(n)}$'s, namely, that (a) each $\mathsf{K}^{(n)}$ has arity $n+1$, and (b) each $\mathsf{K}^{(n)}$ cancels all but its nth argument.

Exercise 6.2.5. Give a simpler proof of lemma 6.2.6. (Hint: Scrutinize the above proof.)

LEMMA 6.2.7. *If $I(M)=I(N)$ in every combinatory algebra, then $M=_{\mathrm{w}}N$ in* CL.

Proof: Let us assume that $M\neq_{\mathrm{w}}N$ in CL. We show that there is a combinatory algebra and an interpretation such that $I(M)\neq I(N)$. Then, by contraposition, the claim of the lemma follows. Consider the algebra that is formed by taking the equivalence classes of CL-terms, and by defining the application operation on them. Since $M\neq_{\mathrm{w}}N$, M and N do not belong to each other's equivalence class. We define I as the identity function, that is, each term is mapped into the equivalence class it generates. It is straightforward to verify that I is an interpretation of CL-terms into a combinatory algebra. Obviously, $I(M)\neq I(N)$. qed

[4]We have been omitting the outside parentheses from algebraic terms (just as we did for CL-terms). In clause (2), they are included to underscore that for $n>2$, $\mathsf{K}^{(n)}$ is right associated: $(\mathsf{K}\cdot(\mathsf{K}\cdot\ldots)\ldots)$.

Equations are easy to handle, and varieties are a well-understood kind of algebras. Moreover, they have important closure properties, for instance, homomorphic images of algebras from a variety belong to the same variety. Sometimes, it might appear that the language of a logic does not contain a connective that leads to an equational algebraization, but there is a way to define equality anyway. The situation is different when we look at CL with $\triangleright_{\mathrm{w}}$ (rather than $=_{\mathrm{w}}$). There is no conjunction or disjunction like connective that would allow us to introduce $=$, and of course, if we focus on $\triangleright_{\mathrm{w}}$, then we do not wish to define equivalence classes via $=_{\mathrm{w}}$.

It is possible to algebraize CL with $\triangleright_{\mathrm{w}}$, that is, to consider the algebra of the inequational calculus for CL. However, the resulting algebra is not characterized by equations. The idea is to place into the same equivalence class not the terms that are weakly equal to each other, but only those terms that *weakly reduce* to each other. For instance, $\mathsf{WI}(\mathsf{WI})$ and $\mathsf{I}(\mathsf{WI})(\mathsf{WI})$ are terms that are not only weakly equal, but also reduce to each other.

DEFINITION 6.2.8. A *combinatorial reduction algebra* is a partially ordered algebra $\mathfrak{A} = \langle A; \leq, \cdot, \mathsf{S}, \mathsf{K} \rangle$ of similarity type $\langle 2,0,0 \rangle$, when the inequations in (1)–(2), and the quasi-inequations in (3)–(4) are true, for all $a,b,c \in A$.

(1) $\mathsf{S} \cdot a \cdot b \cdot c \leq a \cdot c \cdot (b \cdot c)$;

(2) $\mathsf{K} \cdot a \cdot b \leq a$;

(3) $a \leq b$ implies $c \cdot a \leq c \cdot b$;

(4) $a \leq b$ implies $a \cdot c \leq b \cdot c$.

A partially ordered algebra also carries certain tacit assumptions, namely, $\leq$ is a partial order. Sometimes partially ordered algebras are defined to contain everywhere monotone operations. However, we may assume a more general definition, what may be viewed to be the definition of a *tonoid*. (We made the tonicity of $\cdot$ explicit in (3) and (4).[5])

The notion of an interpretation is as in definition 6.2.2, with S, K and $\cdot$ belonging to a combinatorial reduction algebra now.

LEMMA 6.2.9. *If $M \triangleright_{\mathrm{w}} N$ in* CL, *then $I(M) \leq I(N)$ in* every *combinatorial reduction algebra under any interpretation function I.*

Proof: The proof is similar to the proof of lemma 6.2.3. The combinatory axioms for S and K match the inequations in (1) and (2) in definition 6.2.8. The two other conditions are easily seen to hold in view of $\triangleright_{\mathrm{w}}$ being a preorder on the set of CL-terms. The inequational formalization of CL includes the assumptions of an ordered algebra save the antisymmetry of the order relation; thus, the rest of proof is not difficult, and it is left to the reader. qed

[5]For a discussion of notions of ordered algebras and the definition of tonoids, see [65, §§3.9–10].

The algebraization of CL with respect to $\triangleright_{\mathrm{w}}$ is appropriate, as the following lemma shows.

LEMMA 6.2.10. *If $I(M) \leq I(N)$ is every combinatorial reduction algebra under any interpretation function I, then $M \triangleright_{\mathrm{w}} N$.*

Proof: We prove the contraposition of the statement. Let us assume that M does not reduce to N. Let us define $[N] = \{M \colon N \triangleright_{\mathrm{w}} M \wedge M \triangleright_{\mathrm{w}} N\}$. In words, we use in place of an equivalence relation the relation of mutual weak reducibility to generate equivalence classes of CL-terms. From the assumption and the definition, we get that $M \notin [N]$. We have for any CL-term P, that $P \in [P]$. Thus I can be taken as the identity function on terms, or a bit more precisely, $I(P) = [P]$. $I(MN) = [M][N]$, hence the type-lifted application operation may be taken for $\cdot$. It is not difficult to check that $\mathfrak{A} = \langle \{[M] \colon M \text{ is a CL-term }\}; \leq, \cdot, [\mathsf{S}], [\mathsf{K}] \rangle$, where $\leq$ is the obvious partial order on the set of equivalence classes is a combinatorial reduction algebra. Then $I(M) = [M] \nleq [N] = I(N)$ is immediate. qed

Exercise 6.2.6. Verify in detail that the algebra $\mathfrak{A}$ as defined in the above proof is indeed a combinatorial reduction algebra as we claimed.

The model that we describe next is closely related to typed-CL as well as to the canonical construction in the relational models. Indeed, this model can be viewed as a preeminent example of a combinatorial reduction algebra.[6]

In chapter 9, we introduce in detail several typed versions of CL. At this point we do not intend to introduce a formal type system; however, we would like to appeal to the connection between implicational formulas (i.e., simple types) and certain combinators. Implicational formulas are inductively generated from a set of propositional variables by the binary connective $\rightarrow$, which is called implication.

Example 6.2.11. Let us assume that Z^4 is a combinator with axiom $\mathsf{Z}xyzv \triangleright z(xyv)$. We could, perhaps, view Z as an implicational formula so that taking the application operation as modus ponens, $\mathsf{Z}xyzv$ would yield the same formula that $z(xyv)$ does.

Let us take, in place of Z, $(D \rightarrow A \rightarrow B) \rightarrow D \rightarrow (B \rightarrow C) \rightarrow A \rightarrow C$. We take the antecedents of this formula for x, y, z and v, respectively. To get the formula that corresponds to xyz, we detach D and then A from $D \rightarrow A \rightarrow B$ getting B. z is $B \rightarrow C$, and so C is the wff corresponding to the term on the right-hand side, and of course, C is the formula that remains after detaching the antecedents from the formula standing for Z.

After this quick example, we introduce a *partial operation* that takes two formulas $A \rightarrow B$ and C, and returns a formula (if certain conditions can be met), which is the result of a detachment of C' from $A' \rightarrow B'$, where the $'$'s indicate a certain amount of substitution that is sufficient for modus ponens to become applicable. There are pairs

[6] The model comes from Meyer et al. [114], and there it is called "the fool's model." This label might be stunning, but it is an allusion to some ideas of Anselm of Canterbury from the 11th century, and the motivation is more fully explained in the paper itself.

of formulas $A \to B$ and C such that no amount of substitution can make detachment achievable; hence, the operation is partial.[7]

DEFINITION 6.2.12. (MGCI) Let A and B be formulas that do not share any propositional variables. The wff C is a *most general common instance* (*mgci*, for short) of A and B, when C is a substitution instance of both formulas, and all C''s that are common instances of A and B are substitution instances of C.

Mgci's are described in a somewhat complicated fashion, but they are quite simple.

Example 6.2.13. Let the two formulas be p and q. Then p is an mgci of these formulas, so is q. Indeed, any propositional variable would do, but no wff that contains $\to$ is a mgci of them.

The example illustrates why we may use a definite article with mgci's: they all look the same, that is, they are each other's substitution instances when the substitutions are restricted so that they assign distinct propositional variables to distinct propositional variables. Such formulas are called *alphabetic variants* of each other, and the procedure of generating them is called *relettering*.

Example 6.2.14. Let the two formulas be $(p \to p) \to q$ and $s \to (r \to (s \to r))$. In this case, neither formula can be taken to be the mgci, because one of the formulas has an implication in its antecedent with the other having a variable, and the other way around for the consequents. The idea would be to take $r \to (s \to r)$ for q and $p \to p$ for s, but it is not sufficient to use the substitution $[q/r \to (s \to r), s/p \to p]$ on the formulas, and not only because we would end up with formulas that still do not share all their variables: $(p \to p) \to (r \to (s \to r))$ and $(p \to p) \to (r \to ((p \to p) \to r))$. If we compare the resulting formulas, then we can easily see why the purported substitution does not work. We stipulated to change s to $p \to p$; however, we also substituted a wff with an occurrence of s for q.

$[q/r \to ((p \to p) \to r), s/p \to p](p \to p) \to q$ is $(p \to p) \to (r \to ((p \to p) \to r))$, and so is $[q/r \to ((p \to p) \to r), s/p \to p]s \to (r \to (s \to r))$.

Notice that two wff's that have no common propositional variables either have *infinitely many* mgci's or *none*.

Exercise 6.2.7. Create an example with two formulas so that they have no mgci.

Given a pair of formulas, it is always possible to *decide* whether they have an mgci, that is, if they can be *unified*.

The next definition has an algorithmic flavor. In chapter 9, we give a more descriptive definition (that is equivalent to the present one).

DEFINITION 6.2.15. (CONDENSED DETACHMENT) Let $A \to B$ and C be the starting formulas.

[7] Definition 9.1.37, in chapter 9, is an alternative definition of this partial operation.

1. Let C' be a relettering of C so that C has no variables in common with $A \to B$.

2.1. Let D be a mgci of A and C', that is obtained by the substitution σ.

2.2. If there is no mgci for A and C', then there is no formula that results from $A \to B$ and C by condensed detachment.

3.1. $\sigma(B)$ is the result of the condensed detachment of C from $A \to B$.

By the numbering of the steps, we intend to indicate how the algorithm may proceed: 1, 2.1, 3.1 or 1, 2.2.

Example 6.2.16. Let us suppose that we are given the formulas $p \to q \to p$ and $((p \to q) \to p) \to p$. Both of them are implications, hence we could take either for the major premise of modus ponens. We surely have to reletter one of them though. $((r \to s) \to r) \to r$ is a relettering of the latter formula, which is often called *Peirce's law*.

Let the first formula be the major premise. Its antecedent is a propositional variable; hence, we simply have to substitute the second wff for p: $(((r \to s) \to r) \to r) \to q \to ((r \to s) \to r) \to r$. The condensed detachment is successful in the sense that it results in a formula, namely, $q \to ((r \to s) \to r) \to r$.

Now let us reverse the roles of the formulas. Then we want to find a common substitution for $(r \to s) \to r$ and $p \to q \to p$. We saw in example 6.2.14 that, in principle, it is not an obstacle to finding a common instance that the formulas have compound formulas in different places in their antecedent or consequent. But these two formulas have more structure than the previous ones, because of the repeated occurrences of some propositional variables in them. It appears that we have to make $p = r \to s$ and $r = q \to p$. (The constrains obtained from an attempted unification are often written in the form of equations. We follow this practice here using the "$=$" symbol in the meta-language.) The equational form lure us into replacing r in the first equation according to the second one: $p = (q \to p) \to s$. We could proceed the other way around, and get an equation for r: $r = q \to r \to s$. Either way, we end up with an equation in which a variable occurs on one side of the equation and it also occurs on the other side — but there as a proper subformula. However, this situation is excluded, which means that the condensed detachment of the so-called positive paradox from Peirce's law does not yield a formula.

Mgci's are not unique in the sense that any relettering of an mgci is an mgci. In the case of condensed detachments, which are successful, the use of the definite article in "the result" is to be understood similarly. Strictly speaking, there is no unique formula that results, because the renaming of variables is allowed, and we have not defined a deterministic algorithm that would produce for each formula exactly one relettering. However, we wish to stress that a relettering is a very special kind of substitution, which *does not identify* distinct variables and *does not increase* the complexity of the formula.

Certain combinators are associated with implicational formulas and certain implicational formulas are associated with combinators. (The matching though is not total for either set.) Section 9.1 gives a detailed motivation, in the form of a natural deduction calculus, for type assignment. The natural deduction calculus shows how this association between formulas and combinators emerges straightforwardly. Suffice it to say here that, if we are thinking about the combinatory basis $\{\mathsf{S},\mathsf{K}\}$, then the formulas that are assigned to some combinators are exactly the theorems of the implicational fragment of intuitionistic logic. If we consider other bases, especially, bases that are not combinatorially complete, then the set of implicational formulas is a proper subset of those theorems, for example, it may be the set of theorems of the implicational fragment of a relevant logic.

The slogan "formulas-as-types" originated in the context of typing combinators and λ-terms. Here we do not assign formulas to combinators (or in general, to CL-terms) using a proof system; rather, we will interpret CL-terms as sets of formulas. The next definition emphasizes the latter point of view.

DEFINITION 6.2.17. (SETS OF SUBSTITUTION INSTANCES) Let A be an implicational formula. The set of all *substitution instances* of A is denoted by $\{A\}_s$. That is, $\{A\}_s = \{B\colon B \text{ is } \sigma(A)\}$, where σ is a substitution.

There is no standard notation for sets of formulas that are substitution instances of a wff — except by resorting to the use of meta-variables. We added the subscript $_s$ to avoid confusion with the singleton set containing A. We give a list of combinators and sets of formulas associated to them.

Combinator	Set of wff's
S	$\{(p \to q \to r) \to (p \to q) \to p \to r\}_s$
S'	$\{(p \to q) \to (p \to q \to r) \to p \to r\}_s$
K	$\{p \to q \to p\}_s$
B	$\{(p \to q) \to (r \to p) \to r \to q\}_s$
B'	$\{(p \to q) \to (q \to r) \to p \to r\}_s$
W	$\{(p \to p \to q) \to p \to q\}_s$
I	$\{p \to p\}_s$
C	$\{(p \to q \to r) \to q \to p \to r\}_s$
T	$\{p \to (p \to q) \to q\}_s$
CII	$\{((p \to p) \to q) \to q\}_s$
KI	$\{p \to q \to q\}_s$
J	$\{(p \to q \to q) \to p \to q \to p \to q\}_s$

The set $\{p \to q \to p\}_s$ contains all the reletterings of the formula, and all the reletterings of $p \to p \to p$, as well as all formulas that are obtained by substitutions that increase the complexity of the formula. For instance, $(q \to q) \to r \to q \to q$,

$q \to (q \to q) \to q$ and $(p \to r \to s) \to (r \to q \to q) \to p \to r \to s$ are all elements of $\{p \to q \to p\}_s$.

It so happens that some formulas, which have been given names during the historical development of logic, are principal types of some combinators that we introduced earlier. We gave a list of combinators and sets of formulas without any motivation. However, a clear motivation emerges from typed CL, what we introduce in chapter 9.[8]

We have listed nearly all the combinators that appeared in chapter 1. A combinator that is missing is M. WI defines M, and if we think about W, that is, *contraction*: $(p \to p \to q) \to p \to q$, as the major premise and I, that is, *self-identity*: $p \to p$, as the minor premise in an application of condensed detachment, then we can see that there should be no wff associated to M.[9]

There is a "mismatch of levels" between the rule condensed detachment, which is applied to pairs of formulas, and the combinators, which are sets of formulas. However, we can define an operation on sets of formulas that is derived from condensed detachment and is similar to condensed detachment.

DEFINITION 6.2.18. Let X and Y be sets of wff's. The *condensed detachment* of X and Y, denoted by $X \cdot Y$, is defined as

$$X \cdot Y = \{C \colon \exists A \in X\, \exists B \in Y.\, \mathbf{D}(A,B) = C\},$$

where $\mathbf{D}(A,B)$ is the wff (if exists) that is obtained by condensed detachment from A and B.

We call this operation condensed detachment too, because no confusion can arise — precisely because the two operations are applicable to different kinds of objects. This definition does not assume that either set of formulas, X or Y, comprises substitution instances of some single formula.

Example 6.2.19. Let $X = \{(p \to p \to q) \to p \to q\}$ and $Y = \{p \to p, p \to q \to p\}$. Then $r \to r \in X \cdot Y$, and this happens not because $p \to p \in Y$, but because $\mathbf{D}((p \to p \to q) \to p \to q, p \to q \to p) = r \to r$.

LEMMA 6.2.20. *Let us consider $\{A\}_s$ and $\{B\}_s$. $\{A\}_s \cdot \{B\}_s = \{C\}_s$, provided that the condensed detachment of the two sets is not empty.*

Proof: Let us assume that $E \in \{A\}_s \cdot \{B\}_s$. Then there are $C_1 \to C_2 \in \{A\}_s$ and $D \in \{B\}_s$ such that $E = \mathbf{D}(C_1 \to C_2, D)$. We may assume that $C_1 \to C_2$ and D do not share any propositional variables, because the reletterings of these wff's are also elements of the sets. Hence we can suppose that the reletteringin the application of condensed detachment is relegated to the choice of elements from $\{A\}_s$ and

[8] For instance, the notion of principal types is introduced in definition 9.1.23. Roughly speaking, the principal type of a combinator is the generator of a principal cone in the set of wff's when substitution is viewed as an ordering.

[9] To be more precise, *no purely implicational* wff should get attached to M. See section 9.2 for types with further constants.

$\{B\}_s$. By the definition of condensed detachment, there is a substitution σ_1 such that $\sigma_1(C_1) = \sigma_1(D)$, and E is $\sigma_1(C_2)$.

Now, if there are substitution instances of A and B that permit a "successful" application of the condensed detachment rule, then condensed detachment is successful on A and B themselves. The A is $A_1 \to A_2$ and $\sigma_0(A_1 \to A_2) = C_1 \to C_2$. If $\mathbf{D}(A,B) = F$, then $\sigma_2(A_1) = \sigma_2(B)$ and $\sigma_2(A_2) = F$. $\sigma_2(A_1)$ is the mgci of A_1 and B; therefore, $\sigma_1(C_1)$ is a substitution instance of $\sigma_2(A_1)$. Similarly, $\sigma_1(D)$ is a substitution instance of $\sigma_2(B)$. If $\sigma_3(\sigma_2(A_1)) = \sigma_1(C_1)$, then $\sigma_3(\sigma_2(A_2)) = \sigma_1(C_2)$, which means that $E = \sigma_3(F)$. qed

The situation in the proof may be depicted as follows.

$$
\begin{array}{ccccc}
\mathbf{D}(A_1 \to A_2, B) & = & F & \xleftarrow{\sigma_2} & A_2 \\
 & & \sigma_3 \big\downarrow & & \big\downarrow \sigma_0 \\
\mathbf{D}(C_1 \to C_2, D) & = & E & \xleftarrow[\sigma_1]{} & C_2
\end{array}
$$

Exercise 6.2.8. The above proof depends on the fact that the substitutions compose into substitutions. Prove this claim. (Hint: Substitutions are total functions on the set of propositional variables.)

DEFINITION 6.2.21. The *fool's model* for CL is $\mathfrak{M}_{\mathfrak{F}} = \langle T, \subseteq, \cdot \rangle$, where $T = \{\{A\}_s : A$ is a wff$\} \cup \{\emptyset\}$, $\subseteq$ is set inclusion and $\cdot$ is condensed detachment on sets of formulas. An *interpretation* I is a function from the set of CL-terms into T such that the interpretation of a combinator Z is $\{\mathrm{pt}(\mathsf{Z})\}_s$.

This model is not as abstract as a combinatorial reduction algebra where the elements and the operation are not circumscribed beyond the requirement that they satisfy (1)–(4). By the notation T, we intend to allude to "types;" pt(Z) (if exists) is a concrete formula and the $\cdot$ operation is also quite concrete.

LEMMA 6.2.22. $\mathfrak{M}_{\mathfrak{F}}$ *is a* combinatorial reduction algebra.

Proof: **1.** $\langle T, \subseteq \rangle$ is obviously a partially ordered set. We have to verify that $\cdot$ is a total operation. Totality is unproblematic, because $\emptyset$ is included into T. Lemma 6.2.20 provides that a successful condensed detachment between elements of T is an element of T.

2. We will look at clause (2) now. Let us assume that a and b are $\{A\}_s$ and $\{B\}_s$, respectively. K is interpreted as $\{p \to q \to p\}_s$. We may assume that A does not contain p or q, and so by lemma 6.2.20, the result of condensed detachment gives the interpretation of $\mathsf{K} \cdot a$ to be $\{q \to A\}_s$. A similar reasoning leads us to conclude that $\mathsf{K} \cdot a \cdot b$ is $\{A\}_s$, that is, (2) holds.

If either a or b is interpreted as $\emptyset$, then the interpretation of $\mathsf{K} \cdot a \cdot b$ is $\emptyset$ too, and $\emptyset \subseteq X$, for any $X \in T$. (We leave the rest of the proof as an exercise.) qed

Exercise 6.2.9. Complete the previous proof. In other words, show that (1) and (3)–(4) from definition 6.2.8 hold in $\mathfrak{M}_{\mathfrak{F}}$. (Hint: (3) and (4) are nearly obvious.)

Step **2** in the above proof provides a clear indication of why the above model cannot be modified to yield a combinatory algebra in the sense of definition 6.2.1. The prototypical example of a pair of formulas such that the second cannot be unified with the antecedent of the first is $\langle (p \to p \to q) \to p \to q, r \to r \rangle$. This means that the same phenomenon occurs on sets of formulas — with the difference that condensed detachment is turned into a total operation on sets of implicational formulas, by the inclusion of $\emptyset$ into T. $\emptyset$ is the bottom element in T, that is, $\emptyset \subseteq X$, for all elements of T.

The addition of the bottom element to T helps to make $\cdot$ a *total* operation on T. However, at the same time, the addition of the bottom element allows us to raise the question whether the operation is *normal* in the presence of the bottom element. The answer is that the operation $\cdot$ is normal, that is, $X \cdot \emptyset = \emptyset = \emptyset \cdot X$. This is not really a surprise, because the operation on sets is defined by an *existentially* quantified formula involving condensed detachment on formula, which is known not to be a total operation (in view of the example we just mentioned). The somewhat unfortunate consequence of normality is that $\mathsf{K} \cdot a \cdot b \geq a$ cannot be guaranteed in this model. If b happens to be interpreted by $\emptyset$, then we get $\emptyset$ on the left-hand side of the inequation. But a's interpretation in no way depends on b's interpretation, that is, the set on the right-hand side of the $\geq$ need not be empty.

LEMMA 6.2.23. *There is a* fool's model *for* CL *with a basis* $\mathfrak{B}$ *that comprises proper combinators without a cancellator, when each* $\mathsf{Z}_i \in \mathfrak{B}$ *has a principal type associated to it.*

Proof: Let $\mathfrak{B}$ be $\{\mathsf{Z}_1, \ldots, \mathsf{Z}_n\}$, with $\mathrm{pt}(\mathsf{Z}_1), \ldots, \mathrm{pt}(\mathsf{Z}_n)$ being the associated implicational formulas. $\mathfrak{M}_{\mathfrak{F}} = \langle T; \cdot \rangle$ with the interpretation function I such that $I(\mathsf{Z}_i) = \{\mathrm{pt}(\mathsf{Z}_i)\}_s$ is a model of CL with basis $\mathfrak{B}$.

The construction is clearly *not* a combinatory algebra, not even a combinatorial reduction algebra. Because of the stipulation that there are no cancellators in $\mathfrak{B}$, K, for example, is neither an element of $\mathfrak{B}$ nor definable from $\mathfrak{B}$. However, we have the following for any interpretation function. If $\mathsf{Z}_i^m x_1 \ldots x_m \rhd M$, then $I(\mathsf{Z}_i x_1 \ldots x_m) = I(M)$.

The lack of combinatorial completeness precludes these $\mathfrak{M}_{\mathfrak{F}}$'s from satisfying condition (2) in definition 6.2.1, and (1) may not hold either depending on the concrete $\mathfrak{B}$. However, for each $\mathfrak{B}$ we could define a suitably modified notion of a combinatory algebra with a restricted set of distinguished elements and corresponding equations involving them.

The assumption that the implicational formula associated to Z_i is the principal type of the combinator ensures that $I(\mathsf{Z}_i) \cdot I(x_1) \cdot \ldots \cdot I(x_m) = I(M)$ whenever the $I(x_j)$'s are not empty.[10]

We have to scrutinize what happens if some of the $I(x_j)$'s are empty. Let $I(x_l) = \emptyset$, for instance. Then $I(\mathsf{Z}_i) \cdot I(x_1) \cdot \ldots \cdot I(x_l) = \emptyset$, hence all further applications of $\cdot$

[10]In chapter 9, we discuss principal types in detail, and we also show that there is a close connection between the $E \to$ rule and condensed detachment. Here we take it for granted that the principal type of a combinator ensures this equation when the condensed detachment on formulas is successful.

result in $\emptyset$. On the other hand, x_l occurs in M, because Z_i is not a cancellator. Let x_lN be a subterm of M. (The case when Nx_l is a subterm of M is similar.) Then $I(x_l)\cdot I(N)=\emptyset$, and if there are further applications of $\cdot$, then those yield $\emptyset$ too. That is, $I(M)=\emptyset$. It may be somewhat uninteresting that the result is $\emptyset$, but $\emptyset$ results on both sides, which is sufficient for the equality to hold. qed

The "fool's model" interprets CL without types, though the types figure into the interpretation itself. Just as not all combinators are typable with simple types, as we will see in detail in section 9.1, not all combinators are interpreted here by a nonempty set of types. Typability of all CL-terms may be achieved by adding conjunction and constant truth as type constructors, and then the model can be turned into a model of $=_{\mathrm{w}}$. We return to models involving types in section 6.5.

6.3 Encoding functions by numbers

Abstract mathematical structures such as abstract algebras are excellent interpretations of CL — whether with $=_{\mathrm{w}}$ or $\triangleright_{\mathrm{w}}$. However, it may be desirable to have more *concrete interpretations* for CL, which is, after all, one of the general theories of functions. Chapter 3 has shown how to use various combinators to represent functions such as the constant zero function, projections, successor, and further on, addition, multiplication, etc.

Theoretically speaking, it might not be completely definite what natural numbers are, but practically, we all seem to be very comfortable to deal with natural numbers. They are wholesome, some of them are triangular or square numbers, and some of them are even perfect. Therefore, it is especially pleasing that CL can be interpreted using natural numbers — with just a bit of structure added.

The set of subsets of the set of natural numbers is often denoted by $\mathcal{P}(\mathbb{N})$ or by $\mathcal{P}(\omega)$.[11] The model of CL, that we present in this section, is derivatively denoted by $\mathcal{P}\omega$ or $\mathcal{P}_\omega$, and called the *graph model*.

As any power set of some set, $\mathcal{P}(\mathbb{N})$ is a *Boolean algebra*, hence a *lattice*, moreover a *complete* lattice.[12] This suggests — in the light of the next section — that $\mathcal{P}(\mathbb{N})$ may be a suitable set to define a model on.

The power set of $\mathbb{N}$ contains many ($2^{\aleph_0}$-many, to be precise) sets, but all they can be easily categorized into two types of sets: *finite* and *infinite*. Not only infinite sets are "much bigger" than finite sets — they contain infinitely many more elements — but $\mathcal{P}(\mathbb{N})$ itself is too large to be matched with the set of all finite subsets of $\mathbb{N}$. The next definition places a topology on $\mathbb{N}$. For the sake of completeness, we included the definition of a topology into section A.6.

---

[11] ω is an alternative notation for $\mathbb{N}$, which may imply that the set of natural numbers is taken with its usual order $\leq$ as a poset, or with $<$ as an ordinal.

[12] For definitions of these concepts, see section A.6.

DEFINITION 6.3.1. (SCOTT TOPOLOGY ON $\mathbb{N}$) Let the set of *open sets* be denoted by $\mathcal{O}_S$, which is defined by the basis comprising the set of increasing sets generated by a finite subset of $\mathbb{N}$. Clauses (1) and (2) together define $\mathcal{O}_S$.

(1) $s^{\uparrow} = \{X \subseteq \mathbb{N} \colon s \subseteq X\}$ is an *increasing set* that is generated by a finite subset, whenever s is finite.

(2) $\mathcal{O}_S = \tau(\{s^{\uparrow} \colon s \in \mathcal{P}(\mathbb{N}) \wedge s \text{ is finite}\})$.

The idea behind choosing the finite subsets of $\mathbb{N}$ as the elements generating the cones of subsets of $\mathbb{N}$ is that a finite subset represents some *finite* amount of *positive information*. The basis has cardinality $\aleph_0$, because there are only so many finite subsets of $\mathbb{N}$.

Finite subsets also allow an alternative characterization of *continuity*.

DEFINITION 6.3.2. (CONTINUITY) Let f be a function of type $\mathcal{P}(\mathbb{N}) \longrightarrow \mathcal{P}(\mathbb{N})$. f is *continuous* whenever

$$f(X) = \bigcup\{f(s) \colon s \subseteq X \wedge s \text{ is finite}\}.$$

The value of a continuous function (in the just defined sense) is completely determined by the values the function takes on finite sets that are subsets of its argument. Roughly speaking, this means that with continuous functions there is no need to worry about infinite subsets of $\mathbb{N}$ in calculations.

In the context of topologies, which contain open and closed sets, functions that ensure that the inverse image of an open set is open are of particular interest. These functions are called *continuous*, and it is remarkable that the functions that are continuous according the above definition are continuous, and vice versa. (See definition A.6.9 and theorem A.6.10.)

Exercise 6.3.1. Prove the claim, that is, theorem A.6.10.

The finite subsets of $\mathbb{N}$ can be *enumerated*, and the particular enumeration chosen is crucial to the whole composition of the model. For the workings of the model, the enumeration of the finite subsets of $\mathbb{N}$ should be in harmony with the enumeration of ordered pairs over the natural numbers.

DEFINITION 6.3.3. (ENUMERATION OF $\mathbb{N} \times \mathbb{N}$) Consider the natural numbers placed on points with integer coordinates in the top-right quadrant of the coordinate system as illustrated below. There is an obvious 1–1 correspondence between the sets $\mathbb{N}$ and $\mathbb{N} \times \mathbb{N}$, which is the *enumeration of ordered pairs of natural numbers* that we use.

We picture only the $\langle x,y\rangle$ points with both coordinates nonnegative. The pairs in the leftmost column and in the bottom row are to be thought to be on the y- and x-axes, respectively. (We omit the lines that would represent the axes.)

35							
27	34						
20	26	33					
14	19	25	32				
9	13	18	24	31			
5	8	12	17	23	30		
2	4	7	11	16	22	29	
0	1	3	6	10	15	21	28 ...

The informal idea is to fill the quadrant with natural numbers via filling in the short diagonals bottom up. The ordered pairs are the pairs of $\langle x,y\rangle$ coordinates and they may be thought of as a function.

Example 6.3.4. The illustration above makes it easy to see the value of some of the ordered pairs with small numbers: $\langle 5,0\rangle = 15$ and $\langle 2,4\rangle = 25$. However, we might want to be able to have a more efficient way to calculate, let us say, $\langle 7,11\rangle$ than to draw a big picture of the coordinate system and fill in sufficiently many short diagonals.[13] The following formula is another way to calculate the value of a pair.

$$\frac{(x+y)(x+y+1)}{2}+y$$

Exercise 6.3.2. Verify for a couple of pairs from the above illustration that the formula gives the correct numbers. What is $\langle 7,11\rangle$?

Exercise* 6.3.3. Prove that the formula is a function and gives the number obtainable by successively filling the short diagonals. (Hint: The length of the short diagonals is 1, 2, $3,\ldots$ and the sum of the natural numbers up to n is $\frac{1}{2}n(n+1)$.)

The possibility to enumerate the pairs should not be surprising, though the elegance of this particular enumeration may be flattering. An important difference of this enumeration from the enumeration of the positive rational numbers, which we gave earlier, is that now we have an *invertible* function.[14] In other words, $\langle\, ,\, \rangle$ is an *injective function.*

Exercise 6.3.4. Describe an algorithm that for any natural number n, gives a pair $\langle x,y\rangle$, where x and y are the coordinates of the number n.

The finite subsets of $\mathbb{N}$ may be enumerated in various ways. The enumeration that we need will be calculated through the *binary* representation of numbers and using the identity function as an enumeration of $\mathbb{N}$.

[13]Drawing a big chart might not feel like computation, but it is an *effective procedure*, and it is a suitable method to determine the value of the $\langle\, ,\, \rangle$ function. In sum, it is computation.

[14]A sample enumeration of the positive rational numbers may be found in section 3.1. Although the idea behind the enumeration there is to fill in short diagonals, we did not aim at a bijection, and every positive rational occurs $\aleph_0$-many times in that whole enumeration.

Recall that the difference between the decimal and the binary representation of numbers is mainly what the places stand for: 101_{10} can be expanded as $1 \cdot 10^2 + 0 \cdot 10^1 + 1 \cdot 10^0$, whereas 101_2 is $1 \cdot 2^2 + 0 \cdot 2^1 + 1 \cdot 2^0$. To put it differently, in the decimal notation the places indicate (from right to left) $1, 10, 100, 1000, \ldots$, similarly, in the binary notation they stand for $1, 2, 4, 8, 16, \ldots$. Of course, the decimal notation needs and permits more symbols too, because $2, 3, \ldots, 9$ are all less than 10, whereas in binary notation only 0 and 1 are less than 2.

The natural numbers written in binary notation are just (finite) nonempty *strings* over the alphabet $\{0,1\}$. The identity enumeration of natural numbers is the same as the *lexicographic ordering* of all the binary strings that begin with 1.

Example 6.3.5. Let us consider the numbers $7, 8, 9, 10$ and 11. In binary notation these are the strings 111, 1000, 1001, 1010 and 1011. 7_2 comprises only 1's, however, shorter than any of the four other strings (i.e., 111 is only three symbols, whereas the others contain four symbols). 9_2 precedes 10_2, because in the former the second 1 from the left is the rightmost symbol unlike in 10_2. (Of course, we assume that the alphabet $\{0,1\}$ itself comes with the ordering $0 \leq 1$.)

Every binary string can be thought of as a characteristic function of a set. To turn each finite sequence of 0's and 1's into a characteristic function of a set, we can consider each string prefixed with a denumerable sequence of 0's. Now, we apply these functions defined on $\mathbb{N}$ — with the symbol at the end of the sequence (i.e., the rightmost symbol) being the first argument. Each padded string gives a finite subset of $\mathbb{N}$.

Example 6.3.6. 0_{10} looks the same as 0_2: 0. The infinite binary sequence corresponding to 0_2 is

$$\ldots 0000000000000000000000000000,$$

which means that the finite set s_0 has no elements, that is, $s_0 = \emptyset$.

A slightly more interesting example is $26_{10} = 11010_2$, which produces the sequence

$$\ldots 0000000000000000000000011010,$$

which, in turn, means the $0 \notin s_{26}$, neither are 2 or numbers strictly greater than 4 elements of s_{26}. Thus $s_{26} = \{1,3,4\}$.

The enumeration of the finite subsets may be pictured in a coordinate system too. The first 36 finite subsets are listed below.

$\{0,1,5\}$							
$\{0,1,3,4\}$	$\{1,5\}$						
$\{2,4\}$	$\{1,3,4\}$	$\{0,5\}$					
$\{1,2,3\}$	$\{0,1,4\}$	$\{0,3,4\}$	$\{5\}$				
$\{0,3\}$	$\{0,2,3\}$	$\{1,4\}$	$\{3,4\}$	$\{0,1,2,3,4\}$			
$\{0,2\}$	$\{3\}$	$\{2,3\}$	$\{0,4\}$	$\{0,1,2,3\}$	$\{1,2,3,4\}$		
$\{1\}$	$\{2\}$	$\{0,1,2\}$	$\{0,1,3\}$	$\{4\}$	$\{1,2,4\}$	$\{0,2,3,4\}$	
$\emptyset$	$\{0\}$	$\{0,1\}$	$\{1,2\}$	$\{1,3\}$	$\{0,1,2,3\}$	$\{0,2,4\}$	$\{2,3,4\}$

In the case of models that are constructed from terms, we do not have to worry about having a term of the form xx — at least once we have gotten accustomed to seeing such CL-terms. Similarly, there is no reason why a term $a \cdot a$ in an abstract algebra should raise eyebrows. However, if we want to define a function to represent application with $\mathcal{P}(\mathbb{N})$ as its domain, then somehow the definition should be flexible to capture self-application too. If we simply look at a function in a set-theoretic sense, then it is a set of ordered tuples, which is never applicable to itself.

Of course, every number may be seen as an ordered pair (even as an ordered n-tuple), and as a finite subset of natural numbers. The maps between the various data types are 1–1 correspondences, that is, they are total onto invertible functions, which means that one can pick which set of objects to take. Choosing one or the other set invites us — due to our familiarity with numbers and sets — to affix to them the operations, which are most frequently used with those kinds of objects: $+$, $/$ or $\cap$, $\cup$, etc.

Sets comprising different kinds of objects are completely exchangeable when the operations (or even the relations) on one of them have analogues on the other, that is, there is an *isomorphism* between the sets. The latter means that from a certain point of view there is no way to distinguish between those sets despite the fact that from some other angle the objects and operations differ. Part of representation theory and duality theorems concern finding 1–1 maps between sets of objects and between sets of operations on those objects.

To model CL, we have to find sets — certain subsets of $\mathbb{N}$ — to represent combinators, and to define an operation that behaves like the application operation in CL.

DEFINITION 6.3.7. ($\mathfrak{f}$ FROM A SUBSET) Let $X \subseteq \mathbb{N}$. $\mathfrak{f}(X)$ is of type $\mathcal{P}(\mathbb{N}) \longrightarrow \mathcal{P}(\mathbb{N})$, and for any $Y \subseteq \mathbb{N}$, gives

$$\mathfrak{f}(X)(Y) = \{m\colon \exists s_n.\, s_n \subseteq Y \wedge \langle n,m\rangle \in X\},$$

where s_n is a finite subset of $\mathbb{N}$.

$\mathfrak{f}$ is an operation on subsets of $\mathbb{N}$, which turns the subset X into a function. The role of the finite subsets is to ensure that $\mathfrak{f}$ is a continuous function.

LEMMA 6.3.8. *The function* $\mathfrak{f}\colon \mathcal{P}(\mathbb{N}) \longrightarrow (\mathcal{P}(\mathbb{N}) \longrightarrow \mathcal{P}(\mathbb{N}))$ *is* continuous.

Proof: Let us assume that $X \subseteq \mathbb{N}$, and $X = \bigcup_{i \in I} X_i$ for some index set I. By the definition of $\mathfrak{f}$, for any $Y \subseteq \mathbb{N}$, $\mathfrak{f}(X)(Y) = \{m\colon \exists s_n.\, s_n \subseteq Y \wedge \langle n,m\rangle \in X\}$. The usual definition of $\cup$, and the assumption about the X_i's gives that the latter set equals to $\bigcup_{i \in I}\{m\colon \exists s_n.\, s_n \subseteq Y \wedge \langle n,m\rangle \in X_i\}$. The expression after $\bigcup_{i \in I}$ defines $\mathfrak{f}(X_i)(Y)$, that is, we have $\bigcup_{i \in I} \mathfrak{f}(X_i)(Y)$. Therefore, $\mathfrak{f}(X) = \bigcup_{i \in I} \mathfrak{f}(X_i)$, which means — according to definition 6.3.2 — that $\mathfrak{f}$ is continuous, as we intended to show. qed

The possibility to consider a set of natural numbers as a function hints at the idea of using $\mathfrak{f}$ itself as the representation for application.

DEFINITION 6.3.9. Let X and Y be subsets of $\mathbb{N}$. The *application* of X to Y is $\mathfrak{f}(X)(Y)$.

Given the above definition, we can think of interpreting terms as sets of natural numbers. However, just as combinators are *special* CL-terms, so should be the sets that represent them. We start with the identity combinator I, which has as its axiom $\mathsf{I}x \rhd x$.

Example 6.3.10. The variable x can be interpreted as any subset of $\mathbb{N}$ — including $\emptyset$ and $\mathbb{N}$ itself. $\mathsf{I}\emptyset = \emptyset$. The only subset of $\emptyset$ is $\emptyset$, and this is s_0. Then from the definition of $\mathfrak{f}(\mathsf{I})(\emptyset)$, we get that $m \in \emptyset$ iff $s_0 \subseteq \emptyset \wedge \langle 0,m \rangle \in \mathsf{I}$. Of course, $m \in \emptyset$ is always false, hence no number, which looks like in its ordered pair form as $\langle 0,m \rangle$ can be an element of I. If we look back at the small initial fragment of the enumeration of the natural numbers in the coordinate system, then we see that $0,2,5,9,14,20,27,35 \notin \mathsf{I}$. (There are infinitely many more natural numbers that are not elements of I, but we did not picture those.)

If we exclude the leftmost column of natural numbers from I, then we may or may not have the right set for I. Perhaps, $\langle 1,1 \rangle \notin \mathsf{I}$ either.

Exercise 6.3.5. Find an $X \subseteq \mathbb{N}$ that can be used to show (via calculating $\mathfrak{f}(\mathsf{I})(X)$) that $\langle 1,1 \rangle \notin \mathsf{I}$.

To discover a more complete description of the set of numbers that belong to I, let us consider that $\mathfrak{f}(\mathsf{I})(Y)$ should be Y, for any $Y \subseteq \mathbb{N}$. Then we have that $m \in Y$ iff $\exists s_n . s_n \subseteq Y \wedge \langle n,m \rangle \in \mathsf{I}$. For each Y, $\emptyset \subseteq Y$, for each $m \in Y$, $\{m\} \subseteq Y$ and $Y \subseteq Y$ — possibly, with other subsets of Y in between. (Of course, Y may not be an s_n.)

If $s_n \subseteq Y$, then for all $m \in s_n$, $m \in Y$, hence it should be that $\langle n,m \rangle \in \mathsf{I}$. $Y = \mathbb{N}$ is a possibility, and then for all n, $s_n \subseteq Y$. Then, from the previous consideration, we get that

$$\mathsf{I} = \{\langle n,m \rangle : m \in s_n\}.$$

We could have thought that I should be the diagonal starting with $\langle 0,0 \rangle$, or perhaps, with a slight shift from $\langle 0,0 \rangle$. An initial segment of I contains the following pairs: $\langle 1,0 \rangle$, $\langle 2,1 \rangle$, $\langle 3,0 \rangle$, $\langle 3,1 \rangle$, $\langle 4,2 \rangle$, $\langle 5,0 \rangle$, $\langle 5,2 \rangle$, $\langle 6,1 \rangle$, $\langle 6,2 \rangle$, $\langle 7,0 \rangle$, $\langle 7,1 \rangle$, $\langle 7,2 \rangle$, $\langle 8,3 \rangle$, $\langle 9,0 \rangle$ and $\langle 9,3 \rangle$.

The corresponding numbers are not in increasing order, which is a consequence of the set of all finite subsets of $\mathbb{N}$ forming a poset that is not linearly ordered, whereas $\mathbb{N}$ is linearly ordered by its natural ordering. However, we can consider the same subset of I with the elements listed as $\{1,6,7,11,15,23,28,29,30,37,38,45,47,69,81\}$. From the definition of I above, it is clear that I is an *infinite* set of numbers, thus the listed elements constitute only a small subset of I.

The other combinators are also subsets of the set of natural numbers. Let us consider K. It should be interpreted by a set of natural numbers so that applied to any $X,Y \in \mathcal{P}(\mathbb{N})$, the result is X. If $\mathfrak{f}(\mathsf{K})(X)$ is applied to Y, then, by the definition of $\mathfrak{f}$, we get that the result is $\{m\colon \exists s_n . s_n \subseteq Y \wedge \langle n,m \rangle \in \mathfrak{f}(\mathsf{K})(X)\}$. But $m \in \mathfrak{f}(\mathfrak{f}(\mathsf{K})(X))(Y)$ iff $m \in X$, which suggests that $\mathfrak{f}(\mathsf{K})(X)$ should contain all pairs of the form $\langle n,m \rangle$ where $m \in X$, because Y, hence the s_n's are arbitrary. Now some of our visual expectations are vindicated: $\mathsf{K} \cdot X$ looks like a collection of constant functions; for each $m \in X$, a constant function taking the value m.

This gives an informal description of $\mathfrak{f}(\mathsf{K})(X)$, but we would like to get to the numbers that are in K. Although the pairs correspond to natural numbers, it is easier to keep them in their pair form for now. $\mathfrak{f}(\mathsf{K})(X)=\{\,j\colon \exists s_i.\, s_i\subseteq X\wedge\langle i,j\rangle\in\mathsf{K}\,\}$. We know that the j's are $\langle n,m\rangle$'s with $m\in X$. Again X need not be finite, but if it were, then we would get that K contains $\langle i,\langle n,m\rangle\,\rangle$ iff $m\in s_i$. But for any set we can look at its finite subsets, thus we get that

$$\mathsf{K}=\{\,\langle i,\langle n,m\rangle\,\rangle\colon m\in s_i\wedge n\in\mathbb{N}\,\}.$$

K is a cancellator, which is reflected by the "unrestricted" n.

Example 6.3.11. Let X be $\{2\}=s_4$ — for a simple example. $2\in s_4$, therefore, $\langle 4,\langle n,2\rangle\,\rangle\in\mathsf{K}$, for any n. Let us also assume that $Y=\{0,3\}=s_9$. The two subsets of s_4 are $\emptyset$ and $\{2\}$, that is, s_0 and s_4. (Since the empty set has no elements, K's definition excludes all the pairs of the form $\langle 0,j\rangle$.) $\mathfrak{f}(\mathsf{K})(s_4)=\{\,\langle n,2\rangle\colon n\in\mathbb{N}\,\}$.

The subsets of Y are s_0, s_1, s_8 and s_9, which select four pairs from $\mathfrak{f}(\mathsf{K})(s_4)$: $\langle 0,2\rangle$, $\langle 1,2\rangle$, $\langle 8,2\rangle$ and $\langle 9,2\rangle$. Then $2\in\mathsf{K}\cdot X\cdot Y$; indeed, 2 is the only element of this set. This simple example illustrates how a set of numbers can stand for K.

The triples that are elements of K can be viewed as natural numbers. For instance, $1,4,13$ and 34 are elements of K.

Exercise 6.3.6. Find a handful of other numbers that are elements of K. (Hint: The numbers may get large quickly depending on i and n.)

Giving a description of a set of numbers that represent I is, probably, the easiest, and K is not difficult either. The above examples and the definitions of the corresponding sets also provide hints as to the shape of these sets.

Exercise 6.3.7. Find some numbers that belong to S, then define the set that represents S in this model.

Once we have at least K and S, we have a model of CL.

DEFINITION 6.3.12. An *interpretation* of CL is a function I such that (1)–(3) holds.

(1) $I(x)\subseteq\mathbb{N}$;

(2) $I(\mathsf{K})=\mathsf{K}$ and $I(\mathsf{S})=\mathsf{S}$;

(3) $I(MN)=\mathfrak{f}(I(M))(I(N))$.

LEMMA 6.3.13. *If* $M\rhd_{\mathrm{w}}N$ *in* CL, *then* $I(M)\subseteq I(N)$, *for all interpretations* I.

Proof: The interpretation of $\rhd_{\mathrm{w}}$ by $\subseteq$ guarantees that the reflexivity and transitivity of $\rhd_{\mathrm{w}}$ holds. The combinatory axioms hold, because of the choice of the sets of numbers that represent them. The only remaining step is to ensure that $I(M)\subseteq I(N)$ implies $I(MP)\subseteq I(NP)$ as well as $I(PM)\subseteq I(PN)$.

Let us assume that $I(M)\subseteq I(N)$, and $I(P)=Y$. By the definition of $\mathfrak{f}$, $I(MP)=\{m\colon \exists s_n.\, s_n\subseteq Y\wedge\langle n,m\rangle\in I(M)\}$. If $m\in I(MP)$, then — assuming $s_n\subseteq Y$ —

$\langle n,m\rangle \in I(M)$, hence $\langle n,m\rangle \in I(N)$ and $m \in I(NP)$. $s_n \subseteq Y$ is a common expression in $I(MP)$ and $I(NP)$; therefore, this reasoning may be repeated for any $m \in I(MP)$. (For the last step of the proof, see the next exercise.) qed

Exercise 6.3.8. Finish the proof, that is, verify that $I(M) \subseteq I(N)$ implies $I(PM) \subseteq I(PN)$. (Hint: Notice that there is an "asymmetry" in the definition of $\mathfrak{f}$, which means that this case slightly differs from $I(MP) \subseteq I(NP)$.)

The application operation $\mathfrak{f}$ takes pairs of natural numbers in the first argument and depending on the second argument returns the second element of the pair. Although there is a 1–1 correspondence between the natural numbers and ordered pairs of them, the second coordinate is certainly not in 1–1 correspondence with the natural numbers.

Exercise 6.3.9. Show — using finite subsets of $\mathbb{N}$ — that $X \not\subseteq Y$ does not imply that $XZ \not\subseteq YZ$ for all Z. Show that $ZX \not\subseteq ZY$ does not follow either.

The model also can be taken to model weak equality.

LEMMA 6.3.14. *If $M =_{\text{w}} N$ in* CL, *then $I(M) = I(N)$ for all interpretation functions I.*

Proof: Some of the steps are very much like the steps in the proof for $\triangleright_{\text{w}}$, and so we skip them. Let us assume that $M =_{\text{w}} N$. By the hypothesis of induction, $I(M) = I(N)$. Further, let P be a CL-term that is interpreted by Y, that is, $I(P) = Y$. Then $\mathfrak{f}(Y)(I(M)) = \{m\colon \exists s_n. s_n \subseteq I(M) \wedge \langle n,m\rangle \in Y\}$. $I(M) = I(N)$ obviously implies that for all s_n that is a finite subset of $\mathbb{N}$, $s_n \subseteq I(M)$ iff $s_n \subseteq I(N)$. Hence $\{m\colon \exists s_n. s_n \subseteq I(M) \wedge \langle n,m\rangle \in Y\} = \{m\colon \exists s_n. s_n \subseteq I(N) \wedge \langle n,m\rangle \in Y\}$, and so $I(PM) = I(PN)$. (The rest of the proof is easy and left as an exercise.) qed

Exercise 6.3.10. Think about all the steps of the preceding proof. What did we omit? Give all the details of the omitted steps.

Exercise 6.3.11. Consider two subsets of $\mathbb{N}$, let us say, X and Y. Show that there are sets P and Q such that $\mathfrak{f}(P)(X) = \mathfrak{f}(P)(Y)$ and $\mathfrak{f}(Q)(X) = \mathfrak{f}(Q)(Y)$.

The model we outlined in this section is not the only model of CL that can be constructed from natural numbers and finite sets of natural numbers. The enumerations may be thought of as *encodings*, and it has been shown that such enumerations come in tandem, that is, they cannot be randomly paired. Also this particular encoding is not unique, though it may be the most often presented in the literature.

6.4 Domains

We often think of functions as operations on some set of numbers. The previous section showed that sets of natural numbers may be viewed as functions via viewing each number as a pair of natural numbers as well as a finite set of natural numbers. The set-theoretical understanding of functions is that they are sets of ordered tuples.

Obviously, an ordered pair is never identical to either of its elements, because they are different types of objects. On the other hand, 1–1 correspondences and isomorphisms can work magic: one set of objects can stand in for another, and the other way around. The concreteness of natural numbers meant that it was relatively straightforward to find or define suitable 1–1 correspondences. However, if we want to consider sets that might contain elements other that numbers, then correspondences have to be provided in a different way.

Domain theory, as it became called later, started with Scott's work in the late 1960s (see Scott [131]).[15] Some of the ideas behind this theory are similar to the ideas that appeared in the $\mathcal{P}\omega$ model. For example, finite amounts of information and continuous functions play a crucial role.

DEFINITION 6.4.1. (DIRECTED SUBSETS) Let $\langle D, \sqsubseteq \rangle$ be a *poset* (i.e., a set with a partial order $\sqsubseteq$). $X \subseteq D$ is a *directed* subset of D iff $X \neq \emptyset$ and for all $x_1, x_2 \in X$, there is a $y \in X$ such that $x_1 \sqsubseteq y$ and $x_2 \sqsubseteq y$. The *set of directed subsets of* D is denoted by $\mathcal{P}(D)^{\bar{\wedge}}$.

A directed set contains an *upper bound* for any pair of its elements. For instance, all finite subsets of $\mathbb{N}$ are bounded because the largest number in the set is an upper bound for all the elements (because $\mathbb{N}$ is totally ordered).

Example 6.4.2. Let us utilize the enumeration of finite subsets from the previous section. The set $\{s_1, s_2, s_3, s_4\}$ is *not* a directed set (with the partial order $\subseteq$), because $s_3 = \{0, 1\}$ and $s_4 = \{2\}$ are not subsets of each other or of the other elements. The pairs $\langle s_1, s_2 \rangle$, $\langle s_1, s_3 \rangle$ and $\langle s_2, s_3 \rangle$ have upper bounds in the set, but pairs with s_4 in it, do not. The situation could be remedied by throwing in the set s_7 (or one of the infinitely many other sets s_n, for which $s_7 \subseteq s_n$, like s_{23}).

Completeness has several meanings and usages. The next notion of completeness is closely related to the algebraic notion.

DEFINITION 6.4.3. (CPOS) Let $\langle D, \sqsubseteq \rangle$ be a poset. D is a *complete partial order*, a CPO, when (1) and (2) are true.

(1) There is a $\bot \in D$ such that for all $d \in D$, $\bot \sqsubseteq d$.

(2) If $X \in \mathcal{P}(D)^{\bar{\wedge}}$, then there is an $x \in D$ such that $\text{lub}(X) = x$, that is, x is the least upper bound of X in D.

It might seem that we are unfamiliar with CPOs, and some of them are indeed not exactly commonly used sets.

[15] The term "domain" is used for a set of objects that can be arguments of a function. Here this term is used not just to denote a set of objects, but also a certain structure on that set.

Example 6.4.4. The set $\mathbb{N}$ is not a CPO with the usual $\leq$ ordering. The reason is that *all* subsets of $\mathbb{N}$ are directed subsets, but no infinite subset of $\mathbb{N}$ has an upper bound. Consider $\mathbb{N}$ itself or the set $\mathbb{E}$ of odd numbers. Neither $\mathbb{N}$ nor $\mathbb{E}$ has an upper bound. However, the linear ordering of $\mathbb{N}$ ensures that $n \leq m$ or $m \leq n$ for all n and m, thus n or m is an upper bound of $\langle n, m \rangle$. Every infinite subset of natural numbers is unbounded, because the natural numbers are discretely ordered. Thus there is plenty of sets ($2^{\aleph_0}$-many), which are directed but fail to have an upper bound, let alone a least upper bound.

The example pinpoints the "problem" with $\mathbb{N}$. It is easy to turn the set upside down, that is, we can take $\langle \mathbb{N}, \geq \rangle$. Then every subset contains an upper bound, and the upper bound is — again due to the linear ordering — the least one in $\mathbb{N}$. This structure is still not a CPO though. Now the missing component is the bottom element. We can easily add ω with the stipulation that for all $n \in \mathbb{N}$, $n \leq \omega$ (and $\omega \leq \omega$ too). Under the converse of $\leq$, ω becomes the bottom element. Adding an extra element into a set in order to obtain a set bounded from below (or from above) is a frequently used trick — not only in the construction of a CPO.

Exercise 6.4.1. Check that adding ω does the trick by itself. That is, verify that $\langle \mathbb{N} \cup \{\omega\}, \leq \rangle$ is a CPO.

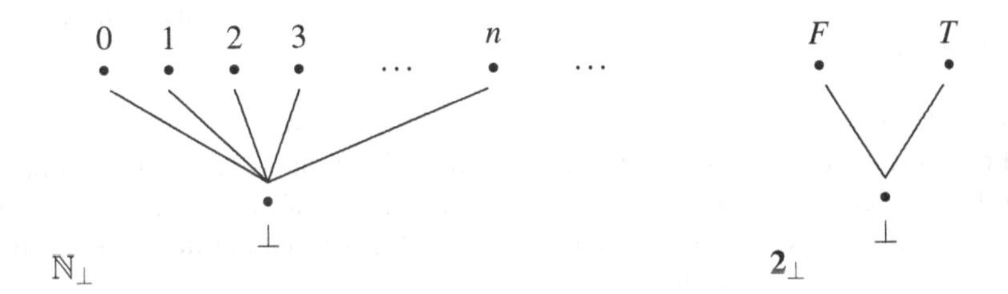

The natural numbers may be viewed with a little amendment as a CPO in a different way.

Example 6.4.5. The set $\mathbb{N} \cup \{\perp\}$ with $x \sqsubseteq y$ iff $x = \perp$ or $x = y$ is a CPO. The set $\mathbb{N}_\perp$ and its partial order are depicted above (on the left). Notice that the usual $\leq$ ordering is dropped now.

Another CPO that is constructed from a familiar structure in the same way is $\langle \{T, F, \perp\}, \sqsubseteq \rangle$. The capital letters T and F stand for the two (classical) truth values "true" and "false," and $\perp$ is an extra element added. This CPO is depicted above (on the right). Notice that the usual ordering $F \leq T$, (or $0 \leq 1$ in $\mathbf{2}$, the two element Boolean algebra) is omitted. In this CPO, the partial ordering can be interpreted as an *information ordering*, and then $\perp$ indicates the lack of information.

Recall that we said that the completeness here is related to the algebraic notion of completeness. The *power set* of any set is a lattice with $\cap$ and $\cup$ (intersection and union) being the meet and join. Moreover, $\mathcal{P}(X)$ is always a *complete* lattice, because intersections and unions of infinitely many sets exist in $\mathcal{P}(X)$. The partial order is $\subseteq$, and since $\mathcal{P}(X)$ is a lattice, another way to look at $\cup$ and $\cap$ as lub and glb (greatest lower bound). Power sets include a least element $\emptyset$, which can be taken to be $\bot$.

CPOs will be the sets of objects into which the objects of CL are interpreted. The objects of CL operate on each other, thus merely having a set of objects will not be sufficient as a model of CL.

DEFINITION 6.4.6. (MONOTONE AND CONTINUOUS FUNCTIONS) Let $\langle C, \sqsubseteq \rangle$ and $\langle D, \sqsubseteq \rangle$ be CPOs. A function f of type $f\colon C \longrightarrow D$ is *monotone* iff $c_1 \sqsubseteq c_2$ implies $f(c_1) \sqsubseteq f(c_2)$.

Let f be a monotone function from $\langle C, \sqsubseteq \rangle$ into $\langle D, \sqsubseteq \rangle$. f is *continuous* iff for all $X \in \mathcal{P}(C)^{\bar{\wedge}}$, $f(\text{lub}X) = \text{lub} f[X]$, where $f[X]$ is the image of the set X with respect to the function f.

We had definition 6.3.2 in the previous section that concerned continuity of a function on subsets of natural numbers. $\mathcal{P}(\mathbb{N})$ is a CPO, and so we might wonder whether there is a connection between the two definitions, or even the previous definition is a special instance of the new one.

LEMMA 6.4.7. *Let f be a function from the* CPO $\langle \mathcal{P}(\mathbb{N}), \subseteq \rangle$ *into* $\langle \mathcal{P}(\mathbb{N}), \subseteq \rangle$. *$f$ is* continuous *in the sense of definition 6.3.2* just in case *it is* continuous *in the sense of definition 6.4.6.*

Proof: We prove the "if and only if" statement in two steps.
1. From left to right, we start with showing that f is monotone if it is continuous according to definition 6.3.2. Let $X \subseteq Y$. For any finite subset s_n, if $s_n \subseteq X$, then $s_n \subseteq Y$. Thus if $f(s_n) \in f(X)$, then $f(s_n) \in f(Y)$. By the definition of $\cup$, then $f(X) \subseteq f(Y)$, as we had to show.

Beyond monotonicity f has to commute with lub on directed sets. Let then $\mathbb{X}$ be a directed set of subsets of $\mathbb{N}$. $\mathbb{X}$ is directed and lub$(\mathbb{X})$ exists in $\mathcal{P}(\mathbb{N})$, which is a CPO. We denote lub$(\mathbb{X})$ by Y; $Y \in \mathbb{X}$. We may assume that $\mathbb{X}$ comprises some X_i's: $\mathbb{X} = \{X_i \colon i \in I\}$, where I is an index set. From the stipulation that Y is least upper bound follows that $X_i \subseteq Y$, for each i. Then $f(X_i) \subseteq f(Y)$ by monotonicity, which ensures that $\text{lub} f[\mathbb{X}] \subseteq f(Y)$. If the converse inclusion fails to hold, then $f(Y) \subsetneq \text{lub} f[\mathbb{X}]$, which means, by the definition of union (that is the lub on $\mathcal{P}(\mathbb{N})$), that for some X_i, $f(X_i) \nsubseteq f(Y)$, but that is not the case. Thus $f(Y) = \text{lub} f[\mathbb{X}]$.
2. From right to left, we assume that f is continuous in the sense of definition 6.4.6. Let $X \subseteq \mathbb{N}$. We consider a directed set $\mathbb{X}$, namely, $\mathbb{X} = \{s_n \colon s_n \subseteq X\}$. lub$(\mathbb{X}) = X$, thus $f(X) = \text{lub} f[\mathbb{X}]$. Because of the shape of $\mathbb{X}$, this implies that $f(X) = \bigcup\{f(s_n) \colon s_n \subseteq X\}$, as we intended to prove. qed

Having established a connection with the previous model, we proceed with the more general structures. Not all CPOs are isomorphic to $\mathcal{P}(\mathbb{N})$, thus from now on we mean continuity in the sense of definition 6.4.6. In order to boost our understanding of continuous functions on CPOs, we look at some examples.

Example 6.4.8. Let C be **4** (see below), which is a CPO, and let D be an arbitrary CPO.[16]

All the *monotone* functions from C into D are continuous. Notice that if C is finite, then every directed set X contains its own lub. Thus, $f[\text{lub}(X)] = f(x)$ for some $x \in X$. Then it is obvious that the monotonicity of f implies that $f(x) = \text{lub}(f[X])$.

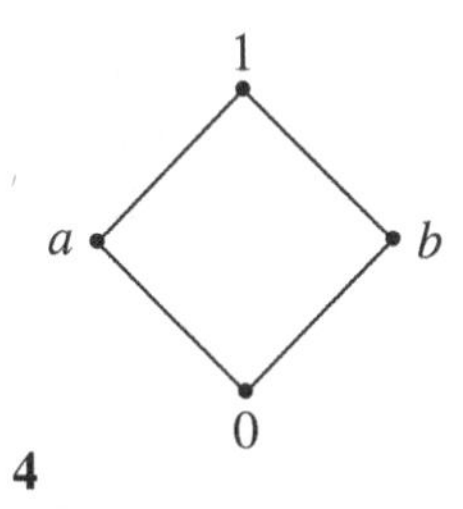

4

Example 6.4.9. We have seen above $\mathbb{N}$ (without its usual ordering $\leq$) supplanted with a $\bot$. Let D be again an arbitrary CPO. All *monotone* functions from $\mathbb{N}_\bot$ into D are continuous. There are many directed sets on $\mathbb{N}_\bot$, but the variation in their shape is very limited. A directed subset of $\mathbb{N}_\bot$ contains either one element, or two elements where one of the elements is $\bot$. Then continuity is obvious.

The next example shows that monotonicity, in general, is *not sufficient* to guarantee that the function is continuous.

Example 6.4.10. In example 6.4.1, we mentioned that topping $\mathbb{N}$ with an element, for instance, ω, creates a structure, which is a CPO. However, there are functions from $\mathbb{N} \cup \{\omega\}$ into a CPO D that are monotone but not continuous. For the sake of concreteness, let us take **2**, which a complete lattice, hence a CPO, in place of D. Let f be defined so that it separates the natural numbers from ω, that is, $f(n) = 0$ for each $n \in \mathbb{N}$, whereas $f(\omega) = 1$. The set of even numbers $\mathbb{E}$ is directed, and $f(\text{lub}(\mathbb{E})) = f(\omega) = 1$. On the other hand, $\text{lub}(f[\mathbb{E}]) = \text{lub}\{0\} = 0$.

Monotonicity is sufficient for continuity if there are *no ascending chains* — like $\mathbb{N}$ itself — in the CPO.

Exercise* 6.4.2. Prove the last statement. (Hint: An ascending chain is a directed set, which has to have a lub in the whole carrier set of the CPO.)

Lastly, there is a way to construct a CPO with a continuous function from any poset with a function.

Example 6.4.11. Let $\langle C, \sqsubseteq \rangle$ be a poset, and let f be of type $f : C \longrightarrow D$, where D is a set. $\mathcal{P}(C)$ as well as $\mathcal{P}(D)$ are CPOs, and there is a natural way *to lift* f into the

[16] **4** is a well-known distributive lattice that can be turned into a Boolean algebra or a De Morgan lattice depending on whether it is complemented by an order-inverting operation.

power sets. For $X \subseteq C$, let the function F be defined as

$$F(X) = \{f(c) \colon c \in X\}.$$

In other words, $F(X)$ is $f[X]$. F is easily seen to be *monotone*. F is also *continuous*, because lub is $\bigcup$ on power sets. Obviously, $F(\bigcup_{i \in I} X_i) = \bigcup_{i \in I} F(X_i) = \bigcup F[X_{i \in I}]$.

So far we looked at sets with an order and functions from or into those sets. To create structures, in which correspondences between sets and functions on those sets can be found, we need a partial order on functions too.

DEFINITION 6.4.12. (ORDERING OF FUNCTIONS) Let f and g be functions from the CPO $\langle C, \sqsubseteq \rangle$ into the CPO $\langle D, \sqsubseteq \rangle$.

$$f \sqsubseteq g \quad \text{iff} \quad \forall x \in C.\, f(x) \sqsubseteq g(x).$$

We have already used $\sqsubseteq$ without indices as the partial order of several CPOs, and now we added yet another use. Context determines uniquely which order is meant.

LEMMA 6.4.13. *The set of all continuous functions between a pair of* CPOs *is a* CPO.

Proof: **1.** $f(x) \sqsubseteq f(x)$ always holds, which means that $\sqsubseteq$ on functions is reflexive. Transitivity and antisymmetry can be established similarly from those properties for $\sqsubseteq$ on D together with the observation that the defining clause is universally quantified.
2. There is a function $\bot$ such that $\bot(c) = \bot$. This function is obviously continuous, and $\bot \sqsubseteq d$ for all d's in D.
3. Let us suppose that $F = \{f_i \colon i \in I\}$ is a directed set of continuous functions. We define g to be lub(F) as follows. For all $x \in C$, $g(c) = \text{lub}\{f_i(c) \colon i \in I\}$. $f_i \sqsubseteq g$ is immediate. Let a distinct function g' be a purported "real" least upper bound of F. Then $g' \sqsubseteq g$ but not the other way around. Therefore, there is a $c \in C$ such that $g'(c) \sqsubseteq g(c)$, but not $g(c) \sqsubseteq g'(c)$. Then there is a f_i such that $f_i \not\sqsubseteq g'$, which contradicts to g''s being an upper bound of F.

We also have to prove that the function g is continuous. Let $X \subseteq C$ and $X \in \mathcal{P}(C)^{\bar{\wedge}}$. lub$(X)$ exists in C, let us say, it is x. We have to show that $g(x) = \text{lub}(g[X])$. $g(x) = \text{lub}_{i \in I}\{f_i(x)\}$. Each f_i is continuous by our assumption about F, hence $f_i(x) = \text{lub}\, f_i[X]$, that is, $f_i(x) = \text{lub}_{j \in J} f_i(x_j)$, where J is the index set for X. Then we have $g(x) = \text{lub}_{i \in I} \text{lub}_{j \in J} f_i(x_j)$, which is the same as $\text{lub}_{j \in J} \text{lub}_{i \in I} f_i(x_j)$. However, $\text{lub}_{i \in I} f_i(x_j) = g(x_j)$, thus $g(x) = \text{lub}_{j \in J} g(x_j)$. The latter is merely another way of writing lub$(g[X])$, which means that g is continuous, as we wanted to prove. qed

In chapter 3, we saw that CL has the expressive power of computable functions, and one way to solve recursive equations is to apply the fixed point combinator. Theorem 2.3.1 is a fixed point theorem for monotone functions on complete lattices. The analogous theorem for CPOs is the following.

LEMMA 6.4.14. (FIXED POINTS IN CPOS) *Let f be a continuous function from the* CPO $\langle D, \sqsubseteq \rangle$ *into itself.* f *has* a least fixed point d_f *in* D, *that is,*

(1) $f(d_f) = d_f$, *and*

(2) *for all* $d' \in D$ *such that* $f(d') = d'$, $d_f \sqsubseteq d'$.

Proof: D has $\perp$ with the property that $\perp \sqsubseteq d$, and also $\perp \sqsubseteq f(\perp)$. f is monotone (because continuous), hence $f(\perp) \sqsubseteq f(f(\perp))$, and in general, $f^n(\perp) \sqsubseteq f^{n+1}(\perp)$. (The superscript indicates the iteration of f.) The set $X = \{f^n(\perp) \colon n \in \mathbb{N}\}$ is a directed set, thus its lub, d_f is in D, by the definition of a CPO. $f(\text{lub}(X)) = \text{lub}(f[X])$, because f is continuous. Notice that $\text{lub}(X) = \text{lub}(f[X])$, because of the way X is defined. That is, $f(d_f) = d_f$.

If d is a fixed point of f, then $f^{n+1}(d) = f^n(f(d)) = f^n(d)$, since $f(d) = d$. Also, $f^n(\perp) \sqsubseteq f^n(d)$, for any $d \in D$, including a d that is a fixed point of f. Thus d_f is indeed the least element in the set of fixed points of f. qed

A function might have more than one fixed point, but some do not have more than one. The least fixed point has a featured spot among all the potentially existing fixed points: its existence is guaranteed, by the above lemma. Working with the least fixed point provides a certain uniformity when moving from one CPO to another CPO.

There is quite a distance between the $\mathcal{P}\omega$ model from the previous section and arbitrary CPOs. In particular, the functions considered were determined by their behavior on *finite subsets* of $\mathbb{N}$.[17] The following definition generalizes this aspect of $\mathcal{P}\omega$ to CPOs that may not contain any natural numbers or their sets.

DEFINITION 6.4.15. Let $\langle D, \sqsubseteq \rangle$ be a CPO. $d \in D$ is a *finite element* iff for all $X \in \mathcal{P}(D)^{\bar{\wedge}}$, if $d \sqsubseteq \text{lub}(X)$, then there is a $y \in X$ such that $d \sqsubseteq y$. The *set of all finite elements* of D is denoted by **fin**.

To see the rationale behind the label "finite," let us consider $\langle \mathbb{N} \cup \{\omega\}, \leq \rangle$. A directed subset X is finite (with $\text{lub}(X) = \max(X)$) or infinite (with $\text{lub}(X) = \omega$). If $n \in \mathbb{N}$, then it is obvious that all directed subsets X with $\text{lub}(X) \geq n$ contain an element suitable for y. If we take ω, then $\omega \leq \text{lub}(\mathbb{N})$, however, no $n \in \mathbb{N}$ satisfies $\omega \leq n$. In sum, all the natural numbers are finite elements, whereas ω is not.

DEFINITION 6.4.16. Let $\langle D, \sqsubseteq \rangle$ be a CPO. D is *algebraic* iff for all $d \in D$, both (1) and (2) hold.

(1) $\{x \colon x \in \mathbf{fin} \wedge x \sqsubseteq d\} \in \mathcal{P}(D)^{\bar{\wedge}}$,

(2) $d = \text{lub}\{x \colon x \in \mathbf{fin} \wedge x \sqsubseteq d\}$.

In words, the CPO D is algebraic whenever the finite elements below each d form a directed set, the lub of which is d itself. Thinking about $\langle \mathbb{N} \cup \{\omega\}, \leq \rangle$ again for a moment, each $n \in \mathbb{N}$ clearly satisfies (1) and (2) above. $\mathbb{N}$ is the set **fin** in this CPO, and $\text{lub}(\mathbb{N}) = \omega$, as needed.

[17] Scott emphasized from early on the ideas of approximation and computing with finite pieces of information — see, e.g., Scott [131].

DEFINITION 6.4.17. (DOMAINS) Let $\langle D, \sqsubseteq \rangle$ be an algebraic CPO. D is *a domain* iff $\text{card}(\mathbf{fin}) = \aleph_0$, that is, D has $\aleph_0$-many finite elements.

This is a good place to note that the term "domain" is not used fully uniformly in the literature (beyond the common use of the term as the set of arguments of a function).[18] Our aim here is merely to recall how models for untyped CL can be obtained based on certain posets. There is an abundance of interesting results about domains; however, we focus our consideration on CPOs.

There are various ways to move between CPOs and to combine CPOs. Lemma 6.4.13 guarantees that the set of continuous functions from the CPO D into itself, that is sometimes denoted by $[D \longrightarrow D]$, is a CPO. Again, recalling an idea from $\mathcal{P}\omega$, we would like to set up a correspondence between D and $[D \longrightarrow D]$.

DEFINITION 6.4.18. (RETRACTION) Let $\langle C, \sqsubseteq \rangle$ and $\langle D, \sqsubseteq \rangle$ be CPOs. A pair of functions $\langle i, j \rangle$ is *a retraction* whenever $i \colon C \longrightarrow D$ and $j \colon D \longrightarrow C$, and $j \circ i = \text{id}_C$. That is, the function composition of i with j is the identity function on C. C is called *a retract* of D by i and j.

The requirement that the composition of i and j is identity implies that i is injective, that is, i embeds C into D. $j \upharpoonright i[C]$ is also injective, although j may not be injective on all of D. In general, $i \circ j \subseteq \text{id}_D$. Such a pair $\langle i, j \rangle$ of functions is an *embedding–projection* pair, which is an alternative terminology used instead of "retraction" adopted from category theory.

Having the above definition we want to define concretely the retraction when the two CPOs are a CPO and the space of its continuous functions. We take $\mathbb{N}_\perp$ to be D_0, and denote by D_{n+1} the continuous functions of D_n. That is, for any n, we set $D_{n+1} = [D_n \longrightarrow D_n]$. All the D_n's are CPOs, since we started with a CPO.

Exercise 6.4.3. Prove that, if we start with $\mathbb{N}_\perp$, then all the function spaces $[D_n \longrightarrow D_n]$ are domains (not just CPOs).

DEFINITION 6.4.19. ($\langle i_n, j_n \rangle$) The *retractions* $\langle i_n, j_n \rangle$ between the D_n's are defined inductively. The elements of the nth retraction have types $i_n \colon D_n \longrightarrow D_{n+1}$ and $j_n \colon D_{n+1} \longrightarrow D_n$. (The subscripts show which D the element belongs to.[19])

(1) $i_0(d_0) = \lambda x_0. d_0$;

(2) $j_0(f_1) = f_1(\perp_0)$;

(3) $i_{n+1}(d_{n+1}) = i_n \circ d_{n+1} \circ j_n$;

(4) $j_{n+1}(f_{n+2}) = j_n \circ f_{n+2} \circ i_n$.

---

[18] For instance, in the volume Gierz et al. [73], a domain is not required to have $\aleph_0$-many finite elements (which are called way-below elements there).

[19] The first glance at the subscripts might suggest that something is wrong with (4). However, it should be kept in mind that only the retraction is defined here; the d's and f's are already defined.

In clause (1), we used λ which should be understood as a symbol in the meta-language rather than in the λ-calculus. The content of this clause is simply that every element of D_0 is taken into D_1 as the *constant function* with that element as its value. (We could have used here combinators informally, and write $\mathsf{K}d$ instead.)

The second clause specifies that for a function from D_1 we take the element that the function yields on $\perp_0$, which is the least element of D_0. Each function in D_1 is continuous, hence monotone, and so the effect of this clause is taking the least value each function in D_1 takes as the representative of that function in D_0. The following picture shows the moves.

$$\begin{array}{ccccc} \mathbb{N}_\perp & = & D_0 & d_0 & f_1(\perp_0) \\ & & i_0 \downarrow \uparrow j_0 & \downarrow & \uparrow \\ [D_0 \longrightarrow D_0] & = & D_1 & \lambda x.d_0 & f_1 \end{array}$$

Exercise 6.4.4. Verify that i_0 and j_0 are continuous functions.

Given $d \in D_{n+1}$, the corresponding function $i_{n+1}(d)$ of type $D_n \longrightarrow D_n$ is the composition of j_n with d and then with i_n. j_n takes an element e of D_{n+1} into an element $j_n(e)$ of D_n, which makes d applicable to that element. The value $d(j_n(e))$ — due d's type — is an element of D_n, hence, i_n maps that element into an element of D_{n+1}.

The definition of j_{n+1} involves a similar transition between D_n's. $j_{n+1}(f_{n+2})$ should give a function in D_{n+1}, which takes arguments from D_n. To calculate $j_{n+1}(f_{n+2})(d_n)$, we insert i_n, that is, $i_n(d_n)$. The latter is an element of D_{n+1}, and so f is applicable. The result is in the same D_{n+1}; an application of j_n takes that element into an element in D_n as needed. The last two clauses may be visualized as follows.

$$\begin{array}{ccc} D_{n+1} & d_{n+1} & j_n \circ f_{n+2} \circ i_n \\ i_{n+1} \downarrow \uparrow j_{n+1} & \downarrow & \uparrow \\ D_{n+2} & i_n \circ d_{n+1} \circ j_n & f_{n+2} \end{array}$$

The functions defined are all *continuous*. To prove this claim inductively, it is useful to prove first the following lemma.

LEMMA 6.4.20. *Continuous functions on* CPOs *compose into* a continuous function.

Proof: Let f and g be continuous functions, $f\colon C \longrightarrow D$ and $g\colon D \longrightarrow E$. We show that $g \circ f$ is continuous. Let $X \in \mathcal{P}(C)^{\overline{\wedge}}$. Then $f(\mathrm{lub}(X)) = \mathrm{lub}(f[X])$. The continuity of f implies its monotonicity, which in turn, implies that $f[X] \in \mathcal{P}(D)^{\overline{\wedge}}$. $\mathrm{lub}(f[X]) \in D$, and by g's continuity, $g(\mathrm{lub}(f[X])) = \mathrm{lub}(g[f[X]])$. However, $g[f[X]] = \{e\colon \exists x \in X. e = g \circ f(x)\}$. $g(\mathrm{lub}(f[X])) = g(f(\mathrm{lub}(X)))$, which is further $g \circ f(\mathrm{lub}(X))$. To sum up, we have that for a directed subset X of C, $g \circ f(\mathrm{lub}(X)) = \mathrm{lub}(g \circ f[X])$. qed

LEMMA 6.4.21. (CONTINUITY OF RETRACTIONS) *For all n, each of the elements of $\langle i_n, j_n \rangle$ is* continuous.

Proof: The proof is by induction on n, which is the index of the functions in the retraction. Exercise 6.4.4 concerned the base case, $n = 0$. To show that i_{n+1} is continuous, we appeal to the preceding lemma and the definition of i_{n+1}. There are three functions in the definition: $i_n\colon D_n \longrightarrow D_{n+1}$, $j_n\colon D_{n+1} \longrightarrow D_n$ and $d_{n+1}\colon D_n \longrightarrow D_n$. All these functions are between CPOs, and d_{n+1} is known to be continuous. i_n and j_n are continuous by the hypothesis of the induction. Then so is i_{n+1}. (j_{n+1} can be proved to be continuous in a similar fashion.) qed

The collection of all the D_n's together with the retractions $\langle i_n, j_n\rangle$ form an *inverse system*. The inverse limit, denoted by $D_\infty = \underleftarrow{\lim}(D_n, \langle i_n, j_n\rangle)$ (or simply by $\underleftarrow{\lim} D_n$), is the space in which a model of CL exists.

The elements of D_∞ can be viewed as infinite sequences of elements, for instance, $d_\infty = \langle d_0, d_1, \ldots, d_n, \ldots\rangle$, where $d_0 \in D_0$, $d_1 \in D_1$, etc. As a concrete example, $\bot_\infty = \langle \bot_0, \bot_1, \ldots, \bot_n, \ldots\rangle$. There is a natural ordering emerging on D_∞ via the coordinate-wise ordering. That is, $d_\infty \sqsubseteq_\infty d'_\infty$ iff $d_n \sqsubseteq_n d'_n$, for all n. Then it can be shown that $\langle D_\infty, \sqsubseteq_\infty\rangle$ is a CPO.

The limit construction gives, for each n, a retraction $\langle i_n^\infty, j_n^\infty\rangle$ such that $i_n^\infty\colon D_n \longrightarrow D_\infty$ and $j_n^\infty\colon D_\infty \longrightarrow D_n$. Informally, i^∞ returns an element of the infinite sequence, whereas the j^∞ yields an infinite sequence. $i_n^\infty(d_\infty) = d_n$ and

$$j_n^\infty(d_n) = \langle j_{n\mapsto 0}(d_n), \ldots, j_{n\mapsto n-1}(d_n), \mathsf{I}_n(d_n), i_{n\mapsto n+1}(d_n), i_{n\mapsto n+2}(d_n), \ldots\rangle,$$

where I_n is the identity function on D_n. Furthermore, $j_{n\mapsto n-1}$ is j_n, $i_{n\mapsto n+1}$ is i_n, and the other i's and j's are compositions of the one-step i's and of the one-step j's that we have defined previously.

D_∞ can be used as a model of CL, if we can find suitable elements to stand for the combinators S and K, and we can define a binary operation to represent application.

DEFINITION 6.4.22. (CL MODEL IN D_∞) Application and the combinators S and K are defined as in (1)–(3). (I_1 abbreviates the element of D_1 that is $\lambda x_0.x_0$. The subscripts indicate to which D_n's the elements belong.)

(1) $\mathsf{S}_\infty = \langle \bot_0, \mathsf{I}_1, j_2(\mathsf{S}_3), \mathsf{S}_3, \mathsf{S}_4, \ldots, \mathsf{S}_n, \ldots\rangle$, where

$\mathsf{S}_n = \lambda x_{n-1}\, y_{n-2}\, z_{n-3}.x_{n-1}(j_{n-3}(z_{n-3}))(y(z))$.[20]

(2) $\mathsf{K}_\infty = \langle \bot_0, \mathsf{I}_1, \mathsf{K}_2, \mathsf{K}_3, \mathsf{K}_4, \ldots, \mathsf{K}_n, \ldots\rangle$, where

$\mathsf{K}_n = \lambda x_{n-1}\, y_{n-2}.\, j_{n-2}(x_{n-1})$.

(3) $d_\infty \cdot d'_\infty = \mathrm{lub}_{n\in\mathbb{N}}\, j_n^\infty(d_{n+1}(d'_n))$.

DEFINITION 6.4.23. A function I from CL-terms into $\langle D_\infty, \cdot, \mathsf{S}_\infty, \mathsf{K}_\infty\rangle$ is an *interpretation* iff it satisfies (1)–(3).

[20]Notice that we use a few extra pairs of parentheses compared to CL or the λ-calculus. For instance, $j_{n-3}(z_{n-3})$ is the application of j to z, in the sense of the usual notation for functions.

(1) $I(x) \in D_\infty$;

(2) $I(\mathsf{S}) = \mathsf{S}_\infty$ and $I(\mathsf{K}) = \mathsf{K}_\infty$;

(3) $I(MN) = I(M) \cdot I(N)$.

The definition allows variables to be interpreted by any elements of D_∞, but requires the combinators and the application operation to have a fixed interpretation.

LEMMA 6.4.24. *If* $M =_{\mathrm{w}} N$ *in* CL, *then* $I(M) = I(N)$ *in* D_∞.

Exercise* 6.4.5. Prove the lemma.

In this model, if $I(MP) = I(NP)$ for all P, then $I(M) = I(N)$, that is, the model is *extensional.*

Exercise 6.4.6. Prove the latter claim. (Hint: The proof might be extracted from the proof of the previous lemma.)

6.5 Models for typed CL

In this section we will be concerned with typed CL. Some systems of typed CL are introduced in chapter 9.[21] The intersection type system $\mathrm{CL}_\wedge$ is slightly more complicated than the type system that contains only $\rightarrow$. However, the additional expressive capability of the language of types is clearly a plus. Later, in chapter 9, we will see that all CL-terms are typable in $\mathrm{CL}_\wedge$. Now we show that $\mathrm{CL}_\wedge$ allows CL-terms to be interpreted in a straightforward way as sets of types.

The type constructors in $\mathrm{CL}_\wedge$ are $\rightarrow$, $\wedge$ and the constant ω. The latter is a top element in the set of types (i.e., it is like ω in $\mathbb{N} \cup \{\omega\}$ rather than the elements of $\mathbb{N}$). Types are thought of as *formulas*.

The type system in itself amounts to the logic $B^T_\wedge$ (where T is the usual notation for the Church constant "truth," which is like ω).

DEFINITION 6.5.1. (AXIOMS FOR $B^T_\wedge$) The axioms and rules for $B^T_\wedge$ are as follows. ($A, B, C, \ldots$ are arbitrary formulas.)

(A1) $A \rightarrow A$

(A2) $(A \wedge B) \rightarrow A$

(A3) $(A \wedge B) \rightarrow B$

[21] We suggest reading this section in parallel with section 9.2, or skipping this section now and moving onto section 6.6. The reader might return to this section as well as the concrete operational model at the end of section 6.2 after reading chapter 9.

(A4) $((A \to C) \wedge (A \to B)) \to (A \to (B \wedge C))$

(A5) $A \to T$

(A6) $T \to (T \to T)$

(R1) $A,\ A \to B \Rightarrow B$

(R2) $A,\ B \Rightarrow A \wedge B$

(R3) $A \to B \Rightarrow (C \to A) \to (C \to B)$

(R4) $A \to B \Rightarrow (B \to C) \to (A \to C)$

A formula A is a *theorem* of $B_\wedge^T$ iff there is a finite sequence of wff's — called a proof — ending with A in which each wff is either an instance of an axiom or obtained from previous formula(s) by an application of a rule.

$B_\wedge^T$ is quite usual as an axiom system, and the notion of a proof is uncomplicated. The presence of $\wedge$ as a type constructor is crucial in ensuring the typability of terms like xx. ω is a universal type that can be assigned to any term. In the spirit of the "formulas-as-types" paradigm, we use "type" and "wff" synonymously here.

At the end of section 6.2, we have seen an interpretation of CL-terms by sets of types, however, that modeling was restricted to $\triangleright_{\mathrm{w}}$. Having more types — that is the real advantage of having more type constructors — allows us to give a model for $=_{\mathrm{w}}$, that is, for expansion too.

DEFINITION 6.5.2. (THEORIES OF TYPES) A set of wff's X is *a theory* when (1)–(3) hold of X.

(1) $X \neq \emptyset$;

(2) if $A, B \in X$ then $A \wedge B \in X$;

(3) if $A \in X$ and $A \to B$ is a theorem then $B \in X$.

The set of types has a natural *preorder* on it: $A \leqslant B$ iff $A \to B$ is a theorem of $B_\wedge^T$. The reflexivity of this relation is obvious due to (A1), and its transitivity follows by the rules (R3) and (R4). As a counterexample to antisymmetry, we can take $A \wedge B$ and $B \wedge A$, which imply each other, but distinct as wff's. We can consider wff's that imply each other together, and then we have a partial order (on the set of equivalence classes of wff's). If we apply this usual trick, then we get *a meet semilattice with a top* from the word algebra of wff's.

Talking about algebras, more specifically, lattices and meet semilattices, certain subsets of elements are called *filters*. A filter is an upward closed set of elements that is also closed under the (binary) meet operation. Sometimes filters are defined not to be empty, and we would want that version of the definition here. There is a 1–1 correspondence between theories and filters, though they are not the same object. There are certain advantages to the less discriminative view of types, because partial orders are easier to treat than quasi-orders. This may be an explanation why in some of the literature the model (or its close variants) are called *filter λ-model.* To complete the model though, we have to have an operation that can stand in for the application operation.

DEFINITION 6.5.3. (FUSION) Let X and Y be theories. The fusion operation, denoted by $\circ$, is defined as

$$X \circ Y = \{B \colon \exists A. A \to B \in X \wedge A \in Y\}.$$

This operation is very similar to the condensed detachment operation from section 6.2. The definition says that $A \to B \in X$ and $A \in Y$, that is, B is obtained by detachment. A difference between the operations is that we do not allow the unification of the formulas in X and Y via substitution. However, substitution is an admissible rule in $B^T_\wedge$, because the axioms are formulated as axiom schemas.

The name "fusion" for $\circ$ comes from the name of a connective that is included in some formulations of B (and other relevance logics). Arguably, detachment is best formulated using fusion, which is an operation closely related to $\to$. ($\to$ is a residual of $\circ$ in the algebra of the logic B.)

The structure that arises from theories together with the fusion operation is a combinatory algebra in the sense of definition 6.2.2. This time we want to interpret not simply the terms of CL, but *type assignments*, or even type assignments given a (consistent) set of assumptions. This means that we have to have an interpretation both for CL-terms and types.

DEFINITION 6.5.4. Let $\mathfrak{A} = \langle D, \cdot, \mathsf{S}, \mathsf{K} \rangle$ be a combinatory algebra, and let I and v be functions that interpret atomic terms and propositional variables.[22]

(1) $I(\mathsf{K}) = \mathsf{K}$, $I(\mathsf{S}) = \mathsf{S}$ and $I(\mathsf{I}) = \mathsf{S} \cdot \mathsf{K} \cdot \mathsf{K}$;

(2) $I(x) \in D$;

(3) $I(MN) = I(M) \cdot I(N)$;

(4) $v(p) \in \mathcal{P}(D)$;

(5) $v(T) = D$;

(6) $v(A \to B) = \{d_1 \colon \forall d_2. d_2 \in v(A) \Rightarrow d_1 \cdot d_2 \in v(B)\}$;

(7) $v(A \wedge B) = v(A) \cap v(B)$.

In a model $\mathfrak{M} = \langle \mathfrak{A}, I, v \rangle$, the type assignment $M \colon A$ is *true*, denoted by

$$\mathfrak{M} \vDash M \colon A \quad \text{iff} \quad I(M) \in v(A).$$

[22]The extensions of these functions to all the CL-terms, and to all the types, respectively, are also denoted by I and v.

A set of type assignments is true whenever each element is true. A type assignment with an assumption set is *valid*, denoted by

$$\Delta \vDash M\colon A \quad \text{iff} \quad \forall \mathfrak{M}.\, \mathfrak{M} \vDash \Delta \text{ implies } \mathfrak{M} \vDash M\colon A.$$

The notation $\Delta \vdash_{\mathrm{CL}_\wedge} M\colon A$ (or $\Delta \vdash M\colon \tau$ in chapter 9, where we use Greek letters for types) means that given the type assignments of variables in Δ, $\mathrm{CL}_\wedge$ *proves* $M\colon A$. The notation $\Delta \vDash M\colon A$, in the above definition, on the other hand, stands for a *semantic* statement that is spelled out on the right-hand side of the iff. The latter contains a universal quantifier, and it should be noted that $\mathfrak{A}$ does not determine uniquely either I or v. Thus the definition of validity involves three universal quantifiers: for all combinatory algebras, for all interpretations of CL-terms (in the algebra), and for all interpretations of types (in the algebra).

In the "fool's model," CL-terms were interpreted by types, and we already mentioned that there is a similarity between the fusion operation on theories and condensed detachment. Notice, however, that clause (6) interprets the implicational wff's via the operation in the combinatory algebra (that is the analogue of the application operation on CL-terms).

The calculus $\mathrm{CL}_\wedge$ allows us to prove expressions of the form $\Delta \vdash M\colon A$. Roughly speaking, we would like to show that the two relations $\vdash$ and $\vDash$ coincide.

LEMMA 6.5.5. (SOUNDNESS OF $\mathrm{CL}_\wedge$) *If $\Delta \vdash M\colon A$ is provable in $\mathrm{CL}_\wedge$, then $\Delta \vDash M\colon A$.*

Proof: **1.** The axiom id is obviously valid, and so is axiom ω. For the combinatory axioms, we have to take a look at the interpretation of the types that are assigned to them.

Let us take $A \to B \to A$. $d_1 \in v(A \to B \to A)$ iff for all $d_2 \in v(A)$, $d_1 \cdot d_2 \in v(B \to A)$. Further, $d_1 \cdot d_2 \in v(B \to A)$ just in case, for all d_3, if $d_3 \in v(B)$ then $(d_1 \cdot d_2) \cdot d_3 \in v(A)$. If we combine the two conditions, then we get that for d_1 to be an element of $v(A \to B \to A)$, it must be the case that for all d_2's in $v(A)$ and d_3's in $v(B)$, $d_1 \cdot d_2 \cdot d_3$ has to be an element of $v(A)$. There is no constraint linking $v(B)$ to $v(A)$ — except that they are subsets of the same set D. Thus given $d_2 \in v(A)$, K is an element that guarantees $\mathsf{K} \cdot d_2 \cdot d_3 \in v(A)$. The function I is fixed on the constants, that is, $I(\mathsf{K}) = \mathsf{K}$, where the K on the left is the combinator, and the K on the right is the constant element of $\mathfrak{A}$. Clearly, $I(\mathsf{K}) \in v(A \to B \to A)$ meaning that $\vDash \mathsf{K}\colon A \to B \to A$.

The interpretation of $A \to A$ is $\{d_1 \colon \forall d_2 . d_2 \in v(A) \Rightarrow d_1 \cdot d_2 \in v(A)\}$. We have S and K as distinguished elements of D, and they behave with respect to $\cdot$ in $\mathfrak{A}$ just like the identically denoted combinators do in CL. In other words, $\mathsf{S} \cdot \mathsf{K} \cdot \mathsf{K} \cdot d = d$, because $\mathfrak{A}$ is a combinatory algebra. Then $\mathsf{S} \cdot \mathsf{K} \cdot \mathsf{K} \in v(A \to A)$. Since SKK defines I in CL, we have above that $I(\mathsf{I}) = \mathsf{S} \cdot \mathsf{K} \cdot \mathsf{K}$, which is sufficient to prove that $I(\mathsf{I}) \in v(A \to A)$, that is, $\vDash \mathsf{I}\colon A \to A$.

The case of the S axiom is left as an exercise.

2. Next we look at the $E \to$ rule. The premises provide, by the hypothesis of induction, that $\Delta_1 \vDash M\colon A \to B$ and $\Delta_2 \vDash N\colon A$. The applicability of the rule means that Δ_1 and Δ_2 do not assign different formulas to one and the same variable. Then for all I and v, Δ_1 is true if Δ_2 is true and the other way around. If $\vDash \Delta_1, \Delta_2$, then both $I(M) \in v(A \rightarrow B)$ and $I(N) \in v(A)$. By the definition of $v(A \rightarrow B)$, $I(M) \cdot I(N)$, that is, $I(MN)$ is an element of $v(B)$. This is exactly what we need for the validity of $\Delta_1, \Delta_2 \vDash MN \colon B$.

3. Lastly, we consider an $E\wedge$ rule. Let us assume that $\Delta \vDash M \colon A \wedge B$. With the proviso that Δ is true, this means that $I(M) \in v(A \wedge B)$. We know that $v(A \wedge B) = v(A) \cap v(B)$; hence, by the usual definition of $\cap$, $I(M) \in v(A)$ is immediate. (Again, we leave the rest of this step for an exercise.) qed

Exercise 6.5.1. Describe the set that is the interpretation of $(A \rightarrow B \rightarrow C) \rightarrow (D \rightarrow B) \rightarrow (A \wedge D) \rightarrow C$.

Exercise 6.5.2. Finish step **3** in the above proof. (Hint: There are two more $\wedge$ rules that we have not considered.)

To prove the converse of the claim of the above lemma, we select a particular combinatory algebra, as well as certain functions I and v.

DEFINITION 6.5.6. Let $\mathfrak{A}_{\mathcal{F}}$ be $\langle \mathcal{F}, \cdot, \mathbf{S}, \mathbf{K}, \mathbf{I} \rangle$, where the components are as follows.

(1) $\mathcal{F}$ is the *set of theories* of wff's (or equivalently, the set of filters on the algebra of $B_\wedge^T$);

(2) $F_1 \cdot F_2 = \{B \colon \exists A. A \in F_2 \wedge A \rightarrow B \in F_1\}$;

(3) $\mathbf{S} = [\{E \colon E = \sigma((A \rightarrow B \rightarrow C) \rightarrow (D \rightarrow B) \rightarrow (A \wedge D) \rightarrow C)\})$;

(4) $\mathbf{K} = [\{E \colon E = \sigma(A \rightarrow B \rightarrow A)\})$;

(5) $\mathbf{I} = [\{E \colon E = \sigma(A \rightarrow A)\})$.

The notation $E = \sigma(G)$ is to indicate that E is a substitution instance of G. The mismatched-looking $[\ )$ forms a *theory* from a set of wff's.[23]

LEMMA 6.5.7. $\mathfrak{A}_{\mathcal{F}}$ *is a* combinatory algebra.

Proof: We have to prove that $\cdot$ is an operation on filters, as well as, that the selected filters, $\mathbf{S}$, $\mathbf{K}$ and $\mathbf{I}$ behave as expected.

1. Let F_1 and F_2 be filters, and let us assume that A_1 and A_2 are in $F_1 \cdot F_2$. By (2), for A_1, $\exists B_1 \in F_2. B_1 \rightarrow A_1 \in F_1$, and similarly for A_2, $\exists B_2 \in F_2. B_2 \rightarrow A_2 \in F_1$. By our starting assumption, F_1 and F_2 are filters, hence $B_1 \wedge B_2 \in F_2$. In $B_\wedge^T$, it is provable that $\rightarrow$ is antitone in its first argument place, in particular, $(B_1 \rightarrow A_1) \rightarrow (B_1 \wedge B_2) \rightarrow A_1$ and $(B_2 \rightarrow A_2) \rightarrow (B_1 \wedge B_2) \rightarrow A_2$. Therefore, the consequents of these implications are also elements of F_1. F_1 is a filter too, and so $((B_1 \wedge B_2) \rightarrow A_1) \wedge ((B_1 \wedge B_2) \rightarrow A_2) \in F_1$. $B_\wedge^T$ proves the latter wff to be equivalent to $(B_1 \wedge B_2) \rightarrow (A_1 \wedge A_2)$. Then, by the definition of $\cdot$, $A_1 \wedge A_2 \in F_1 \cdot F_2$.

[23]This notation is widely used in the algebraic literature for *filters*.

For the other defining feature of filters, let us assume that $A \in F_1 \cdot F_2$ and $A \to B$ is a theorem. The definition of $\cdot$ implies that $C \to A \in F_1$ and $C \in F_2$ for some formula C. By rule (R3), $(C \to A) \to C \to B$ follows, thus $C \to B \in F_1$. But this implies that $B \in F_1 \cdot F_2$, again, by (2), that is, by the definition of $\cdot$.

2. We consider the easiest constant $\mathbf{I}$. (The exercise below asks you to complete this step of the proof.) The easy direction involves showing that $F \subseteq \mathbf{I} \cdot F$, for any $F \in \mathfrak{F}$. Let $A \in F \in \mathfrak{F}$. By the definition of $\mathbf{I}$ in (5), $A \to A \in \mathbf{I}$ too, hence $A \in \mathbf{I} \cdot F$.

For the converse, let $F \in \mathfrak{F}$, and $A \in \mathbf{I} \cdot F$. Then there is a $B \in F$ such that $B \to A \in \mathbf{I}$. The definition of $\mathbf{I}$ provides that $\bigwedge_{i=1}^{n}(C_i \to C_i) \to B \to A$ for some n. $B_\wedge^T$ proves every instance of $C_i \to C_i$, hence also their conjunction. Then by rule (R1), $B \to A$ is a theorem, thus by the definition of theories (or filters), $A \in F$. qed

Exercise 6.5.3. Verify that $\mathbf{S}$ and $\mathbf{K}$ behave as expected.

In order to be able to prove the completeness of $\mathrm{CL}_\wedge^T$, we also have to define I and v appropriately. We know that $\mathfrak{A}_\mathfrak{F}$ is a combinatory algebra with the elements of the carrier set being filters of formulas. The usual canonical interpretation in various nonclassical logics takes a wff and maps that into a proposition, which is a set of theories (of a suitable kind). Thus we have to find an element in those propositions, that is, a theory that can be taken to interpret the term that can be assigned the formula (represented by the proposition).

LEMMA 6.5.8. *The set of formulas that can be assigned to a* CL*-term in* $\mathrm{CL}_\wedge$ *is a* theory. *In other words,* $\{A : \Delta \vdash M : A\} \in \mathfrak{F}$.

Proof: **1.** First of all, ω (i.e., T) can be assigned to any formula, therefore, $\{A : \Delta \vdash M : A\} \neq \emptyset$. (Notice that this is in contrast with the interpretation of CL-terms in the last model in section 6.2.)

Let $B, C \in \{A : \Delta \vdash M : A\}$. Then both $\Delta \vdash M : B$ and $\Delta \vdash M : C$, hence by rule $I\wedge$, $\Delta \vdash M : B \wedge C$. The latter means that $B \wedge C \in \{A : \Delta \vdash M : A\}$.

Next, let $B \in \{A : \Delta \vdash M : A\}$ and $A \to B$ a theorem of $B_\wedge^T$. Rule up guarantees that $\Delta \vdash M : B$ as well. qed

The proof is a nice illustration of how the addition of the top element to the set of types and the new type constructor $\wedge$ provide a nonempty and well-shaped set of types for all CL-terms.

Actually, if Δ is the least set such that $\Delta \vdash M : A$ is provable in $\mathrm{CL}_\wedge$, then $\{A : \Delta \vdash M : A\}$ is the theory generated by $\{B : \exists x.\, x : B \in \Delta\}$.

Exercise 6.5.4. Prove the previous claim. (Hint: There are two $\wedge$ elimination rules.)

DEFINITION 6.5.9. The *interpretation* of a wff A is $v(A) = \{F \in \mathfrak{F} : A \in F\}$.

Let I be a function assigning elements of $\mathfrak{F}$ to variables, $x, y, z, \ldots$. Let Δ' be a modification of Δ where $x_i : A_i$ when $A_i \in I(x_i)$. Then the definition of $I(M)$ provides that $I(M) = \{A : \Delta' \vdash M : A\}$. (The interpretation of the constants is unchanged; that is, they are certain selected filters.)

LEMMA 6.5.10. *The function I as defined above is an* interpretation *of* CL*-terms into the combinatory algebra* $\mathfrak{A}_{\mathcal{F}}$.

Proof: We are only concerned with the interpretation of the binary application operation. Let us assume that $A \in I(N)$ and $A \to B \in I(M)$. By the definition of $\cdot$, $B \in I(M) \cdot I(N)$. But an application of $E \to$ gives $MN \colon B$ (with the combined assumption set), hence $B \in I(MN)$, as we need.

To prove the converse, we have to make sure that $\Delta \vdash MN \colon B$ implies that for some A, $A \to B \in I(M)$ and $A \in I(N)$. The axioms contain atomic terms, hence we only have to scrutinize the rules of $\mathrm{CL}_\wedge$. Obviously, $E \to$ gives what we need.

If $\Delta \vdash MN \colon B$ is by $E\wedge$, then by hypothesis, $\Delta \vdash MN \colon B \wedge C$ implies $\Delta \vdash M \colon A \to (B \wedge C)$ and $\Delta \vdash N \colon A$, which is sufficient for $A \in I(N)$. $(A \to (B \wedge C)) \to A \to B$ is a $B^T_\wedge$ theorem; hence, $A \to B \in I(M)$, as desired. The case of the other $E\wedge$ rule is alike. If $\Delta \vdash MN \colon B$ is by $I\wedge$, then B is $C \wedge D$. By the hypothesis of the induction, we have that $\Delta \vdash MN \colon C$ implies $E \to C \in I(M)$ and $E \in I(N)$, as well as $\Delta \vdash MN \colon D$ implies $F \to D \in I(M)$ and $F \in I(N)$. The two implicational types imply $(E \wedge F) \to (C \wedge D)$ in $B^T_\wedge$. Given the assumptions provided by the hypothesis of the induction, we can obtain a common type assigned to M and a common type assigned to N in the following way.

$$\dfrac{\dfrac{\Delta \vdash M \colon E \to C \quad \Delta \vdash M \colon F \to D}{\Delta \vdash M \colon (E \wedge F) \to (C \wedge D)} \qquad \dfrac{\Delta \vdash N \colon E \quad \Delta \vdash N \colon F}{\Delta \vdash N \colon E \wedge F}}{\Delta \vdash MN \colon C \wedge D}$$

This shows that $(E \wedge F) \to (C \wedge D) \in I(M)$ and $C \wedge D \in I(N)$ as we need.

The last two rules to consider are mon and up. The former is straightforward. If $\Delta \vdash MN \colon B$ is by up, then $A \leq B$ for some A, and by hypothesis, $C \to A \in I(M)$ and $C \in I(N)$. However, $A \leq B$ implies $C \to A \leq C \to B$, thus we have

$$\dfrac{\dfrac{\Delta \vdash M \colon C \to A \qquad C \to A \leq C \to B}{\Delta \vdash M \colon C \to B} \qquad \Delta \vdash N \colon C}{\Delta \vdash MN \colon B}.$$

Thus $C \to B \in I(M)$, which is sufficient to establish the claim. qed

LEMMA 6.5.11. *If $\Delta \vDash M \colon A$, then $\Delta \vdash M \colon A$ is* provable in $\mathrm{CL}_\wedge$.

Proof: The algebra of theories (i.e., $\mathfrak{A}_{\mathcal{F}}$, as defined above) is a combinatory algebra by lemma 6.5.7. We have proved that a particular function I is an interpretation of CL-terms in this algebra. v can be taken to be the canonical interpretation of types as propositions. Thus $\langle \mathfrak{A}_{\mathcal{F}}, I, v \rangle$ is a model of $\mathrm{CL}_\wedge$.

Let us assume that $\mathrm{CL}_\wedge$ does not prove $\Delta \vdash M \colon A$. Then $A \notin I(M)$, which means that A is not a member of a particular theory (or filter) from $\mathcal{F}$. However, $v(A)$ contains only filters of which A is an element. This gives $I(M) \notin v(A)$, that is, $\mathfrak{M} \vDash M \colon A$ is not true. The definition of I ensures that $\mathfrak{M} \vDash \Delta$; that is, $\langle \mathfrak{A}_{\mathcal{F}}, I, v \rangle$ makes the type assignments in the assumption set true without making $M \colon A$ true. By definition 6.5.4, $\Delta \vDash M \colon A$ is not valid. qed

The intersection type assignment system $\mathrm{CL}_\wedge$ allows a type to be assigned to any CL-term. However, without $\wedge$ and T, not all CL-terms are typable.[24] The logic of types in $\mathrm{CL}_\wedge$ is $B^T_\wedge$, and it is known that the extension of the implicational logic $B_\rightarrow$ by conjunction and top is *conservative*. The conservativity of the extension may be shown to carry over to $\mathrm{CL}_\wedge$ when that is viewed as an extension of $\mathrm{CL}_\rightarrow$.

LEMMA 6.5.12. *The above model is* complete *for* $\mathrm{CL}_\rightarrow$.

Exercise* 6.5.5. Use the conservativity of $B^T_\wedge$ over $B_\rightarrow$ to show the conservativity of $\mathrm{CL}_\wedge$ over $\mathrm{CL}_\rightarrow$, and thereby, give a proof of this lemma.

Another way to utilize the existence of a model for $\mathrm{CL}_\wedge$ is to eliminate the type assignments. Then we get *a model of pure* CL. We built the model above piece by piece, and so we can leave off parts of the construction. The core results pertaining to the possibility of this deletion are lemma 6.5.7 and lemma 6.5.8.

The previous section considered CL models in an abstract setting. $\mathcal{F}$, the set of filters (or theories) forms a lattice; moreover, a complete lattice with intersection as meet and with the least covering filter of filters as join.[25] Clearly, $\langle \mathcal{F}, \subseteq \rangle$ is a CPO with $\{T\}$ being the least element in $\mathcal{F}$. Furthermore, this model is a domain.

Exercise 6.5.6. Recall the definition 6.4.17. Give a detailed proof that $\langle \mathcal{F}, \subseteq \rangle$ is a CPO, and also a domain.

In section 9.2, we briefly mention, a *union type* assignment system CL_+. The type inferences in CL_+ prove exactly the theorems of the logic known as B^T_+.[26]

DEFINITION 6.5.13. (POSITIVE B^T) The positive fragment of the logic B^T is axiomatized by adding axioms (A7)–(A10) to the axiomatization of $B^T_\wedge$ (in definition 6.5.1).

(A7) $A \rightarrow (A \vee B)$

(A8) $A \rightarrow (B \vee A)$

(A9) $((A \rightarrow B) \wedge (C \rightarrow B)) \rightarrow (A \vee C) \rightarrow B$

(A10) $(A \wedge (B \vee C)) \rightarrow ((A \wedge B) \vee (A \wedge C))$

The axioms are easily seen to be analogues of the type inference axioms that were added to $\mathrm{CL}_\wedge$ to obtain CL_+. The standard logical notions of B^T_+ — such as proof and theoremhood — are defined as usual.

The Stone representation of Boolean algebras or the completeness of classical logic uses *prime* filters. Prime theories (prime filters) have the property that, if they

---

[24] See section 9.1 for simply typed systems of CL.

[25] The lattice $\mathcal{F}$ is not always a chain like the filters on $\mathbb{Q}$ are, which are well-known from the Dedekind completion of $\mathbb{Q}$. This explains why union cannot be taken for join, in general. For other ways to define the "covering," see, for example, [25, §9.4].

[26] A subscript $+$ is often used to indicate the positive fragment of a logic, which means the fragment that does not contain negation (and constants like F or f).

contain a disjunction (a join), then they contain a component from the disjunction (the join). Fundamentally, this is the reason why intersection and union on sets of prime filters can represent meet and join of a distributive lattice.

To make a long story short, from the semantical studies of nonclassical logics, mainly of relevance logics, it is well-known that *sets of prime theories* can be used to give a model for various positive relevance logics. But it is just as well-known that the set of prime theories is not closed under a $\cdot$ -like operation — unlike the set of theories is. In other words, we cannot simply replace theories by prime theories in the previous model.

DEFINITION 6.5.14. (HARROP THEORIES) The set of *Harrop formulas*, hwf, is defined by (1)–(3).

(1) $A \in \text{hwf}$ if A is an atomic formula;

(2) $A \wedge B \in \text{hwf}$ if $A, B \in \text{hwf}$;

(3) $A \to B \in \text{hwf}$ if $B \in \text{hwf}$ and $A \in \text{wff}$.

A set of wff's X is a *Harrop theory* iff X is a theory in which for all formulas $A \in X$, there is a Harrop formula B such that $B \to A$ is a theorem of B_+^T. The set of Harrop theories is denoted by $\mathcal{H}$.

The application operation is defined as before, that is, as in definition 6.5.6. Before we prove that the Harrop theories provide closure under the application operation, we state a lemma that gives the crucial step in the proof of the next lemma.

LEMMA 6.5.15. (BUBBLING LEMMA FOR B_+^T) *Let M and J be finite subsets of $\mathbb{N}$. Then $\bigwedge_{m \in M} A_m \to B_m \leq \bigvee_{j \in J} C_j \to D_j$ just in case*

$$\exists j \in J \colon C_j \leq \bigvee_{m \in M} A_m \wedge \forall K. K \subsetneq M \Rightarrow (C_j \leq \bigvee_{k \in K} A_k \vee \bigwedge_{i \in M-K} B_i \leq D_j).$$

Exercise* 6.5.7. Prove the bubbling lemma for B_+^T.

LEMMA 6.5.16. *The set $\mathcal{H}$ is* closed under $\cdot$ *, the application operation.*

Proof: Let us assume that $F_1, F_2 \in \mathcal{H}$, and $B \in F_1 \cdot F_2$. Then there is $A \in F_2$ with $A \to B \in F_1$. At this point, we do not know whether $B \in \text{hwf}$ or not. For $F_1 \cdot F_2$ to be a Harrop theory, we have to show that there is a C (possibly, B itself) such that $C \in \text{hwf}$ and $C \to B$ is a theorem.

$A \in F_2 \in \mathcal{H}$, hence there is a $D \in \text{hwf}$ such that $D \to A$ is a theorem. By suffixing, $(A \to B) \to D \to B$, therefore, $D \to B \in F_1$. Then there is an hwf $E \in F_1$ such that $E \to D \to B$ is a theorem. Since E is hwf, it can be taken to be of the form $(\bigwedge_{m \in M} E_m \to F_m) \wedge \bigwedge_{k \in K} p_k$, where the F_m's themselves are hwf's. Then by the bubbling lemma, there is a finite subset of M, let us say J, for which $D \to \bigwedge_{j \in J} E_j$ and $\bigwedge_{j \in J} F_j \to B$. We need a hwf such that that formula implies B, and we already have that $\bigwedge_{j \in J} F_j$ implies B. All the F_j's are hwf's, and J is finite. The conjunction of hwf's is an hwf, by the definition above. Since $D \in F_1$, then $\bigwedge_{j \in J} E_j \in F_1$ too. Then $\bigwedge_{j \in J} F_j \in F_1 \cdot F_2$, which shows that $F_1 \cdot F_2$ is a Harrop theory, as we intended to prove. qed

Harrop theories are quite special among all the sets of formulas — they provide the desired closure under $\cdot$ for a model. However, there are *not enough many* of them, in general. In other words, they cannot provide an isomorphic representation for CL. We do not pursue further the problem of modeling CL based on Harrop theories. Instead we conclude this section with a couple of exercises.

Exercise 6.5.8. In order to obtain a model from CL, suitable sets of formulas has to be selected to interpret the combinators. Work out those details of the model, and prove soundness.

Exercise ** **6.5.9.** Determine why completeness is not provable using Harrop theories. Fix the problem. (Hint: This is a completely open-ended exercise.)

6.6 Relational models

The models for typed CL showed an interaction between structures which included a binary operation and types. We also saw a connection between CL and the minimal relevance logic B. Many nonclassical logics have been interpreted starting with a *relational structure* rather than an algebra.

These sorts of semantics were first introduced by Kripke in the late 1950s for normal modal logics. The usual modal operators, $\Box$ and $\Diamond$, are unary operations in (the algebra of) modal logics. However, the set of maximally consistent sets of formulas (in the sense of classical logic) is not closed under the operations that correspond to these modal operators. Kripke's idea was to use a *binary relation* instead, which later on, became known as the "accessibility relation" between "possible worlds." Taking the *existential image* operation determined by the relation, we can obtain an operation on sets of possible worlds that shares tonicity and distributivity properties with $\Diamond$, possibility.

The lack of closure under an operation is a recurring problem; we have already encountered this difficulty in the case of B^T_+ and the $\cdot$ operation. Relevance logics are typically not extensions of classical logic. However, prime filters on the algebra of these logics perform adequately in the modeling of conjunction and disjunction — just like ultrafilters (or maximally consistent sets of formulas) do in normal modal logics. The (former) prime filters though are not maximal with respect to not being identical to the carrier set of the algebra. The *relational paradigm* that originated in the context of logics with unary operators was adapted to relevance logics by Dunn, Meyer and Routley. Later on, it was generalized and extended to other nonclassical logics by them and others.[27]

[27] For a comprehensive and state-of-the-art exposition see [25], which is completely devoted to this approach (and it also contains an extensive bibliography).

We have *combinatory algebras* and *combinatorial reduction algebras* that were introduced in definition 6.2.1 and definition 6.2.8, respectively. The relational semantics for various nonclassical logics can be viewed as *representations* of the algebras of those logics. Having isolated the algebras of CL, we can consider representations of combinatory and combinatorial reduction algebras. First, we assume that we have a combinatorial reduction algebra — with constants S and K; then, we note some necessary modifications for a semantics for CL with $=_{\mathrm{w}}$.

DEFINITION 6.6.1. (STRUCTURES) A *structure* (or *frame*) is $\mathfrak{F} = \langle U, \leq, R, \mathbf{s}, \mathbf{k} \rangle$, where the structure satisfies (f1)–(f4).[28]

(f1) $U \neq \emptyset, \quad \leq \subseteq U^2, \quad R \subseteq U^3, \quad \mathbf{s}, \mathbf{k} \in U, \quad R{\downarrow}\text{-}{\uparrow}$;

(f2) $\alpha \leq \alpha, \quad \alpha \leq \beta \wedge \beta \leq \gamma. \Rightarrow \alpha \leq \gamma, \quad \alpha \leq \beta \wedge \beta \leq \alpha. \Rightarrow \alpha = \beta$;

(f3) $\exists u, v. R\mathbf{s}\alpha u \wedge Ru\beta v \wedge Rv\gamma\delta. \Rightarrow \exists u, v. R\alpha\gamma u \wedge R\beta\gamma v \wedge Ruv\delta$;

(f4) $\exists u. R\mathbf{k}\alpha u \wedge Ru\beta\delta. \Rightarrow \alpha \leq \delta$.

Informally, we have a partially ordered set of *situations* (or possible worlds). In extensions of classical logic, for instance, in normal modal logics, the situations may be thought of as *complete* and *consistent* descriptions of states of the world, and accordingly, they are called possible worlds. Here we cannot assume that the descriptions are complete, which motivates both the label "situations" for them and the stipulation of a partial order on the set of situations. U, the set of situations is assumed not to be empty, as usual.

The relation linking some of the situations is R. If R were binary, then it would be natural to call it an accessibility relation. However, R is *ternary*, and so it is perhaps, more natural to view R as a *compatibility* relation on triples of situations, or even as a *reachability* relation; combining a pair of situations, from which the situation in the last argument of R may be reached. R is stipulated to interact with the ordering on the situations, in particular, R is *antitone* (decreasing) in its first argument place, and it is *monotone* (increasing) in its last argument place.

$\mathbf{s}$ and $\mathbf{k}$ are elements of U — they will be involved in the modeling of the constants S and K of the combinatorial reduction algebra, as the notation is intended to suggest. These distinguished elements are required to satisfy their respective conditions in (f3) and (f4).

We intend to use a certain kind of subsets — *cones* — of the set of all situations to represent CL-terms.[29] Given a structure with U, the set of all cones on U is denoted by $\mathcal{P}(U)^{\uparrow}$.

To obtain an interpretation of CL-terms, we augment a structure with an interpretation function v that is stipulated to conform to certain conditions.

[28]The term "frame" is sometimes used for a certain class of Heyting algebras. Our usage more closely follows the use of this term is part of the literature on modal logics.

[29]Definition A.9.11 in section A.9 introduces cones in semilattices. The same definition can be adopted for our frames, because the definition does not depend on the existence of greatest lower bounds.

DEFINITION 6.6.2. (MODELS) Let $\mathfrak{F}$ be a structure as defined above. A *model* is $\mathfrak{M} = \langle \mathfrak{F}, v \rangle$, where v has to satisfy (v1)–(v4).

(v1) $v(x) \in \mathcal{P}(U)^{\uparrow}$;

(v2) $v(\mathsf{S}) = \mathbf{s}^{\uparrow} \in \mathcal{P}(U)^{\uparrow}$;

(v3) $v(\mathsf{K}) = \mathbf{k}^{\uparrow} \in \mathcal{P}(U)^{\uparrow}$;

(v4) $v(MN) = \{\gamma\colon \exists\alpha,\beta.\, \alpha \in v(M) \wedge \beta \in v(N) \wedge R\alpha\beta\gamma\}$.

The inequation $M \geq N$ is *valid* iff $v(M) \subseteq v(N)$ for all models on all structures.

LEMMA 6.6.3. *If* $v(M), v(N) \in \mathcal{P}(U)^{\uparrow}$, *then so is* $v(MN)$.

Proof: Let us assume that $\gamma \in v(MN)$ and $\gamma \leq \delta$. Clause (v4) above means that $\exists\alpha,\beta.\, \alpha \in v(M) \wedge \beta \in v(N) \wedge R\alpha\beta\gamma$. After instantiating the quantifiers and eliminating conjunction, we have $R\alpha\beta\gamma$. R is monotone in its third argument place, and so we get that $R\alpha\beta\delta$ holds too. Then by conjunction and existential quantifier introduction, $\delta \in v(MN)$, as we had to show. qed

THEOREM 6.6.4. *If* $M \rhd_{\mathrm{w}} N$ *then* $M \geq N$ *is* valid.

Proof: We prove that $v(\mathsf{K}MN) \subseteq v(M)$ and that function application is interpreted by an operation that is monotone in both argument places. (See the next exercise for a step of the proof that we do not detail.)

1. Let $\delta \in v(\mathsf{K}MN)$, that is, by (v4), $\exists\gamma \in v(\mathsf{K}M)\, \exists\beta \in v(N).\, R\gamma\beta\delta$. Further, $\exists\varepsilon \in v(\mathsf{K})\, \exists\alpha \in v(M).\, R\varepsilon\alpha\gamma$. After quantifier and conjunction elimination, we have $R\varepsilon\alpha\gamma$. $v(\mathsf{K}) = \mathbf{k}^{\uparrow}$, that is, $\mathbf{k} \leq \varepsilon$. Then also $R\mathbf{k}\alpha\gamma$, because R is antitone in its first argument place. According to (f4), $\alpha \leq \delta$, and since $v(M) \in \mathcal{P}(U)^{\uparrow}$, $\delta \in v(M)$. This is sufficient to show that $v(\mathsf{K}MN) \subseteq v(M)$.

2. Let us assume that $v(M) \subseteq v(N)$. If $\delta \in v(MP)$, then we have to show that $\delta \in v(NP)$. By clause (v4), $\exists\alpha \in v(M)\, \exists\beta \in v(N).\, R\alpha\beta\delta$. Existential quantifier and conjunction elimination give $\varepsilon \in v(M)$. But by the starting assumption, $\varepsilon \in v(N)$, thus retracing the steps backward, we have that $\exists\alpha,\beta.\, \alpha \in v(N) \wedge \beta \in v(P) \wedge R\alpha\beta\gamma$, that is, $\delta \in v(NP)$. The definition of function application on upward-closed sets is completely symmetric in its two arguments. Then it is obvious that $v(PM) \subseteq v(PN)$ may be established based on the same assumption $v(M) \subseteq v(N)$. qed

Exercise 6.6.1. Prove that a model satisfies $v(\mathsf{S}MNP) \subseteq v(MP(NP))$.

In order to be able to prove the converse of the claim, we have to define a structure that can make sufficiently many distinctions. We define a so-called *canonical structure*, which is constructed from CL-terms. We need the following notion.

DEFINITION 6.6.5. (DOWNWARD-DIRECTED CONES) Let $\langle X, \leq \rangle$ be a poset, and $Y \subseteq X$. Y is a *downward-directed cone* on X iff

(1) $x \in Y$ and $x \leq y$ imply $y \in Y$;

(2) $Y \neq \emptyset$;

(3) $x, y \in Y$ imply $\exists z \in Y. z \leq x \wedge z \leq y$.

The set of *all downward-directed cones* on X is denoted by $\mathcal{P}(X)^{\uparrow\veebar}$, or if X is clear from the context, then simply by $\mathfrak{C}^{\veebar}$.

These special subsets are the analogues of filters on lattices. (Cf. definition A.9.13.) Informally speaking, downward directedness in necessary, because S is a duplicator. The effect of a duplicator with respect to types is identification, and in the completeness proof, they have a similar impact. A downward-directed cone provides identification by ensuring that there is an element that is less than the two given elements.

DEFINITION 6.6.6. (CANONICAL STRUCTURE AND MODEL) The *canonical structure* is $\mathfrak{F}_c = \langle \mathfrak{C}^{\veebar}, \subseteq, R_c, [\mathsf{S}), [\mathsf{K}) \rangle$, where R_c is defined by (c1).

(c1) $R_c\alpha\beta\gamma \;\Leftrightarrow\; \forall M, N. M \in \alpha \wedge N \in \beta. \Rightarrow (MN) \in \gamma$.

The *canonical model* is $\mathfrak{M}_c = \langle \mathfrak{F}_c, v_c \rangle$, where $v_c(M) = \{C \in \mathfrak{C}^{\veebar} : M \in C\}$.

The components of $\mathfrak{F}_c$ are of appropriate type. $[\mathsf{S})$ and $[\mathsf{K})$ are the principal cones generated by S and K, respectively.[30]

LEMMA 6.6.7. (CANONICAL STRUCTURE IS A STRUCTURE) *The canonical structure is* a structure *in the sense of definition 6.6.1.*

Proof: We prove one of the conditions, namely, that $[\mathsf{K})$ behaves as needed. Let us assume that $R_c[\mathsf{K})\alpha u$ and $R_c u\beta\delta$, as well as $M \in \alpha$. $\mathsf{K} \in [\mathsf{K})$, hence $\mathsf{K}M \in u$. $\beta \neq \emptyset$, and if $N \in \beta$, then $\mathsf{K}MN \in \delta$. The axiom of K guarantees that $\mathsf{K}MN \leq M$, hence, $M \in \delta$ too.

We leave two further steps from the proof as exercises. qed

Exercise 6.6.2. Prove that R_c has the tonicity that was stipulated for the compatibility relation in (f1) in definition 6.6.1.

Exercise 6.6.3. Show that (f3) is true in the canonical structure. (Hint: It is essential in this step that the cones are not arbitrary but downward-directed cones.)

THEOREM 6.6.8. (COMPLETENESS) *If $v(M) \subseteq v(N)$ in all models on all structures, then $M \geq N$.*

Proof: It is obvious that the constants get interpretations that satisfy conditions (v2) and (v3), respectively. It remains to show that v_c is a homomorphism for function application on cones, which is the content of the next exercise.

The last step of the proof is to show that, if $M \geq N$ is false, then the two terms may be separated. $[M)$ is a downward-directed cone and it cannot contain N, because $M \geq N$ is false. $v_c(M)$ contains $[M)$, but $v_c(N)$ does not. qed

[30] The order relation is $\leq$, which is generated by $\geq$ or equivalently, $\triangleright_w$.

Exercise 6.6.4. Prove that $v_c(MN)$ satisfies (v4).

Lastly, we note a modification that is not completely obvious if the $=_w$ relation is to be modeled. Function application in combinatory logic would not be a normal operation if there would be a bottom element in the set of CL-terms. In a model, a bottom element appears simply as $\emptyset$. Since function application is not a normal operation, though it is modeled by an existential image operation, $\emptyset$ must be excluded from $\mathcal{P}(U)^{\uparrow}$.

Of course, the conditions characterizing **s** and **k** have to be strengthened to biconditionals.[31]

[31]Further details on relational semantics for CL and structurally free logics, see, for example, [66], [26], [14] and [16].

Chapter 7

Dual and symmetric combinatory logics

The application of a numerical function f to arguments $x_1,\ldots,x_n$ is usually written as $f(x_1,\ldots,x_n)$, which is an example of *prefix notation* where the function symbol comes first. There are other notational conventions used, such as $+$ being placed between its arguments in *infix notation*, or the exponent appearing as a superscript. In the case of these functions, we think of numbers as a different sort of object than the sort the functions belong to. This distinction is reinforced by the set-theoretical understanding of functions. However, we saw in chapter 6 that the natural numbers themselves may be seen as functions, and at the same time, computable functions on $\mathbb{N}$ may be viewed as natural numbers. This suggests that $f(n)$ could be thought of as an application of n to f — not only the other way around, an application of f to n.

Dual and *symmetric combinators* are functions that are applicable to arguments from the right-hand side, or from either side. The introduction of these new types of constants changes CL and its properties.

7.1 Dual combinators

In certain areas of mathematics, it is not unusual that an operation may be applied to an object in more than one way or from more than one "direction."

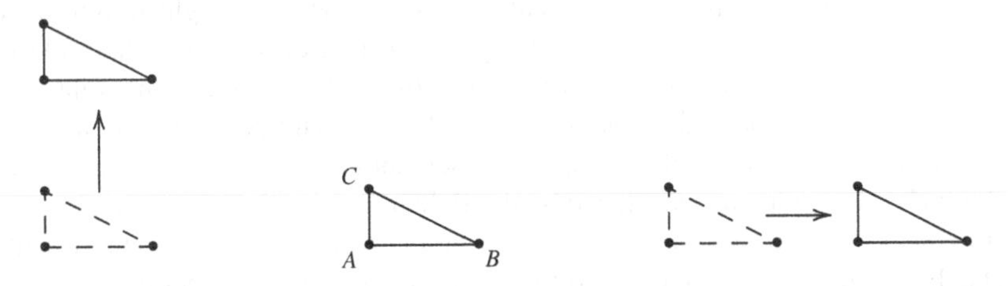

Ordinate and abscissa parallel transformations

A simple geometrical example can show that a polygon in a two-dimensional coordinate system can be moved in parallel with either of the two axes, which in general, results in different locations. The diagram above illustrates this idea.

Universal algebra considers operations abstractly. The label "universal" shows that the operational structures investigated do not need to be thought of as operations on some set of numbers.

The function application operation of CL has a certain similarity to group *multiplication*, and it is even more similar to the binary operation of a groupoid. Group multiplication is not commutative — unless the group is Abelian. A groupoid multiplication is not even associative. A multiplication operation of a non-Abelian group or groupoid may be residuated, when there are two further operations appropriately related to multiplication. These other operations are called *residuals*. For example, a combinatorial reduction algebra from definition 6.2.8 would be residuated if there were two further binary operations $\rightarrow$ and $\leftarrow$ satisfying

$$b \leq c \leftarrow a \quad \text{iff} \quad a \cdot b \leq c \quad \text{iff} \quad a \leq b \rightarrow c.$$

The $\rightarrow$ operation resembles $\rightarrow$ from typed CL, which we introduce in chapter 9. An immediate question then is what corresponds to types that can be generated when $\leftarrow$ is included.

A connective like $\leftarrow$ is quite well-known in nonclassical logics. (Sometimes another symbol, $\backslash$ is used for essentially the same connective.) $\leftarrow$ was introduced into sequent calculi, which do not contain structural rules.[1] These sequent calculi, nowadays, are often called Lambek calculi, and they do not stipulate their fusion operation, $\circ$, to be commutative. $\backslash$ and $/$ cannot be proved to coincide in either the associative or in the nonassociative Lambek calculus. This sort of connection between $\rightarrow$, $\leftarrow$, and the combinators and dual combinators is pursued further in section 7.3 below.

The idea of having functions that apply from the right-hand side is not as far-fetched as it first might appear. Among programming languages, APL stands out as an early example that, in effect, uses suffix and infix notation for functions, because the parsing of strings and the execution of commands proceeds from the right — unlike in many other programming languages. Some other languages later also used suffix notation for functions.[2]

It is well-known that infix notation may be replaced by a parentheses-free *prefix notation* when each function has a fixed (and known) arity. A version of the prefix notation for the connectives of classical logic is still widely taught in introductory logic courses. It could be thought that the parentheses-free notation is helpful to students who fail to properly parenthesize strings of symbols to make them into wff's; but as it turns out, they have just as much difficulty with parsing or writing wff's in prefix notation. The prefix notation for sentential connectives is called "Polish notation," because it was introduced by the Polish logician J. Łukasiewicz.

In *object-oriented programming*, there is an interaction between methods and objects. For instance, a method called the replacement of a subobject might be applicable to both strings and trees. Of course, those objects are likely to have different

[1] See Lambek [95] and Lambek [96]. Newer expositions may be found in many places — e.g., in Bimbó and Dunn [25, Ch. 5].

[2] For instance, some Hewlett-Packard printers used to work with such a language.

low-level implementations, that is, they are different data types. The replacement then is carried out differently too. However, when **`replacement(a, b)`** is evaluated, then the objects **`a`** and **`b`** are the ones that determine how **`replacement`** will be performed.

These examples provide ample motivation for us to consider functions that are applicable from the right-hand side within CL.

We adopt the notational convention that *dual combinators* are denoted by *lowercase* sans serif letters. The duals of the already introduced combinators will be denoted by the same (but small) letter.

Dual combinators are *constants* just as combinators are. Having only dual combinators among the constants has some interest, but it is quite easy to see that this leads to a notational variant of CL. Unless we explicitly state otherwise, we assume that there are some combinators and some dual combinators in the set of constants. The definitions of terms and subterms are unchanged — see definitions 1.1.1 and 1.1.4.

Exercise 7.1.1. Consider the set of CL terms that are generated by the duals of some combinators together with an infinite set of variables. Define a translation between this new set of CL-terms and the set generated by the matching combinators together with variables.

We could use the same symbol $\triangleright$ in axioms of dual combinators, for example, to write $x\mathsf{i} \triangleright x$. However, it will not cause confusion if we use the "dual symbol" $\triangleleft$, which will emphasize that the dual combinators apply from the right. We use $\triangleright$ and $\triangleleft$, just as we might use with equivalent meaning $\geq$ and $\leq$, when we switch the numbers too. The *axioms* for some dual combinators are as follows. We place the dual combinator as the head of a redex on the right-hand side of $\triangleleft$, which is also in harmony with the use of ${}_1\triangleleft$ as a symbol for one-step expansion in previous chapters.

$$x \triangleleft y(x\mathsf{k}) \qquad zyx \triangleleft z(y(x\mathsf{b})) \qquad zy(zx) \triangleleft z(y(x\mathsf{s}))$$
$$xx \triangleleft x\mathsf{m} \qquad y(zx) \triangleleft z(y(x\mathsf{c})) \qquad y(yx) \triangleleft y(x\mathsf{w})$$
$$xyx \triangleleft x\mathsf{y} \qquad zxy \triangleleft z(y(x\mathsf{b}')) \qquad zx(zy) \triangleleft z(y(x\mathsf{s}'))$$

If we would consider only dual combinators as constants, then we could dualize the convention about omitting parentheses from right-associated sequences. However, we typically want to include both types of constants into one system. We cannot have in place though both conventions without introducing ambiguity. We wish to keep CL-terms *unambiguous*. This is the reason why the terms on the right-hand side of the $\triangleleft$'s contain parentheses — when the combinator is at least binary — unlike redexes in axioms for combinators.

THEOREM 7.1.1. (UNIQUE DECOMPOSABILITY) *Every term is* uniquely decomposable, *that is, every term has a* unique formation tree.

Proof: Formation trees for terms are introduced in section A.1, and they closely reflect the way in which terms are put together from their components.

A term is obtained from one of the clauses in definition 1.1.1. If the term M is a constant — a combinator or a dual combinator — then M is one of the two types of constants, but never of both. If M is a variable, then M is a variable in one way. In each of these cases, M is an atomic rather than a complex term, and belongs to exactly one category. If M is obtained by the last clause in definition 1.1.1, then M must be of the form (NP). By the hypothesis of induction, N and P are uniquely decomposable, and then M is uniquely decomposable too. qed

There are objects in mathematics that are not uniquely decomposable. For instance, strings from definition A.1.18 do not have a unique formation tree, despite the fact that each formation tree of a string has the same leaves. It is possible to define strings so that each string can be composed in only one way. This is rarely done, probably, because this causes the connection between the set of strings and a monoid to be lost (or at least to become hidden).

Exercise 7.1.2. Form a string (not a CL-term!) that contains each of $\{\mathsf{B},\mathsf{k},x,\mathsf{C},\mathsf{w}\}$. Draw all the possible formation trees for the string. (Hint: You might wish first to calculate the number of trees.)

CL-terms are *not strings*, even when dual combinators are included into the language, and theorem 7.1.1 ensures that dual combinators do not introduce any ambiguity into the terms themselves.

Definition 1.3.1 is unchangeably applicable to redexes that are headed by a combinator. However, we need to expand the notion of redexes to incorporate dual combinators.

DEFINITION 7.1.2. (REDEXES) If Z^n is a combinator and z^m is a dual combinator, then the terms

$$\mathsf{Z}M_1\ldots M_n \quad \text{and} \quad (N_m\ldots(N_1\mathsf{z})\ldots)$$

are *redexes*, and the displayed occurrences of Z and z are the *heads* of the redexes, respectively. The terms $M_1,\ldots,M_n$ are the *arguments* of Z; similarly, $N_1,\ldots,N_m$ are the *arguments* of z.

Example 7.1.3. $\mathsf{I}x$ is a redex, and so is $x\mathsf{i}$. On the other hand, iI is not a redex, whereas Ii is two redexes. I is the head of one of the redexes with argument i, and i is the head of the other redex with I as its argument.

This is perhaps, the simplest example of a term in which two redexes overlap — indeed, as CL-terms coincide — but neither is wholly inside of an argument of the other one.

Exercise 7.1.3. Find terms in which there is a redex headed by the combinator S and there is a redex headed by the dual combinator c and the two redexes are the same as CL-terms.

Exercise 7.1.4. Find terms with redexes headed by W and b' such that the redexes overlap, however, they do not coincide as terms.

Exercise 7.1.5. Find terms with redexes headed by S' and t such that the redexes overlap, however, neither redex is wholly inside of an argument of the other constant.

Looking at the axioms of the identity combinator and its dual, it appears that we should get two different results from Ii depending on which constant is applied to the other as its argument. This intuition mirrors the informal motivations we put forward earlier. We make the idea formal in the next definition, which expands definition 1.3.2.

DEFINITION 7.1.4. (ONE-STEP REDUCTION) Let Z be an n-ary combinator or dual combinator that is a primitive. Let the axiom of Z (with variables $x_1,\ldots,x_n$) yield the term P. If N is a CL-term with an occurrence of a redex that has Z as its head and $M_1,\ldots,M_n$ as the arguments, and N' is the replacement of that redex occurrence by the term by $[x_1/M_1,\ldots,x_n/M_n]P$, then N *one-step reduces* to N'. This is denoted by $N \rhd_1 N'$.

Notice that sizewise Z is in between a combinator and a dual combinator. That is because the definition of one-step reduction does not distinguish between combinators and dual combinators; the difference between those constants surfaces in the shape of the redexes they form.

Strictly speaking, definition 1.2.2 — the definition of substitution — is not applicable to the present case. However, the incorporation of dual combinators is quite straightforward, and so instead of repeating the definition, we assume that Z can be viewed as a little too tall small letter — with a little squinting. The addition of dual combinators does not lead to significant changes in the way how *substitution* and *replacement* are preformed. Therefore, we use the earlier definitions and notions with the obvious necessary modifications tacitly stipulated.

The notion of *weak reduction* is unchanged, that is, it is the transitive, reflexive closure of one-step reduction. Of course, the set of CL-terms now is expanded, and so a term might contain a redex with a dual combinator as its head.

Example 7.1.5. Let us consider the term $x(M_1(M_2(M_3\mathsf{b})))$, where M_1, M_2 and M_3 are terms. The subterm $M_1(M_2(M_3\mathsf{b}))$ is a redex with the head b. Thus we have that $x(M_1(M_2(M_3\mathsf{b}))) \rhd_1 x(M_1M_2M_3)$ and so $x(M_1(M_2(M_3\mathsf{b}))) \rhd_{\mathrm{w}} x(M_1M_2M_3)$.

Depending on the shape of the terms M_1, M_2 and M_3, there might be other redexes in the term in the above example. We saw in example 7.1.3 that a term, Ii that contains overlapping redexes headed by a combinator and by a dual combinator, can reduce to two terms that do not reduce to each other.

We continue to use $=_{\mathrm{w}}$ as the notation for *weak equality*, and this relation is the transitive symmetric reflexive closure of $\rhd_1$ as before. (Cf. definition 1.3.14.) Previously, S and K turned out not to be weakly equal; were I included as a primitive, $\mathsf{I} \neq_{\mathrm{w}} \mathsf{S}$ and $\mathsf{I} \neq_{\mathrm{w}} \mathsf{K}$ would be provable too. That is, all the primitive constants that we introduced with distinct axioms were not equal in the sense of $=_{\mathrm{w}}$.

The case of the term Ii suggests that the situation might have changed with the inclusion of the dual combinators. Ii shows immediately that $\mathsf{I} =_{\mathrm{w}} \mathsf{i}$.

First, we prove an easy lemma that becomes important now.

LEMMA 7.1.6. *Let M be a term over the combinatory basis $\{\mathsf{I}\}$, that is, the only atomic terms in M are* I*'s.*[3] $M \triangleright_{\mathrm{w}} \mathsf{I}$.

Dually, if $M \in \{\mathsf{i}\}^{\oplus}$, then $M \triangleright_{\mathrm{w}} \mathsf{i}$.

Proof: If M is I, then obviously, $M \triangleright_{\mathrm{w}} \mathsf{I}$. Otherwise, M is (M_1M_2). By the hypothesis of the induction, $M_1 \triangleright_{\mathrm{w}} \mathsf{I}$ and $M_2 \triangleright_{\mathrm{w}} \mathsf{I}$, hence $(M_1M_2) \triangleright_{\mathrm{w}} \mathsf{II}$. But of course, $\mathsf{II} \triangleright_{\mathrm{w}} \mathsf{I}$. We leave writing out the details of the dual case to the reader. qed

The next lemma shows why the reductions of CL-terms built from identities is of any interest now.

LEMMA 7.1.7. (IDENTITIES EQUATE) *Let $\mathfrak{B}$, a combinatory basis contain at least I and an n-ary proper dual combinator z_1. Then $\mathsf{I} =_{\mathrm{w}} \mathsf{z}_1$. Furthermore, $\mathsf{z}_2^m \in \mathfrak{B}$ implies $\mathsf{z}_1 =_{\mathrm{w}} \mathsf{z}_2$, as long as z_2 is proper.*

Dually, if $\mathsf{i}, \mathsf{Z} \in \mathfrak{B}$, where Z is proper, then $\mathsf{Z} =_{\mathrm{w}} \mathsf{i}$, and Z weakly equals *to all* proper combinators *in $\mathfrak{B}$.*

$\mathsf{i}, \mathsf{I} \in \mathfrak{B}$ *implies that* all proper constants *in $\mathfrak{B}$ are* weakly equal.

Proof: The crux of the proof is the CL-term $\mathsf{I} \ldots \mathsf{I}\mathsf{z}_1$, where all the arguments of z_1 are filled by I. Having performed n one-step reductions, we have that $\mathsf{I} \ldots \mathsf{I}\mathsf{z}_1 \triangleright_{\mathrm{w}} \mathsf{z}_1$. On the other hand, $\mathsf{I} \ldots \mathsf{I}\mathsf{z}_1 \triangleright_{\mathrm{w}} M$, where M is a term that has only I as its atomic term (possibly, occurring repeatedly). Now that term weakly reduces to I. $\mathsf{I} =_{\mathrm{w}} \mathsf{z}_1$ as the result.

Next, it is sufficient to note that the only features of z_1 that figured into the proof are that z_1 is a dual and proper combinator. This means that the same steps — possibly, with m I's instead of n I's — can be repeated for a z_2. $\mathsf{I} =_{\mathrm{w}} \mathsf{z}_1$ and $\mathsf{I} =_{\mathrm{w}} \mathsf{z}_2$ give that $\mathsf{z}_1 =_{\mathrm{w}} \mathsf{z}_2$. (The rest of the proof is easy and left as an exercise.) qed

Exercise 7.1.6. Finish the previous proof.

Such a sweeping equating of constants should raise the alarm with respect to the consistency of the set of CL-terms with combinators and dual combinators.

LEMMA 7.1.8. *For any terms M and N from the set of terms over the combinatory basis $\{\mathsf{S}, \mathsf{K}, \mathsf{k}, \mathsf{S}\}$ (with or without a denumerable set of variables), $M =_{\mathrm{w}} N$. That is, both $\{\mathsf{S}, \mathsf{K}, \mathsf{S}, \mathsf{k}\}^{\oplus}$ and $(\{\mathsf{S}, \mathsf{K}, \mathsf{k}, \mathsf{S}\} \cup \{x_i \colon i \in \mathbb{N}\})^{\oplus}$ are* inconsistent *with respect to the weak equality relation.*

Proof: Let us consider the term $\mathsf{K}x(y\mathsf{k})$. Obviously, $\mathsf{K}x(y\mathsf{k}) \triangleright_1 x$ and $\mathsf{K}x(y\mathsf{k}) \triangleright_1 y$. By the definition of the weak equality as the transitive symmetric reflexive closure of $\triangleright_1$, $x =_{\mathrm{w}} y$. We can substitute any term for x and y, that is, we may consider the term $\mathsf{K}M(N\mathsf{k})$ instead of the previous one. The two one-step reductions lead the M and N, respectively. Therefore, $M =_{\mathrm{w}} N$. qed

[3]The set of terms over a set of terms is given in definition A.1.20. The inclusion of dual combinators does not impact the definition except that the set of constants is potentially larger.

Both example 7.1.3 and lemma 7.1.8 show that the Church–Rosser property no longer holds for the set of all CL-terms given a combinatorially complete basis. Moreover, the mere inclusion of K and k suffices for all terms to turn out weakly equal to one another. We can pursue these themes further. Something much stronger can be proved concerning the Church–Rosser property about sets of terms with dual combinators.

THEOREM 7.1.9. (NO CONFLUENCE) *Let $\mathfrak{B}$ be a combinatory basis that contains at least one combinator and at least one dual combinator, both of which are proper. There is a term $M \in (\mathfrak{B} \cup \{x_i : i \in \mathbb{N}\})^{\oplus}$ such that $M \triangleright_{\rm w} N_1$ and $M \triangleright_{\rm w} N_2$, but* there is no term *P to which both N_1 and N_2 reduce.*

Before we turn to the proof itself, we give some definitions and explanations about the proof. We have not emphasized so far that the *classification of combinators* in section 1.2 can be extended straightforwardly to include dual combinators as well. We already called i a dual identity combinator, and we can call m a dual duplicator, etc.

Exercise 7.1.7. Write out the axiomatic characterizations for dual combinators that are identities, associators, cancellators, permutators, duplicators and regulars. (Hint: The convention concerning omitting parentheses is unchanged, which may be a bit confusing.)

The categorization of combinators and dual combinators is handy in the proof of the theorem. The next definition introduces a new category of combinators and dual combinators, which will be found surprisingly useful in the proof. This class of combinators and dual combinators is almost a subcategory of the cancellators. Incidentally, the combinators in this category are exactly those that were used in chapter 3 to represent the *projection functions*.

DEFINITION 7.1.10. (VERY SHORT COMBINATORS) If Z is an n-ary combinator with axiom $\mathsf{Z}x_1 \ldots x_n \triangleright x_i$ (where $1 \leq i \leq n$), then Z is a *very short combinator*. Similarly, z^n with axiom $(x_n \ldots (x_1 \mathsf{z}) \ldots) \triangleright x_i$ is a *very short dual combinator*.

The proof of theorem 2.1.14 used the notion of simultaneous one-step reduction, that we denoted by $\triangleright_{p1}$. This relation on CL-terms was confluent, and its transitive closure is $\triangleright_{\rm w}$. The claim of theorem 2.1.14 was a *universal statement*, which applied to each and every CL-term.

The claim that we wish to prove now is an *existential statement*. Notice that it is obviously false that we can demonstrate the lack of confluence starting with an *arbitrary* term.

A variable x, for instance, weakly reduces only to itself, and $x \triangleright_{\rm w} x$ prevents us from using x to demonstrate the lack of confluence. This fact, of course, chould have been foreseen, because the term x is "not new," so to speak. x is in the set of CL-terms when the only constants are combinators, and that set of the CL-terms is confluent. This observation leads us to look for terms that are *new* in the set of CL-terms when both combinators and dual combinators are permitted as constants.

Thus all CL-terms that do not have an occurrence of a dual combinator fall short of demonstrating the lack of confluence.

A little bit of thought — and perhaps, exercise 7.1.1 — leads us to the conclusion that all CL-terms that are void of at least one combinator, whether they do or do not contain a dual combinator, are also unsuitable to prove the failure of the Church–Rosser property.

So far, we established that a term that could show the lack of confluence must have at least a combinator and at least a dual combinator in it. We can still further delineate the set of terms that cannot exhibit confluence with respect to weak reduction. However, we delay doing that until definition 7.1.14 after the proof of theorem 7.1.9 and theorem 7.1.11.

To sum up, we have to select terms carefully to be able to show that after two weak reductions the resulting terms do not reduce to a common term. A further difficulty is that we are not considering a concrete combinatory basis like $\{\mathsf{S},\mathsf{K},\mathsf{k},\mathsf{S}\}$ or $\{\mathsf{S},\mathsf{s}\}$, but we made a claim concerning *all bases* $\mathfrak{B}$ such that $\{\mathsf{Z}^n,\mathsf{z}^m\}\subseteq\mathfrak{B}$, where $m,n\in\mathbb{N}$, and $\mathsf{Z}^m,\mathsf{z}^n$ are proper constants.

Instead of the elegant proof for theorem 2.1.14, which relied on confluence being preserved under transitive closure, now we get a lengthy *proof by cases*. The description of the cases is where we make use of the categorization of the combinators. We have only one category in which the reduct of a redex is fully described, namely, the category of identity combinators and their duals. A way to look at the following proof is that the versatility of our constants together with the existential character of the claim is what *complicates* the proof. After all the preliminary remarks about the context and the style of the proof, we turn to the proof itself.

Proof: All the cases involve *two overlapping redexes* one of which has a combinator as its head, whereas the other has a dual combinator as its head. A complete proof — including all the cases — is given in [15]. Here we include only a couple of cases for the sake of illustration.

Case 1. Let us assume that Z^n is a *very short combinator* with axiom $\mathsf{Z}x_1\ldots x_n\rhd x_i$. We have assumed that there is a dual combinator in the basis, let us say z^m. As always, there is a denumerable set of variables available to us. For the sake of transparency, we use both x's and y's. The x's fill in most of the arguments of Z and the y's are placed into most of the arguments of z. The overlap between the redexes implies that one of the arguments of Z is a complex term, which contains z and possibly some y's, whereas one of the arguments of z is a complex term, which contains Z and possibly some x's. In particular, we arrange the overlap to occur in the i*th argument place* of Z. The next term and the following two one-step reductions show the lack of confluence.

$$\mathsf{Z}x_1\ldots(y_{m-1}\ldots(y_1\mathsf{z})\ldots)_i\ldots x_n$$

$$\rhd_1\ (y_{m-1}\ldots(y_1\mathsf{z})\ldots) \qquad\qquad M\ {}_1\!\lhd$$

$M\in\{\mathsf{Z}x_1\ldots x_{i-1},y_{m-1},\ldots,y_1,x_{i+1},\ldots,x_n\}^{\oplus}$, and depending on what kind of dual combinator z is, the shape of M varies. But we do not need to concern ourselves with the properties of z at all. It is sufficient to note that z does not occur in M, because z is a proper dual combinator, whereas z occurs in $(y_{m-1} \dots (y_1\mathsf{z})\dots)$. Moreover, the latter term is not a redex. This means that confluence could obtain only if $M \rhd_{\mathrm{w}} (y_{m-1} \dots (y_1\mathsf{z})\dots)$ could happen. But even if M contains a redex, that redex is headed by an occurrence of Z, which is a proper combinator that cannot introduce z via $\rhd_{\mathrm{w}}$.

Case 2. We may assume that Z^n is *not a very short combinator*, which means that $\mathsf{Z}x_1 \dots x_n \rhd M$, where $\ell(M) \geq 2$. Of course, $M \in \{x_1, \dots, x_n\}^{\oplus}$ as before. For this particular case, we assume that *M has a subterm (x_ix_j)*, where $i \neq j$.

If Z is not very short, then inevitably, there is at least one subterm that is formed from two variables. Those may be occurrences of distinct variables (like x_i and x_j) or of one variable (like x_k). (Some cases in the proof, which we will not recall here, deal with combinators that yield a term that has only one or more subterms of the form (x_kx_k).)

We also assume that Z is *not a duplicator for i*. Of course, we tacitly assume that $1 \leq i, j \leq n$. The index i in (x_ix_j) is the same i for which argument Z is not a duplicator. Lastly, we assume that $N \lhd (y_m \dots (y_1\mathsf{z})\dots)$. We consider a term — followed by two one-step reductions.

$$\mathsf{Z}x_1 \dots (y_{m-1} \dots (y_1\mathsf{z})\dots)_i \dots x_n$$

$$\rhd_1 [(x_ix_j)/((y_{m-1} \dots (y_1\mathsf{z})\dots)x_j)]M \qquad\qquad ([y_m/\mathsf{Z}x_1 \dots x_{i-1}]N)x_{i+1} \dots x_n \,{}_1\!\lhd$$

Again, the reduction of the redex that has z as its head leads to a term in which there are no occurrences of z, because z is a proper combinator. However, the term on the left, $[(x_ix_j)/((y_{m-1} \dots (y_1\mathsf{z})\dots)x_j)]M$ is such that it has a nonredex occurrence of z. That is, z occurs in the term, but there is no redex in the term. It is obvious that the two one-step reductions resulted in distinct terms, and no confluence obtains in this case.

Case 3. For this case we continue to assume that Z is *not a very short combinator* and that M, the term in the axiom of Z, has a *subterm of the form (x_ix_j)* with $i \neq j$. We also assume that Z is a *duplicator for i*. (We can make this assumption, because in the previous case we have considered that Z is not a duplicator for i.)

Our assumption about z is that it is *not a cancellator for its last argument*, that is, if $(y_m \dots (y_1\mathsf{z})\dots) \rhd N$ is z's axiom then y_m occurs in N.

Let us consider the same term as in the previous case.

$$\mathsf{Z}x_1 \dots (y_{m-1} \dots (y_1\mathsf{z})\dots)_i \dots x_n$$

$$\rhd_1 \dots ((y_{m-1} \dots (y_1\mathsf{z})\dots)x_j) \dots \qquad\qquad N \,{}_1\!\lhd$$

The subterm that is displayed on the left is intended to emphasize that z has at least one nonredex occurrence in that term. x_i has more that one occurrence in M, the reduct in the axiom of Z. However, the complex term $(y_{m-1} \dots (y_1\mathsf{z})\dots)$ remains intact even if further $\rhd_1$ steps are possible — due to our supposition that z is not a final cancellator. To see this, consider that if $(y_{m-1} \dots (y_1\mathsf{z})\dots)x_j$ is a subterm of a redex occurrence of z, then it is necessarily a subterm of the last argument of the dual combinator. This means that no further reductions — if there are any possible — can result in a term that has no occurrence of z.

On the other hand, the term N does not contain any occurrences of z, but it contains at least one occurrence of Z. We have not stipulated anything about z except that it does not cancel its mth argument; therefore, we do not know whether N reduces to a term that has no occurrences of Z or not. However that is unimportant. Z is a proper combinator and so no matter if N does or does not reduce further, it surely does not reduce to a term that contains z.

Case 4. Now we change our assumption about z, and further refine the assumptions concerning Z.

Let Z be, as before, *not a very short combinator*, but let Z be a *duplicator for its ith* argument place. Again, $\mathsf{Z}x_1 \ldots x_n \rhd M$ with a *subterm* (x_ix_j), where $i \neq j$, in M is stipulated. M might have several occurrences of (x_ix_j), but we assume that at least one of the occurrences of (x_ix_j) is not in a subterm of the form $((\ldots(x_ix_j)\ldots)x_i)$.

For z, we assume that it is a *cancellator for its mth* argument. (This possibility is complementary to the assumption in the previous case.) We consider a term — together with two one-step reductions.

$$\mathsf{Z}x_1 \ldots (y_{m-1} \ldots (y_1\mathsf{z})\ldots)_i \ldots x_n$$

$$\rhd_1 \ldots ((y_{m-1} \ldots (y_1\mathsf{z})\ldots)x_j)\ldots \qquad\qquad N\,{}_1\lhd$$

All three terms, the one we started with, the term on the left, and the one on the right look very similar to those in the previous case. However, they are not the same terms. (The similarity is apparent, and it is due to the partial specification of Z and z, as well as of their reducts.) Despite the differences, the reasoning that allows us to conclude the failure of confluence in this case is like the preceding one.

We may suppose that the displayed subterm on the left is *the* subterm that is not in the last argument of an occurrence of z. Then even if further reductions of redexes headed by z are possible, the exhibited occurrence of z remains in any further terms that can result. Once again, N has no occurrence of z, because z is a proper combinator, and z does not occur in an argument of z in the term we started with. This shows that the two terms cannot reduce to a common term.

These sample cases show some of the typical features of the structure of this proof. The rest of the proof may be found in the paper we cited above. qed

Undoubtedly, a proof that would be shorter and would have fewer cases and subcases would be more elegant and more easy to comprehend.

Exercise* 7.1.8.** Find a better arrangement of the possibilities or find a completely different but more elegant proof for the theorem. (Hint: This is a completely open-ended exercise, which might turn out to be the size of a huge project.)

Our theorem appears to have a certain limitation: we only considered proper combinators and proper dual combinators. We know that the fixed point combinator is important, and we also know that it is definable from proper combinators. In the proof of theorem 7.1.9, however, we relied on the possibility of placing terms into various argument places of combinators. Thus it is *not immediate* that if the fixed point combinator or the dual fixed point combinator are primitive, that is, included into or constitute the combinatory basis then the Church–Rosser property should invariably fail.

THEOREM 7.1.11. (NO CONFLUENCE WITH FIXED POINT CONSTANTS) *Let $\mathfrak{B}$ be a combinatory basis that contains at least one combinator and at least one dual combinator, and at least one element of $\mathfrak{B}$ is a fixed point constant, that is,* $\mathsf{Y} \in \mathfrak{B}$ *or* $\mathsf{y} \in \mathfrak{B}$. *Then there is a term* $M \in (\mathfrak{B} \cup \{x_i \colon i \in \mathbb{N}\})^{\oplus}$ *such that* $M \rhd_{\mathrm{w}} N_1$ *and* $M \rhd_{\mathrm{w}} N_2$, *but* there is no term P *to which both* N_1 *and* N_2 *reduce.*

Proof: The combinatory bases that satisfy the stipulation in the above statement are "more concrete," so to speak, than those in 7.1.9. We know that at least one of the combinators is a fixed point combinator or its dual. Thus we need to consider only three cases, and two of those are each other's duals, hence we omit one of them.

Case 1. Let us assume that we have Y *in* $\mathfrak{B}$, as well as a *proper dual combinator* z^m. The term and the two one-step reductions we consider are

$$\mathsf{Y}(y_{m-1}\ldots(y_1\mathsf{z})\ldots)$$

$$\rhd_1 (y_{m-1}\ldots(y_1\mathsf{z})\ldots)(\mathsf{Y}(y_{m-1}\ldots(y_1\mathsf{z})\ldots)) \qquad M\,{}_1\!\lhd\,.$$

The term M does not contain any occurrences of z, because z is a proper combinator. On the other hand, the first occurrence of the subterm $(y_{m-1}\ldots(y_1\mathsf{z})\ldots)$ is not a redex and it does not overlap with the redexes $(\mathsf{Y}(y_{m-1}\ldots(y_1\mathsf{z})\ldots))$. The latter implies that the first occurrence of $(y_{m-1}\ldots(y_1\mathsf{z})\ldots)$ remains a subterm of all terms to which $(y_{m-1}\ldots(y_1\mathsf{z})\ldots)(\mathsf{Y}(y_{m-1}\ldots(y_1\mathsf{z})\ldots))$ reduces. This establishes that the Church–Rosser property does not hold.

Case 2. Now we consider the fully concrete combinatory basis $\{\mathsf{Y},\mathsf{y}\}$. (Given **Case 1** and its dual, the only remaining possibility is that there are no proper combinators or dual combinators in $\mathfrak{B}$; hence, $\mathfrak{B}$ contains only the fixed point constants.) We would like to form a term that comprises two redexes, one headed by Y and the other by y. Both Y and y are unary, and so there are not many terms to choose from. Indeed, we can create the following term only.

$$\mathsf{Yy}$$

$$\rhd_1 \mathsf{y}(\mathsf{Yy}) \qquad \mathsf{YyY}\,{}_1\!\lhd$$

Of course, neither term is in weak normal form. Indeed, both terms have infinite one-step reduction sequences. Some of the terms to which $\mathsf{y}(\mathsf{Yy})$ reduces are as follows.

$$\mathsf{y}(\mathsf{y}(\mathsf{Yy})), \quad \mathsf{y}(\mathsf{YyY}), \quad \mathsf{y}(\mathsf{y}(\mathsf{y}(\mathsf{Yy}))), \quad \mathsf{y}(\mathsf{YyYY}), \quad \mathsf{y}(\mathsf{y}(\mathsf{Yy})\mathsf{Y})$$

Some of the terms to which YyY reduces are

$$\mathsf{YyYY}, \quad \mathsf{YyYYY}, \quad \mathsf{y}(\mathsf{Yy})\mathsf{Y}, \quad \mathsf{y}(\mathsf{y}(\mathsf{Yy}))\mathsf{Y}, \quad \mathsf{y}(\mathsf{YyY})\mathsf{Y}.$$

It might be confusing that all these terms are built from y's of various sizes with parentheses inserted here and there. However, a little thought and some careful parsing of the terms show that $\mathsf{y}(\mathsf{Yy})$ reduces to terms that are of the form $\mathsf{y}M$ (with various terms in place of M), whereas YyY reduces to terms that are of the form $N\mathsf{Y}$ (with different complex terms in place of N). This latter fact is slightly disguised by the convention of omitting parentheses from left-associated terms, which to start with, allows us to write YyY in lieu of $(\mathsf{Yy})\mathsf{Y}$. qed

There are infinitely many — $\aleph_0$-many, to be precise — terms that cannot be used to demonstrate the lack of confluence. Given that there are just as many CL-terms, it may *appear* as if "practically all" CL-terms are confluent. But, with a little thought, it is quite obvious that in each case in both proofs above, there are $\aleph_0$-many terms that can show that the Church–Rosser property fails.

Exercise 7.1.9. Substantiate the claim in the preceding sentence. (Hint: Just as it is easy to delineate a denumerable set of confluent terms, it is easy to delineate a denumerable set of nonconfluent terms provided we have already found one nonconfluent term.)

We might reiterate that all the terms that we used in the proofs of theorems 7.1.9 and 7.1.11 had at least two redexes such that a dual redex overlapped with a redex. Unfortunately, the absence of a pair of overlapping redexes in a term — one headed by a combinator, another headed by a dual combinator — is not sufficient to guarantee confluence "in the long run."

Example 7.1.12. Let us consider the CL-term $\mathsf{KK}xy(z((z\mathsf{k})\mathsf{k}))$. The term contains two redexes: $\mathsf{K}x$ and $(z((z\mathsf{k})\mathsf{k}))$. Obviously, these two subterms *do not overlap* in $\mathsf{KK}xy(z((z\mathsf{k})\mathsf{k}))$; the term is not even composed of these two subterms as its immediate subterms. However, $\mathsf{KK}xy(z((z\mathsf{k})\mathsf{k})) \rhd_1 \mathsf{K}y(z((z\mathsf{k})\mathsf{k}))$, and also $\mathsf{KK}xy(z((z\mathsf{k})\mathsf{k})) \rhd_1 \mathsf{KK}xy(z\mathsf{k})$. Both these terms contain two redexes, but this time the redexes overlap. Furthermore, one-step reductions of both terms can yield $\mathsf{K}y(z\mathsf{k})$. However, the latter term not only contains overlapping redexes, but it is the "archetypical" CL-term, which involves a combinator and a dual combinator and can be used to show nonconfluence and even inconsistency. Of course, $\mathsf{K}y(z\mathsf{k}) \rhd_{\mathrm{w}} y$ and $\mathsf{K}y(z\mathsf{k}) \rhd_{\mathrm{w}} z$.

The *lack of overlap* between a redex and a dual redex is not sufficient condition *for confluence*, when the term contains a combinator and a dual combinator. On the other hand, the *overlap* between a redex and a dual redex is not a sufficient condition for the *lack of confluence*.

Exercise 7.1.10. Find a term in which K and k head overlapping redexes, nonetheless, the reduction sequences of the term converge.

A slight modification of the term in example 7.1.12 can give us a term, for instance, $\mathsf{KK}xy(z(z\mathsf{k}))$, which eventually — in some sense — reduces to exactly one term, namely, y. This term is not very different from the CL-term $\mathsf{KK}xy(\mathsf{K}zz)$, which does not contain any dual combinators. $\mathsf{KK}xy(z(z\mathsf{k})) \rhd_{\mathrm{w}} \mathsf{K}y(z(z\mathsf{k}))$ and $\mathsf{KK}xy(\mathsf{K}zz) \rhd_{\mathrm{w}}$

$\mathsf{K}y(\mathsf{K}zz)$. In the resulting terms, the redexes $z(z\mathsf{k})$ and $\mathsf{K}zz$, respectively, are fully inside the argument of the initial K.

There are terms, which do not involve cancellators and still, they do not show the lack of confluence despite containing overlapping redexes.

<u>Example</u> 7.1.13. Let us consider Z_1^2 and z_2^2 with axioms $\mathsf{Z}_1x_1x_2 \rhd x_1x_2x_1$ and $x_2(x_1\mathsf{z}_2) \rhd x_2x_1x_1$. Both constants are duplicators for their first argument, and additionally, Z_1 is a permutator, whereas z_2 is an associator. The following term and two pairs of one-step reductions show that the term reduces to the same term no matter which redex is reduced first.

$$\mathsf{Z}_1x(x\mathsf{z}_2)$$

$$\rhd_1 x(x\mathsf{z}_2)x \rhd_1 xxxx \qquad\qquad xxxx\ {}_1\!\lhd\ \mathsf{Z}_1xxx\ {}_1\!\lhd$$

The starting term has two overlapping redexes, and neither redex is wholly inside of an argument of a constant. Admittedly, we choose the constants carefully, and this time, we filled the arguments by the same variable x. However, there seems to be no easy way to exclude terms like $\mathsf{Z}_1x(x\mathsf{z}_2)$ or to give a general characterization of such terms.

Exercise 7.1.11. Find a term with occurrences of Z_1, z_2 and x (only), which can show the lack of confluence.

Exercise 7.1.12. Find other terms, with converging reduction sequences in which overlapping redexes are headed by combinators and dual combinators that were not mentioned in the last couple of examples.

Exercise* 7.1.13. Determine whether it is possible to construct a term with overlapping redexes in which the arguments of the combinator are filled out with x's, and the arguments of the dual combinator are filled out with y's (except at the place of the head of the other redex), so that the term still reduces to the same term eventually.

It would be good to have a characterization of terms that do not have the diamond property with respect to $\rhd_w$. As we tried to illustrate, there is a "gray area" in which we do not have concepts of shape that would allow us to capture exactly *the boundary* between CL-terms that do and do not lead to nonconfluence. Of course, we have the notion of the diamond property itself and so together with $\rhd_w$ we have one kind of characterization. But we would like to have another characterization that is independent of those notions. We can give a description of a set of terms that surely have the diamond property — even though we know that this characterization is not exhaustive in the sense that some CL-terms that cannot be used to show the failure of the diamond property do not belong to this set.

DEFINITION 7.1.14. (SAFE TERMS) Let M be a CL-term. M is a *safe term* iff (1) or (2) is true of M.

(1) M has no redex occurrences of a combinator or of a dual combinator; or

(2) M has only redex occurrences of combinators or dual combinators, and for any two redexes,

(a) they do not overlap; or

(b) if N_1 and N_2 are two overlapping redexes in M, then either N_1 is fully inside of an argument of the head of N_2, or the other way around.

The terms we started with in each of the cases in the proofs of theorems 7.1.9 and 7.1.11 were not safe, which should be the case given our motivation for defining safe terms. However, the CL-term from example 7.1.13 is not safe either.

The next lemma ensures that some sets of CL-terms that we already considered are subsets of the set of safe terms.

LEMMA 7.1.15. *All* CL-*terms that do not contain a dual combinator are* safe. *Dually, all* CL-*terms that do not contain a combinator (though perhaps contain dual combinators) are* safe.

Proof: Clearly, many CL-terms, which do not have occurrences of dual combinators, fall under clause (1), because they are atomic. Or if they are not atomic, the combinator that occurs in them is not the head of a redex.

If a CL-term contains just one redex, then (2) obtains vacuously. If there are at least two redexes, then they are either pairwise disjoint, or some of them overlap. The former situation obviously satisfies (a) in (2). We really need to scrutinize the case when there is an overlap. If there are only combinators (but no dual combinators) in a CL-term, then a redex can completely coincide only with itself. In other words, if M and N are *two* redexes, then one is a proper subterm of the other.

Now let M and N be two redexes. They are both headed by a combinator, let us say, by Z_1^n and Z_2^m, respectively. Thus M is $\mathsf{Z}_1 M_1 \ldots M_n$ and N is $\mathsf{Z}_2 N_1 \ldots N_m$, and both terms are left-associated. Without loss of generality, we may stipulate that N is a proper subterm of M. Then N is one of the M_i's or N is inside of one of the M_i's, which falls under (b) in (2). Indeed, it is well-known that redexes can overlap only this way in CL without dual combinators. The dual case is similar; hence, we omit the details. qed

Example 7.1.16. All terms — whether atomic or not — that have only variables as their atomic subterms are safe, because in such a term there is no occurrence, hence there is no redex occurrence of a combinator or of a dual combinator.

The term $\mathsf{KK}xy(z((z\mathsf{k})\mathsf{k}))$ from example 7.1.12 is not safe, because some occurrences of the constants do form redexes, but some others do not. The difficulty is that after some $\triangleright_1$ steps the combinator and the dual combinator that originally headed no redexes became heads of overlapping redexes (which do not satisfy (b) in (2)).

The terms $\mathsf{KK}xy(z(z\mathsf{k}))$ and $\mathsf{KK}xy(\mathsf{K}zz)$ are not safe either. However, if we drop the first occurrence of K from either term, then we get two safe terms, namely, $\mathsf{K}xy(z(z\mathsf{k}))$ and $\mathsf{K}xy(\mathsf{K}zz)$. Another potential modification to turn the original CL-terms into safe one is by restructuring the terms as $\mathsf{K}(\mathsf{K}xy)(z(z\mathsf{k}))$ and $\mathsf{K}(\mathsf{K}xy)(\mathsf{K}zz)$. The former two terms' nf is xz, whereas the latter two ones' nf is x.

Safe terms can be characterized by appealing to how they can be constructed. Accordingly, the definition is similar to, but not the same as the definition of CL-terms in general.

DEFINITION 7.1.17. (SAFE TERMS, 2) The set of *safe terms* is defined as the union of the sets of terms obtained by (1)–(4) (the set of whole terms), and by (5)–(6) (the set of nonredex terms), respectively.

(1) If x is a variable, then x is a whole term;

(2) if Z^n is a combinator, and $M_1,\ldots,M_n$ are whole terms, then $\mathsf{Z}M_1\ldots M_n$ is a whole term;

(3) if z^n is a dual combinator, and $M_1,\ldots,M_n$ are whole terms, then the term $(M_n\ldots(M_1\mathsf{z})\ldots)$ is a whole term;

(4) if M and N are whole terms, then (MN) is a whole term.

(5) If M is a variable or a constant, then M is a nonredex term;

(6) if M,N are nonredex terms, and M is not of the form $\mathsf{Z}^nP_1\ldots P_{n-1}$, and N is not of the form $(P_{n-1}\ldots(P_1\mathsf{z}^n)\ldots)$, then (MN) is a nonredex term.

The crucial clauses are (2) and (3) which require that all the arguments of a combinator or of a dual combinator are filled out by safe terms.

Exercise 7.1.14. Show that definition 7.1.14 and definition 7.1.17 yield the same subset of CL-terms. (Hint: The two inductive parts of definition 7.1.17 generate sets that have some common elements.)

It may be interesting to note that the complement of the set of safe terms is *not closed under* $\triangleright_{\mathrm{w}}$. We can demonstrate this by finding just one CL-term, which is not safe, but reduces to a safe one. $\mathsf{K}x\mathsf{K}y(z(z\mathsf{k})\mathsf{k})$ is such a term: $\mathsf{K}x\mathsf{K}y(z(z\mathsf{k})\mathsf{k}) \triangleright_1 xy(z(z\mathsf{k})\mathsf{k})$.

A more interesting question is if the set of safe terms is closed under $\triangleright_{\mathrm{w}}$.

LEMMA 7.1.18. (SAFE TERMS REDUCE TO SAFE TERMS) *If M is a safe term that contains occurrences of only proper or fixed point combinators and dual combinators, then $M \triangleright_{\mathrm{w}} N$ implies that N is* a safe term.

Proof: Clearly, we can easily dispense with all the safe terms that are safe due to clause (1) in definition 7.1.14. Otherwise, if $M \triangleright_{\mathrm{w}} N$, then either M is N, or there are $\triangleright_1$ steps linking M to N. If M is N, then the claim is obviously true. Let us assume that M' resulted by n $\triangleright_1$ steps from M. The assumption that M' itself is safe. We may also assume that N is yet another $\triangleright_1$ step away from M' — otherwise, N is safe due to previous considerations.

If the redex reduced in getting from M' to N is a stand-alone redex, then the claim is obviously true: this $\triangleright_1$ step replaces a subterm with a subterm leaving the rest of the term intact, and the new subterm can contain a redex only if the head is a fixed point constant. However, in that case the new redex is like the old one, hence the term remains safe if it was.

The more involved cases are the following two. M' may contain a redex P that contains inside some of its arguments further redexes, or P itself may be inside an argument of another redex. The reduction of P replaces that subterm with one that is itself safe, because P was safe. A one-step reduction cannot alter the internal structure of the arguments of a constant. In fact, it does not matter whether P is within a redex or contains redexes in its arguments.

Replacing the term that is an argument in a redex with another term does not destroy the redex. Thus N is safe too. qed

We note that the definition of simultaneous one-step reduction (definition 2.1.6) can be adopted as it is for dual combinatory logic. We simply assume that the notion of redexes is the one in definition 7.1.2, which includes dual redexes too. The simultaneous one-step reduction is unchangeably situated between one-step reduction and weak reduction.

LEMMA 7.1.19. *If* $M \triangleright_1 N$*, then* $M \triangleright_{p1} N$*. If* $M \triangleright_{p1} N$*, then* $M \triangleright_{p1} P_1 \ldots P_{n-1} \triangleright_{p1} P_n \triangleright_{p1} N$ *for some* n*, or* M *is* N*.*

Proof: The first claim is obvious. For the second claim, we note that $\triangleright_{p1}$ is reflexive, and if $M \triangleright_{p1} N$ by reflexivity, then M is N. If there is exactly one redex reduced, then the claim is again obvious.

Lastly, for the truth of the claim, it is sufficient to observe that if the redexes do not overlap, then their respective one-step reductions do not affect each other. Therefore, they can be preformed in a sequence too. qed

It may be worthwhile to point out that even if the simultaneous one-step reduction of a set of nonoverlapping redexes produces a nonsafe term, the $\triangleright_{p1}$ step can be *emulated* by a series of one-step reductions. This is the reason why we did not stipulate that M is safe as a condition in the lemma.

We can now turn to proving that the set of safe terms has the Church–Rosser property. Theorems 7.1.9 and 7.1.11 showed that trying to select a subset of combinators and dual combinators in order to ensure that the generated set of CL-terms has the Church–Rosser property is futile. The concept of safe terms carves out a proper subset of CL-terms along different lines: it restricts how terms can be constructed.

THEOREM 7.1.20. *If* $\mathbb{T}$ *is the set of safe* CL*-terms generated by a combinatory basis* $\mathfrak{B}$ *comprising proper and fixed point constants only, then for any* $M \in \mathbb{T}$*,* $M \triangleright_w N_1$ *and* $M \triangleright_w N_2$ *imply that* there is a $P \in \mathbb{T}$ *such that* $N_1 \triangleright_w P$ *and* $N_2 \triangleright_w P$*.*

Proof: $\mathbb{T}$ is the set of safe terms. $M \in \mathbb{T}$, hence by lemma 7.1.18, N_1 and N_2 are also safe. If P with the property $N_1 \triangleright_w P$ and $N_2 \triangleright_w P$ is found, then it follows that P is safe, and also that $P \in \mathbb{T}$.

We have reaffirmed that the $\triangleright_{p1}$ relation is sandwiched between one-step and weak reduction. Thus we only have to show that $\triangleright_{p1}$ has the Church–Rosser property on safe terms. This step is similar to the proof of lemma 2.1.13 — except that we consider only safe terms.

1. Let us assume that M is atomic. Then N is M, which is sufficient for the diamond property to hold.
2. Now let us assume that M is complex, for instance, (M_1M_2). If M does not contain redexes (e.g., because all occurrences of all combinators and dual combinators are in nonredex position), then $M \triangleright_{p1} M$ only. Hence, confluence follows — exactly as in case **1**. The same is true when there are redexes in M, but the set of selected redexes is empty.

There are two subcases that we have to scrutinize. M itself may be a redex and may be reduced in one of the $\triangleright_{p1}$ moves, or all the redexes are inside of M_1 and M_2. It is important to note that M cannot be a redex in two ways, that is, M cannot be of the form $\mathsf{Z}^n N_1 \ldots N_{n-1} M_2$ and at the same time of the form $M_1(Q_{m-1} \cdots (Q_1 \mathsf{z}^m)\ldots)$. This possibility is excluded by the stipulation that M is a safe term. If M is of the form $\mathsf{Z}^n N_1 \ldots N_{n-1} M_2$, then it is excluded — for the same reason of being a safe term — that M is also of the form $M'(Q_{m-1} \cdots (Q_1 \mathsf{z}^m)) \ldots M'_{i+1} \cdots M'_n$. The dual case, when M is a dual redex and it contains a redex not fully confined to one of the arguments of the head of the dual combinator, is excluded as well.
2.1 If M is a redex with a combinator as the head, then M is $\mathsf{Z}^n N_1 \ldots N_n$, that is, M_1 is $\mathsf{Z}^n N_1 \ldots N_{n-1}$ and M_2 is N_n. All other redexes in the term are completely inside of one of the N's, because M is safe. The one-step reductions of nonoverlapping redexes from the N_i's yield some N'_i's. (For notational convenience, we write N'_i's for all i, with the understanding that if no redex was reduced in one of the N_i's then N'_i really is N_i.) Then if $\mathsf{Z} N_1 \ldots N_n \triangleright_1 N$ and each $N_i \triangleright_{p1} N'_i$, then $N \triangleright_{p1} P$ as well as $\mathsf{Z} N'_1 \ldots N'_n \triangleright_1 P$. The common term is $[N_n/N'_n, \ldots, N_1/N'_1]N$. (The dual case when M is headed by a dual combinator is similar.)
2.2 If M is not a redex, or it is not a redex that is reduced in the $\triangleright_{p1}$ step, then all the redexes are subterms of either M_1 or M_2. We know that since M is safe, so are M_1 and M_2. Having apportioned the redexes between M_1 and M_2 for both reductions $M \triangleright_{p1} N_1$ and $M \triangleright_{p1} N_2$, we have that $M_1 \triangleright_{p1} N_{11}$ and $M_2 \triangleright_{p1} N_{12}$, and $M_1 \triangleright_{p1} N_{21}$ and $M_2 \triangleright_{p1} N_{22}$. The $N_{11} \triangleright_{p1} P_1$ and $N_{21} \triangleright_{p1} P_1$, and similarly, $N_{21} \triangleright_{p1} P_2$ and $N_{22} \triangleright_{p1} P_2$, by hypothesis. M_1 and M_2 do not overlap; therefore, $N_1 \triangleright_{p1} P_1P_2$ and $N_2 \triangleright_{p1} P_1P_2$. The term P_1P_2 is a suitable P. This concludes the proof that $\triangleright_{p1}$ has the Church–Rosser property on safe terms.

As before, we know that transitive closure is a monotone operation, and it preserves confluence. Then by lemma 7.1.19, $\triangleright_{\mathrm{w}}$ has the Church–Rosser property on the set of safe terms. qed

Exercise*7.1.15. Give a conceptually different proof of the previous theorem. (Hint: Try to make the alleged similarity between the proof of lemma 2.1.13 and the above proof rigorous based on example 7.1.16.)

It might appear that the restriction to safe terms is too strong, and it indeed may be. However, it is not clear how to enlarge the set of safe terms without risking the loss of confluence. One idea could be that it is sufficient to require that M *reduces to a safe term*. On a second glance, this is clearly not sufficient. For instance, the term $\mathsf{KK}xy(z(z\mathsf{k})\mathsf{k})$ *does* reduce to a safe term, namely, to y. But $\mathsf{KK}xy(z(z\mathsf{k})\mathsf{k}) \triangleright_{\mathrm{w}} y$ and

$\mathsf{KK}xy(z(z\mathsf{k})\mathsf{k}) \rhd_{\mathrm{w}} z$, that is, *there is a safe term* to which $\mathsf{KK}xy(z(z\mathsf{k})\mathsf{k})$ reduces, but this does not show confluence.

Another idea could be that the real problem with terms like $\mathsf{KK}xy(z(z\mathsf{k})\mathsf{k})$, which is not a safe term but which is also not a term with overlapping redexes and dual redexes, is that there are one-step reductions that lead to a nonsafe term that contains a redex and a dual redex, which overlap. We can expand the set of safe terms to the set of *mild* terms. Before that we give a name to the CL-terms that we indisputably want to separate out, because all the starting terms in the proofs of theorems 7.1.9 and 7.1.11 belong to this set.

DEFINITION 7.1.21. (TANGY TERMS) M is a *tangy* CL-term iff there is a redex and a dual redex in M that overlap, and neither redex is completely inside of an argument of the head of the other redex.

The set of safe and tangy CL-terms are *disjoint* sets, however, their union is not the set of CL-terms — as long as there are combinators and dual combinators in the language. The informal idea is that tangy terms are likely to reduce to terms that will not further converge. Example 7.1.13 shows that this is not always the case, however, nontangy terms can lead to nonconfluence only if reductions pass through a tangy term.

DEFINITION 7.1.22. (MILD TERMS) M is a *mild* CL-term iff M does not reduce to a tangy term.

Example 7.1.23. The term $\mathsf{KK}xy(z(z\mathsf{k})\mathsf{k})$ is *not mild* but *not tangy* either. We have seen that this term reduces to a tangy term, hence it is not mild.

A term that looks very much like the previous one, $\mathsf{K}x(\mathsf{K}y)(z\mathsf{k}(z\mathsf{k}))$ is not safe, because some occurrences of the constants are not in a head position within a redex. The term is not tangy, because the two redexes in the term do not overlap. This term is mild because the only possible $\rhd_1$ steps lead to $\mathsf{K}x(\mathsf{K}y)z$ and $x(z\mathsf{k}(z\mathsf{k}))$. The latter terms clearly cannot reduce to tangy terms, hence these three terms are all mild.

A difference between the safe and tangy terms on one hand, and the mild terms on the other, is that given a term, it can be determined by analyzing the shape of the term if it belongs to either of the first two categories, whereas reductions may have to be considered in order to determine if a term is mild.

Exercise* 7.1.16. Describe a procedure which will generate all the terms that have to be scrutinized to determine if a given term is mild or not.

LEMMA 7.1.24. *The complement of the set of tangy terms is* not closed *under weak reduction. The set of mild terms is* closed *under weak reduction.*

Exercise 7.1.17. Prove the lemma. (Hint: The proof is nearly immediate from the definitions of the tangy and mild terms together with some examples we considered.)

Mild terms are indeed well-behaved: they do not cause divergence, let alone inconsistency.

LEMMA 7.1.25. *The simultaneous one-step reduction on the set of mild terms* has the Church–Rosser property.

Proof: The structure of the proof is similar to the proof of theorem 7.1.20. However, we group the cases differently — for the sake of transparency.

1. If either or both of the terms N_1, N_2 resulting from M by $\triangleright_{p1}$ are M, then the truth of the claim is immediate.

2. If N_1 and N_2 are the same term, then again, N_1 is sufficient for the truth of the claim.

3. Let M, N_1 and N_2 be distinct, but of course, $M \triangleright_{p1} N_1$ and $M \triangleright_{p1} N_2$. It follows that M is of the form M_1M_2, for some M_1 and M_2. The set of redexes in each $\triangleright_{p1}$ step is pairwise disjoint, hence $M \triangleright_{w} N_1$ and $M \triangleright_{w} N_2$. Then N_1 and N_2 are mild terms too, moreover, all terms (if there are any) between M and N_1, N_2 — obtained by $\triangleright_1$ steps — are also mild.

We consider the shape of (M_1M_2). M itself may be a redex or a dual redex (but not both), and this may be the sole (dual) redex reduced in one of the two initial $\triangleright_{p1}$ reductions. However, then all other redexes and dual redexes must be inside of the arguments of the head, because M is a mild term. This case is like **2.1** in the proof of theorem 7.1.20.

This situation is essentially the same as when M is safe.

It may be that M is not a redex or a dual redex. (Or M may be a redex or dual redex, but the $\triangleright_{p1}$ steps do not include this (dual) redex in the set of nonoverlapping redexes that are reduced in either $\triangleright_{p1}$ step. We will proceed with the weaker assumption, namely, without assuming that M is a redex or a dual redex.) All the nonoverlapping redexes are inside M_1 and M_2. M is a mild term, hence M_1 and M_2 are mild as well. Clearly, $M_1, M_2 \triangleright_{p1} N_1$ means that N_1 is $N_{11}N_{12}$, and similarly, $M_1M_2 \triangleright_{p1} N_2$ implies that N_2 is $N_{21}N_{22}$. There is a P_1 to which N_{11} and N_{21} reduce via $\triangleright_{p1}$, and similarly, there is a P_2, such that $N_{21} \triangleright_{p1} P_2$ and $N_{22} \triangleright_{p1} P_2$. Therefore, (P_1P_2) is a term that ensures confluence. qed

Exercise** **7.1.18.** Find a characterization for mild terms without quantifying over all the terms to which a CL-term reduces or show that no such description is possible. (Hint: A completely new approach might be needed to understand how redexes can be created and how they can interact.)

Exercise** **7.1.19.** Give an exhaustive characterization of CL-terms that have only converging reduction series. (Hint: The question is to be understood as asking for a solution of the potential difficulty which is to find a way to define this set without appealing to all the reducts or to all the reduction sequences of a term.)

7.1.1 Inequational and equational theories

Dual combinatory logic also can be presented in the form of equational and inequational theories. The lack of confluence may or may not be an undesirable property — depending on the context in which the CL-terms are used. If the Church–Rosser property holds with respect to some reduction relation, then the set of CL-terms is consistent with that relation. However, if the Church–Rosser property *does not hold* for a reduction relation, then it *does not follow* that the set of CL-terms with that relation added is inconsistent. Of course, in the latter situation, consistency does not follow either.

We start with *inequational systems.*

We first take the combinatory basis $\mathfrak{B}_1 = \{\mathsf{S},\mathsf{K},\mathsf{k},\mathsf{s}\}$ and a denumerable set of variables. The choice of the combinatory basis affects only the set of the axioms, and this is the most powerful basis we can choose.

Let us reconsider the axioms and rules that we introduced in chapter 5 together with two new axioms. We denote this system by $IQ_{\mathfrak{B}_1}$.

$$x \geq x \quad \text{id}$$

$$\frac{M \geq N}{[x_1/P_1,\ldots,x_n/P_n]M \geq [x_1/P_1,\ldots,x_n/P_n]N}\ \text{sub}$$

$$\frac{M \geq P \quad P \geq N}{M \geq N}\ \text{tr}$$

$$\frac{M \geq N}{PM \geq PN}\ \text{m}_l \qquad \frac{M \geq N}{MP \geq NP}\ \text{m}_r$$

$$\mathsf{S}xyz \geq xz(yz)\ \ \mathsf{S} \qquad z(y(x\mathsf{s})) \geq zy(zx)\ \ \mathsf{s}$$

$$\mathsf{K}xy \geq x\ \ \mathsf{K} \qquad y(x\mathsf{k}) \geq x\ \ \mathsf{k}$$

The new axioms are those for s and k. It may be interesting to note that previously m_l and m_r expressed that terms reduce in argument places just as they do standing alone, whereas m_r expressed that arguments preserve reduction between functions. Now the two rules are are completely alike, because there are functions that apply from the right-hand side.

The notion of a *proof* is as before. We can define inconsistency to be the provability of $M \geq N$, for all M and N.

LEMMA 7.1.26. (VARIABLE CONSERVATION) *If $M \geq N$ is provable in the inequational calculus $IQ_{\mathfrak{B}_1}$, then x occurs in N only if x* occurs in M.

Proof: The proof is by structural induction on proofs.

1. There are five axioms in $IQ_{\mathfrak{B}_1}$. The claim is quite obvious for the identity axiom, as well as for the S and s axioms. Let us consider $\mathsf{K}xy \geq x$. Only x occurs on the right-hand side of $\geq$, and x *does* occur on the left, as needed. The axiom for k is similar, in fact, with respect to the occurrences of variables it does not matter from which side the constant is applicable.

2. In the case of the rules, we suppose that the claim holds for all the premises. Let us consider the substitution rule first. If x occurs in $[x_1/P_1,\ldots,x_n/P_n]N$, then there are two possible ways how this could happen. x could have been introduced by a P_i, or x could have remained in N after the substitutions.

If x occurs in P_i (where $1 \leq i \leq n$), then x_i must have had at least one occurrence in N, otherwise, x would not occur in the term that results from the substitution. By the hypothesis of the induction, x_i occurs in M too. Since P_i is substituted in M for x_i, x occurs in $[x_1/P_1, \ldots, x_n/P_n]M$.

The other case is even more straightforward: if x occurs in N after the substitution because it occurred in N, then x is not one of the x_i's. By hypothesis, x occurs in M, and due to its distinctness from the x_i's, it is retained after the substitution.

The monotonicity rules are easy to verify, and we leave that to the reader.

Let us quickly glance at the transitivity rule. If x occurs in N, then by hypothesis, x occurs in P, and further it occurs in M. This completes the proof. qed

The calculus $IQ_{\mathfrak{B}_1}$ can prove that $\mathsf{K}x(y\mathsf{k}) \geq x$ and $\mathsf{K}x(y\mathsf{k}) \geq y$. Here are the derivations.

$$\frac{\mathsf{K}xy \geq x}{[y/y\mathsf{k}]\mathsf{K}xy \geq x} \qquad\qquad \frac{z(y\mathsf{k}) \geq y}{\mathsf{K}x(y\mathsf{k}) \geq y}$$

On the left-hand side, we used the substitution $[y/y\mathsf{k}]$ and did not carry out the substitution to make it obvious. On the right-hand side, we started with a variant of the k axiom, which itself could be obtained by $[y/z]$, and we performed the $[z/\mathsf{K}x]$ substitution on the bottom line. Both proofs utilize substitution, but the two proofs start with *different axioms.*

The two derivations also make explicit that one of the constants is "inactive" in each proof, because its axiom is not used in obtaining the bottom inequality.

LEMMA 7.1.27. There is no CL-term M *such that for all* N, $M \geq N$ *is provable in* $IQ_{\mathfrak{B}_1}$.

Proof: The claim is almost a consequence of lemma 7.1.26. We only have to add that M — as any CL-term — is finite, that is, $\ell(M) \in \mathbb{N}$. For the sake of concreteness, let us assume that $\ell(M) = n$. We can choose $n+1$ distinct variables: $x_1, \ldots, x_{n+1}$. Were $M \geq x_i$ provable for each i, where $1 \leq i \leq n+1$, by lemma 7.1.26, x_i would occur in M. However, this contradicts $\ell(M) = n$. qed

The inequational presentation of dual CL is not more and not less powerful than the presentation where we start with CL-terms, and we use substitution and replacement to characterize relations on the set of the CL-terms.

Exercise* 7.1.20. Prove that if $M \rhd_{\mathrm{w}} N$ in the set of CL-terms over the combinatory basis $\mathfrak{B}_1$, then $IQ_{\mathfrak{B}_1}$ proves $M \geq N$, and vice versa.

We can extend the $IQ_{\mathfrak{B}_1}$ calculus by *extensionality rules*. We denote the resulting system by $IQ^{\eta}_{\mathfrak{B}_1}$.

$$\frac{Mx \geq Nx \quad x \notin \mathrm{fv}(MN)}{M \geq N}\,\mathrm{ext}_r \qquad\qquad \frac{xM \geq xN \quad x \notin \mathrm{fv}(MN)}{M \geq N}\,\mathrm{ext}_l$$

Exercise 7.1.21. Recall that for any x, $\mathsf{SKK}x \rhd_w x$ and $\mathsf{SK(KK)}x \rhd_w x$. Is $\mathsf{SKK} \geq \mathsf{SK(KK)}$ or $\mathsf{SK(KK)} \geq \mathsf{SKK}$ provable in $IQ^{\eta}_{\mathfrak{B}_1}$?

One of the uses of extensionality is to define constants. For instance, B with axiom $\mathsf{B}xyz \rhd x(yz)$ is definable as $\mathsf{S(KS)K}$.

Exercise 7.1.22. Suppose that an axiom for B is added to $IQ^{\eta}_{\mathfrak{B}_1}$. What does it mean in the extended system that $\mathsf{S(KS)K}$ defines B?

Exercise 7.1.23. In section 1.3, we defined the notion of strong reduction (i.e., $\rhd_s$). What is the relationship between provability in $IQ^{\eta}_{\mathfrak{B}_1}$ and $\rhd_s$?

The weak equality relation between CL-terms over the combinatory basis $\mathfrak{B}_1$ is the *total relation*. We choose another combinatory basis, which comprises the combinators of the $\lambda\mathsf{I}$-calculus and its duals. That is, $\mathfrak{B}_2 = \{\mathsf{B}, \mathsf{I}, \mathsf{W}, \mathsf{C}, \mathsf{c}, \mathsf{w}, \mathsf{i}, \mathsf{b}\}$. The equational calculus $EQ_{\mathfrak{B}_2}$ is by replacing some of the axioms of the equational calculus for CL with new ones.[4]

$$x = x \quad \text{id}$$

$$\frac{M = N}{[x_1/P_1, \ldots, x_n/P_n]M = [x_1/P_1, \ldots, x_n/P_n]N} \;\text{sub}$$

$$\frac{M = P \quad P = N}{M = N} \;\text{tr} \qquad \frac{M = N}{N = M} \;\text{sym}$$

$$\frac{M = N}{PM = PN} \;\text{m}_l \qquad \frac{M = N}{MP = NP} \;\text{m}_r$$

$$\mathsf{B}xyz = x(yz) \;\; \mathsf{B} \qquad zyx = z(y(x\mathsf{b})) \;\; \mathsf{b}$$

$$\mathsf{I}x = x \;\; \mathsf{I} \qquad x = x\mathsf{i} \;\; \mathsf{i}$$

$$\mathsf{W}xy = xyy \;\; \mathsf{W} \qquad y(yx) = y(x\mathsf{w}) \;\; \mathsf{w}$$

$$\mathsf{C}xyz = xzy \;\; \mathsf{C} \qquad y(zx) = z(y(x\mathsf{c})) \;\; \mathsf{c}$$

In the absence of cancellators, we can be sure that our system is consistent due to the following lemma.

LEMMA 7.1.28. *If $M = N$ is provable in $EQ_{\mathfrak{B}_2}$, then x occurs in M* just in case *it occurs in N.*

[4]Equality is symmetric, which makes it convenient to place redexes with dual combinators on the right-hand side again.

Exercise 7.1.24. Prove this lemma. (Hint: The structure of the proof is like that of lemma 7.1.26.)

Although we have consistency, $EQ_{\mathfrak{B}_2}$ is slightly unusual. Recall that lemma 7.1.7 showed that the presence of the identity constants conflates constants.

Exercise 7.1.25. Construct proofs in $EQ_{\mathfrak{B}_2}$ that show that $\mathsf{I} = \mathsf{c}$ and $\mathsf{W} = \mathsf{i}$ are provable.

Exercise 7.1.26. Construct a proof in $EQ_{\mathfrak{B}_2}$ that show that $\mathsf{B} = \mathsf{b}$.

These observations — together with properties of the bracket abstraction definable by $\mathsf{B}, \mathsf{I}, \mathsf{W}$ and C — suggest that $EQ_{\mathfrak{B}_2}$ can be equivalently defined to contain *one polyadic* constant $\flat$. The identity axiom and the rules remain the same, and the axioms for $\flat$ are

$$\flat x_1 \ldots x_n = M \qquad \text{and} \qquad M = (x_n \ldots (x_1 \flat)\ldots),$$

where $n > 0$ and M contains at least one (but possibly multiple) occurrences of each of the x's, and it may contain occurrences of $\flat$ itself. We denote this calculus as $EQ_\flat$.

Exercise* 7.1.27. Prove that $EQ_{\mathfrak{B}_2}$ and $EQ_\flat$ are equivalent. (Hint: The two calculi have different languages; thus first you should come up with a suitable notion of equivalence.)

$EQ_{\mathfrak{B}_2}$ may be extended to an *extensional* calculus by including ext_l and ext_r (with $=$ in place of $\geq$). Let us denote the new calculus by $EQ^\eta_{\mathfrak{B}_2}$.

Exercise 7.1.28. Find examples of provable equations $M = N$ in $EQ^\eta_{\mathfrak{B}_2}$ such that $M =_{\text{w}} N$ is false.

Exercise* 7.1.29. In view of the previous exercise, the relation on CL-terms determined by provability in $EQ^\eta_{\mathfrak{B}_2}$ is a proper subset of $=_{\text{w}}$. Give a characterization, perhaps somewhat like the definition of $=_{\text{w}\zeta}$, of this new relation.

7.2 Symmetric combinators

Intuitionistic logic is sometimes called *constructive*, because of its interpretation by constructions or proofs. The original motivation in Brouwer's works behind intuitionism as an approach to mathematics also relied on ideas about how mathematical objects are constructed.

Constructions and proofs are both effective, that is, they have an algorithmic flavor. Partly because intuitionism started off as a criticism of classical logic, and partly because how mathematicians and logicians in the classical tradition reacted to intuitionism, classical logic has become to be perceived as *nonconstructive*.

Lemma 9.1.35 in chapter 9 epitomizes the relationship between proofs of type assignments for combinators over the basis $\{\mathsf{S},\mathsf{K}\}$ and intuitionistic logic, more precisely, its implicational fragment.

Symmetric λ-calculus was introduced by Barbanera and Berardi in [5] to establish a connection between classical propositional logic as a set of types and a λ-calculus.[5] In intuitionistic logic, there is a certain *asymmetry* between formulas that are of the form $\neg\varphi$ and formulas that do not begin with a negation: the former formulas are equivalent to their double negations, the latter, in general, are not. There is no similar distinction in classical logic, that is, the double negation law is applicable in both directions to all formulas. Without going into further details, we point out that if function application is symmetric, then the symmetry between the formulas φ and $\neg\varphi$ is restored.

Symmetric λ-calculus defines the set of Λ-terms just as other Λ's do — see definition 4.1.1. The symmetric character of function application operation, which is the invisible operation between terms signaled by their juxtaposition, becomes apparent once redexes are defined.

DEFINITION 7.2.1. (β- AND η-REDEXES) The Λ-terms $M\lambda x.N$ and $(\lambda x.N)M$ are β-*redexes*. If $x \notin \mathrm{fv}(N)$, then $\lambda x.Nx$ and $\lambda x.xN$ are η-*redexes*.

Other notions like *one-step β-reduction*, *β-reduction*, etc., are defined as before, but assuming that the new redexes are incorporated.

Example 7.2.2. $y\lambda x.x$ is in $\beta\eta$ nf in (nonsymmetric) Λ, but it is a β-redex in the symmetric λ-calculus: $y\lambda x.x \rhd_{1\beta} y$.

The slogan "everything is a function" takes on a new meaning in the symmetric λ-calculus. A λ-abstract $\lambda x.M$ can occur as a subterm of a β normal form, if $\lambda x.M$ is the whole term itself.

Closed Λ-terms are called *combinators in* Λ, and in the symmetric λ-calculus they apply to arguments both from left (as usual), but also from right. For instance, $(\lambda xy.yx)(\lambda zvw.v(zw))$ yields $\lambda y.y(\lambda zvw.v(zw))$ in one β-reduction step, but it also gives $\lambda vw.v((\lambda xy.yx)w)$. Neither of these terms is in nf, and we further get $\lambda y.(\lambda vw.v(yw))$ and $\lambda vw.v(\lambda y.yw)$, and then $\lambda vw.vw$. The two resulting Λ-terms are clearly not α-equivalent.

Symmetric λ-calculus straightforwardly leads to the idea of considering *symmetric constants* in CL. Of course, we suppose that function application is symmetric too, that is, there is no information built-in into this binary operation about the function–argument roles of its arguments. Constants that behave alike on the left-hand side and on the right-hand side of an operation are not a rarity. For instance, 1 is the

[5] The types in lemma 9.1.35 are simple types, whereas those for symmetric λ-calculus involve the Boolean type constructors $\bot$, $\wedge$ and $\vee$.

left–right identity for $\cdot$ (i.e., for multiplication). Multiplication on $\mathbb{R}$ is commutative, which means that 1 must behave this way. However, the commutativity of the underlying operation is not a requirement for a constant to be "symmetric" in this sense. A *monoid* contains an operation that may be called *concatenation*, which has a left–right identity.[6] Some monoids are *Abelian*, for instance meet semilattices (see definition A.9.7), whereas others are not (see definition A.1.18). By stipulating that function application is symmetric, we do not stipulate that it is a commutative operation.

A symmetric combinator may be thought of as a combinator *amalgamated* with its dual into one constant. Symmetric versions of the combinators S and K are denoted as $\widehat{\mathsf{S}}$ and $\widehat{\mathsf{K}}$. Similarly, placing a $\widehat{\ }$ over a sans serif letter will indicate that the constant is for a symmetric combinator. The set of CL-terms is defined as before — but with only symmetric combinators included as constants. In order to be able to state axioms for symmetric combinators concisely, we define an operation on CL-terms. The notation we use reveals a certain analogy between this dual operation and the converse of a binary relation.

DEFINITION 7.2.3. (DUAL TERMS) The *dual of a* CL-*term* M is recursively defined by (1)–(3).

(1) $x^{\smile}$ is x, whenever x is a variable;

(2) $\widehat{\mathsf{Z}}^{\smile}$ is $\widehat{\mathsf{Z}}$, whenever $\widehat{\mathsf{Z}}$ is a constant;

(3) $(MN)^{\smile}$ is $(N^{\smile}M^{\smile})$.

The dual of a left-associated CL-term is a right-associated CL-term, and vice versa.

Exercise 7.2.1. Define dual terms for CL with combinators and dual combinators. (Hint: Of course, many definitions can be given, however, there is one definition that is a straightforward adaptation of the previous definition.)

For symmetric combinators we give an axiom with a left-associated term, and stipulate that the dual of both terms is also an axiom for the constant. For example, $\widehat{\mathsf{K}}xy \rhd x$; the duals are $y(x\widehat{\mathsf{K}})$ and x, so we have $y(x\widehat{\mathsf{K}}) \rhd x$ too. For $\widehat{\mathsf{S}}$ the axiom that we state explicitly is $\widehat{\mathsf{S}}xyz \rhd xz(yz)$, and we have $z(y(x\widehat{\mathsf{S}})) \rhd zy(zx)$ implicitly. As a rule, for the well-known combinators that we introduced in section 1.2, their symmetric cousins have the same axiom with a left-associated term plus the dual. (Of course, we could use completely different notation, because S and $\widehat{\mathsf{S}}$ are simply two constants.)

The notions of redexes and one-step reductions are defined as previously (with some obvious adjustments for the symmetric constants).

LEMMA 7.2.4. *If* $\widehat{\mathsf{Z}}^n$ *is the head of a redex (where* $n > 1$*), then that redex is either* a left-associated term *or* a right-associated term, *but* not both.

[6]See definition A.9.3.

Exercise 7.2.2. Prove the lemma. (Hint: left- and right-associated terms are defined in section A.1.)

The lemma underscores a useful observation: each concrete redex behaves like either a redex with a combinator or as a redex with a dual combinator. In other words, a symmetric combinator can form redexes in two ways, but never in one concrete term. Incidentally, the latter also holds for $n = 1$, though if $\ell(M) = 2$, then the term can be viewed both as a left- or as a right-associated term.

CL and Λ are closely related, but the differences between them resurface, and perhaps, most vigorously in the symmetric case.

Exercise 7.2.3. The symmetric Λ-term $(\lambda xy.\,yx)\widehat{\ }(\lambda zvw.\,v(zw))$, which we considered above, can be thought of in symmetric CL as $\widehat{\mathsf{T}}\widehat{\mathsf{B}}'$. What terms can be obtained by $\triangleright_{\mathrm{w}}$ when $\widehat{\mathsf{T}}\widehat{\mathsf{B}}'$ is supplied with sufficiently many arguments? (Hint: A combinator in CL requires as many arguments as its arity for $\triangleright_{\mathrm{w}}$ steps.)

Exercise*7.2.4. Develop translations between symmetric CL and symmetric Λ. (Hint: The previous exercise may point at new difficulties.)

Perhaps, one could expect that if dual combinators led to a widespread lack of confluence, then symmetric combinators do so even more often. Symmetric combinators in general lead to the failure of the Church–Rosser property. However, the ability of a symmetric constant to form a redex in two directions not only adds potential $\triangleright_{\mathrm{w}}$ steps at the upper tip of the diamond, so to speak, but also raises the chance of finding more reductions some of which might converge. In other words, there is a similarity between dual and symmetric CL, but their differences are greater than it may seem at first.

THEOREM 7.2.5. (NO CONFLUENCE) *Let $\mathfrak{B}_1$ be a combinatory basis containing symmetric constants. If there are at least two proper constants in $\mathfrak{B}_1$, then $\triangleright_{\mathrm{w}}$* does not have the Church–Rosser property *on the set of terms generated by $\mathfrak{B}_1$ together with a denumerable set of variables.*

Before we turn to the proof, it is useful to point out that the statements in theorems 7.1.9 and 7.2.5 are *independent*, and so are their proofs. The general structure of the proofs is similar though, because both proofs are based on the analysis of the structure of CL-terms that contain at least two overlapping redexes. That is, we will have to consider several cases again. We assume that the classification of combinators and dual combinators is extended to include symmetric combinators in a straightforward way.

Proof: We give a few examples of the cases.[7]
Case 1. Let us assume that one of the two proper symmetric constants that are in $\mathfrak{B}_1$, let us say $\widehat{\mathsf{Z}}_1$, is a *very short combinator*. In particular, $\widehat{\mathsf{Z}}_1^n x_1 \ldots x_n \triangleright x_i$ (where $1 \leq i \leq n$). We consider the following CL-term and two one-step reductions. (We

[7] The complete proof may be found in [17].

use similar conventions concerning the choice of variables, placing the reducts, etc., as we did in the proof of theorem 7.1.9. Unless we state otherwise, the arity of $\widehat{\mathsf{Z}}_1$ is n and the arity of $\widehat{\mathsf{Z}}_2$ is m.)

$$\widehat{\mathsf{Z}}_1 x_1 \ldots (y_{m-1} \ldots (y_1 \widehat{\mathsf{Z}}_2) \ldots)_i \ldots x_n$$

$$\rhd_1 (y_{m-1} \ldots (y_1 \widehat{\mathsf{Z}}_2) \ldots) \qquad\qquad M\, {}_1\lhd$$

The term on the left contains an occurrence of $\widehat{\mathsf{Z}}_2$. The arity of a symmetric combinator does not change via one-step reductions, which means that the term is not a redex. The term on the right-hand edge does not contain any occurrence of $\widehat{\mathsf{Z}}_2$, because it is a proper constant. $\widehat{\mathsf{Z}}_1$ is also proper, therefore, even if further $\rhd_1$ steps are possible starting with M, no further reducts will have an occurrence of $\widehat{\mathsf{Z}}_2$. The lack of confluence is obvious.

Case 2. We may assume that neither $\widehat{\mathsf{Z}}_1$ nor $\widehat{\mathsf{Z}}_2$ is very short. For this case, we assume that both are *unary*. This greatly simplifies the choice of the term we wish to consider.

$$\widehat{\mathsf{Z}}_1 \widehat{\mathsf{Z}}_2$$

$$\rhd_1 M \qquad\qquad N\, {}_1\lhd$$

We do not know the exact shapes of M and N, however, we do not need to know them to see that the diamond property fails. By assumption, $\widehat{\mathsf{Z}}_1$ and $\widehat{\mathsf{Z}}_2$ are distinct and both proper. M has occurrences of $\widehat{\mathsf{Z}}_2$ only, and N has occurrences of $\widehat{\mathsf{Z}}_1$ only. It is clear that no confluence can obtain.

Case 3. Next we consider when one of the two constants $\widehat{\mathsf{Z}}_1$ and $\widehat{\mathsf{Z}}_2$ is *not unary*, for the sake of definiteness we suppose that $\widehat{\mathsf{Z}}_2^{\geq 2}$. For $\widehat{\mathsf{Z}}_1$, we assume that for some $i \neq j$, $(x_i x_j)$ *is a subterm* of the reduct of $\widehat{\mathsf{Z}}_1$, when $\widehat{\mathsf{Z}}_1$'s argument places are filled out with x's.[8] Furthermore, we also assume that an occurrence of $(x_i x_j)$ is not a subterm of a term of the form $(\ldots(x_i x_j)\ldots)x_i$. We consider a term likethe term in **case 1**.

$$\widehat{\mathsf{Z}}_1 x_1 \ldots (y_{m-1} \ldots (y_1 \widehat{\mathsf{Z}}_2) \ldots)_i \ldots x_n$$

$$\rhd_1 \ldots ((y_{m-1} \ldots (y_1 \widehat{\mathsf{Z}}_2) \ldots) x_j) \ldots \qquad\qquad M\, {}_1\lhd$$

The CL-term on the right-hand side does not contain an occurrence of $\widehat{\mathsf{Z}}_2$, whereas the term on the left does. By placing $(x_i x_j)$ (with the term $(y_{m-1} \ldots (y_1 \widehat{\mathsf{Z}}_2) \ldots)$ substituted for x_i) within a pair of $\ldots$, we intend to emphasize that that occurrence cannot disappear from the term even if further $\rhd_1$ steps are possible. Again, because both $\widehat{\mathsf{Z}}_1$ and $\widehat{\mathsf{Z}}_2$ are proper, no confluence can obtain. (The rest of the proof may be found in the paper mentioned earlier.) qed

The next three theorems further emphasize the differences between dual and symmetric combinatory logics.

[8] A careful reader might have noticed that the last assumption about $\widehat{\mathsf{Z}}_1$ implies that $\widehat{\mathsf{Z}}_1$ is not unary (either), which does not complement the previous case. Other cases, which are not included here, deal with the situation when $\widehat{\mathsf{Z}}_1$ is unary.

THEOREM 7.2.6. *Let $\widehat{\mathsf{Z}}$ be a proper symmetric combinator with arity $n \geq 2$, which is a cancellator for its nth argument. Weak reduction* lacks the Church–Rosser property *on the set of* CL-*terms generated by $\widehat{\mathsf{Z}}$ and $2 \cdot (n-1)$ variables.*

Proof: Let us consider the term and the one-step reductions:

$$\widehat{\mathsf{Z}}x_1 \ldots x_{n-1}(y_{n-1} \ldots (y_1\widehat{\mathsf{Z}})...)$$

$$\triangleright_1 M \qquad\qquad N\, {}_1\!\triangleleft .$$

M is formed of x's only, whereas N is the dual term of M, but with all x_i's replaced by y_i's. Neither term contains a combinator, thus the lack of confluence is obvious. qed

THEOREM 7.2.7. *Let $\widehat{\mathsf{Z}}$ be a proper symmetric combinator of arity $n \geq 2$ with axiom $\widehat{\mathsf{Z}}x_1 \ldots x_n \triangleright x_n M$. Weak reduction* lacks the diamond property *on the set of* CL-*terms generated by $\widehat{\mathsf{Z}}$ and $2 \cdot (n-1)$ variables.*

Exercise 7.2.5. Prove the theorem. (Hint: The constraint on the arity of $\widehat{\mathsf{Z}}$ is essential.)

In the case of dual combinatory logic, one constant by itself — whether that constant is a combinator or a dual combinator — cannot cause the failure of confluence. Of course, a symmetric constant can be thought of as two constants in one, and so the fact that one symmetric combinator can lead to nonconfluence is not astonishing.

Theorem 7.1.11 showed that the fixed point constants also cause nonconfluence in the dual setting. Not so when the fixed point constant is symmetric.

THEOREM 7.2.8. (CHURCH–ROSSER PROPERTY WITH FIXED POINT CONSTANT) *Let $\mathfrak{B}_2$ be the combinatory basis $\{\widehat{\mathsf{Y}}, \widehat{\mathsf{I}}\}$. The relation $\triangleright_{\mathrm{w}}$ on the set of* CL-*terms generated by the basis $\mathfrak{B}_2$ and a denumerable set of variables* has the Church–Rosser property.

Proof: The proof that $\triangleright_{\mathrm{w}}$ has the Church–Rosser property is by structural induction on the starting term. The only case that is different — indeed, special — for the current set of CL-terms, is when we have a complex term and it is contains two overlapping redexes. Both $\widehat{\mathsf{I}}$ and $\widehat{\mathsf{Y}}$ are unary, hence there are four distinct terms that can be formed; two of them are duals of each other. We consider one of the latter two.

$$\widehat{\mathsf{I}}\widehat{\mathsf{Y}}$$

$$\triangleright_1 \widehat{\mathsf{Y}} \qquad\qquad \widehat{\widehat{\mathsf{I}}\widehat{\mathsf{Y}}}\widehat{\mathsf{I}}\, {}_1\!\triangleleft$$

The left-hand-side term is in nf, hence confluence can obtain if the right-hand-side term can be reduced to $\widehat{\mathsf{Y}}$. But, of course, $(\widehat{\widehat{\mathsf{I}}\widehat{\mathsf{Y}}})\widehat{\mathsf{I}}$ does reduce to $\widehat{\mathsf{Y}}$. qed

Exercise 7.2.6. There are two terms that are not duals of the term we considered in the proof. Finish the proof by showing that confluence obtains in those cases.

Despite the somewhat prolific character of symmetric constants in forming redexes, the notion of safe terms is secure for symmetric CL. (We assume the obvious modifications in definition 7.1.14 which consists in replacing all constants with symmetric combinators.) The reason is lemma 7.2.4, which ensures — together with the positive arity of all combinators — that once sufficiently many arguments have been supplied to a symmetric constant, it is — in effect — has been reduced to a combinator or a dual combinator.

THEOREM 7.2.9. (SAFE TERMS ARE SAFE) *Let $\mathbb{T}$ be a set of safe terms built from a denumerable set of variables and a set of proper symmetric combinators.* $\triangleright_w$ *has the* Church–Rosser property *on* $\mathbb{T}$.

The other related notions from section 7.1.1 — tangy and mild terms — can be straightforwardly adapted for CL with symmetric combinators. In view of theorem 7.2.8, however, it is somewhat less clear that those concepts are the most suitable ones if we try to delineate classes of CL-terms on which $\triangleright_w$ is and is not confluent.

Exercise 7.2.7.** Find a proper superset of the set of safe terms such that weak reduction retains the Church–Rosser property.

7.2.1 Inequational and equational theories

Symmetric CL also can be presented in the form of inequational and equational theories. The lack of confluence for $\triangleright_w$ does not prevent us from using a functionally complete combinatory basis. Let $\mathfrak{B}_1 = \{\widehat{\mathsf{S}}, \widehat{\mathsf{K}}\}$. The *inequational* calculus for symmetric CL based on $\mathfrak{B}_1$ is denoted by $IQ_{\mathfrak{B}_1}$. In section 7.1.1, we gave inequational and equational calculi for dual CL. The possibility to apply constants from the right is a common new feature of dual and symmetric CL compared to CL with combinators only. The set of axioms includes id, together with the following four axioms. The rules are sub, tr, m_r and m_l. (We list only $\widehat{\mathsf{S}}_l$, $\widehat{\mathsf{S}}_r$, $\widehat{\mathsf{K}}_l$ and $\widehat{\mathsf{K}}_r$ here; the rest of the axioms and rules is as in $IQ_{\mathfrak{B}_1}$ in section 7.1.1.)

$$\widehat{\mathsf{S}}xyz \geq xz(yz) \;\; \widehat{\mathsf{S}}_l \qquad z(y(x\widehat{\mathsf{S}})) \geq zy(zx) \;\; \widehat{\mathsf{S}}_r$$

$$\widehat{\mathsf{K}}xy \geq x \;\; \widehat{\mathsf{K}}_l \qquad y(x\widehat{\mathsf{K}}) \geq x \;\; \widehat{\mathsf{K}}_r$$

We do not introduce any conventions about axioms for symmetric combinators in an inequational logic. The two axioms per symmetric combinator make it clear that the axiom subscripted with $_l$ is like the axiom for the combinator denoted by the same letter; the axiom subscripted with $_r$ is like the axiom for the dual combinator denoted by the same letter.

The analogues of lemmas 7.1.26 and 7.1.27 are true for $IQ_{\mathfrak{B}_1}$.

Exercise 7.2.8. State and prove a lemma that parallels lemma 7.1.26, but concerns the symmetric calculus $IQ_{\mathfrak{B}_1}$.

Exercise 7.2.9. State and prove a lemma for the symmetric $IQ_{\mathfrak{B}_1}$ calculus that is an analogue of lemma 7.1.27 for the dual calculus $IQ_{\mathfrak{B}_1}$.

Our aim in formulating the inequational calculus is to capture the relation $\triangleright_w$ on the set of CL-terms over the basis $\mathfrak{B}_1$ together with a denumerable set of variables.

LEMMA 7.2.10. *If* $M \triangleright_w N$, *then* $IQ_{\mathfrak{B}_1}$ proves $M \geq N$, *and* the other way around.

Proof: From left to right, we proceed according to the definition of $\triangleright_w$. $M \triangleright_w M$ for any term M. We start a corresponding proof in $IQ_{\mathfrak{B}_1}$ by the identity axiom, and apply the rule of substitution with M for x, which gives $M \geq M$.

To simulate a one-step reduction, we recall that a redex is a subterm of a CL-term. Therefore, in the formation tree of the term, the redex corresponds to a subtree. A redex is headed by a constant followed by sufficiently many terms (from left to right or from right to left); let us say, the terms are $M_1, \ldots, M_n$. We can start a proof by the $_l$ or $_r$ axiom for the constant, and then we can apply substitution to $x_1 \ldots x_n$. This yields an inequality that represents the redex and the reduct with M's in place of x's in $\triangleright_1$. The left and right monotonicity rules applied (finitely many times) give the term with the redex on the left, followed by $\geq$, and then by the term with the reduct.

Now let us assume that $M \triangleright_w N$ holds due to the transitive closure of $\triangleright_1$, that is there is some CL-term P, such that $M \triangleright_w P$ and also $P \triangleright_w N$, where the lengths of the one-step sequences between M and P, and between P and N are strictly less than the length of the sequence between M and N. Then by hypothesis, $M \geq P$ and $P \geq N$ are provable in $IQ_{\mathfrak{B}_1}$. The transitivity rule is exactly what captures in the inequational calculus successive $\triangleright_1$ steps and that gives $M \geq N$.

For the right-to-left direction, let us assume that $M \geq N$ is provable. The base case is that $M \geq N$ is an axiom. $x \triangleright_w x$ is a special instance of the identity relation on the set of CL-terms, which is a subset of $\triangleright_w$, by definition. The axioms for combinators are special instances of $\triangleright_1$, which is also a subset of $\triangleright_w$, by definition. The whole term is the redex, and no application of substitution is needed, because the axioms in $IQ_{\mathfrak{B}_1}$ are formulated using variables.

In the induction step, one of the rules must have been applied. If the rule of substitution has been applied, we appeal to the fact that there is a proof of the same inequation in which the substitution rule has been applied to an axiom. If the axiom is identity, then the resulting inequation is an instance of the identity relation on the set of CL-terms. If the axiom involves a constant, then the resulting inequation is an instance of $\triangleright_1$.

The transitivity rule is unproblematic: $N \triangleright_w P$ and $P \triangleright_w M$ together imply $N \triangleright_w M$. For the monotonicity rules we note that $P \triangleright_w P$ for all CL-terms. It $M \triangleright_w N$, by assumption, then both $(MP) \triangleright_w (NP)$ and $(PM) \triangleright_w (PN)$ follow. qed

Exercise* 7.2.10. Prove that in $IQ_{\mathfrak{B}_1}$, for any proof, there is a proof with the same end inequation in "substitutional normal form." We let the latter mean that the sub rule is applied to axioms only.

$IQ_{\mathfrak{B}_1}$ can be extended by *extensionality rules*. They are the rules ext_r and ext_l as in $IQ^{\eta}_{\mathfrak{B}_1}$ in section 7.1.1.

Exercise 7.2.11. Suppose that $IQ_{\mathfrak{B}_1}$ has been extended to $IQ^{\eta}_{\mathfrak{B}_1}$ as described. Find a characterization of the provable inequations as a superset of the $\triangleright_{\mathrm{w}}$ relation. (Hint: Exercises 7.1.21 and 7.1.22 are somewhat related to this question.)

Another direction to extend our calculus is to make the inequality relation symmetric, which leads to an *equational calculus*. This time we assume that the combinatory basis is $\mathfrak{B}_2$, which includes the constants $\widehat{\mathsf{J}}$ and $\widehat{\mathsf{I}}$. We do not repeat the axiom and the rules from $EQ_{\mathfrak{B}_2}$ from section 7.1.1 that do not involve any combinators. The axioms for the elements of $\mathfrak{B}_2$ are the following four. (The axioms for combinators and dual combinators are, of course, not included now.)

$$\widehat{\mathsf{J}}xyzv = xy(xvz)\ \ \widehat{\mathsf{J}}_l \qquad\qquad z(vx)(yx) = v(z(y(x\widehat{\mathsf{J}})))\ \ \widehat{\mathsf{J}}_r$$

$$\widehat{\mathsf{I}}x = x\ \ \widehat{\mathsf{I}}_l \qquad\qquad x = x\widehat{\mathsf{I}}\ \ \widehat{\mathsf{I}}_r$$

There is no cancellator in the combinatory basis. This means that we can be sure of the consistency of the whole system.

LEMMA 7.2.11. *There are* CL-*terms* M *and* N *such that the equation* $M = N$ *is* not provable *in* $EQ_{\mathfrak{B}_2}$. *Therefore,* $EQ_{\mathfrak{B}_2}$ *is* consistent.

Exercise 7.2.12. Prove the lemma. (Hint: A suitable modification or adaptation of lemma 7.1.26 may be useful here.)

The proof of the previous lemma makes the next statement self-evident.

COROLLARY 7.2.11.1. *There is* no CL-term M *that is equal to all* CL-*terms* N.

The inclusion of $\widehat{\mathsf{I}}$ into $\mathfrak{B}_2$ leads to the indistinguishability of the constants with respect to provable equations in which they occur — as it happened in the equational dual calculus. This motivates the following exercises.

Exercise*7.2.13. Prove that for any $\widehat{\mathsf{Z}}_1, \widehat{\mathsf{Z}}_2 \in \mathfrak{B}_2$, if $\widehat{\mathsf{Z}}_1 M_1 \ldots M_n = N$ is provable, then $\widehat{\mathsf{Z}}_2 M_1 \ldots M_n = N$ is provable too.

Exercise*7.2.14. Extend $EQ_{\mathfrak{B}_2}$ to an extensional calculus, and compare the relation emerging from provable equations to $=_{\mathrm{w}}$. (Hint: If the relations do not coincide, give a characterization of the new relation.)

Exercise 7.2.15.** Investigate various combinatory bases with symmetric combinators that do not include $\widehat{\mathsf{I}}$ from the point of view of provable equalities between terms. (Hint: Recall that $\{\mathsf{J}, \mathsf{I}\}$ is sufficient to define all noncancellative combinators, but without I the interdefinability between combinators becomes much more subtle — both in $\{\widehat{\mathsf{J}}\}$ and in other bases such as $\{\widehat{\mathsf{B}}, \widehat{\mathsf{C}}, \widehat{\mathsf{W}}\}$.)

7.3 Structurally free logics

Sequent calculi divide rules into two groups: *connective rules*, which always introduce a connective, and *structural rules*, which change the structure of the sequent. It has become standard practice — following Curry — to label some structural rules with combinators. For example, the exchange rule can be likened to C, the regular permutator. In the case of *LK*, the sequent calculus for classical logic, these labels can be taken to be useful pointers that the principal type schema of the combinator involves an application of the structural rule bearing the label. However, the structural connective of *LK* (i.e., ,) is associative, and so there is a certain *mismatch* between , and function application.

The idea of using a more ramified structural connective goes back at least to a sequent calculus introduced by Dunn in [63], and a binary structural connective prominently featured in the approach taken in [11]. The idea of introducing combinators into a sequent calculus (although in a somewhat limited fashion) seems to have first occurred in [109].

This section introduces *structurally free logics* that integrate some of the features of *substructural logics* and *type assignment systems*.[9] We will consider a class of sequent calculi — rather than just one calculus — hence, we refer to these systems together as *LC**. Sequent calculi are *inherently modular* (if appropriately formulated) in the sense that the omission or the addition of a connective means dropping or including a couple of rules. The isolation of the groups of rules from one another was originally intended even for classical logic, where the connectives are strongly linked by interdefinability results. The separation of groups of rules is more of a necessity in many substructural logics, where connectives are seldom definable from each other. In all logics, there are meta-theoretical advantages springing from the isolation of groups of rules. In line with this approach, we introduce rules for combinators and dual combinators so that they remain independent of each other.

The *language* of *LC* may contain the connectives $\wedge$ (*conjunction*), $\vee$ (*disjunction*), $\rightarrow$ (*implication*), $\leftarrow$ (*co-implication*) and $\circ$ (*fusion*). The constants may include some of the usual Church and Ackermann constants, $\boldsymbol{T}$ (*extensional truth* or *triviality*), $\boldsymbol{F}$ (*extensional falsity* or *absurdity*), and $\boldsymbol{t}$ (*intensional truth*), $\boldsymbol{f}$ (*intensional falsity*). Of course, we also allow the inclusion of *combinators, dual combinators* and *symmetric combinators*. The intensional truth constant is closely related to I, and to some extent also to i and $\widehat{\mathsf{I}}$; however, they are *not* the same.

DEFINITION 7.3.1. (WELL-FORMED FORMULAS (WFF'S)) The set *wff*, that is, *the set of well-formed formulas* is inductively defined from a set of propositional variables $\mathbb{P}$, by clauses (1)–(4).

[9]Structurally free logics were introduced in [66], and further developed and investigated in [26], [14], [19] and [25]. Type assignment systems with simple types and intersection types are presented in chapter 9.

(1) $\mathbb{P} \subseteq \text{wff}$;

(2) $\{\boldsymbol{T}, \boldsymbol{F}, \boldsymbol{t}, \boldsymbol{f}\} \subseteq \text{wff}$;

(3) if c is a combinator, a dual combinator or a symmetric combinator, then $c \in \text{wff}$;

(4) if $A, B \in \text{wff}$ and $* \in \{\wedge, \vee, \leftarrow, \circ, \rightarrow\}$, then $(A * B) \in \text{wff}$.

The combinators are atomic formulas and to this extent they are similar to propositional variables and to the other constants introduced in (2). The idea of making *combinators into wff's* is completely new in structurally free logics. In chapter 9, we will see that typing or type assignment never allows a type to become a term or vice versa; the set of CL-terms and the set of type variables are fully disjoint in those frameworks. Typing and type assignment are ways to put together certain types and certain terms creating a new category of expressions. In structurally free logics, propositional variables are somewhat like *variables* in pure CL, but they are also somewhat like the *type variables* in typed CL.

The four constants introduced in (2) have varying importance with respect to the LC^* logics. $\boldsymbol{t}$ is *vital* to the formulation of these logics as sequent calculi. This constant is closely related to, though in general, distinct from I, which may or may not be included in the set of constants. (We may think of $\boldsymbol{t}$ as the always present alter ego of I.) $\boldsymbol{T}$ and $\boldsymbol{F}$ are wff's that, respectively, are implied by and imply all the wff's in the language. $\boldsymbol{f}$ has a limited importance in LC^*, because we do not allow more than one formula to occur on the right-hand side of a sequent. We do not include $+$, the dual of $\circ$, and we completely omit $\boldsymbol{f}$ from now on.

Exercise 7.3.1. Work out what rules or axioms are needed to add $\boldsymbol{f}$ to LC^*, and what the impact of the addition is.

The original sequent calculus of Gentzen for classical logic included a comma (,) in both the *antecedent* and the *succedent* of a sequent. We define the LC^* logics as *right-singular* sequent calculi; that is, only one wff and no *structural connective* can occur on the right-hand side of a sequent. To emphasize that the structural connective on the left-hand side of a sequent is related to $\circ$ — rather than to $\wedge$ — we will use ;. We do not need a second structural connective; however, the usage of ; here is in harmony with the notation in sequent calculi that utilize two structural connectives. (Such sequent calculi were first introduced by Dunn in [63] for R_+.)

DEFINITION 7.3.2. (STRUCTURES) The set *str*, the *set of structures*, is inductively defined by (1)–(2).

(1) $\text{wff} \subseteq \text{str}$;

(2) if $\mathfrak{A}, \mathfrak{B} \in \text{str}$, then $(\mathfrak{A}; \mathfrak{B}) \in \text{str}$.

The parentheses that surround $\mathfrak{A}; \mathfrak{B}$ in the second clause mean that ; is not a polyadic operation and it is not an associative operation either, because we do not postulate additional equivalences between structures. We denote structures by $\mathfrak{Gothic\ letters}$ (as in clause (2)).

Structures could be viewed as *trees* — just like CL-terms. (Some formation trees for CL-terms are given in example A.1.6 in section A.1.) A difference is that the leaves of a tree are now formulas and the binary operation ; replaces function application, which causes binary branching in the formation tree.

A formula may occur in several places in a structure. We may assume that we can identify a particular occurrence of a formula in a structure. Similarly, we can identify a particular *occurrence of a structure* in a structure. We rarely have to make fully explicit a concrete structure, and so for us, it is important that the identification is possible, but not what exact mechanism is chosen to do the job. We will use the notation $\mathfrak{A}[\mathfrak{B}]$ to indicate that a particular occurrence of the structure $\mathfrak{B}$ within the structure $\mathfrak{A}$ has been selected. This means that $\mathfrak{B}$ is a subtree of $\mathfrak{A}$, possibly, comprising only a leaf.

We may *replace* a subtree by a tree, that is, a structure by a structure; the result is obviously a structure again. To simplify the notation that would emerge by analogy from the replacement of a subterm by a CL-term in a CL-term, we omit the structure that is replaced when context can provide that information. The full notation would look like $[\mathfrak{B}/\mathfrak{C}]\mathfrak{A}$, or with the new term located next to a selected structure $\mathfrak{A}[\mathfrak{B}/\mathfrak{C}]$. Instead we use $\mathfrak{A}[\mathfrak{C}]$, when the replacement occurs in a limited context, for instance, within a rule. If $\mathfrak{A}[\mathfrak{B}]$ figures into the premise of a rule, and we have $\mathfrak{A}[\mathfrak{C}]$ in the lower sequent, then we always assume that $\mathfrak{C}$ has been inserted exactly where the particular chosen occurrence of $\mathfrak{B}$ was in the premise. (In the case of the $\vee\vdash$ rule, we assume that all three bracketed structures occur in the same place in the formation tree of $\mathfrak{A}$.) It is hopefully obvious that we really do not need a more precise procedure, such as indexing, to identify an occurrence of $\mathfrak{B}$ in $\mathfrak{A}$ for the purpose of formulating the rules. (To substantiate the claim that occurrences of structures can be precisely identified, we give further details in section A.7.)

DEFINITION 7.3.3. (SEQUENTS IN LC^*) If $\mathfrak{A}$ is a structure and A is a well-formed formula, then $\mathfrak{A}\vdash A$ is *a sequent.*

We list the *axioms* and *rules* "by groups." Once we will have accumulated all of them, we will specify which rules are to be omitted if a restricted vocabulary is considered.

$$A\vdash A \quad \text{id}$$

$$\frac{\mathfrak{A}[A]\vdash C}{\mathfrak{A}[A\wedge B]\vdash C}\ \wedge\vdash \qquad \frac{\mathfrak{A}[B]\vdash C}{\mathfrak{A}[A\wedge B]\vdash C}\ \wedge\vdash \qquad \frac{\mathfrak{A}\vdash A \quad \mathfrak{A}\vdash B}{\mathfrak{A}\vdash A\wedge B}\ \vdash\wedge$$

$$\frac{\mathfrak{A}[A]\vdash C \quad \mathfrak{A}[B]\vdash C}{\mathfrak{A}[A\vee B]\vdash C}\ \vee\vdash \qquad \frac{\mathfrak{A}\vdash A}{\mathfrak{A}\vdash A\vee B}\ \vdash\vee \qquad \frac{\mathfrak{A}\vdash B}{\mathfrak{A}\vdash A\vee B}\ \vdash\vee$$

$$\frac{\mathfrak{A}[A;B] \vdash A}{\mathfrak{A}[A \circ B] \vdash A} \circ\vdash \qquad \frac{\mathfrak{A} \vdash A \quad \mathfrak{B} \vdash B}{\mathfrak{A};\mathfrak{B} \vdash A \circ B} \vdash\circ$$

$$\frac{\mathfrak{A} \vdash A \quad \mathfrak{B}[B] \vdash C}{\mathfrak{B}[A \to B;\mathfrak{A}] \vdash C} \to\vdash \qquad \frac{\mathfrak{A};A \vdash B}{\mathfrak{A} \vdash A \to B} \vdash\to$$

$$\frac{\mathfrak{A} \vdash A \quad \mathfrak{B}[B] \vdash C}{\mathfrak{B}[\mathfrak{A};B \leftarrow A] \vdash C} \leftarrow\vdash \qquad \frac{A;\mathfrak{A} \vdash B}{\mathfrak{A} \vdash B \leftarrow A} \vdash\leftarrow$$

These rules are not unique to the LC^* calculi; that is, some of the rules appear in some other logics too. The *identity axiom* is often formulated for all formulas A, as above, rather than for propositional variables p. There is no thinning built into this axiom. (All versions of the thinning rule correspond to cancellators, and we may or may not want to include such combinators. Even if want to have a thinning kind of rule in some of the LC^* calculi, we want to be sure that *a trace of an application* of the rule is retained in the sequent in the form of a combinator.)

The *rules for conjunction* and *disjunction* are the so-called structure-free, or additive rules. They differ from their counterpart in Gentzen's LK, because the structures in the premises of the two premise rules are stipulated to be the same.

The latter 6 rules show the connections both between ; and $\circ$, and between the two implications and $\circ$.

The three noncombinatory constants each add a *rule* or an *axiom*.

$$\boldsymbol{F} \vdash A \;\; \boldsymbol{F}\vdash \qquad \frac{\mathfrak{A}[\mathfrak{B}] \vdash A}{\mathfrak{A}[\boldsymbol{t};\mathfrak{B}] \vdash A} \; \boldsymbol{t}\vdash \qquad \mathfrak{A} \vdash \boldsymbol{T} \;\; \vdash\boldsymbol{T}$$

Before we describe the rules for the combinators, we introduce a piece of new notation to make the general form of the rules look more agreeable. Definition A.1.20 defined a set of terms that can be obtained from a starting set of terms.

DEFINITION 7.3.4. (GENERATED TERMS AND STRUCTURES) Let $\mathbb{T} = \{t_1, \ldots, t_n\}$ be a set of CL-terms. If M is a CL-term from the set $\{t_1, \ldots, t_n\}^{\oplus}$, that is, the set of CL-terms that is generated by the set of terms $\mathbb{T}$ according to definition A.1.20, then we occasionally use $(\!| t_1, \ldots, t_n |\!)$ as a *notation* for the CL-term M.

Let $\{\mathfrak{A}_1, \ldots, \mathfrak{A}_n\}$ be a set of structures. If $\mathfrak{B}$ is a structure from the set $\{\mathfrak{A}_1, \ldots, \mathfrak{A}_n\}^{\oplus}$, that is, the set of structures similarly generated by $\{\mathfrak{A}_1, \ldots, \mathfrak{A}_n\}$ according to definition A.7.6, then we occasionally use $(\!| \mathfrak{A}_1, \ldots, \mathfrak{A}_n |\!)$ as a notation for the structure $\mathfrak{B}$.

An advantage of using the $(\!| \;\; |\!)$ notation is that we can skip the cumbersome circumscription of the set to which M or $\mathfrak{B}$ belong. At the same time, the notation does not require that we give a precise shape to M or $\mathfrak{B}$. By introducing the notation in parallel for CL-terms and structures, we intend to stress the similarities between the two concepts. We transplant the convention of omitting parentheses from left-associated CL-terms to left-associated structures.

We give an exhaustive list of rules for concrete combinators and dual combinators.

$$\frac{\mathfrak{A}[\mathfrak{B}]\vdash A}{\mathfrak{A}[\mathsf{I};\mathfrak{B}]\vdash A}\ \mathsf{I}\vdash \qquad \frac{\mathfrak{A}[\mathfrak{B}]\vdash A}{\mathfrak{A}[\mathfrak{B};\mathsf{i}]\vdash A}\ \mathsf{i}\vdash$$

$$\frac{\mathfrak{A}[\mathfrak{B};(\mathfrak{C}\ \mathfrak{D})]\vdash A}{\mathfrak{A}[\mathsf{B};\mathfrak{B};\mathfrak{C};\mathfrak{D}]\vdash A}\ \mathsf{B}\vdash \qquad \frac{\mathfrak{A}[\mathfrak{D};\mathfrak{C};\mathfrak{B}]\vdash A}{\mathfrak{A}[\mathfrak{D};(\mathfrak{C};(\mathfrak{B};\mathsf{b}))]\vdash A}\ \mathsf{b}\vdash$$

$$\frac{\mathfrak{A}[\mathfrak{C};(\mathfrak{B}\ \mathfrak{D})]\vdash A}{\mathfrak{A}[\mathsf{B}';\mathfrak{B};\mathfrak{C};\mathfrak{D}]\vdash A}\ \mathsf{B}'\vdash \qquad \frac{\mathfrak{A}[\mathfrak{D};\mathfrak{B};\mathfrak{C}]\vdash A}{\mathfrak{A}[\mathfrak{D};(\mathfrak{C};(\mathfrak{B};\mathsf{b}'))]\vdash A}\ \mathsf{b}'\vdash$$

$$\frac{\mathfrak{A}[\mathfrak{B};\mathfrak{D};\mathfrak{C}]\vdash A}{\mathfrak{A}[\mathsf{C};\mathfrak{B};\mathfrak{C};\mathfrak{D}]\vdash A}\ \mathsf{C}\vdash \qquad \frac{\mathfrak{A}[\mathfrak{C};(\mathfrak{D};\mathfrak{B})]\vdash A}{\mathfrak{A}[\mathfrak{D};(\mathfrak{C};(\mathfrak{B};\mathsf{c}))]\vdash A}\ \mathsf{c}\vdash$$

$$\frac{\mathfrak{A}[\mathfrak{C};\mathfrak{B}]\vdash A}{\mathfrak{A}[\mathsf{T};\mathfrak{B};\mathfrak{C}]\vdash A}\ \mathsf{T}\vdash \qquad \frac{\mathfrak{A}[\mathfrak{B};\mathfrak{C}]\vdash A}{\mathfrak{A}[\mathfrak{C};(\mathfrak{B};\mathsf{t})]\vdash A}\ \mathsf{t}\vdash$$

$$\frac{\mathfrak{A}[\mathfrak{B};\mathfrak{C};\mathfrak{C}]\vdash A}{\mathfrak{A}[\mathsf{W};\mathfrak{B};\mathfrak{C}]\vdash A}\ \mathsf{W}\vdash \qquad \frac{\mathfrak{A}[\mathfrak{C};(\mathfrak{C};\mathfrak{B})]\vdash A}{\mathfrak{A}[\mathfrak{C};(\mathfrak{B};\mathsf{w})]\vdash A}\ \mathsf{w}\vdash$$

$$\frac{\mathfrak{A}[\mathfrak{B};\mathfrak{B}]\vdash A}{\mathfrak{A}[\mathsf{M};\mathfrak{B}]\vdash A}\ \mathsf{M}\vdash \qquad \frac{\mathfrak{A}[\mathfrak{B};\mathfrak{B}]\vdash A}{\mathfrak{A}[\mathfrak{B};\mathsf{m}]\vdash A}\ \mathsf{m}\vdash$$

$$\frac{\mathfrak{A}[\mathfrak{B};\mathfrak{D};(\mathfrak{C};\mathfrak{D})]\vdash A}{\mathfrak{A}[\mathsf{S};\mathfrak{B};\mathfrak{C};\mathfrak{D}]\vdash A}\ \mathsf{S}\vdash \qquad \frac{\mathfrak{A}[\mathfrak{D};\mathfrak{C};(\mathfrak{D};\mathfrak{B})]\vdash A}{\mathfrak{A}[\mathfrak{D};(\mathfrak{C};(\mathfrak{B};\mathsf{s}))]\vdash A}\ \mathsf{s}\vdash$$

$$\frac{\mathfrak{A}[\mathfrak{C};\mathfrak{D};(\mathfrak{B};\mathfrak{D})]\vdash A}{\mathfrak{A}[\mathsf{S}';\mathfrak{B};\mathfrak{C};\mathfrak{D}]\vdash A}\ \mathsf{S}'\vdash \qquad \frac{\mathfrak{A}[\mathfrak{D};\mathfrak{B};(\mathfrak{D};\mathfrak{C})]\vdash A}{\mathfrak{A}[\mathfrak{D};(\mathfrak{C};(\mathfrak{B};\mathsf{s}'))]\vdash A}\ \mathsf{s}'\vdash$$

$$\frac{\mathfrak{A}[\mathfrak{B}]\vdash A}{\mathfrak{A}[\mathsf{K};\mathfrak{B};\mathfrak{C}]\vdash A}\ \mathsf{K}\vdash \qquad \frac{\mathfrak{A}[\mathfrak{B}]\vdash A}{\mathfrak{A}[\mathfrak{C};(\mathfrak{B};\mathsf{k})]\vdash A}\ \mathsf{k}\vdash$$

$$\frac{\mathfrak{A}[\mathfrak{B};\mathfrak{C};(\mathfrak{B};\mathfrak{E};\mathfrak{D})]\vdash A}{\mathfrak{A}[\mathsf{J};\mathfrak{B};\mathfrak{C};\mathfrak{D};\mathfrak{E}]\vdash A}\ \mathsf{J}\vdash \qquad \frac{\mathfrak{A}[\mathfrak{D};(\mathfrak{E};\mathfrak{B});(\mathfrak{C};\mathfrak{B})]\vdash A}{\mathfrak{A}[\mathfrak{E};(\mathfrak{D};(\mathfrak{C};(\mathfrak{B};\mathsf{j})))]\vdash A}\ \mathsf{j}\vdash$$

$$\frac{\mathfrak{A}[\mathfrak{B};(\mathsf{Y};\mathfrak{B})]\vdash A}{\mathfrak{A}[\mathsf{Y};\mathfrak{B}]\vdash A}\ \mathsf{Y}\vdash \qquad \frac{\mathfrak{A}[\mathfrak{B};\mathsf{y};\mathfrak{B}]\vdash A}{\mathfrak{A}[\mathfrak{B};\mathsf{y}]\vdash A}\ \mathsf{y}\vdash$$

$$\frac{\mathfrak{A}[\mathfrak{D};\mathfrak{B};\mathfrak{C}]\vdash A}{\mathfrak{A}[\mathsf{V};\mathfrak{B};\mathfrak{C};\mathfrak{D}]\vdash A}\ \mathsf{V}\vdash \qquad \frac{\mathfrak{A}[\mathfrak{C};(\mathfrak{B};\mathfrak{D})]\vdash A}{\mathfrak{A}[\mathfrak{D};(\mathfrak{C};(\mathfrak{B};\mathsf{v}))]\vdash A}\ \mathsf{v}\vdash$$

Note that all the rules for combinators are *left-introduction rules*. A combinator can be introduced on the right by the identity axiom only, and afterward it may become a subformula of a formula. (This is true for the constant ***t*** too.) We have not listed any rules that would explicitly mention a *symmetric* combinator. We think of a symmetric combinator as the merger of a combinator and its dual pair. Thus instead of repeating the rules for I and i, for instance, we stipulate that $\widehat{\mathsf{I}}$, the symmetric identity combinator, comes with two introduction rules on the left — like the ones for

I and i, except that those constants are replaced by $\widehat{\mathsf{I}}$. In general, for any symmetric combinator $\widehat{\mathsf{Z}}$, there are *two* left-introduction rules, which are like the rules for Z and z.

The concrete combinators and dual combinators in the rules above have been introduced earlier. However, there are infinitely many proper combinators, and just as many dual and symmetric combinators that are not listed above. We formulate *rule schemas* that allow us to generate a concrete rule for any such constant from its axiom.

Let Z be an n-ary proper combinator. Using the notation $(\!|\ \ |\!)$, we can state the axiom for Z as:

$$\mathsf{Z}x_1 \ldots x_n \rhd (\!| x_1, \ldots, x_n |\!) .$$

The axioms for dual and symmetric proper combinators can be similarly rewritten. Then the rules in LC^* look like:

$$\frac{\mathfrak{A}[(\!| \mathfrak{B}_1, \ldots, \mathfrak{B}_n |\!)] \vdash A}{\mathfrak{A}[\mathsf{Z}; \mathfrak{B}_1; \ldots; \mathfrak{B}_n] \vdash A}\ \mathsf{Z}\vdash \qquad \frac{\mathfrak{A}[(\!| \mathfrak{B}_1, \ldots, \mathfrak{B}_n |\!)] \vdash A}{\mathfrak{A}[\mathfrak{B}_n; (\ldots; (\mathfrak{B}_1; \mathsf{z})\ldots)] \vdash A}\ \mathsf{z}\vdash .$$

The premise appears to be identical in both rules; however, this does not mean that the structure bracketed in $\mathfrak{A}$ should be the same. Rather, the $(\!|\ \ |\!)$ notation is sufficiently liberal to cover both terms, one of which is the dual of the other (when z is the dual variant of Z).[10]

The notion of *a proof* is usual: a proof is a tree in which all the leaves are instances of axioms, and the sequents in the other nodes of the tree are obtained by applications of the rules. The root is the sequent proved. We depicted the formation trees with their roots at the top. The trees that are proofs in a sequent calculus have their root at the bottom.

Each sequent can be turned into a formula, which means that provable sequents can be viewed as formulas. First of all, a sequent is not a formula, because it may contain the structural connective ;. But the rule $\circ\vdash$ fuses bits of structures into formulas. Second, a sequent is not a formula because it has two parts with $\vdash$ in the middle. However, every sequent of the form $A \vdash B$ can be turned into a formula; more precisely, it can be almost turned into a formula. The $\boldsymbol{t}\vdash$ rule introduces $\boldsymbol{t}$ on the left: $\boldsymbol{t};A \vdash B$. Then the $\vdash\rightarrow$ rule is applicable, yielding $\boldsymbol{t} \vdash A \rightarrow B$. If the latter sequent is provable in LC^*, we say that the wff $A \rightarrow B$ is provable or a *theorem* of LC^*. The constant $\boldsymbol{t}$ is a sort of "placeholder" that fills in what would be an empty left-hand side in a sequent calculus like *LK*. (To justify that the formula $A \rightarrow B$ can be equivalently viewed as the sequent we started with, we will need the cut rule.)

A significant difference of LC^* from the classical and intuitionistic sequent calculi is that there are absolutely *no structural rules* in LC^*. More precisely, there are no structural rules that are applicable to all structures, but we introduce three rules that

[10] We do not use $\smile$ on $(\!|\ \ |\!)$, because the latter notation was not intended to specify the shape of a term completely; thus, $(\!| \mathfrak{B}_1, \ldots, \mathfrak{B}_n |\!)^{\smile}$ would not be meaningful.

involve $\boldsymbol{t}$. Also, the combinatory rules are intended to *replace* the structural rules.

$$\frac{\mathfrak{A}[\boldsymbol{t};\boldsymbol{t}] \vdash A}{\mathfrak{A}[\boldsymbol{t}] \vdash A}\; t^{\rightarrow}\vdash \qquad \frac{\mathfrak{A}[\boldsymbol{t};\mathfrak{B};\mathfrak{C}] \vdash A}{\mathfrak{A}[\boldsymbol{t};(\mathfrak{B};\mathfrak{C})] \vdash A}\; t^{\circ}\vdash \qquad \frac{\mathfrak{A}[\boldsymbol{t};\mathfrak{B};\mathfrak{C}] \vdash A}{\mathfrak{A}[\mathfrak{B};(\boldsymbol{t};\mathfrak{C})] \vdash A}\; t^{\leftarrow}\vdash$$

A wff by itself is a structure, and so these rules are *special instances* of structural rules, which are not included into LC^* in their general form. The rule $\boldsymbol{t}^{\rightarrow} \vdash$ is a very special contraction rule: it allows us to collapse adjacent $\boldsymbol{t}$'s into a single $\boldsymbol{t}$. The second rule, $\boldsymbol{t}^{\circ} \vdash$ moves $\boldsymbol{t}$ that is in front of a bigger structure inside of that structure. Lastly, the third rule does the opposite: $\boldsymbol{t}^{\leftarrow} \vdash$ allows us to push $\boldsymbol{t}$ from inside of a structure, moreover, in the process, $\boldsymbol{t}$ can jump over a structure toward the right.

There is one more rule that is *paramount* for a sequent calculus to be sensible. The rule is called the *single cut rule*. The appropriate formulation for LC^* is

$$\frac{\mathfrak{A} \vdash A \quad \mathfrak{B}[A] \vdash B}{\mathfrak{B}[\mathfrak{A}] \vdash B}\; \text{cut.}$$

The cut rule expresses the transitivity of the consequence relation, and a special case of cut corresponds to the rule of modus ponens. To see transitivity in the cut rule, we consider the premises $\mathfrak{A} \vdash A$ and $A \vdash B$ — the result is $\mathfrak{A} \vdash B$. For modus ponens, we take an instance of the cut with premises $\boldsymbol{t} \vdash A$ and $A \vdash B$. The latter can be viewed as $\boldsymbol{t} \vdash A \to B$, where the implication is explicit. The application of the cut gives $\boldsymbol{t} \vdash B$, which asserts the formula resulting from A and $A \to B$ by modus ponens.

The cut rule has to be included into a sequent calculus unless it can be shown to be an admissible rule. A rule is *admissible* when its addition to the calculus does not increase the set of provable formulas (i.e., theorems). The cut rule is *not a derivable rule* in LC^*, because given $\mathfrak{A} \vdash A$ and $\mathfrak{B}[A] \vdash B$ there is no way to apply some of the other rules to obtain $\mathfrak{B}[\mathfrak{A}] \vdash B$. Nonetheless, we have the following.

THEOREM 7.3.5. (CUT THEOREM FOR LC^*) *The single cut rule is* admissible *in* LC^*.

The proof of this theorem is lengthy and we only sketch some of the ideas in section A.7. We note though that the cut rule has more than one form; some of those variants are equivalent in some calculi in the presence of structural rules. Famously, in *LK* an expulsive version called *mix* is admissible too, and the first proof of the admissibility of single cut proceeded through the proof of the admissibility of mix. An LC^* calculus may include contraction-like rules, in which case the following form of the cut rule can be shown to be admissible by fewer inductions than the admissibility of the single cut can be shown.

$$\frac{\mathfrak{A} \vdash A \qquad \mathfrak{B}\langle A\rangle \vdash B}{\mathfrak{B}\langle\mathfrak{A}\rangle \vdash B}\; \text{cut}$$

The angle brackets indicate at least one but possibly more than one occurrence of the structure A in the structure $\mathfrak{B}$. The convention about replacement for the square bracket notation is inherited by the $\langle\ \rangle$'s; that is, $\mathfrak{B}\langle\mathfrak{A}\rangle$ is the result of replacing the selected occurrences of A by $\mathfrak{A}$.

The cut rule is a remarkable and important rule. If the cut rule is included in a calculus, then the *subformula property* cannot be demonstrated (unless, of course, the cut is admissible).

DEFINITION 7.3.6. (SUBFORMULA PROPERTY) A sequent calculus has the *subformula property* iff a provable sequent has a proof in which each formula that occurs in a sequent in the proof is a subformula of the proven sequent.

The admissibility of the cut rule in a sequent calculus implies the subformula property. A perhaps more obvious consequence of the admissibility of the cut is that the cut need not be added separately, because it is "tacitly" already there.

The cut rule, when looked at without taking into account its admissibility, does not support the subformula property, because the wff A disappears from the proof. However, the admissibility of the cut rule means precisely that the arbitrariness of A *in proofs* is apparent.

Cut can speed up proofs, and even if practical or complexity type considerations are not taken into account, cut is necessary in order to have a calculus with the *replacement property*. To illustrate the idea, let us consider classical propositional logic in which $A \wedge B$ is provably equivalent to $B \wedge A$. The truth of the replacement theorem means that no matter where $A \wedge B$ occurs in a wff, that occurrence can be replaced by $B \wedge A$ without affecting the provability or validity of the wff. In *LC**, for instance, $\boldsymbol{t} \circ \boldsymbol{t}$ and $\boldsymbol{t}$ are provably equivalent; hence, we want a sequent that contains one of them to be provable just in case the same sequent with the other in that spot is provable.

THEOREM 7.3.7. *LC** *with all the rules listed is* consistent.

Proof: Let us assume the contrary. Then $p \vdash q$ is provable, where p and q are propositional variables. However, the admissibility of the cut rule guarantees that $p \vdash q$ has a cut-free proof, in which all the formulas occurring in the proof are subformulas of $p \vdash q$. It is easy to verify by looking at all the rules, that there is no way to prove $p \vdash q$ — as long as the propositional variables p and q are distinct. qed

*LC** contains many rules, and we mentioned that further rules might be added using the rule schemas $\mathsf{Z}\vdash$ and $\mathsf{z}\vdash$. Having such a big system, we might consider if we want to omit some of the rules, and what happens, if we do.

First of all, we may consider the *LC** calculi with only some of the constants (combinators, dual combinators or symmetric combinators).

Second, we may omit some of the connectives, for instance, $\wedge$ and $\vee$. Another advantage of the cut theorem (theorem 7.3.5) is that conservativity readily follows.

Approaching the *LC** calculi from another prespective, we may start with a small calculus containing the axioms and rules id, $\circ\vdash$, $\vdash\circ$, $\to\vdash$, $\vdash\to$, $\boldsymbol{t}\vdash$, $\boldsymbol{t}^{\to}\vdash$ and $\boldsymbol{t}^{\circ}\vdash$ together with the combinatory rules $\mathsf{I}\vdash$, $\mathsf{B}\vdash$, $\mathsf{B}'\vdash$ and $\mathsf{W}\vdash$. Let us denote this calculus as LC_1.

LEMMA 7.3.8. *If M is a combinator over the set $\{\mathsf{B},\mathsf{B}',\mathsf{I},\mathsf{W}\}$ and typable by the simple type A, then LC_1* proves *$M^\circ \vdash A$. The converse is also true; that is, if A is a* purely implicational formula, *and $M^\circ \vdash A$ is provable in LC_1, then A is a simple type of M.*

By M° we denote the wff that is obtained by replacing juxtaposition in the CL-term M by fusion. LC^* combines types and combinators by treating all of them as formulas, thus the claim should not be very surprising.

Exercise* 7.3.2. Prove the lemma. (Hint: Type assignment systems are treated in chapter 9; those with simple types are in section 9.1.)

<u>Example</u> 7.3.9. The CL-term $\mathsf{BB'W}$ can be assigned the type $(p \to p \to q) \to ((q \to r) \to p \to r)$. We give a proof of the sequent $\mathsf{B};\mathsf{B}';\mathsf{W};p \to p \to q;q \to r;p \vdash r$, which after a few more easy steps can be turned into the desired sequent.

$$\dfrac{\dfrac{\dfrac{\dfrac{p \vdash p \qquad \dfrac{p \vdash p \qquad \dfrac{q \vdash q \qquad r \vdash r}{q \to r;q \vdash r}}{q \to r;(p \to q;p) \vdash r}}{q \to r;(p \to p \to q;p;p) \vdash r}}{\dfrac{q \to r;(\mathsf{W};p \to p \to q;p) \vdash r}{\mathsf{B}';(\mathsf{W};p \to p \to q);q \to r;p \vdash r}}}{\mathsf{B};\mathsf{B}';\mathsf{W};p \to p \to q;q \to r;p \vdash r}$$

A nearly obvious result is the following.

LEMMA 7.3.10. *If M is an n-ary proper combinator over the basis $\{\mathsf{B},\mathsf{B}',\mathsf{I},\mathsf{W}\}$ such that $Mx_1\ldots x_n \rhd_{\mathrm{W}} N$, then $M^\circ;\mathfrak{B}_1;\ldots;\mathfrak{B}_n \vdash ([x_1/\mathfrak{B}_1,\ldots,x_n/\mathfrak{B}_n]N)^\circ$ is* provable *in LC_1.*

The lemma in effect says that the combinatory rules faithfully reproduce one-step reductions in CL. The reduction steps are blended together by the combinatory rules.

<u>Example</u> 7.3.11. Let us consider the CL-term $\mathsf{BB'W}$ again, which is a ternary combinator. $\mathsf{BB'W}xyz \rhd_{\mathrm{W}} y(xzz)$, and LC_1 proves $\mathsf{B} \circ \mathsf{B}' \circ \mathsf{W};\mathfrak{B}_1;\mathfrak{B}_2;\mathfrak{B}_3 \vdash \mathfrak{B}_2^\circ \circ (\mathfrak{B}_1^\circ \circ \mathfrak{B}_3^\circ \circ \mathfrak{B}_3^\circ)$. To simplify the proof, we turn the $\mathfrak{B}$'s into B's, that is, instead of structures, we take formulas. (The next exercise — with sufficiently many $\circ\vdash$ steps — shows that this is harmless.)

$$\vdots$$

$$\dfrac{\dfrac{\dfrac{\dfrac{\dfrac{B_2 \vdash B_2 \qquad (B_1;B_3);B_3 \vdash (B_1 \circ B_3) \circ B_3}{B_2;(B_1;B_3;B_3) \vdash B_2 \circ ((B_1 \circ B_3) \circ B_3)}}{B_2;(\mathsf{W};B_1;B_3) \vdash B_2 \circ ((B_1 \circ B_3) \circ B_3)}}{\mathsf{B}';(\mathsf{W};B_1);B_2;B_3 \vdash B_2 \circ ((B_1 \circ B_3) \circ B_3)}}{\mathsf{B};\mathsf{B}';\mathsf{W};B_1;B_2;B_3 \vdash B_2 \circ ((B_1 \circ B_3) \circ B_3)}}{\mathsf{B} \circ \mathsf{B}' \circ \mathsf{W};B_1;B_2;B_3 \vdash B_2 \circ ((B_1 \circ B_3) \circ B_3)}$$

Exercise 7.3.3. We used in the previous example that for any structure $\mathfrak{A}$, $\mathfrak{A} \vdash \mathfrak{A}^\circ$ is provable in LC_1 (hence in LC^*). Prove this for LC_1. (Hint: Use structural induction on the construction of $\mathfrak{A}$.)

Exercise 7.3.4. Prove lemma 7.3.10. (Hint: You may want to use the result from the previous exercise.)

So far we have illustrated that LC_1 can emulate combinatory reduction as well as type assignment with simple types. However, LC_1 can assign types to combinators that do not have simple types — despite the fact that LC_1 does not contain the extensional connectives. The presence of the fusion operation allows us to form compound types, which resemble the two CL-terms in the axiom of the combinator.

<u>Example</u> 7.3.12. Let us consider the CL-term WI or simply M. This term is not typable, because xx, the reduct of Mx is not typable due to the fact that there is no implicational formula that is identical to its antecedent. Given the above lemma and example, it is easy to see that LC_1 proves $\mathsf{W} \circ \mathsf{I} \vdash \mathfrak{B}^\circ \to (\mathfrak{B} \circ \mathfrak{B})^\circ$. Here is the proof (with the $\mathfrak{B}$'s fused into B's).

$$\dfrac{\dfrac{\dfrac{\dfrac{\dfrac{B \vdash B}{\mathsf{I}; B \vdash B} \quad B \vdash B}{\mathsf{I}; B; B \vdash B \circ B}}{\mathsf{W}; \mathsf{I}; B \vdash B \circ B}}{\mathsf{W} \circ \mathsf{I}; B \vdash B \circ B}}{\mathsf{W} \circ \mathsf{I} \vdash B \to (B \circ B)}$$

LC_1 can be conservatively extended to LC_2 by $\wedge$ and $\vee$. This gives further possibilities to prove sequences that can be likened to type assignments in an extended type assignment system.

Exercise* 7.3.5. Investigate the connection between LC_2 and the intersection type assignment system (with the basis $\{\mathsf{B}, \mathsf{B}', \mathsf{I}, \mathsf{W}\}$).

As the next extension, we may consider the inclusion of rules for the dual combinators $\mathsf{b}, \mathsf{b}', \mathsf{i}$ and w.

Exercise 7.3.6. Which formulas could be thought to replace the simple types for dual combinators?

Exercise* 7.3.7. Give an exhaustive characterization of the formulas that may appear in provable sequents on the right-hand side of $\vdash$ when the left-hand side contains only a constant or fused constants.

There is a lot more that could be added here about structurally free logics, their connections to type assignment systems, their models, etc. We hope that this brief section gave the flavor of these calculi and the sorts of problems and of some results that emerge from this approach on the edge of proof theory and CL.

Chapter 8

Applied combinatory logic

We have emphasized that CL is a *powerful theory*. Thus it is not surprising that combinatory logic has been applied in the foundations of mathematics. The approach that intended to turn CL into a suitably general framework is called *illative combinatory logic*. We devote the first section to this topic.

The origin of CL is the question of the eliminability of bound variables and the problem of substitution. We give a detailed illustration how the incorporation of combinators into a first-order language allows the complete elimination of bound variables.

These examples of applications of CL in mathematics, philosophy and logic are not exhaustive. Some other applications that we do not detail here include compilers and type systems for functional programming languages. Depending on where we draw the line between pure and applied CL, some of the earlier topics could be viewed as applications too, for example, the formalization of recursive functions or the type assignment systems.

8.1 Illative combinatory logic

One of the great intellectual themes of the early 20th century was the *quest for an overarching theory* that could serve as the foundations for mathematics. The approach based on combinatory logic has been termed illative combinatory logic.

CL — inevitably — contains constants. So far, we had combinators, and then dual and symmetric combinators. Illative combinatory logic calls for the inclusion of other sorts of constants that can be thought of as symbols for logical and mathematical operations and relations. For example, constants for connectives such as implication, negation, conjunction and disjunction may be included together with constants for quantifiers. Other constants may stand for equality, set membership and the subset relation.

Illative combinatory logic *extends* pure combinatory logic, where the latter is thought of as the theory of weak equality. The inclusion of *new constants*, of course, expands the set of CL-terms.

Example 8.1.1. P is the symbol for $\supset$, that is, for implication. By definition 1.1.1, constants are CL-terms. With the extended set of constants, the set of CL-terms is extended too.

$\mathsf{P}xx$ is thought of as $x \supset x$, and $\mathsf{P}(\mathsf{P}xy)(\mathsf{P}(\mathsf{P}yz)(\mathsf{P}xz))$ is $(x \supset y) \supset ((y \supset z) \supset (x \supset z))$. This formula is theorem of classical logic (with formulas in place for x, y and z), and it is the principal type schema of B'.[1]

The use of P is reminiscent of C, which is used in prefix notation for implication. A difference is that C is used in a parentheses-free manner. The next three expressions are "essentially the same."

(1) $\mathsf{P}(\mathsf{P}xy)(\mathsf{P}(\mathsf{P}yz)(\mathsf{P}xz))$

(2) $CCpqCCqrCpr$

(3) $A \to B \to (B \to C) \to A \to C$

(1) is an illative CL-term. (2) is a formula in prefix notation with p, q and r standing for propositional variables. (3) is the simple type schema of B' with some parentheses omitted according to association to the right. Notice that (1) is similar to (2) in respect to the locations where P appears (instead of C), but it includes parentheses to indicate the arguments of P (in accordance with the convention about delimiting arguments by parentheses in CL-terms).

The inclusion of constants would be more or less pointless if they would be simply symbols, which are not completely unlike meta-variables except that no substitution is possible for them. Accordingly, the set of weak equalities is enriched by *assertions* and *rules* that are specific for terms of predetermined shapes. In fact, if the identity is represented by a constant, then weak equalities may be viewed as a proper subset of assertions too; then all the theorems of the system are of one kind.

DEFINITION 8.1.2. Let the set of constants be expanded by P. Further, let the set of CL-terms be defined as in 1.1.1, where Z ranges over $\{\mathsf{S}, \mathsf{K}, \mathsf{P}\}$. The set of *theorems* comprises *equalities* (in the extended language) provable in $EQ_{\mathfrak{B}_1}$ with the combinatory axioms restricted to S and K, and *assertions* obtainable from the assertions (A1)–(A3) by rules (R1)–(R2).[2]

(A1) $\mathsf{P}M(\mathsf{P}NM)$

(A2) $\mathsf{P}(\mathsf{P}M(\mathsf{P}NP))(\mathsf{P}(\mathsf{P}MN)(\mathsf{P}MP))$

(A3) $\mathsf{P}(\mathsf{P}(\mathsf{P}MN)M)M$

(R1) $\mathsf{P}MN$ and M imply N

(R2) M and $M = N$ imply N

[1]Simply typed CL is introduced in the next chapter, in section 9.1, with $\to$ for implication. Implicational fragments of various logics are briefly recounted in section A.9.

[2]We use $M, N, P, \ldots$ as before to denote meta-terms. This incorporates the effect of substitution into the assertions and rules without postulating a separate substitution rule.

We denote this system by $\mathfrak{I}\mathfrak{l}$, and call it the *minimal illative combinatory logic*.

The axioms together with (R1) are equivalent — with M and N thought of as formulas rather than CL-terms — to the implicational fragment of classical logic.

The notion of a proof for assertions is the same as in an axiomatic calculus. Namely, a *proof* is a finite sequence of assertions in which each assertion is an instance of (A1)–(A3) or obtainable from previous elements of the sequence by an application of one of the rules (R1) and (R2).

For our purposes the following theorems are of particular importance. We list them together with their usual name.

(1) $\mathsf{P}MM$ [self-implication]

(2) $\mathsf{P}(\mathsf{P}M(\mathsf{P}MN))(\mathsf{P}MN)$ [contraction]

Exercise 8.1.1. Show that (1) and (2) are theorems. (Hint: To prove (2), it may be useful to prove $\mathsf{P}(\mathsf{P}M(\mathsf{P}NP))(\mathsf{P}N(\mathsf{P}MP))$ first.)

LEMMA 8.1.3. *The system* $\mathfrak{I}\mathfrak{l}$ *is* inconsistent.

Proof: The proof of the lemma can be carried out by utilizing *Curry's paradox*. We use the later version of the paradox now. In order to keep CL-terms to readable size, we freely use some of the well-known combinators (which are definable from S and K). We can form the CL-term $\mathsf{C}(\mathsf{B}(\mathsf{WB})\mathsf{P})y$, where y is a variable. $\mathsf{C}(\mathsf{B}(\mathsf{WB})\mathsf{P})yx \triangleright_{\mathrm{w}} \mathsf{P}x(\mathsf{P}xy)$, hence the equality of the latter two CL-terms is provable. Then, of course,

$$\mathsf{C}(\mathsf{B}(\mathsf{WB})\mathsf{P})y(\mathsf{Y}(\mathsf{C}(\mathsf{B}(\mathsf{WB})\mathsf{P})y)) \triangleright_{\mathrm{w}} \mathsf{P}(\mathsf{Y}(\mathsf{C}(\mathsf{B}(\mathsf{WB})\mathsf{P})y))(\mathsf{P}(\mathsf{Y}(\mathsf{C}(\mathsf{B}(\mathsf{WB})\mathsf{P})y))y).$$

Y is the fixed point combinator, hence, we know that $\mathsf{Y}M =_{\mathrm{w}} M(\mathsf{Y}M)$ for any M, including $\mathsf{C}(\mathsf{B}(\mathsf{WB})\mathsf{P})y$. We abbreviate $\mathsf{C}(\mathsf{B}(\mathsf{WB})\mathsf{P})y$ by M. The displayed weak reduction is of the form $M(\mathsf{Y}M) \triangleright_{\mathrm{w}} \mathsf{P}(\mathsf{Y}M)(\mathsf{P}(\mathsf{Y}M)y)$. Hence, $\mathsf{Y}M = \mathsf{P}(\mathsf{Y}M)(\mathsf{P}(\mathsf{Y}M)y)$ is a provable equation. We construct the following sequence of equations and assertions.[3]

1. $\mathsf{P}(\mathsf{P}(\mathsf{Y}M)(\mathsf{P}(\mathsf{Y}M)y))(\mathsf{P}(\mathsf{Y}M)y)$ [theorem, by exercise 8.1.1]
2. $\mathsf{Y}M = \mathsf{P}(\mathsf{Y}M)(\mathsf{P}(\mathsf{Y}M)y)$ [provable equation]
3. $\mathsf{P}(\mathsf{Y}M)(\mathsf{P}(\mathsf{Y}M)y)$ [by (R2) from 1. and 2.]
4. $\mathsf{Y}M$ [by (R2) from 3. and 2.]
5. $\mathsf{P}(\mathsf{Y}M)y$ [by (R1) from 3. and 4.]
6. y [by (R1) from 5. and 4.]

[3]We insert the equations for transparency; they can be left out to obtain a sequence of assertions only, which conforms to the definition of proofs.

This means that y, which is a variable, is a theorem. Clearly, we could have used a meta-term everywhere in the proof or any concrete CL-term — nothing depends on the fact that y is a variable. Then it is immediate that all CL-terms are theorems, which means that $\mathfrak{I}\mathfrak{l}$ is absolutely inconsistent. qed

Curry's original paradox starts with a different CL-term. Let the meta-term M stand for $\mathsf{CP}N$, where N is a meta-term. $\mathsf{CP}Nx \rhd_{\mathrm{w}} \mathsf{P}xN$, and so $\mathsf{CP}N(\mathsf{Y}(\mathsf{CP}N)) \rhd_{\mathrm{w}} \mathsf{P}(\mathsf{Y}(\mathsf{CP}N))N$. Of course, $\mathsf{Y}(\mathsf{CP}N) \rhd_{\mathrm{w}} \mathsf{CP}N(\mathsf{Y}(\mathsf{CP}N))$. That is, using M, we have that $\mathsf{Y}M = \mathsf{P}(\mathsf{Y}M)N$. Again we construct a proof (with some equations restated and inserted) to show that N is a theorem.

1. $\mathsf{P}(\mathsf{P}(\mathsf{Y}M)(\mathsf{P}(\mathsf{Y}M)N))(\mathsf{P}(\mathsf{Y}M)N)$ [theorem, by exercise 8.1.1]
2. $\mathsf{Y}M = \mathsf{P}(\mathsf{Y}M)N$ [provable equation]
3. $\mathsf{P}(\mathsf{P}(\mathsf{Y}M)(\mathsf{Y}M))(\mathsf{P}(\mathsf{Y}M)N)$ [by (R2) from 1. and 2.]
4. $\mathsf{P}(\mathsf{Y}M)(\mathsf{Y}M)$ [theorem, by exercise 8.1.1]
5. $\mathsf{P}(\mathsf{Y}M)N$ [by (R1) from 3. and 4.]
6. $\mathsf{Y}M$ [by (R2) from 5. and 2.]
7. N [by (R1) from 5. and 6.]

Again, N is a meta-term, hence all CL-terms are theorems of $\mathfrak{I}\mathfrak{l}$.

These paradoxes depend only on the presence of two formulas in the form of CL-terms, namely, $\mathsf{P}(\mathsf{P}M(\mathsf{P}MN))(\mathsf{P}MN)$ and $\mathsf{P}MM$. An immediate thought is to leave out (at least) the first one of these CL-terms from among the CL-terms that are assertions or theorems. $\mathsf{P}(\mathsf{P}M(\mathsf{P}MN))(\mathsf{P}MN)$ is a theorem of many nonclassical logics: from intuitionistic logic to the logic of entailment and the logic of relevant implication. The implicational fragment of a logic without contraction is *quite weak* in the sense that many theorems, with which we are familiar, are no longer provable. Then (A2) could hardly be an axiom, and neither could $\mathsf{P}(\mathsf{P}MN)(\mathsf{P}(\mathsf{P}M(\mathsf{P}NP))(\mathsf{P}MP))$ be an axiom if $\mathsf{P}MM$ is included.

One could think that economizing on the implicational part of a logic may be compensated by the inclusion of other connectives and assertions for them. In general, there is no guarantee that this approach leads to a consistent system.[4]

Let us assume that the set of constants is $\{\mathsf{S},\mathsf{K},\mathsf{P},\mathsf{N}\}$. The informal interpretation of N is that it is *negation*. We consider the following two additions to $\mathfrak{I}\mathfrak{l}$ — an *axiom* and a *rule*, which are (A4) and (R3). ((A1)–(A3) can be weakened as much as desired, because now we only rely on rule (R1).) We denote the modified system by $\mathfrak{I}\mathfrak{l}'$.

(A4) $\mathsf{P}(\mathsf{P}M(\mathsf{N}M))(\mathsf{N}M)$

[4]Variations on Curry's paradox have been investigated in the literature. We recall only one new variant from among the paradoxes considered in [20].

(R3) M and $\mathsf{N}M$ imply N

(A4) is a so-called reductio principle, because it can be read as "if A implies not-A (i.e., its own negation), then not-A." (R3) expresses that a contradiction implies anything.

The CL-term $\mathsf{W}(\mathsf{B}(\mathsf{CP})\mathsf{N})$ does not have any occurrences of any variables, and we denote this term by Q. $\mathsf{W}(\mathsf{B}(\mathsf{CP})\mathsf{N})x \triangleright_{\mathrm{w}} \mathsf{P}x(\mathsf{N}x)$. We can construct a sequence of equations and assertions to prove N as 1–8:

1. $\mathsf{P}(\mathsf{P}(\mathsf{YQ})(\mathsf{N}(\mathsf{YQ})))(\mathsf{N}(\mathsf{YQ}))$ [instance of (A4)]
2. $\mathsf{YQ} = \mathsf{Q}(\mathsf{YQ})$ [provable equation]
3. $\mathsf{Q}(\mathsf{YQ}) = \mathsf{P}(\mathsf{YQ})(\mathsf{N}(\mathsf{YQ}))$ [provable equation]
4. $\mathsf{YQ} = \mathsf{P}(\mathsf{YQ})(\mathsf{N}(\mathsf{YQ}))$ [provable equation]
5. $\mathsf{P}(\mathsf{YQ})(\mathsf{N}(\mathsf{YQ}))$ [by (R2) from 1. and 3.]
6. YQ [by (R2) from 5. and 4.]
7. $\mathsf{N}(\mathsf{YQ})$ [by (R1) from 5. and 6.]
8. N [by (R3) from 6. and 7.]

N is an arbitrary meta-term; therefore, $\mathfrak{I}\mathfrak{l}'$ is inconsistent.

Given the three examples of inconsistency proofs for illative systems, we may try to find the commonalities in them. An inevitable part of the constructions is the presence of *constants* (beyond combinators), which come with some axioms and rules added to pure CL as the theory of weak equality on CL-terms. The constants, which are typically considered, and their characterization are guided by some connectives from a logic.

However, there is another crucial element that comes from pure CL itself: combinatorial completeness. In each case, we started with a CL-term (with or without a free variable) that we denoted by M or Q. In fact, these terms are the results of *abstracting out a variable* in place of which we inserted the fixed point of the CL-term itself. For instance, the last construction really starts with $\mathsf{P}x(\mathsf{N}x)$, which can be seen to be an instance of a subterm of axiom (A4). By combinatorial completeness, there is a combinator that reduces to $\mathsf{P}x(\mathsf{N}x)$, when applied to x, and we denoted this combinator by Q. We know also that from a combinatorially complete basis the *fixed point combinator* is definable. Then YQ is a CL-term.

Combinatorial completeness in an illative system has a certain similarity to *naive comprehension* in (naive) set theory. The naive comprehension principle permits a set to be formed given any property φ, that is, $\{x\colon \varphi(x)\}$ is a set. To reinforce this similarity, let us consider *Russell's paradox*. Let R stand for $\{x\colon x \notin x\}$. $R \in R$ if and only if R satisfies the defining property for R, that is, $R \notin R$. But A iff $\neg A$ is a classical contradiction, which implies anything.

We already have all the components in $\mathfrak{I}\mathfrak{l}'$ to reconstruct this paradox.

1. $\mathsf{N}(\mathsf{YN}) = \mathsf{YN}$ [provable equation]

2. $\mathsf{P}(\mathsf{P}(\mathsf{YN})(\mathsf{N}(\mathsf{YN})))(\mathsf{N}(\mathsf{YN}))$ [instance of (A4)]

3. $\mathsf{P}(\mathsf{P}(\mathsf{YN})(\mathsf{YN}))(\mathsf{N}(\mathsf{Y}N))$ [by (R2) from 2. and 1.]

4. $\mathsf{P}(\mathsf{YN})(\mathsf{YN})$ [theorem, by exercise 8.1.1]

5. $\mathsf{N}(\mathsf{YN})$ [by (R1) from 3. and 4.]

6. YN [by (R2) from 5. and 1.]

7. N [by (R3) from 5. and 6.]

The CL-term on line 4 is an instance of the same theorem that figured into Curry's original paradox. $A \supset A$ is an axiom or a theorem of nearly every extant logic, because it guarantees a very elementary property of $\supset$, namely, reflexivity. The latter is essential to the algebraizability of a logic.

Informally, the equation on line 1 may already look problematic, because the two CL-terms in the equation are exactly those from which anything may be concluded in some logics; notably, both in intuitionistic logic and in classical logic.

"Solutions to Russell's paradox" in set theory typically restrict the formation of sets, for example, by giving a series of axioms that describe allowable ways of forming sets. The so-called *Zermelo–Fraenkel set theory* (ZF) and the so-called *von Neumann–Bernays–Gödel set theory* (NBG) follow this approach. Neither ZF nor NBG contains naive comprehension.

Another approach that aims at guaranteeing consistency of set theory restricts the logic that is used but keeps the naive comprehension principle. Given this view, the importance of Curry's paradoxes is that they make clear that negation is not needed to arrive at inconsistency.

Illative combinatory logic took a similar turn as set theory. The dominant idea has been that the categories of expressions have to be separated from each other by the introduction of constants that function like predicates do in a first-order language. For example, H may be used for propositions, that is, $\mathsf{H}M$ means that M is a CL-term that stands for a proposition. Such sorting of CL-terms into categories may be likened to the elimination of sorts in first-order logic. This may be accomplished if there are unary predicates such that they are true of the elements of exactly one sort. Incidentally, in set theories that incorporate ur-elements into the domain, sorting or a special predicate S (which is true of sets, but not of ur-elements) is typically used in a similar fashion.

We are not going to go into the details of further developments in illative combinatory logic, which itself could be a topic of a whole book.

8.2 Elimination of bound variables

Schönfinkel introduced combinators as part of the procedure that accomplishes the elimination of bound variables.

Variables are clearly important in a *first-order language*. We already saw that variables and variable binding is important in Λ. Free variables in first-order formulas may be thought to formalize pronouns such as "this" or "it," when used deictically or anaphorically in a natural language sentence. Thus, they play a less significant role in formal sciences than bound variables do.

The main function of variables is to exhibit *variable binding relationships* precisely, for instance, to indicate whether an argument is existentially or universally quantified. Another function of variables (both free and bound) is to show that certain arguments are *identified*.

Example 8.2.1. Let P be a 2-place predicate, and let a be a name constant. We may think that P means "strictly less than," and a is 0. We may form a series of simple sentences — assuming that both quantifiers are in the language. We list some wff's and their intended informal meaning.

(1) $P(a,a)$ — "0 is strictly less than 0."

(2) $\forall x. P(x,a)$ — "Everything is strictly less than 0."

(3) $\exists x. P(x,a)$ — "Something is strictly less than 0."

(4) $\forall x. P(a,x)$ — "0 is strictly less than everything."

(5) $\exists x. P(a,x)$ — "0 is strictly less than something."

(6) $\forall x\, \forall y. P(x,y)$ — "Everything is strictly less than everything."

(7) $\exists x\, \exists y. P(x,y)$ — "Something is strictly less than something."

(8) $\forall x\, \exists y. P(x,y)$ — "Everything is strictly less than something."

(9) $\exists x\, \forall y. P(x,y)$ — "Something is strictly less than everything."

None of these wff's is logically valid. If we suppose that the quantifiers can range over some concrete set of numbers, for instance, $\mathbb{N}$ or $\mathbb{Z}$ is the domain of interpretation, then we could consider whether the sentences are true or false. This would clearly separate some of the wff's that are alike except that their quantifiers are different: (3) is true and (2) is false if the domain is $\mathbb{Z}$.

The last two English sentences are ambiguous, though they have a preferred reading, which is what is expressed by the wff next to which they are placed.

The usual convention in the interpretation of first-order wff's is that syntactically distinct variables such as x and y *may* but *need not* be interpreted as denoting the same individual object. Thus in order to express that two argument places of one predicate or several predicates should denote the same object can be achieved by re-using a variable. (10) and (11) are not equivalent to any of (1)–(9). For example, (3) and (7) are true if the domain of the interpretation is $\mathbb{N}$, but the alike (11) is false.

(10) $\forall x. P(x,x)$ — "Everything is strictly less than itself."

(11) $\exists x. P(x,x)$ — "Something is strictly less than itself."

In Λ, bound variables can be renamed. The renaming of bound variables yields α-equality, which is part of β-reduction and all the other relations on Λ-terms that are supersets of β-reduction. α-equality is well motivated by the idea that λ is an abstraction operator; if Λ allows (as it is usual) unrestricted function application, then α-equality (in some form) is an inevitable part of Λ.

Renaming of bound variables is permitted in first-order logic too — despite the fact that the quantifiers do not create functions from expressions. The above list of wff's might seem "systematic," however, we could have used wff's with other variables. For example, (10) could have been $\forall z. P(z,z)$. The latter wff and the one that actually appears in (10) are distinct wff's, however, they are logically equivalent.

Some wff's are *logically valid* or they are *theorems*, provided that we have a suitable formalization. In a philosophical tradition, where logic is considered a unique and fully general framework of correct reasoning, theorems of logic are viewed as formulas that have a very special status among all the wff's. The wff $\forall x. x = x$ expresses that all objects are self-identical; so does the wff $\forall y. y = y$. This shows that bound variables are auxiliary to the *logical form of formulas*, in general, but also of theorems, in particular.

The logical form of formulas would be made explicit, if the bound variables would be absent from the wff. But of course, simply omitting the bound variables will not work. If the variables are "freed" by dropping the quantifiers, then the meaning of the formula is changed (together with the formula itself); if the variables themselves are omitted (too), then the wff turns into an expression that is no longer a well-formed formula. Even if we would create a new category of expressions specifically to express the logical form of wff's by combinations of predicates, connectives and quantifiers, we would still run into a problem.

Example 8.2.2. Let us assume that we have two one-place predicates N and Z, plus a binary predicate G. The formula

$$\forall x. Z(x) \supset \exists y. N(y) \wedge G(y,x)$$

can be easily given an interpretation. For instance, with a suitable interpretation of the predicates and having a domain comprising the real numbers, the wff could mean that for any integer there is greater natural number.

If we omit the quantifiers, then we get $Z(x) \supset (N(y) \wedge G(y,x))$. A somewhat artificial translation of this wff is "If this is an integer, then that is a greater natural number." The designation of x as "this" and of y as "that" is quite arbitrary. The expressive capability of English is limited when it comes to rendering open formulas as natural language statements.

If we omit only the variable occurrences that are bound, then we get $\forall\ .Z(\ ) \supset \exists\ .N(\ ) \wedge G(\ ,\ )$. This expression with "holes" in it is clearly not a wff. A more acute problem is that the expression is *ambiguous* — exactly, because the variables no longer show the binding relationships.

Classical logic is teeming with interdefinabilities, which allows easy definitions of normal forms. There are *conjunctive*, *disjunctive* and *prenex* normal forms (among others). The bound variables could be eliminated without altering or diminishing the meaning of a formula, if a suitable normal form like equivalent of a formula could be found.

Schönfinkel solved this normalization sort of problem by introducing the *combinators* and the NEXTAND operator. We first explain what the new operator is, then we turn to the way the combinators are applied in the language of first-order logic to achieve the elimination of bound variables.

In section A.8, we outlined functional completeness for classical sentential logic. Either of the two connectives $\dagger$ and $\mid$ are alone expressive enough for sentential logic — as lemma A.8.6 states. However, even if all the connectives are replaced by their definitions using, let us say $\dagger$, there is still a need for at least one quantifier. This would mean that the set of logical constants is at least a doubleton.

The logical essence of a first-order formula can be expressed by a single operator which binds a variable and is binary (i.e., it forms a formula from two formulas). The *new operator* is denoted by $\mid$, and the variable that is bound by $\mid$ will appear in the superscript, for instance, as $\mid^x$. The presence of $\mid$ means that beyond atomic formulas, that is, formulas that contain a predicate symbol with their arguments filled out by terms, we need only the following rule $(6')$ to form formulas. If wff's are defined as in A.8.1, then (6) and (7) may be replaced by $(6')$.

$(6')$ If x is a variable, and A and B are wff's, then $(A \mid^x B)$ is a wff.

Quantification in first-order logic is more often than not defined without the requirement that the variable immediately following the quantifier occurs free in the formula which is the scope of the quantifier. $\mid$ is completely similar in this respect: $(6')$ does not require or imply that A or B contain x free, however, if either does, then we consider all the free occurrences of x in A and B to be bound in $(A \mid^x B)$.

The name NEXTAND intends to suggest that the intended interpretation of $\mid$ is "it is not the case that there is an x such that both A and B." This can be captured by a formula: $\neg\exists x\,(A \wedge B)$. The latter is equivalent in first-order logic to the formula $\forall x \neg(A \wedge B)$. The symbol $\mid$ has been used for NAND; hence, yet another way to write the same formulas is as $\forall x\,(A \mid B)$, which may explain the $\mid^x$ notation.

A *special case* of $\forall x \neg(A \wedge B)$ is when A and B are one-place predicates with x in their argument place. The formula expresses that the extensions of the two predicates are disjoint. A likely more often used wff to express this is $\neg\exists x\,(A(x) \wedge B(x))$; or even $\forall x\,(A(x) \supset \neg B(x))$. Among the so-called categorical statements from syllogistic, this is the "universal negative" statement, often abbreviated as E. Such statements correspond to natural language sentences such as "No apples are bananas" or "There are no venomous tuataras." Of course, A and B may be complex formulas, and they may contain predicates of arbitrary arity. For instance, "No Apple computer ships with a Windows operating system" could be formalized by a wff with a more complex structure to reflect the complexity of the natural language sentence.

Schönfinkel has shown that the $|$ operator itself can express all the connectives and quantifiers of a first-order language. It is well-known nowadays that $|$ (the sentential connective) by itself can express all the truth-functional connectives of classical logic. The $|$ operator can be shown to express $\neg$ and $\wedge$ similarly, once we will have noted that binding a variable that has no occurrences in the component formulas is harmless.

LEMMA 8.2.3. (NEXTAND SUFFICES) *Let A be a formula in a first-order language. There is* a formula A' *that is logically equivalent to A and* contains *no connectives or quantifiers except* the $|$ operator.

Proof: We show that if the first-order language contains $|$ as defined using $\forall$, $\neg$ and $\wedge$, then there is a formula A' as specified for any formula A. Strictly speaking, we consider a definitional extension of the first-order language in which A is given. However, considering a richer language is unproblematic, and indeed, the usual way to prove claims concerning expressibility. We divide the proof into three steps: one for each of $\neg$, $\wedge$ and $\forall$. Since it is well-known that any first-order formula can be expressed using only these connectives and quantifier, then it will be immediate that $|$ itself is similarly sufficient.

1. Let A be of the form $\neg B$. Without loss of generality, we may assume that $x \notin \mathrm{fv}(A)$.[5] $\neg B$ is logically equivalent to $\neg(B \wedge B)$ by the idempotence of $\wedge$. If $x \notin \mathrm{fv}(A)$, then $x \notin \mathrm{fv}(B)$ and $x \notin \mathrm{fv}(\neg(B \wedge B))$ either. Then $\neg(B \wedge B)$ is logically equivalent to $\forall x \neg(B \wedge B)$. The latter is equivalent, by the definition of $|$, to $B |^x B$.

2. Let us assume that A is $B \wedge C$. $B \wedge C$ is logically equivalent to its double negation $\neg\neg(B \wedge C)$. Once again, there is a variable that has no free occurrences in the latter wff. We do not have any specific assumptions about B or C containing occurrences of some variables; hence, we may assume that $x \notin \mathrm{fv}(B \wedge C)$. Provided this assumption, we know that $\neg(B \wedge C)$ is equivalent to $\forall x \neg(B \wedge C)$. In the context of $\neg$, we can replace — according to the replacement theorem — $\neg(B \wedge C)$ by $\forall x \neg(B \wedge C)$. Then $\neg\neg(B \wedge C)$ is equivalent to $\neg(B |^x C)$. This wff is of the form like A in case **1**. We could select x again as the variable that is not free in $B |^x C$, since the formula contains one bound occurrence of that variable only. However, to make the structure of the whole formula clearer, we assume that $y \notin \mathrm{fv}(B \wedge C)$. There are always infinitely many suitable variables to choose from. Then finally, $B \wedge C$ is logically equivalent to $(B |^x C) |^y (B |^x C)$.

3. Lastly, let A be $\forall x.B$, where x may be a free variable of B. Otherwise, the universal quantifier may be omitted, because it is vacuous. B is logically equivalent

[5]We use the notation fv for formulas now with a similar meaning as for λ-terms. The definition of the set of free variables for a formula is straightforward; hence, we omit its details.

to $\neg(\neg B \wedge \neg B)$ by double negation and a De Morgan law. Then A is equivalent to $(\neg B \,|^x\, \neg B)$. As in the previous cases, we can choose a variable that is not free in B, let us say, $y \notin \mathrm{fv}(B)$. By case **1**, A is further equivalent to $(B \,|^y\, B) \,|^x\, (B \,|^y\, B)$. This completes the proof. qed

Exercise 8.2.1. Work out the definitions of some other logical constants such as $\vee$, $\supset$, $\equiv$ and $\exists$.

Exercise 8.2.2. Define the two binary connectives $|$ and $\dagger$ using the NEXTAND operator. (Hint: $|$ should be obvious at this point.)

There has been a certain interest in combining connectives and quantifiers. Universally quantified statements are often conditionals. For instance, the sentence "All topological spaces over a set X contain the set X as an element of the set of opens" will straightforwardly formalize into a formula that is universally quantified, and the main connective in the scope of the quantifier will be a $\supset$. Probably, this observation motivated the integration of $\supset$ with one or more $\forall$'s.[6]

Schönfinkel chose to combine the connective $|$ with $\forall$. However, it is possible to combine either of $\dagger$ and $|$ with either of $\forall$ and $\exists$.

Exercise 8.2.3. Define an operator based on $\dagger$ with existential quantification. Is this operator sufficient like NEXTAND? (Hint: Either give a proof of sufficiency, or provide a counterexample that shows inexpressibility.)

It is obvious that there are *infinitely many truth functions*. It is fairly well-known that there are no zeroary or unary, and there are exactly two binary connectives that can express all the truth functions. A less well-known fact is that there are 56 *ternary connectives*, and of course, infinitely many connectives of higher arity having this property. Every connective that gives rise to a functionally complete singleton set of connectives can be combined with either $\forall$ or $\exists$ and the resulting operator is by itself sufficient.

Exercise* 8.2.4. Find a ternary operator that is sufficient to express all wff's of first-order logic, and prove that it is indeed sufficient.

We do not go deeper into the exploration of problems of functional completeness and sufficiency of operations.[7] Rather, we pose a somewhat open-ended question in the following exercise.

Exercise* 8.2.5. Schönfinkel's idea of eliminating the bound variables could be based on a different operator that is sufficient by itself. Is there an operator the use of which is advantageous? (Hint: Some applications of first-order logic in computer science or translations of natural and meta-language sentences may lead — typically or exclusively — to a limited class of wff's. Then questions about the length of the resulting expressions and efficiency can arise.)

[6] See, for example, Curry and Feys [53, p. 86] and Church [46, §28, §30].

[7] These issues have been investigated in [24].

Schönfinkel introduced the operator $|$, which clearly reduces the number of logical constants, but does not allow the elimination of bound variables.

Definitions A.8.1 and A.8.2 stipulated a denumerable set of individual variables. However, we may as well assume that the variables are ordered by their index set $\mathbb{N}$. This suggest that we consider wff's that contain a full initial segment of the list of variables. For instance, if x is the first variable, and y is the second one, then no formula can have y as a bound variable unless they have occurrences of the variable x too. Given that the renaming of bound variables is permitted in first-order logic, the requirement that bound variables occur in a certain order is not a real restriction on the set of formulas from which variables are eliminated. In other words, for every formula of first-order logic, there is a wff that is logically equivalent to the former and satisfies the sequentiality requirement on its bound variables.

We can go yet another step further, and we may assume that no variables can occur both free and bound. Having two denumerable sets of variables, one of which supplies only free, the other of which provides only bound variables may complicate certain definitions such as substitution, but it is theoretically unproblematic. This could guarantee that bound variables are eliminable from a class of wff's in a normal form, and the reconstruction of the formula with variable bindings is unique.[8]

To reproduce the "essence" or *formal structure* of wff's we will not limit our consideration to strings of symbols that are wff's. Indeed, this should be expected given that combinators are added into formulas of a first-order language.

Schönfinkel introduced a new constant U, which then may be viewed from two standpoints. First of all, UPQ expresses the disjointness of the extensions of two unary predicates P and Q. For instance, the sentence "No primes are even" (which is, of course, false) may be formalized over the domain of natural numbers by $\neg\exists x.Px \wedge Ex$, where P and E are one-place predicates, respectively. Then $Px \,|^x\, Ex$ is an equivalent formalization in which P and E are followed by x only. UPE expresses the same — except that the variables have been eliminated.

The previous way of thinking about U is illuminating, however, it is not general enough to provide for the elimination of *all* bound variables. It may happen that the two wff's in the scope of a $|$ have different arities, and one may have x as its last argument, whereas the other has y. The upshot is that we want to treat formulas as *syntactic objects*, in particular, almost like CL-terms. Let us assume that X and Y are expressions that may contain predicates, variables, the operator $|$, combinators and U. We know that variables bound by $|$ can be renamed just as the variables bound by $\forall$ (or by λ). Thus we can simplify the requirement that x is not free in X to x does not occur in X; similarly, x may not occur in Y. Then $Xx \,|^x\, Yx$ is rewritten as UXY.

[8]To simplify the presentation and to make it more closely resembling [130], from now on, we omit the parentheses and commas from atomic formulas. These auxiliary symbols could be left in the wff's, but the expressions that result after the elimination of bound variables would be longer. No ambiguity results from this simplification when distinct letters are used for different categories of symbols.

Example 8.2.4. Let us consider the sentence "Everybody in Bloomington knows somebody from Nebraska." In order to keep the structure of the formula and the resulting expression simple, we use B and N as one-place predicates, and K as a two-place one. (The previous two stand for "in Bloomington" and "in Nebraska," whereas K is "knows.") The sentence can be formalized in a usual first-order language as

$$\forall x\,(Bx \supset \exists y\,(Ny \wedge Kxy)).$$

This is equivalent to $\forall x \neg(Bx \wedge \forall y \neg(Ny \wedge Kxy))$, and then to $(Bx \,|^x\, (Ny \,|^y\, Kxy))$. The second part of the wff can be written as $UN(Kx)$, because y does not occur in N or in Kx. U, N, etc., are treated as atomic terms, which means that $\mathsf{B}(UN)Kx$ reduces to the previous $UN(Kx)$. Now both expressions in the argument places of $|^x$ end with x — with no other occurrences of x elsewhere. Thus we can use U again: $UB(\mathsf{B}(UN)K)$ is the resulting expression, which has no bound variables at all.

It is completely accidental that we did not need to utilize many combinators in this example, because the variables happened to be in the right positions. B was needed only to disassociate a predicate and its first argument.

Example 8.2.5. For a more complicated example, let us consider the sentence "Each bike has an owner or a rider." We may formalize this sentence, which is on the surface only slightly more complicated than the previous one, by using the unary predicates B, R and O, together with the binary predicate H. (The three former stand for "bike," "rider" and "owner," whereas the latter stands for "has" with the arguments in the same order as in the natural language sentence.) The wff in usual notation looks like:

$$\forall x\,(Bx \supset \exists y\,((Ry \vee Oy) \wedge Hxy)).$$

This wff can be turned into a wff with $|$ only: $Bx \,|^x\, (((Ry \,|^v\, Ry) \,|^z\, (Oy \,|^w\, Oy)) \,|^y\, Hxy)$. To eliminate v from $Ry \,|^v\, Ry$, we have to introduce v twice, which leads to $\mathsf{K}(Ry)v \,|^v\, \mathsf{K}(Ry)v$. Then we get $U(\mathsf{K}(Ry))(\mathsf{K}(Ry))$. Similarly, $U(\mathsf{K}(Oy))(\mathsf{K}(Oy))$ results from $\mathsf{K}(Oy)w \,|^w\, \mathsf{K}(Oy)w$. To eliminate z, we need to add z to both expressions: $\mathsf{K}(U(\mathsf{K}(Ry))(\mathsf{K}(Ry)))z$ and $\mathsf{K}(U(\mathsf{K}(Oy))(\mathsf{K}(Oy)))z$. We have at this point

$$Bx \,|^x\, (U(\mathsf{K}(U(\mathsf{K}(Ry))(\mathsf{K}(Ry))))(\mathsf{K}(U(\mathsf{K}(Oy))(\mathsf{K}(Oy)))) \,|^y\, Hxy).$$

This expression shows the need for combinators beyond B and K, because multiple occurrences of y have to be collapsed into one. Unless we want to stipulate an infinite number of primitive combinators, we will need not only duplicators, but also permutators to move separate occurrences of a variable next to each other.

Exercise* 8.2.6. Find an expression that eliminates all the bound variables from the preceding example. (Hint: It may be useful to turn $\mathsf{K}(U(\mathsf{K}(Ry))(\mathsf{K}(Ry)))$ into an expression of the form Xy with $y \notin \mathrm{fv}(X)$ as the first step; then the same combinator will work for $\mathsf{K}(U(\mathsf{K}(Oy))(\mathsf{K}(Oy)))$ too.)

The role of the combinators is crystal clear from the example: they allow expressions — starting with formulas — to be brought into a form where U can be applied.

The only logical constant is $|$ and the definition of wff's implies that if the wff is not atomic, then it contains at least one subformula of the form $Py_1 \ldots y_n \,|^x\, Qz_1 \ldots z_m$. This provides a paradigmatic case to illustrate the role of the combinators more abstractly.

1. It may happen that y_n and z_m are both x, and none of the other variables are x. Then no combinators are inserted, and the subformula is replaced by $U(Py_1 \ldots y_{n-1})(Qz_1 \ldots z_{m-1})$.

2. If one of the two variables y_n and z_m is x, but the other is not, then x needs to be added where it is missing. (We also assume that the other y's and z's are not x either.) For the sake of definiteness, let us assume that y_n is x. Then $Qz_1 \ldots z_m$ is replaced by $\mathsf{K}(Qz_1 \ldots z_m)x$ — just like in the bracket abstraction algorithm, when the variable that is being abstracted does not occur in the term.[9]

3. Now it may be that one or more of the y's and z's is x. Then $Py_1 \ldots y_n$ may be viewed as the CL-term in the combinatorial completeness lemma (i.e., lemma 1.3.9). Since we have the expression that is the end result, we would have a certain freedom to choose the order of the arguments, in general; but if we are at the stage where x is being eliminated, then x has to be positioned as the last argument of function. The whole structure of the starting wff will guide the abstraction procedure, because the goal is not simply to find an arbitrary expression but one that allows replacement of $|^x$ by U.

A *great insight* and discovery of Schönfinkel is that all the possible expressions that fall under case **3** or emerge as a result of repeated introduction of U's into the expression can be dealt with by using *finitely many* — really, a handful of — *combinators*. The fact that occurrences of variables must be added, shifted around, and possibly contracted shows some of the same complications that arise in an application of the substitution operation. The problem of the elimination of bound variables is really the same problem as *the problem of substitution*. This explains that while both Schönfinkel and Curry worked on CL, they drew their motivation from (seemingly) different questions.

The idea that manipulations on atomic formulas may allow those formulas to be brought into a standard form from which variable binding may be eliminated is directly reflected by versions of Quine's *predicate functor* approach.[10] The predicates and their arguments are treated as different kinds of expressions, which leads to the apparent simplification that an atomic formula is a string with no parentheses or grouping. Unfortunately, the lack of structure in strings excludes the possibility of selecting finitely many operations to turn atomic formulas into a suitable form. For the sake of comparison, we briefly recall one concrete system that closely resembles one of Quine's. This will make clear that Schönfinkel's combinatory approach is *superior* to predicate functors.

The set of connectives $\{\neg, \wedge\}$ is functionally complete, and every formula of first-order logic is equivalent to a formula in which only these two connectives and

---

[9]See the proof of lemma 1.3.9 and section 4.3 for descriptions of bracket abstraction algorithms.

[10]See [123], [122] and [121] for various implementations of essentially the same ideas.

$\exists$ occur.

Negation can be dealt with one predicate functor. The elimination of bound variables is accomplished by a new operator, which is denoted by ↄ. The insertion of a new variable is the result of padding, which is denoted by ʗ. Collapsing two occurrences of a variable at the beginning of a sequence is achieved by the functor ℸ. The other functors that allow the manipulation of finite *sequences* of variables come in multiples.

F and G stand for possibly complex predicates, which are atomic predicates or constructed from such by applications of predicate functors. Their arity is indicated in the superscript. The list of predicate functors is:

(1) $-F\, x_1 \ldots x_n$ is $\neg(F^n\, x_1 \ldots x_n)$;

(2) ↄ$F\, x_2 \ldots x_n$ is $\exists x_1 . F^n\, x_1 \ldots x_n$;

(3) ʗ$F\, x_0 x_1 \ldots x_n$ is $F^n\, x_1 \ldots x_n$;

(4) ℸ$F\, x_1 x_2 \ldots x_{n-1}$ is $F^n\, x_1 x_1 x_2 \ldots x_{n-1}$;

(5) $F \cap G\, x_1 \ldots x_{\max(m,n)}$ is $F^m\, x_1 \ldots x_m \wedge G\, x_1 \ldots x_n$;

(6) $P_n F\, x_2 \ldots x_n x_1$ is $F^n\, x_1 \ldots x_n$;

(7) $p_n F\, x_2 x_1 x_3 \ldots x_n$ is $F^n\, x_1 \ldots x_n$.

(1) simply changes the location of the negation from a formula to a predicate. ↄ eliminates the first variable, when that is existentially quantified; the resulting complex predicate has arity one less. ʗ functions like a cancellator, in particular, it adds a new first element to a sequence of variables. ℸ acts like the duplicator M, because it collapses two occurrences of a variable in the first and second argument places into one. ℸ reduces the arity of a predicate by one. $\cap$ merges two sequences of variables where they are either the same or one is the initial segment of the other. The arity of the new complex predicate is $\max(m,n)$. The two sorts of permutations in (6) and (7) are the operations that are most apparently must come in multiples. Neither P_n nor p_n changes the arity of the predicate, but both change the order of variables in the sequence that follows the predicate.

Example 8.2.6. In order to give a flavor of how the bound variables are eliminated in this setting, we consider the sentence from example 8.2.4 again. That is, our sentence is "Everybody in Bloomington knows somebody from Nebraska," which is formalized as

$$\forall x (Bx \supset \exists y (Ny \wedge Kxy)).$$

This time we need an equivalent formula with $\exists$, $\neg$ and $\wedge$ only. We can get $\neg\exists x (Bx \wedge \neg\exists y (Ny \wedge Kxy))$. We first have to deal with the subformula $Ny \wedge Kxy$. We may either pad Ny or permute x and y in Kxy. The narrowest scope of a quantifier is that of $\exists y$; therefore, we permute the arguments of Kxy. Then $Ny \wedge P_2Kyx$ can be turned into $N \cap P_2K\, yx$. From $\exists y . (N \cap P_2K\, yx)$, we get ↄ$(N \cap P_2K)\, x$. The negation turns into $-$, which is applied to $\text{ɔ}(N \cap P_2K)$. ɔ has decreased the arity of the complex predicate $N \cap P_2K$, thus $\cap$ is applicable without any obstacles at this point. We get $\neg\exists x\,(B \cap - \text{ɔ}(N \cap P_2K)\, x)$; and by two further steps, we obtain a complex predicate as the results of the elimination of bound variables from the above displayed formula.

$$- \text{ɔ}(B \cap - \text{ɔ}(N \cap P_2K)).$$

Chapter 9

Typed combinatory logic

CL proved to be quite powerful in itself, as we have already seen in chapters 2 and 3. The original motivations to develop CL were not primarily the goal to capture the notion of computable functions, but to eliminate bound variables and to provide a framework for the foundations of mathematics. Curry and his students created *illative combinatory logic*, that we briefly outlined in chapter 8. Schönfinkel had not had a chance to expand the elimination of bound variables to higher-order logics, or to investigate the consequences of the addition to CL of sentential or predicate logic. However, the early systems that were proposed to provide a general framework for the foundations of mathematics proved to be inconsistent. Still, some of the ideas behind the theory of functionality, that was invented by Curry in this context, proved very fruitful both theoretically and in applications.

The present chapter is concerned with some *typed* calculi for CL; thus we will use the terms *type-free* CL or *untyped* CL when we wish to refer specifically to CL without types, that is, to some variant of CL that we considered so far (except in section 6.5). A question that is unanswered in type-free CL, is the characterization of the class of terms that *strongly normalize*, or further, terms that *normalize*. Recall that a term strongly normalizes when all the reduction sequences starting with that term are finite. The Church–Rosser theorem ensures that the order in which redexes are reduced does not matter, and strong normalization guarantees that the term not only has an nf, but the order of reductions cannot prevent us from getting to the nf. Simple types and intersection types provide an answer to these questions in an elegant way.

9.1 Simply typed combinatory logic

CL and the λ-calculi are not the only formalisms that take functions as some (or all) of the objects that there are. *Set theory* and *category theory* also deal with functions. A striking difference between functions in CL (or in λ-calculi) and in the two other theories is that in set theory and in category theory functions always come with a *domain* and a *codomain*.

All functions in all *categories* have a pair of objects associated to them, which are thought of as the domain and codomain of the function. In fact, objects cannot exist without functions, because each object has an identity function on it. The role of the objects as (co)domains becomes clear when the composition of functions is considered: $A \xrightarrow{f} B$ and $B \xrightarrow{g} C$ compose into $f \circ g$ with the pair of objects $\langle A, C \rangle$, however, $g \circ f$ need not exist, because A and C are not supposed to be the same object.

In *set theory*, functions are sets of ordered pairs. We consider for the sake of simplicity unary functions — nothing essential depends on this choice. If we think, for a moment, only about total functions, then the set of the first elements in the pairs is the *domain* of the function, and the function is *not applicable* to anything that is not in that set. Similarly, the *codomain* (or *image*) of a function is determined by the second elements of the pairs. Sometimes it is more convenient to consider functions as *partial* and to allow the domain and codomain of a function to be supersets of the narrowest sets that could be taken as the domain and codomain. For instance, it is common to talk about division as a function on the set rational numbers, that is, to take $/ : \mathbb{Q} \times \mathbb{Q} \longrightarrow \mathbb{Q}$ (or $/ : \mathbb{Q} \longrightarrow \mathbb{Q}^{\mathbb{Q}}$), rather than to be "more precise" and to specify the least sets that can be taken: $\mathbb{Q} \times (\mathbb{Q} - \{0\}) \longrightarrow \mathbb{Q}$. However, an important point in either case is that the division function cannot become an argument of itself. This is quite obvious notationally, because $\frac{3}{/}$ is not even a well-formed expression without further variables or numbers as, for instance, in $\frac{3}{x/y}$. As another quick example, let us consider a connective from sentential logic. As long as $\vee$ (disjunction) is a connective, $\vee$ cannot take numbers as its arguments. This is so despite the temptation induced by the apparent structure of an English sentence such as "x equals to 0 or 1." The string $x = (0 \vee 1)$ is not well-formed formula; neither is $\vee\vee$.

Type-free CL does not place any restriction on the applicability of a function to an object, and since all functions are objects, a function can be applied to itself too. This does not mean that all *strings* in the language of CL are well-formed terms, however, the exceptions are due only to misplaced occurrences of the auxiliary symbols that indicate grouping. $)x(($KK is not a CL-term, but not because K cannot be applied to K, but because the parentheses are not well-balanced and they do not surround complex terms, which is the only permitted way for them to appear in a term.

Typed CL introduces a completely new category of expressions: *types*. The expressions of primary concern are neither CL-terms, nor types on their own, but their pairs.

DEFINITION 9.1.1. (SIMPLE TYPES) Let $\mathbb{P}$ be the set of *basic types*. ($\mathbb{P} = \emptyset$ is not an interesting case, hence we stipulate that $\mathrm{card}(\mathbb{P}) \geq 1$.) The *type constructor* is $\rightarrow$, which is a binary function on types. The set of *simple types* is inductively defined by (1) and (2).

(1) If $\alpha \in \mathbb{P}$, then α is a type;

(2) if τ and σ are types, then $(\tau \rightarrow \sigma)$ is a type.

As a notational convention, we may omit the outside parentheses from types together with parentheses from *right-associated* types.

The last notational convention might seem to be a typo; however, it is not. Once we put types and terms together, it will be clear that omitting parentheses from left-associated terms and from right-associated types concur with each other.

Exercise 9.1.1. Restore the parentheses in the following types. (a) $(\alpha \to \beta) \to (\alpha \to \beta \to \gamma) \to \alpha \to \gamma$, (b) $(\delta \to \gamma) \to (\gamma \to \alpha) \to \delta \to \alpha$, (c) $((\gamma \to \gamma) \to (\alpha \to \varepsilon) \to \delta) \to (\alpha \to \varepsilon) \to (\gamma \to \gamma) \to \delta$.

Exercise 9.1.2. Omit as many parentheses as possible from the types in (a)–(c) — without introducing ambiguity. (a) $(\alpha \to ((\alpha \to \beta) \to \beta))$, (b) $((\beta \to (\delta \to \varepsilon)) \to ((\gamma \to \beta) \to (\gamma \to (\delta \to \varepsilon))))$, (c) $((((\alpha \to (\varepsilon \to \delta)) \to ((\delta \to \varepsilon) \to (\varepsilon \to \alpha))) \to (\alpha \to (\varepsilon \to \delta))) \to (\alpha \to (\varepsilon \to \delta)))$.

The types may be thought of along the lines outlined above, that is, indicating the range of *inputs* and the range of *outputs* for a function. This interpretation is important for functional programming languages. Another interpretation is obtained by considering types as *formulas*, which is suggested by the use of the (short) $\to$ symbol. The *formulas-as-types* approach has proved to be productive not only from the point of view of CL, but also from the point of view of implicational logics.

There are two ways to combine CL-terms and types, which are often called *Church typing* (or *explicit typing*) and *Curry typing* (or *implicit typing*). In the former approach, terms are typed by a type giving rise to a new set of objects: *typed terms*. The second approach takes a slightly more flexible view in that some terms can be assigned *a set of types* rather than a single type, and the set may be neither the empty set, nor a singleton. This explains why the two kinds of typing are sometimes called explicit and implicit typing.

The Church-types can be seen to restore for functions (i.e., CL-terms) their domains and codomains similarly to the set theoretic view of total functions. But it may be thought that this leads to a too detailed typing in which a higher level of abstraction gets lost. Although the identity functions on $\mathbb{Z}^+$ and on $\mathbb{Z}^-$ are *not the same*, they are both restrictions of the identity function on $\mathbb{Z}$. Moreover, they both can be characterized as the set of all pairs of the form $\langle x,x\rangle$, with the only difference being the set over which the variable x ranges. One could even contend that all the functions that are called identity functions — from set theory to universal algebra and category theory — have something in common, and that is exactly what is captured by I in CL.

We arrive at a similar conclusion if we recall that bound variables are merely placeholders and indicators of binding relationships within expressions. Of course, CL does not contain bound variables at all. However, the elimination of bound variables, that led to the invention of CL, was really inspired by this observation concerning, in particular, first-order logic. In other words, bound variables can always be renamed using fresh variables. This step is called α-equality in λ-calculi. In CL, the combinators are constants with no internal structure; but in the λ-calculi, the combinators are *closed λ-terms*, which could suggest that the types associated to the variables that occur in the closed λ-term should not affect (or somehow should affect less seriously) the type associated to the closed term.

The set of CL-terms is defined as before. (See definition 1.1.1.) Now we introduce a new class of well-formed expressions comprising a term and a type.

DEFINITION 9.1.2. The set of *typed terms* is inductively defined as follows.

(1) If M is an atomic term (i.e., a variable or a combinatory constant) and τ is a type, then $M\colon \tau$ is a typed term;

(2) if $M\colon \tau \to \sigma$ and $N\colon \tau$ are typed terms, then $(M\colon \tau \to \sigma\, N\colon \tau)\colon \sigma$ is a typed term.

As a notational convention, we will write $MN\colon \sigma$ for $(M\colon \tau \to \sigma\, N\colon \tau)\colon \sigma$ when the types of M and N can be reconstructed from the given derivation or from the surrounding text.

Typed terms are often denoted by placing the type into the superscript on the term (e.g., M^τ) instead of the notation given above. The notation for types and typed terms, as well as for type assignments (that we introduce below), varies greatly. To mention only one variation, Curry used F in place of $\to$ in a prefix-style notation. Rather than try to give an exhaustive list of typing notations appearing in the (older) literature or in programming languages, we merely note that there is no uniform notation and care has to be taken in each case to determine what the intended interpretation of some notation involving functions, terms, types, etc., is.

The above definition plays a role similar to that of the definition of well-formed formulas in first-order logic. However, that is only the initial step, which is followed by some characterization of provable formulas or logically valid formulas. The typing systems in CL are often presented as *natural deduction calculi*.

Natural deduction systems include *introduction* and *elimination* rules for logical constants, and they may include axioms too.[1] For example, if conjunction, $\wedge$, is a logical connective, then in a natural deduction system, there is a rule that specifies how a formula $A \wedge B$ may be introduced when one or both of the wff's A and B are somehow already in the derivation. Another rule specifies how a conjunction can be used to conclude A or B.

The rules of a natural deduction calculus may be displayed by vertically or by horizontally arranging the premises and the conclusions. We will use the latter style of picturing the rules of typed CL, and so we give as an example the rules for $\wedge$ presented horizontally.

$$\frac{\Delta \vdash A \quad \Delta \vdash B}{\Delta \vdash A \wedge B}\, I\wedge \qquad \frac{\Delta \vdash A \wedge B}{\Delta \vdash A}\, E\wedge \qquad \frac{\Delta \vdash A \wedge B}{\Delta \vdash B}\, E\wedge$$

Capital Greek letters stand for sets of wff's, $\vdash$ indicates derivability, and I and E are labels that, respectively, stand for introduction and elimination. An informal rendering of the $I\wedge$ rule is that, if A and B are both derivable from a set of wff's Δ,

[1] Natural deduction systems (for classical and intuitionistic logics) were introduced by Gentzen [69] with the idea that these systems closely resemble the way how people reason from assumptions to conclusions.

then $A \wedge B$ is derivable from the same set of formulas. The elimination rules exhibit two — very similar — ways to use a formula that is a conjunction in a derivation: if $A \wedge B$ is derivable from Δ, then the same set of wff's Δ suffices for the derivation of either conjunct A and B. The rules for conjunction are, probably, the easiest to understand from among the natural deduction rules, which is one of the reasons we use them here as examples. Also, these rules will reappear in an adapted form in the next section.

Returning to simple types, we note that the only type constructor is $\rightarrow$. The natural deduction rules for $\rightarrow$ in classical logic are:

$$\frac{\Delta, A \vdash B}{\Delta \vdash A \rightarrow B}\ I\!\rightarrow \qquad\qquad \frac{\Delta \vdash A \rightarrow B \quad \Gamma \vdash A}{\Delta, \Gamma \vdash B}\ E\!\rightarrow.$$

The expressions Δ, A and Δ, Γ indicate sets of formulas too, namely, $\Delta \cup \{A\}$ and $\Delta \cup \Gamma$. The two rules for $\rightarrow$ can be viewed as inference steps that correspond to the *deduction theorem* and the rule *modus ponens* in an axiomatic formulation of classical logic.

The natural deduction systems *NK* (for classical logic) and *NJ* (for intuitionistic logic) from Gentzen [69] do not contain any axioms. However, if we are aiming at adding CL-terms to rules like $I \rightarrow$ and $E \rightarrow$, then we have to consider the meaning of the types (i.e., wff's) in relation to the terms. As we already mentioned, the informal idea behind associating $A \rightarrow B$ to a function was that the type describes the domain and codomain of the function, which is the interpretation of the term. Then the $I \rightarrow$ rule should be amended by terms so that from terms with types A and B, a term with type $A \rightarrow B$ is formed. This can be done without difficulty in λ-calculus where λ is an operator that creates a function. However, CL does not contain a λ-operator as a primitive, rather the effect of the λ-operator can be simulated by *bracket abstraction*. As a result, a natural deduction system for typed CL does not have a $I \rightarrow$ rule; rather it contains *axioms* for all those combinators that are taken to be undefined.

We use lowercase Greek letters $\alpha, \beta, \ldots$ as *variables for basic simple types*, and we use lowercase Greek letters $\tau, \sigma, \ldots$ as *variables for simple types*, as in definition 9.1.1. Now — unlike in the sample natural deduction rules from classical logic — we use uppercase Greek letters $\Gamma, \Delta, \ldots$ for sets of *typed terms*, where each term is a *variable*. Δ, Γ is the union of the two sets Δ and Γ, moreover, we call a set of typed terms *univalent* whenever each variable that occurs in the set (as a typed term) has a *unique type*.[2]

DEFINITION 9.1.3. (SIMPLY-TYPED CL, $\mathrm{CL}_{\rightarrow}$) The axioms and the rules of $\mathrm{CL}_{\rightarrow}$, the simply-typed CL, are the following. We assume that the undefined combinators are S and K, and that all assumption sets are univalent.

[2]Sometimes such sets are called consistent; however, "consistent" is an overused term, which would be also misleading here in view of the intersection type system introduced in the next section.

$$\Delta \vdash x\colon \tau \quad \text{if } x\colon \tau \in \Delta \ \text{id}$$

$$\Delta \vdash \mathsf{S}\colon (\alpha \to \beta \to \gamma) \to (\alpha \to \beta) \to \alpha \to \gamma \ \vdash\mathsf{S} \qquad \Delta \vdash \mathsf{K}\colon \alpha \to \beta \to \alpha \ \vdash\mathsf{K}$$

$$\frac{\Delta \vdash M\colon \tau \to \sigma \quad \Gamma \vdash N\colon \tau}{\Delta, \Gamma \vdash MN\colon \sigma}\ E\!\to \qquad \frac{\Delta \vdash M\colon \tau}{[\alpha_i/\sigma_i]_{i \in I}\Delta \vdash M\colon [\alpha_i/\sigma_i]_{i \in I}\tau}\ \text{sub}$$

The *first axiom* is an artifact of the horizontal presentation of the natural deduction system: any assumption is, of course, derivable from the assumption set in which it occurs. The two other axioms give *types to the combinators* S and K. Notice that the types for the combinators in these axioms are specified using basic types.

The rule on the left looks somewhat similar to the formation clause in definition 9.1.2, as well as to the implication elimination rule in NK. (The latter similarity explains its label.) But $E \to$ adds here an important condition for applicability, namely, that $M\colon \tau \to \sigma$ and $N\colon \tau$ are provable from some sets of assumptions.

The last rule is the rule of *substitution*. Since the types for the two combinators S and K in the axioms include the basic types α, β and γ explicitly, we need a substitution rule if we intend to give types to complex combinators. In general, the rule of substitution allows us to obtain typed terms from typed terms, in which the terms look the same as before and the types are alike too but they are (typically) more complex. We adapt the substitution notation for sets of type assignments: $[\alpha_i/\sigma_i]\Delta$ means that for each $x\colon \tau \in \Delta$ the substitution is performed on the type, as in $x\colon [\alpha_i/\sigma_i]\tau$. We kept terms and types as disjoint sets of expression in $\mathrm{CL}_{\to}$, hence no confusion will arise from the extension of the scope of the substitution operation, because a type cannot be substituted in a term (or vice versa).

As we already mentioned a requirement on the assumption sets in the axioms and rules is that they are *all univalent*. This creates a condition for applicability in one case, because Δ, Γ cannot contain any x with τ and σ, if those types are not the same. Typically, one would like to find derivations of typed terms with as small an assumption set as possible. For instance, $\emptyset \vdash \mathsf{S}\colon (\alpha \to \beta \to \gamma) \to (\alpha \to \beta) \to \alpha \to \gamma$ clearly includes the minimal assumption set. (The $\emptyset$ symbol is usually omitted in lieu of a little space on the left-hand side of $\vdash$, and we will occasionally omit $\emptyset$.) There is a parallel here with classical logic, which allows, for instance, the derivation of $(A \wedge B) \vee (A \wedge C)$ from $\{A, B \vee C, E \wedge \neg E\}$, but a more economical premise set (without the plain contradiction $E \wedge \neg E$) produces a more interesting claim about derivability in classical logic.

The notion of a proof in $\mathrm{CL}_{\to}$ is quite standard. A *proof* is a tree, in which each node is of the form $\Delta \vdash M\colon \tau$ (with some Δ, M and τ), the leaves of the tree are instances of the axioms, and all other nodes are justified by an application of one of the two rules. If $\Delta \vdash M\colon \tau$ is the root of a tree that is a proof, then this can be phrased as "given the set of typed variables Δ, the CL-term M is typed by the simple type τ."

Example 9.1.4. Let us consider the terms x and y. If we type them, respectively, as

$\alpha \to \beta$ and α, then we may get the derivation:

$$\frac{\{x\colon \alpha \to \beta\} \vdash x\colon \alpha \to \beta \qquad \{y\colon \alpha\} \vdash y\colon \alpha}{\{x\colon \alpha \to \beta,\ y\colon \alpha\} \vdash xy\colon \beta}.$$

This sample proof includes a small binary tree with three nodes. Notice that the axioms are merely repetitions of an assumption — of the only assumption in the singleton sets. It is usually taken to be self-evident in natural deduction systems that an assumption set implies each of its elements, though in some formulations a *reiteration* rule may be needed too. We use the $\vdash$ symbol to separate the premises from their consequences, which necessitates the inclusion of the identity axiom.

The derivation could be started with *nonminimal* sets of typed terms with the same end result.

$$\frac{\{x\colon \alpha \to \beta,\ y\colon \alpha\} \vdash x\colon \alpha \to \beta \qquad \{x\colon \alpha \to \beta,\ y\colon \alpha\} \vdash y\colon \alpha}{\{x\colon \alpha \to \beta,\ y\colon \alpha\} \vdash xy\colon \beta}$$

Even larger sets of typed variables could be taken, because neither the axioms, nor the rules require that all the variables on the left of the $\vdash$ occur on the right too. Alternatively, we could modify the combinatory axioms $\vdash \mathsf{S}$ and $\vdash \mathsf{K}$ by setting $\Delta = \emptyset$, and the id axiom assigning a type to a variable by setting $\Delta = \{x\colon \tau\}$. With the original rules, $\{x\colon \alpha \to \beta,\ y\colon \alpha,\ z\colon \beta\} \vdash xy\colon \beta$ is provable; however, it would no longer be provable in the modified natural deduction system. The equivalence of the two systems could be restored by adding a *monotonicity* (or *weakening*) rule like

$$\frac{\Delta \vdash M\colon \tau}{\Delta, \Gamma \vdash M\colon \tau},$$

where Δ, Γ is required to be univalent as in the other rules.

Exercise 9.1.3. Prove that the modified natural deduction system with the monotonicity rule added is equivalent to $\mathrm{CL}_{\to}$ from definition 9.1.3.

The types that we assigned to S and K are well-known formulas in logic.[3] $(A \to B \to C) \to (A \to B) \to A \to C$ (with Roman letters standing for wff's) is known as the *self-distribution of implication on the major*. The wff $A \to B \to A$ is called *positive paradox*, because of the counterintuitive meaning of this wff, when the connective $\to$ is interpreted as implication. It may be a good idea to take some familiar formulas as types for combinators, but that is surely not a sufficient reason to select these two wff's (rather than some other wff's) as types for S and K.

Let us recall that the axiom of K is $\mathsf{K}xy \rhd x$. If x is to be typed by α, then the type of $\mathsf{K}xy$ should be α too. Using the $E \to$ rule, which, in effect, says that *application in a term is detachment in the type*, we can see that K should have a type that has two antecedents (i.e., the consequent of the whole wff should be a type with

[3]Of course, we assume a little squinting to help us to see expressions in two languages as similar or the same.

$\rightarrow$), the first of which is α. Since we do not have any information about the type of y, we may assume that that is β (i.e., a type possibly different from α). These pieces completely determine what the type of K is: $\alpha \rightarrow \beta \rightarrow \alpha$. In the case of Church typing, K has many typed "alter egos," and the one with type $\alpha \rightarrow \alpha \rightarrow \alpha$ is a different typed term than the K in the axiom for K. So is the one with type $\beta \rightarrow \gamma \rightarrow \beta$ — despite the fact that (the untyped) K is a component of each of these typed terms. Here is a derivation that shows how we get α for the compound term $\mathsf{K}xy$.

$$\dfrac{\dfrac{\vdash \mathsf{K}\colon \alpha \rightarrow \beta \rightarrow \alpha \qquad \{x\colon \alpha\} \vdash x\colon \alpha}{\{x\colon \alpha\} \vdash \mathsf{K}x\colon \beta \rightarrow \alpha} \qquad \{y\colon \beta\} \vdash y\colon \beta}{\{x\colon \alpha,\ y\colon \beta\} \vdash \mathsf{K}xy\colon \alpha}$$

Exercise 9.1.4. Write out the derivation of $\mathsf{S}xyz\colon \gamma$ step by step. What is the (smallest possible) assumption set for this typed term?

We gave types for only two combinatory constants. Since all the combinators are definable from S and K, their typed versions should be derivable by choosing these combinators with suitable types, and then deriving the types of the complex combinators according to their definition.

Exercise 9.1.5. Try to discover the type of B from its axiom $\mathsf{B}xyz \rhd x(yz)$. Is the type the same as the type obtained from the definition of B and using $\mathrm{CL}_{\rightarrow}$? (Hint: Recall that B is definable as $\mathsf{S}(\mathsf{KS})\mathsf{K}$.)

Exercise 9.1.6. Try to find suitable types for the combinators C, W and J. (Hint: Their axioms are listed in section 1.2.)

Having seen some examples, now we take a look at what happens if we have, for example, $\mathsf{K}\colon \alpha \rightarrow \beta \rightarrow \alpha$ and we (attempt to) apply this typed term to itself. (We mention the basic types α and β explicitly to keep the example uncomplicated.) The type $\alpha \rightarrow \beta \rightarrow \alpha$ should be detached from itself, that is, from $\alpha \rightarrow \beta \rightarrow \alpha$. This is, obviously, impossible, because no implicational wff can be detached from itself. If the type is atomic, then the nonapplicability of detachment is even more conspicuous. Indeed, the Church typing system does not allow a typed term to be applied to itself, which is vividly illustrated by the case of a variable.[4]

$$\dfrac{\{x\colon \delta \rightarrow \varepsilon\} \vdash x\colon \delta \rightarrow \varepsilon \quad \{x\colon \delta \rightarrow \varepsilon\} \vdash x\colon \delta \rightarrow \varepsilon}{\{x\colon \delta \rightarrow \varepsilon\} \vdash xx\colon ???}$$

LEMMA 9.1.5. *Some* CL-*terms* cannot be typed *in the Church typing system* $\mathrm{CL}_{\rightarrow}$. *However, by definition, the type of a typed term (if there is one) is* unique.

Proof: The proof is by the preceding attempted derivation together with the observation that the substitution rule cannot remedy the situation, because the substitution is applied to all the typed terms in the set of assumptions as well as to the typed term on the right-hand side of the $\vdash$. qed

[4]Of course, ??? is not a type; we inserted the question marks to emphasize that there is no suitable simple type for this CL-term.

DEFINITION 9.1.6. The CL-term M is called *typable* by the type τ iff there is a derivation in $\mathrm{CL}_{\rightarrow}$ ending with $\Delta \vdash M \colon \tau$.

The above lemma then says that not all CL-terms are typable. This is an important claim, and it is also important to realize that *not only* terms of the form MM are untypable.

The rule of substitution affects the typed terms on both sides of the turnstile, but on the left-hand side, we can have only variables. Thus even using the Church-style typing we can have the derivation:

$$\dfrac{\dfrac{\vdash \mathsf{K}\colon \alpha \rightarrow \beta \rightarrow \alpha}{\vdash \mathsf{K}\colon (\alpha \rightarrow \beta \rightarrow \alpha) \rightarrow \gamma \rightarrow \alpha \rightarrow \beta \rightarrow \alpha} \qquad \vdash \mathsf{K}\colon \alpha \rightarrow \beta \rightarrow \alpha}{\vdash \mathsf{KK}\colon \gamma \rightarrow \alpha \rightarrow \beta \rightarrow \alpha}.$$

The two occurrences of K in KK are *not* the same typed terms. Reversing the notational convention that we introduced above, the bottom line in the derivation is

$$\vdash (\mathsf{K}\colon \alpha \rightarrow \beta \rightarrow \alpha\ \mathsf{K}\colon (\alpha \rightarrow \beta \rightarrow \alpha) \rightarrow \gamma \rightarrow \alpha \rightarrow \beta \rightarrow \alpha)\colon \gamma \rightarrow \alpha \rightarrow \beta \rightarrow \alpha,$$

which makes obvious that the two K's are not the same typed terms. There are no variables involved in the derivation, which underscores that combinators should be characterizable by the *structure of the type* they can be given rather than by the concrete types their arguments can take (or the concrete types that would occur as types of the term when combinators are viewed as closed λ-terms). If we could "forget" or "overlook" the fact that the two occurrences of K have different types, then it seems that we could type — at least some instances of — terms that are self-applications.

Curry-style type systems differ from the Church-style type systems in how combinators are typed. It is usual to read $M\colon \tau$ as a *type assignment* statement rather than the statement that M has the type τ. We introduced this notation for the typed terms already in order to make a smooth transition to the type assignment systems, which we will focus on from now on.

Variables can be assigned arbitrary types as before, and $x\colon \tau$ is read as "τ is a type assigned to the variable x." A *univalent* assumption set Δ may not contain both $x\colon \sigma$ and $x\colon \tau$ provided that $\sigma \neq \tau$. Combinators and terms that contain combinators are now viewed as *sets of types* with no assumption of univalence (or that the set is a singleton). That is, the derivation of a particular type assignment $M\colon \tau$ means only that τ is one of the types that can be assigned to M (possibly, depending on a set of assumptions). If Z is a combinator, then a derivation may contain $\mathsf{Z}\colon \tau$ and $\mathsf{Z}\colon \sigma$, while $\tau \neq \sigma$.

To make the change in the meaning of $M\colon \tau$ official, we give the next definition (that supersedes definition 9.1.2).

DEFINITION 9.1.7. Let τ be a type and let M be a term. Then $M\colon \tau$ is a *type assignment* in a Curry-style typing system.

If we compare definitions 9.1.2 and 9.1.7, then we see that the latter one is simpler. The former definition paralleled the definition of terms (cf. definition 1.1.1) with an additional condition concerning types, which now emerges from proofs (only).

The definition of the natural deduction system for type-assignment is the earlier definition 9.1.3. Notice that in virtue of the notational convention concerning the hiding of the types of subterms, all the axioms and rules can be easily viewed now as type assignments.

Old terminology in logic, especially in syllogistics, included a distinction between *subjects* and *predicates*. Some of the literature on typed CL inherited and retained this terminology (with modified content): the *term* is called the subject and the *type* is called the predicate. We do not use this terminology, because it would add unnecessary complication without any gain in the understanding of typed terms or type assignments. However, some lemmas and theorems are commonly labeled with reference to this terminology, and we use those labels for an easy identification of results.

LEMMA 9.1.8. (SUBJECT CONSTRUCTION) *Let D be a derivation of the type assignment $\Delta \vdash M\colon \tau$. Let D' be a tree that is obtained by omitting each premise set from the left-hand side of each $\vdash$ together with $\vdash$ itself. Further, all the type assignments (i.e., the $:\tau$'s) are omitted after each term. Lastly, if there is an application of the substitution rule, then one of the two nodes of the tree is omitted too. D' is* the formation tree *of the term M.*

Proof: The proof is by structural induction.

1. If D is $\Delta \vdash M\colon \tau$, then it is an instance of one of the axioms. For example, the root may be $\Delta \vdash \mathsf{K}\colon \alpha \to \beta \to \alpha$. Then Δ, $\vdash$, $:$ and $\alpha \to \beta \to \alpha$ are omitted. K is the formation tree of the combinator K. The two other axioms are similar.

2. Let us assume that D's height is greater than 1. One possibility is that the last step in the derivation is of the form

$$\frac{\Delta \vdash M\colon \tau \to \sigma \qquad \Gamma \vdash N\colon \tau}{\Delta, \Gamma \vdash MN\colon \sigma}.$$

By the inductive hypothesis, D_1' and D_2' obtained from the subtrees rooted by the two premises of the rule are formation trees of M and of N, respectively. The transformation adds a new root with binary branching incorporating the two trees D_1' and D_2', and the new root is MN. Then D' is indeed the formation tree of the latter term.

Now let us assume that the last step in D is an application of the substitution rule. The last but one line in D is $\Delta \vdash M\colon \sigma$, and the transformation gives D_1' which, by the hypothesis of induction, is the formation tree of M. However, the last line in D contains the same term M, and one of the two occurrences of M is omitted from D'. Thus the claim is clearly true. qed

The previous lemma means that in a derivation *all the subterms* of a term appear as terms. Moreover, each subterm appears in the leaves as many times as it occurs in the term at the root. The next lemma can be proved using the previous lemma together with the fact that each node in a derivation contains a type assignment on the right-hand side of the $\vdash$.[5]

LEMMA 9.1.9. *If* $\Delta \vdash M\colon \tau$ *is derivable in* $\mathrm{CL}_{\rightarrow}$, *then* every subterm *of the term* M *is* typable.

For the sake of emphasis, it is useful to re-state the theorem in a contraposed form.

COROLLARY 9.1.9.1. *If* M *is not typable, then* no term N, *of which* M *is a subterm, is typable in* $\mathrm{CL}_{\rightarrow}$.

The application operation is the basic syntactic operation on terms. Given provable type assignments, it parallels modus ponens performed on types. However, the possibility of forming terms from terms is not the only and not the most interesting relation on the set of CL-terms. The paramount relation on CL-terms is *weak reduction*. It is remarkable that types are preserved under this relation, but before we can state and prove that theorem we need the following lemma.

LEMMA 9.1.10. *Let* D *be a derivation of* $\Delta \vdash M\colon \tau$, *and let* N *be a subterm of* M *such that there is a* D', *which is a subtree of* D, *and* D' *ends in* $\Delta' \vdash N\colon \sigma$. *If* $\Delta' \vdash P\colon \sigma$ *is provable, then* $\Delta \vdash [N/P]M\colon \tau$ *is provable as well, where* $[N/P]M$ *is the result of the replacement of the subterm* N, *which corresponds to the occurrence of the term* N *in* D', *by* P.

Proof: The claim of the lemma can be rephrased to say that the *replacement* of a subterm with a term does not affect typability as long as the new term can be typed with the same type given the same set of type assignment assumptions.

The derivation of $\Delta \vdash M\colon \tau$ contains a subtree with a root $\Delta' \vdash N\colon \sigma$, where this N is the occurrence of the subterm that is replaced, by lemma 9.1.8. $P\colon \sigma$ has a derivation with the same assumption set Δ'. We only have to ensure that the replacement of the subtree of N with the tree of P results in a derivation of $\Delta \vdash M\colon \tau$.

If N is M, then the claim is trivially true, because $\Delta \vdash P\colon \sigma$ is a derivation by the hypothesis of the lemma. Otherwise, one of the two rules, $E\rightarrow$ and sub, has been applied to $\Delta' \vdash N\colon \sigma$ in D. The substitution rule does not affect N, and Δ' as well as σ are the same in $\Delta' \vdash N\colon \sigma$ and in $\Delta' \vdash P\colon \sigma$. If the rule that was applied to N is the $E\rightarrow$ rule, then there are two subcases to consider. The part of the derivation in focus might look like

$$\frac{\begin{matrix}\vdots \\ \Delta'' \vdash Q\colon \sigma \rightarrow \chi\end{matrix} \qquad \begin{matrix}\vdots \\ \Delta' \vdash N\colon \sigma\end{matrix}}{\Delta'', \Delta' \vdash QN\colon \chi}.$$

Clearly, the following is a derivation too.

$$\frac{\begin{matrix}\vdots \\ \Delta'' \vdash Q\colon \sigma \rightarrow \chi\end{matrix} \qquad \begin{matrix}\vdots \\ \Delta' \vdash P\colon \sigma\end{matrix}}{\Delta'', \Delta' \vdash QP\colon \chi}$$

[5] The notion of typability is literally the same as before — see definition 9.1.6 — but $\mathrm{CL}_{\rightarrow}$ is the Curry-style type assignment system now.

The other subcase, when $\Delta' \vdash N\colon \sigma$ is the left premise, is similar and we omit its details. qed

The above lemma does not allow us to simply replace a subterm by a term — unlike the replacement in type-free CL. P must be typable, moreover, it must be typable with the same σ as N, given the same set of type assignments for variables.

We have informally suggested that types should be preserved under $\triangleright_w$, when we motivated the particular type appearing in axiom $\vdash \mathsf{K}$. The following theorem shows that the system $\mathrm{CL}_\rightarrow$ is designed so that this invariance holds for all terms.

THEOREM 9.1.11. (SUBJECT REDUCTION) *Let $\Delta \vdash M\colon \tau$ in $\mathrm{CL}_\rightarrow$, and let $M \triangleright_w N$. Then $\Delta \vdash N\colon \tau$ in $\mathrm{CL}_\rightarrow$.*

Proof: The two combinators that we included are S and K. They are both proper; that is, they cannot introduce new variables or constants. If $M \triangleright_w N$, then there is a finite chain of one-step reductions that starts with M and leads to N. Accordingly, the proof is by induction on the length of the sequence of $\triangleright_1$ steps.

1. If M and N are the same terms, which is possible due to the reflexivity of $\triangleright_w$, then the claim is obvious.

2. Let us suppose that the claim is true for a sequence of $\triangleright_1$ steps of length $n \geq 1$. We may assume, for the sake of concreteness, that $M \triangleright_w N' \triangleright_1 N$, and that N' contains a redex of the form $\mathsf{K}Q_1Q_2$. Then N' is $[\mathsf{K}Q_1Q_2/Q_1]N$. By lemma 9.1.8, a derivation of $\Delta \vdash M\colon \tau$ contains a subtree rooted with $\Delta' \vdash \mathsf{K}Q_1Q_2\colon \sigma$, that is, $\mathsf{K}Q_1Q_2$ is typable, and so are Q_1 and Q_2. The typability of M by τ given Δ may not determine uniquely the type of Q_1 and Q_2, but we may suppose that K, Q_1 and Q_2, respectively, can be assigned the types $\sigma \rightarrow \chi \rightarrow \sigma$, σ and χ in accordance with the derivation. $\mathsf{K}Q_1Q_2 \triangleright_w Q_1$, and clearly the reduct has the same type as $\mathsf{K}Q_1Q_2$ has. Since Q_1 and Q_2 are subterms of $\mathsf{K}Q_1Q_2$, if $\Delta' \vdash \mathsf{K}Q_1Q_2\colon \sigma$, then $\Delta' \vdash Q_1 : \sigma$ and $\Delta' \vdash Q_2\colon \chi$ must be provable. $\sigma \rightarrow \chi \rightarrow \sigma$ can be assigned to K, since this is a substitution instance of the type in K's axiom. This establishes that there is a derivation of $\Delta' \vdash \mathsf{K}Q_1Q_2\colon \sigma$ with Q_1 and Q_2 having the stipulated types, and having the assumption set Δ'.

First, we replace the subtree $\Delta' \vdash \mathsf{K}Q_1Q_2\colon \sigma$ in the derivation of $\Delta \vdash M\colon \tau$ by the (potentially new) subtree ending in $\Delta' \vdash \mathsf{K}Q_1Q_2\colon \sigma$. The new derivation of $\Delta \vdash M\colon \tau$ is clearly a derivation, because the roots of the subtrees are the same, namely, $\Delta' \vdash \mathsf{K}Q_1Q_2\colon \sigma$. Now we apply lemma 9.1.10. The new derivation of $\Delta \vdash M\colon \tau$ contains a subtree with root $\Delta' \vdash Q_1 : \sigma$, and now we replace the subtree rooted in $\Delta' \vdash \mathsf{K}Q_1Q_2\colon \sigma$ with this subtree. The resulting tree is a derivation of $\Delta \vdash [Q_1/\mathsf{K}Q_1Q_2]M\colon \tau$.

We leave the case when the head of the redex is S as an exercise. qed

Exercise 9.1.7. Consider the other possible case of a one-step reduction in the above proof, that is, when the redex is headed by S. Write out the details of the case.

Exercise* 9.1.8. In the proof of the last theorem, we mentioned that it may not be possible to recover the types of all subterms uniquely from $\Delta \vdash M\colon \tau$ itself. Construct a pair of derivations that illustrates this phenomenon.

It is useful to state (a special instance of) the claim of the previous theorem in contraposed form.

COROLLARY 9.1.11.1. *If the term M is not typable and M is the result of a one-step reduction of N, then N is* not typable *either. In other words, a term obtained by* one-step expansion *from a nontypable term is not typable either.*

In contrast with this corollary, we might wonder whether a subject expansion theorem could be proved. To put it differently, the question is whether typability is preserved in both directions along the $\triangleright_1$ relation.

LEMMA 9.1.12. *There is a term M, which is not typable but obtained from a typable term N* by expansion. *In other words, there is a nontypable term M, such that $M \triangleright_{\mathrm{w}} N$ and N is typable.*

Proof: We would like to have a simple CL-term to prove this lemma. The term $\mathsf{K}xM$ is not typable, if M is not typable, but $\mathsf{K}xM \triangleright_{\mathrm{w}} x$, and x is clearly typable both in itself, and also as a subterm of $\mathsf{K}xy$, for example. Thus we only have to find a nontypable term (possibly, without an occurrence of x).

The prototypical term that is not typable by simple types is WI, that is, M. Exercise 9.1.6 asked you to try to discover types for three combinators — including W. The type of I is nearly obvious: $\tau \to \tau$, and later we will see that $\mathsf{SKK}\colon \alpha \to \alpha$ is provable in $\mathrm{CL}_{\to}$. The identity combinator takes a term as its argument and returns the same term as its value. If $x\colon \tau$, then we want to have $\mathsf{I}x\colon \tau$; therefore, I should have *self-implication* as its type. The type of W is the formula that is usually called *contraction*: $(\sigma \to \sigma \to \tau) \to \sigma \to \tau$. In order to type WI, we have to detach an instance of $\tau \to \tau$ from the type of W. This means that σ and $\sigma \to \chi$ must be *unified*; that is, we need to instantiate these types so that they become the same type. A bit more precisely, substitutions have to be found for σ and χ such that once they are applied to both σ and $\sigma \to \chi$ the resulting types coincide.

We could try substituting $\sigma \to \chi$ for σ. This would turn σ into $\sigma \to \chi$, which is the other type — before the substitution. However, the same substitution has to be applied to $\sigma \to \tau$ too, and $[\sigma/\sigma \to \chi]\sigma \to \chi$ is $(\sigma \to \chi) \to \chi$. Of course, we could try other substitutions, for instance, we could substitute σ for χ. But the problem really is that σ is a *proper* subformula of $\sigma \to \chi$; hence, *there is no substitution* that can turn the two formulas into the same formula. This fact is often phrased as "an implication is not unifiable with its antecedent."

To see that M is not typable without appealing to its definition as WI, we might look at the reduct of $\mathsf{M}x$, that is, at xx. The latter term is not typable, thus $\mathsf{M}x$ is not typable either, by corollary 9.1.11.1. A variable can be assigned an arbitrary type, hence if $\mathsf{M}x$ is not typable, then M must be the subterm that is not typable.

We can now put together the untypable term that reduces to a typable one. We have $\mathsf{K}x(\mathsf{M}y) \triangleright_1 x$. qed

The proof used the informal procedure of assigning types to terms that we introduced before exercise 9.1.4. In $\mathrm{CL}_{\to}$, we have two combinatory axioms, and the procedure's result for K and S coincide with the types in the axioms $\vdash \mathsf{K}$ and $\vdash \mathsf{S}$.

Indeed, using a definition of W (from $\{S, K\}$), and then performing detachments between minimally suitable instances of the types of S and K, we would get the same type for W as the one mentioned above (modulo relabeling the type letter by letter). Although the procedure always works in the case of proper combinators (if the combinator is typable), in general, there is no guarantee that the types assigned to a term in this fashion yield a type assignment that is derivable in $CL_{\rightarrow}$.

The *fixed point combinator* Y is definable from certain combinatory bases, even from bases that are not combinatorially complete. However, we can take Y to be a primitive combinator in CL with the axiom $\mathsf{Y}x \rhd x(\mathsf{Y}x)$. Let us apply the strategy that we demonstrated before exercise 9.1.4 to see whether we can assign some implicational type to Y. We may assign any type to x, thus let us say, $x\colon \sigma$ and $\mathsf{Y}\colon \sigma \rightarrow \tau$. Then the term on the left-hand side of $\rhd$ has type τ. We would like to assign the same type to the reduct of $\mathsf{Y}x$, and so we want to make the type of x implicational, because the reduct is of the form xM. Let us then modify the previous types by taking σ to be $\sigma_1 \rightarrow \sigma_2$. Then M, that is, $\mathsf{Y}x$ must have type σ_1. That is not a problem, because we can adjust further the types stipulated so far: we simply identify σ_1 and σ_2. The resulting type on the right-hand side of $\rhd$ will be σ, hence we identify τ and σ too.

To summarize what we have gotten about assigning a simple type to Y at this point, we list all the subterms and their assigned types (using only one type χ to avoid confusion with the types above). $x\colon \chi \rightarrow \chi$, $\mathsf{Y}\colon (\chi \rightarrow \chi) \rightarrow \chi$, $\mathsf{Y}x\colon \chi$ and $x(\mathsf{Y}x)\colon \chi$. It appears that we succeeded in assigning a type to Y, but there is an arbitrary choice we made — unlike in the case when we applied this procedure to K. We could have started with $\mathsf{Y}x \rhd_{\mathrm{w}} x(x(\mathsf{Y}x))$. This is not the axiom for Y, however, given theorem 9.1.11, we wish to have $x(x(\mathsf{Y}x))\colon \chi$ as well. According to the type assignments, we have now, $x\colon \chi \rightarrow \chi$ and $x(\mathsf{Y}x)\colon \chi$. This gives $x(x(\mathsf{Y}x))\colon \chi$ as we wanted to have. By induction on the structure of the reducts of $\mathsf{Y}x$, we can prove that each of them gets the same type χ.

It would seem that we can add the axiom $\mathsf{Y}\colon (\alpha \rightarrow \alpha) \rightarrow \alpha$ to $CL_{\rightarrow}$. A problem is that this type for Y is *not derivable* from its definition from $\{S, K\}$ in $CL_{\rightarrow}$ — unlike the types for W, I and B, for example. Another objection to adding such an axiom will emerge when we will see that all terms that are typable in $CL_{\rightarrow}$ have an nf, but $\mathsf{Y}x$ has no nf.

Nonetheless, sometimes, this type is added in typed functional programming languages, where often there are other type constructors beyond $\rightarrow$.[6] To be more precise, a typing rule might be added that allows forming the fixed point of a function.

$$\frac{\Delta \vdash M\colon \tau \rightarrow \tau}{\Delta \vdash \mathsf{Y}M\colon \tau}$$

The result of an addition of a rule like the one above is nearly the same as including an axiom for Y, except that the type assignment $\vdash \mathsf{Y}\colon (\alpha \rightarrow \alpha) \rightarrow \alpha$ is not provable.

[6] See, for example, Pierce [116, p. 144], where fix is introduced.

Exercise 9.1.9. Find and list all the subterms together with their types (if any) of the following two terms each of which defines Y. (a) BM(CBM), (b) BWI(BWB). (Hint: In exercises 9.1.5 and 9.1.6, some of the combinators that you were asked to find types for were B, C and W.)

Exercise 9.1.10. We claim that the term W(BB(B(BW))BC)(BWB) is not typable either. What does this term do? Why is this term not typable? (Hint: Find the subterms in this term that are not typable.)

Lastly, we may also notice that $(\alpha \to \alpha) \to \alpha$ is *not a theorem* of classical logic, let alone of intuitionistic logic. Let α be false, then $\alpha \to \alpha$ is true, but $(\alpha \to \alpha) \to \alpha$ is obviously false according to the truth matrix for $\to$ in classical logic. This strongly suggests that we should not include $\vdash \mathsf{Y}\colon (\alpha \to \alpha) \to \alpha$ as a new axiom into $\mathrm{CL}_{\to}$, and we will not consider this extension of the type assignment system $\mathrm{CL}_{\to}$ here.

LEMMA 9.1.13. *All the simple types that can be assigned to a combinator in* $\mathrm{CL}_{\to}$ *are* theorems of $H_{\to}$, *that is, of the implicational fragment of intuitionistic logic.*

Proof: The statement actually follows from the Curry–Howard isomorphism, which concerns proofs in $H_{\to}$ (not only theorems), and we will prove that theorem later. However, it is useful to record this result now. The formulas $A \to B \to A$ and $(A \to B \to C) \to (A \to B) \to A \to C$ are axiom (schemas) of $H_{\to}$, and $A \to B$ and A imply B is the (only) rule of inference. Substitution is either a primitive rule or it is an admissible rule as a result of the calculus comprising schemas rather than concrete axioms.

A combinator does not contain any variables, hence we may choose all the Δ's to be void in the $\mathrm{CL}_{\to}$ derivation (without destroying the derivation itself). By the subject construction theorem, if $\vdash \mathsf{Z}\colon \tau$ is provable, then the proof is a tree, in which the leaves are instances of the combinatory axioms. The types in those axioms are instances of the axioms in $H_{\to}$, and applications of $E \to$ and sub, the two rules of $\mathrm{CL}_{\to}$, are instances of applications of the two rules in $H_{\to}$, when the CL-terms are omitted (or disregarded). qed

Now we return to $\mathrm{CL}_{\to}$ (without axioms for Y or for similar combinators) and to the issue of typability and type preservation under *expansions*. It is not very surprising that dropping a nontypable term utilizing a cancellator produces a counterexample to preservation of typability under expansion. There is a, perhaps, less obvious way to obtain a typable term from a nontypable one via one-step reduction.

Example 9.1.14. We have elaborated on xx not having a simple type. However, we emphasized that in a Curry-style typing system, certain combinators that are self-applications of a combinator are typable. II is typable, for instance, by $\tau \to \tau$, because the left I can be assigned the type $(\tau \to \tau) \to \tau \to \tau$, and the right I can be assigned the same type $\tau \to \tau$ as the type for II.

Clearly, no combinator Z_{xx} can have a type that has xx in its reduct, because xx is not typable. If there is a subterm xx in the reduct of Z_{xx}, then Z_{xx} is a duplicator.

However, it might happen that the reduct of a redex headed by Z_{xx} is typable, for example, because I occurs in the place of the argument x. (Of course, putting I's in place of the x's may not be sufficient in itself for the typability of the reduct. There might be a y for which Z_{xx} is a duplicator too, etc.)

Example 9.1.15. WIK is not typable, however, $\mathsf{WIK} \triangleright_1 \mathsf{IKK}$, and the latter term is typable as the next (segment of a) derivation makes it transparent. On the top line, we abbreviate $\alpha \to \beta \to \alpha$ by π (in order to fit the derivation onto the page).

$$\dfrac{\dfrac{\begin{matrix}\vdots\\ \vdash \mathsf{I}\colon (\pi \to \gamma \to \pi) \to \pi \to \gamma \to \pi\end{matrix} \qquad \begin{matrix}\vdots\\ \vdash \mathsf{K}\colon \pi \to \gamma \to \pi\end{matrix}}{\vdash \mathsf{IK}\colon (\alpha \to \beta \to \alpha) \to \gamma \to \alpha \to \beta \to \alpha} \qquad \vdash \mathsf{K}\colon \alpha \to \beta \to \alpha}{\vdash \mathsf{IKK}\colon \gamma \to \alpha \to \beta \to \alpha}$$

The type we assigned to IKK here is the same that we assigned earlier to KK.

We could hope that if expansion is limited to combinators that are not cancellators or duplicators then the type, and therefore, typability is preserved. Of course, an immediate problem with such a theorem is that it is not applicable to $\mathrm{CL}_{\to}$ as it is, because S is a duplicator and K is a cancellator. In other words, the theorem would be vacuously true, since it would not cover any expansion steps. Lemma 9.1.12 showed why K has to be excluded from expansion. In example 9.1.15, we used WI and emphasized the presence of a subterm xx; thus it might appear the exclusion of S is overly zealous, since the reduct of $\mathsf{S}xyz$ does not contain an occurrence of xx. Notice, however, that SII defines M just as WI does.

The moral is that not all expansion steps introducing K and S lead to a nontypable term given a typable one, but we have no way, in general, to characterize and separate those that do and those that do not.

To ensure that both the subject reduction and subject expansion theorems are provable and applicable, we may start with a different combinatory basis that does not include a cancellator or a duplicator. (Of course, no such basis is combinatorially complete.)

DEFINITION 9.1.16. The type assignment calculus $\mathrm{CL}(\mathsf{B},\mathsf{C},\mathsf{I})_{\to}$ is defined by the following four axioms and two rules. (Again, all the assumption sets are stipulated to be univalent.)

$$\Delta \vdash x\colon \alpha \quad \text{if } x\colon \alpha \in \Delta \ \text{ id} \qquad\qquad \Delta \vdash \mathsf{B}\colon (\alpha \to \beta) \to (\gamma \to \alpha) \to \gamma \to \beta \ \vdash\!\mathsf{B}$$

$$\Delta \vdash \mathsf{C}\colon (\alpha \to \beta \to \gamma) \to \beta \to \alpha \to \gamma \ \vdash\!\mathsf{C} \qquad\qquad \Delta \vdash \mathsf{I}\colon \alpha \to \alpha \ \vdash\!\mathsf{I}$$

$$\dfrac{\Delta \vdash M\colon \tau \to \sigma \qquad \Gamma \vdash N\colon \tau}{\Delta, \Gamma \vdash MN\colon \sigma}\ E\!\to \qquad\qquad \dfrac{\Delta \vdash M\colon \tau}{[\alpha_i/\sigma_i]_{i\in I}\Delta \vdash M\colon [\alpha_i/\sigma_i]_{i\in I}\tau}\ \text{sub}$$

The notions of *derivability* and *typability* are as before.

THEOREM 9.1.17. (SUBJECT EXPANSION) *Let M and N be terms over the combinatory basis $\{\mathsf{B},\mathsf{C},\mathsf{I}\}$. If $M \triangleright_{\mathrm{w}} N$ and $\Delta \vdash N\colon \tau$ is provable in $\mathrm{CL}(\mathsf{B},\mathsf{C},\mathsf{I})_{\rightarrow}$, then so is $\Delta \vdash M\colon \tau$.*

Proof: The set of terms over $\{\mathsf{B},\mathsf{C},\mathsf{I}\}$ (with a denumerable set of variables) is called the set of *linear terms*. All terms in this set have the property that if $M \triangleright_{\mathrm{w}} N$ then $\mathrm{fv}(M) = \mathrm{fv}(N)$; therefore, it is straightforward to establish a 1–1 correspondence between occurrences of the variables in M and N even if a variable occurs more than once in both M and N. Of course, occurrences of combinators need not be preserved in a similar way, for the obvious reason that a combinator might be the head of a redex.

The subject construction lemma (i.e., lemma 9.1.8) and the replacement lemma (i.e., lemma 9.1.10) are true for $\mathrm{CL}(\mathsf{B},\mathsf{C},\mathsf{I})_{\rightarrow}$ as well.

Let us assume that there is a derivation of $\Delta \vdash N\colon \tau$ in $\mathrm{CL}(\mathsf{B},\mathsf{C},\mathsf{I})_{\rightarrow}$. Because of the linearity of the term N, all the variables in M (if there are any) are in N, and all the variables that occur in N are in Δ too. Therefore, $\Delta \vdash N\colon \tau$ has a derivation in which at each node, a variable is assigned a type in the assumption set iff the variable occurs in the term on the right-hand side of the $\vdash$. Also, the substitution steps (if there are any in the particular derivation of $\Delta \vdash N\colon \tau$) may be replaced by substitution steps on axioms. Then there is a derivation D of $\Delta \vdash N\colon \tau$ such that all the assumption sets except for instances of the id axiom are Δ's.[7]

The rest of the proof is by induction on the length of $\triangleright_{\mathrm{w}}$, that is, on the number of one-step expansions that comprise the converse of $\triangleright_{\mathrm{w}}$ linking N and M. If the length is 0, then the claim is obviously true.

We assume that the claim is true for expansions of length n, and we show that it is true when one more step is added. There are three cases to consider, depending on the combinator that is the head of the newly introduced redex.

Let us assume that I is the head, and P is a subterm of N that is replaced by $\mathsf{I}P$. P is typable, and in fact, $\Delta' \vdash P\colon \sigma$ (where $\Delta' \neq \Delta$ only if P is a variable) is a subtree of the derivation D of $\Delta \vdash N\colon \tau$ (where all the assumption sets have been harmonized). We only have to show that $\mathsf{I}P$ can be typed by σ.

$$\dfrac{\dfrac{\dfrac{\Delta \vdash \mathsf{I}\colon \alpha \rightarrow \alpha}{\Delta \vdash \mathsf{I}\colon \sigma \rightarrow \sigma} \qquad \begin{matrix}\vdots\\ \Delta \vdash P\colon \sigma\end{matrix}}{\Delta \vdash \mathsf{I}P\colon \sigma}}{\dfrac{\vdots}{\Delta \vdash [P/\mathsf{I}P]N\colon \tau}}$$

The above section of a tree is a derivation provided the subtree above the ellipsis is replacing the subtree that has as its root $\Delta' \vdash P\colon \sigma$ in the original derivation D.

The next case is C. Let us assume that the particular subterm that is expanded is $P_1P_2P_3$, which becomes $\mathsf{C}P_1P_3P_2$ after the expansion. Then there is derivation

[7] See lemmas 9.1.19 and 9.1.20 below, in which similar claims for $\mathrm{CL}_{\rightarrow}$ are proved.

of $\Delta \vdash P_1P_2P_3 : \sigma$, which is a subtree of the derivation D of $\Delta \vdash N : \tau$, due to the subject construction lemma. Then there are derivations of $\Delta \vdash P_1P_2 : \tau_1 \to \sigma$ and $\Delta'_1 \vdash P_3 : \tau_1$, as well as of $\Delta'_2 \vdash P_1 : \tau_2 \to \tau_1 \to \sigma$ and $\Delta'_3 \vdash P_2 : \tau_2$. (In the three latter cases, if $\Delta'_i \neq \Delta$, then P_i is a variable. However, $\Delta'_i = \Delta$ if P_i is not a variable, and the assumption sets may be identical even in some cases when P_i is a variable.) We have to show that, given all these subderivations, $\Delta \vdash \mathsf{C}P_1P_3P_2 : \sigma$ is derivable. Let us consider the following segment of a derivation. π_1 and π_2, respectively, abbreviate $(\alpha \to \beta \to \gamma) \to \beta \to \alpha \to \gamma$ and $(\tau_1 \to \tau_2 \to \sigma) \to \tau_2 \to \tau_1 \to \sigma$.

$$\dfrac{\dfrac{\dfrac{\dfrac{\Delta \vdash \mathsf{C} : \pi_1}{\Delta \vdash \mathsf{C} : \pi_2} \quad \Delta'_2 \vdash P_1 : \tau_2 \to \tau_1 \to \sigma}{\Delta \vdash \mathsf{C}P_1 : \tau_1 \to \tau_2 \to \sigma} \quad \Delta'_1 \vdash P_3 : \tau_1}{\Delta \vdash \mathsf{C}P_1P_3 : \tau_2 \to \sigma} \quad \Delta'_3 \vdash P_2 : \tau_2}{\Delta \vdash \mathsf{C}P_1P_3P_2 : \sigma}$$

The above tree is a derivation, and if we replace the subtree with the root $\Delta \vdash P_1P_2P_3 : \sigma$ by the tree above, then we get a derivation of $\Delta \vdash [P_1P_2P_3/\mathsf{C}P_1P_3P_2]N : \tau$.

The last case is left as exercise 9.1.12. qed

Exercise 9.1.11. Verify that lemma 9.1.8 and lemma 9.1.10 are true for $\mathrm{CL}(\mathsf{B},\mathsf{C},\mathsf{I})_\to$ as claimed in the proof of the above theorem.

Exercise 9.1.12. Complete the proof of the subject expansion theorem by writing out the details of the case when the head of the newly introduced redex is B.

The type assignments that we introduced in $\mathrm{CL}(\mathsf{B},\mathsf{C},\mathsf{I})_\to$ for the combinators B,C and I are provable in $\mathrm{CL}_\to$. Thus we could have taken the approach of *extending* $\mathrm{CL}_\to$ by these axioms. In virtue of the replacement lemma, such an expansion would not lead to any new typable terms when B,C and I are replaced by their definitions obtainable from S and K.

LEMMA 9.1.18. *The axioms for* B,C *and* I *from* $\mathrm{CL}(\mathsf{B},\mathsf{C},\mathsf{I})_\to$ *are provable in* $\mathrm{CL}_\to$ *with arbitrary assumption sets, including* $\Delta = \emptyset$.

Exercise 9.1.13. Prove this lemma. (Hint: Choose a definition for each of the three combinators, e.g., SKK for I, etc., and reconstruct the proof tree from the root upward. Alternatively, you can construct the tree top-down, possibly, relabeling some types in the whole tree so that the types in the root match the types in the corresponding axioms.)

Concerning the extension of $\mathrm{CL}_\to$ with the axioms for B,C and I, we note that we can select one of the definitions of these combinators in terms of S and K, perhaps, requiring that the terms are in weak nf, and we can think about the axioms as if B,C and I were replaced by those terms. Then the next claim is obviously true, by lemma 9.1.10 (replacement), and by lemma 9.1.18.

COROLLARY 9.1.18.1. *The addition of the axioms for* B,C *and* I *from* $\mathrm{CL}(\mathsf{B},\mathsf{C},\mathsf{I})_{\rightarrow}$ *to* $\mathrm{CL}_{\rightarrow}$ *is conservative; that is, no new type assignments result for terms over the combinatory basis* $\{\mathsf{S},\mathsf{K}\}$.

Linear terms get their name from *linear logic*, which was introduced by J.-Y. Girard in [74]. The implicational fragment of linear logic is what has been called $BCI_{\rightarrow}$ logic, which was introduced by C. A. Meredith and A. Prior in the early 1960s in [106]. The proof-theoretic interest in $BCI_{\rightarrow}$ logic stems from at least two remarkable properties. If the types of these combinators are taken as axioms with propositional variables, then a severely restricted form of the rule of detachment, called *condensed detachment*, almost suffices in place of detachment and substitution. In fact, the introduction of the rule of condensed detachment led to the invention of $BCI_{\rightarrow}$ logic, though substitution and detachment can be replaced by this rule in certain "weaker" logics too. We will return to this rule later in this section.

Another attractive feature of a logic with the types of B,C and I being theorems is that, in its sequent calculus formulation, sequents may be viewed as *strings* of formulas (due to the *associativity* of the sequent-forming operation). Associativity can be linked to B and C. Furthermore, the order of the formulas is immaterial too because of the *commutativity* of that operation. (The commutativity is the result of the presence of C and I.) These properties lead to quite simple sequent calculi about which meta-theorems are provable without much difficulty.

The possibility to use large assumption sets makes some considerations of provable type assignments easy. However, it is also useful to be able to narrow down that set when necessary.

LEMMA 9.1.19. *Let* M *be a term with* $\mathrm{fv}(M)=X$. *If* M *is typable by* τ, *then there is a proof in* $\mathrm{CL}_{\rightarrow}$ *with root* $\Delta\vdash M\colon\tau$, *where* Δ *does not assign types to any variables other that those in* X. *Therefore,* $x\colon\sigma\in\Delta$, *with some* σ, *iff* $x\in\mathrm{fv}(M)$.

Proof: For the right-to-left direction, let us assume that $x\in\mathrm{fv}(M)$. By lemma 9.1.8, every occurrence of x in M corresponds to a leaf in a proof D of $\Delta'\vdash M\colon\tau$, where Δ' is some assumption set. The existence of a proof D follows from the hypothesis of the lemma that M is typable by τ. In a derivation, the only operation on assumption sets is union, therefore, x is assigned a type by Δ', whatever Δ' is.

The conditional from left to right may be proved by, first, assuming (to the contrary) that $y\colon\sigma$ is included in Δ, though $y\notin\mathrm{fv}(M)$. Again, by the subject construction lemma, no leaf of the derivation is an axiom of the form $\Delta''\vdash y\colon\chi$. All leaves that are instances of the combinatory axioms remain instances of the same axioms with $\emptyset$ on the left-hand side of the $\vdash$. All instances of the id axiom with $x\colon\sigma$ on the right-hand side of the $\vdash$ can be changed to contain only the type assignment $x\colon\sigma$ in Δ. The two rules, $E\rightarrow$ and sub, do not have side conditions, therefore, the previous applications of these rules are applications of the same rules now.

In sum, for any axiom $\Delta'\vdash N\colon\chi$, $\Delta''\vdash N\colon\chi$ is an axiom, where $x\colon\sigma\in\Delta''$ only if $x\in\mathrm{fv}(N)$. Given that the premises of the rules have this property, the conclusion has the same property, either because the term is the same in the premise and in the conclusion, or because if $x\in\mathrm{fv}(N)$, then $x\in\mathrm{fv}(NP)=\mathrm{fv}(PN)$. Thus, indeed, if M

is typable by τ, then there is a derivation with root $\Delta \vdash M\colon \tau$, where Δ assigns types only to the free variables of M. qed

The rule of substitution is very handy when we are constructing a proof. We can start with some axioms, and in order to be able to apply detachment, we can substitute in the types. Substitution is sufficient for the application of that rule, because any pair of terms M and N can be put together into MN. In other words, the conditions for the applicability of the $E\rightarrow$ rule are the univalence of the union of the assumption sets and finding substitutions for the types of the premises so that they turn out to be of the form $\sigma \rightarrow \tau$ and σ. The combination of detachment with just enough substitution allowed is the rule of *condensed detachment* that we already mentioned.

On the other hand, it would be easier to think about derivations if the application of the substitution would be eliminated or at least restricted.

LEMMA 9.1.20. *If there is a derivation of $\Delta \vdash M\colon \tau$ possibly with applications of the substitution rule, then there is a derivation with the same root, in which the substitution rule is applied to axioms only.*

Proof: The proof is by induction on the structure of the derivation.

By lemma 9.1.19, we should not be concerned about those elements of Δ in which there is a variable x that is not free in M. The type assignments to these variables do not affect in an essential way the derivation of $\Delta \vdash M\colon \tau$. At the same time, a variable may be assigned any type, hence any substitution on any disenfranchised variable turns a derivation into a derivation.

The derivation is either an axiom, and then the claim is trivially true, or the last step in the derivation is an application of one of the two rules.

If the last step is an application of the rule of substitution, then the derivation looks like

$$\dfrac{\begin{matrix}\vdots\\ \Delta \vdash M\colon \sigma\end{matrix}}{[\alpha_i/\chi_i]\Delta \vdash M\colon [\alpha_i/\chi_i]\sigma}\,.$$

By inductive hypothesis, there is a derivation of $\Delta \vdash M\colon \sigma$ such that all the substitution steps come right after instances of the axioms. Substitutions compose (by function composition) into a substitution, hence the substitution step in the last step of the derivation may be incorporated into the substitutions after the leaves of the tree.[8]

The other case is when the last step is an application of detachment and the derivation looks like

$$\dfrac{\begin{matrix}\vdots\\ \Delta_1 \vdash M\colon \sigma \rightarrow \tau\end{matrix} \qquad \begin{matrix}\vdots\\ \Delta_2 \vdash N\colon \sigma\end{matrix}}{\Delta_1, \Delta_2 \vdash MN\colon \tau}\,.$$

We may assume that, if $x \notin \mathrm{fv}(MN)$, then x does not occur in $\Delta_1 \cup \Delta_2$ at all, or that it is in Δ_1 only. If there is an occurrence of a variable both in Δ_1 and Δ_2, then

[8]A substitution is a total function from $\mathbb{P}$ into the set of types. The bracket-and-slashes notation displays only a restriction of the substitution function to a (finite) subset of $\mathbb{P}$.

the two occurrences carry the same type assignment in virtue of the requirement of univalence, which is a condition for the applicability of the rule. By inductive hypothesis, there are derivations with the stipulated property ending in $\Delta_1 \vdash M\colon \sigma \to \tau$ and $\Delta_2 \vdash N\colon \sigma$. The application of the rule $E\to$ puts together two trees in each of which the applications of substitution (if there are any) are on the leaves, by the hypothesis of the induction. This completes the proof. qed

Example 9.1.21. Occasionally, a type may completely disappear by the time the final type assignment is reached. The following segment of a derivation illustrates this situation together with some of the potential complications from the proof of the previous lemma.

$$\dfrac{\dfrac{\begin{array}{c}\vdots\\ \{x\colon \gamma,\, y\colon \beta\} \vdash \mathsf{SK}\colon (\beta\to\gamma\to\beta)\to\beta\to\beta\end{array}}{\{x\colon \beta,\, y\colon \alpha\} \vdash \mathsf{SK}\colon (\alpha\to\beta\to\alpha)\to\alpha\to\alpha} \qquad \{x\colon \beta,\, z\colon \alpha\} \vdash \mathsf{K}\colon \alpha\to\beta\to\alpha}{\{x\colon \beta,\, y\colon \alpha,\, z\colon \alpha\} \vdash \mathsf{SKK}\colon \alpha\to\alpha}$$

In the left branch of the tree, a substitution has been carried out; that is, $\{x\colon \beta, y\colon \alpha\} \vdash \mathsf{SK}\colon (\alpha\to\beta\to\alpha)\to\alpha\to\alpha$ is the result of the substitution $[\gamma/\beta,\beta/\alpha]\{x\colon \gamma,\, y\colon \beta\} \vdash \mathsf{SK}\colon [\gamma/\beta,\beta/\alpha](\beta\to\gamma\to\beta)\to\beta\to\beta$.

Clearly, $\mathrm{fv}(\mathsf{SKK}) = \emptyset$, and so the assignments to x, y and z can be omitted. We can similarly reconstruct the missing part of the derivation — provided we aim at having a tree as flat as possible, that is, a tree of minimal height.

$$\dfrac{\dfrac{\vdash \mathsf{S}\colon (\alpha\to\beta\to\gamma)\to(\alpha\to\beta)\to\alpha\to\gamma}{\vdash \mathsf{S}\colon (\beta\to(\gamma\to\beta)\to\beta)\to(\beta\to\gamma\to\beta)\to\beta\to\beta} \qquad \dfrac{\vdash \mathsf{K}\colon \alpha\to\beta\to\alpha}{\vdash \mathsf{K}\colon \beta\to(\gamma\to\beta)\to\beta}}{\vdash \mathsf{SK}\colon (\beta\to\gamma\to\beta)\to\beta\to\beta}$$

The substitutions may be composed to obtain a derivation of $\{x\colon \beta,\, y\colon \alpha\} \vdash \mathsf{SK}\colon (\alpha\to\beta\to\alpha)\to\alpha\to\alpha$. $\Delta = \{x\colon \beta,\, y\colon \alpha\}$ is not minimal, and we can pick types for x and y independently. Indeed, assigning to these disenfranchised variables basic types that do not occur anywhere else in the derivation clarifies the whole derivation. (Again, we abbreviate some of the types to keep the size of the derivation manageable: π stands for $\alpha\to(\beta\to\alpha)\to\alpha$, and τ is $\alpha\to\beta\to\gamma$.)

$$\dfrac{\dfrac{\{x\colon \delta,\, y\colon \varepsilon\} \vdash \mathsf{S}\colon \tau\to(\alpha\to\beta)\to\alpha\to\gamma}{\{x\colon \beta,\, y\colon \alpha\} \vdash \mathsf{S}\colon \pi\to(\alpha\to\beta\to\alpha)\to\alpha\to\alpha} \qquad \{y\colon \alpha\} \vdash \mathsf{K}\colon \pi}{\{x\colon \beta,\, y\colon \alpha\} \vdash \mathsf{SK}\colon (\alpha\to\beta\to\alpha)\to\alpha\to\alpha}$$

Let us summarize the potential modifications we can make to $\mathrm{CL}_{\to}$. We can trim the *set of assumptions* to contain only type assignments for those variables that are free (i.e., occur) in the term on the right-hand side of the $\vdash$. We can also replace the axioms for combinators by *schemas* and drop the substitution rule. Lastly, if desired we might want to add a rule of *weakening* (or *monotonicity*) for the assumption set.

The reformulated $\mathrm{CL}_{\rightarrow}$ looks like as follows.

$$\{x\colon \tau\} \vdash x\colon \tau \ \text{id}$$

$$\vdash \mathsf{K}\colon \pi \rightarrow \sigma \rightarrow \pi \ \vdash\!\mathsf{K} \qquad\qquad \vdash \mathsf{S}\colon (\pi \rightarrow \sigma \rightarrow \chi) \rightarrow (\pi \rightarrow \sigma) \rightarrow \pi \rightarrow \chi \ \vdash\!\mathsf{S}$$

$$\frac{\Delta_1 \vdash M\colon \sigma \rightarrow \tau \qquad \Delta_2 \vdash N\colon \sigma}{\Delta_1, \Delta_2 \vdash MN\colon \tau}\ E\!\rightarrow \qquad\qquad \frac{\Delta_1 \vdash M\colon \tau}{\Delta_1, \Delta_2 \vdash M\colon \tau}\ \text{mon}$$

As before, all the assumption sets are required to be *univalent.*

LEMMA 9.1.22. *There are combinators that can be assigned* more that one type, *and there are types that can be assigned to* more that one combinator.

There are combinators that cannot be assigned a type, and there are types that cannot be assigned to a combinator.

Proof: The first claim is obvious, for example, K can be assigned the type $\alpha \rightarrow \beta \rightarrow \alpha$, $\alpha \rightarrow \alpha \rightarrow \alpha$ as well as $(\alpha \rightarrow \beta) \rightarrow \beta \rightarrow \alpha \rightarrow \beta$, etc. These types are distinct, though they are similar to one another to a certain extent.

To prove the second claim, let us consider SKK and SKK(SKK). The first can be assigned the type $\alpha \rightarrow \alpha$, hence also the type $(\alpha \rightarrow \alpha) \rightarrow \alpha \rightarrow \alpha$. Then $\alpha \rightarrow \alpha$ is a type both for SKK and SKK(SKK). (Cf. example 9.1.14 that contains similar terms and types.)

For the third statement, let us consider SII (where I is, of course, SKK). If we identify π with $\sigma \rightarrow \chi$ in the axiom of S, then the rest of the type after a detachment is $((\sigma \rightarrow \chi) \rightarrow \sigma) \rightarrow (\sigma \rightarrow \chi) \rightarrow \chi$. This is a type for SI, and to get a type for SII, we would need to turn $(\sigma \rightarrow \chi) \rightarrow \sigma$ into $\pi \rightarrow \pi$ or into an instance of $\pi \rightarrow \pi$. However, no implication can be unified with its *antecedent* (or its *consequent*), independently of how complex the subtypes (e.g., σ and χ) are.

Lastly, the type we already mentioned in connection to Y, $(\alpha \rightarrow \alpha) \rightarrow \alpha$ is not a theorem of intuitionistic logic. Therefore, by lemma 9.1.13, $(\alpha \rightarrow \alpha) \rightarrow \alpha$ cannot be assigned as a type to a combinator. qed

The axiom *schemas* for combinators mean that all the types that can be assigned to S and K are *substitution instances* of those schemas.

DEFINITION 9.1.23. A type τ is *a principal type* for a term M iff all the types that can be assigned to M are substitution instances of τ. A principal type of M is denoted by $\mathrm{pt}(M)$.

We assumed an infinite sequence of atomic types, thus the next claim is immediate.

LEMMA 9.1.24. *A term has either* no *or* infinitely many *principal types.*

Exercise 9.1.14. Write out the proof of the lemma. (Hint: Types are defined to have a finite length.)

A variable can be assigned any type. Thus any variable has a principal type α. Type assignments to variables are far more permissive than the principal types for the combinators S and K in their axioms, which makes the principal types for variables somewhat uninteresting. Sometimes the notion of principal types is generalized to principal pairs, which is the principal type of a term containing free variables with respect to a set of type assignment assumptions about those variables. We focus here on *closed terms* only.

A closed term in $\mathrm{CL}_{\rightarrow}$ can contain occurrences of S and K only; that is, it is a combinator. We may ask then if a combinator, which is not necessarily atomic, has or does not have a principal type.

LEMMA 9.1.25. *For every typable combinator Z, there is a type σ such that $\vdash \mathsf{Z}\colon \sigma$ is provable in $\mathrm{CL}_{\rightarrow}$, and for any τ, if $\vdash \mathsf{Z}\colon \tau$ is provable in $\mathrm{CL}_{\rightarrow}$, then τ is a substitution instance of σ. That is, every typable combinator has a* principal type.

Proof: If a term is typable, then there is a derivation of $\vdash \mathsf{Z}\colon \tau$ in $\mathrm{CL}_{\rightarrow}$. By lemma 9.1.8, all the leaves of the derivation tree are instances of the combinatory axioms, because each leaf in the formation tree of Z is S or K.

Now we will show that there is a derivation of $\vdash \mathsf{Z}\colon \sigma$, where σ is the principal type for Z. (Of course, σ may be τ itself or a relabeling of τ.) The core idea is to use at each detachment step instances of types that are as general as possible. In other words, we apply condensed detachment with respect to the terms.

The most general instances of the axioms are the axioms themselves. If $\vdash \mathsf{Z}_1\colon \chi_1 \rightarrow \chi_2$ and $\vdash \mathsf{Z}_2\colon \chi_3$ at some step in the new derivation, then we create substitution instances of $\chi_1 \rightarrow \chi_2$ and χ_3, let us say, $\xi \rightarrow \chi_2'$ and ξ to obtain $\vdash \mathsf{Z}_1\mathsf{Z}_2\colon \chi_2'$. Informally, our strategy is that we identify types or substitute a type of higher degree only if that is necessary for the instances of χ_1 and χ_3 to turn out the same. Formally, we require that for any pair of substitutions s_1 and s_2 such that $s_1(\chi_1) = s_2(\chi_3)$, there is an s_3 with the property that $s_3(\xi) = s_1(\chi_1)$. That is, any other type that could be taken in place of ξ, can be obtained by substitution from ξ.

The step of finding ξ is *unification*, because ξ is an instance of both χ_1 and χ_3. The search for ξ can be described algorithmically, that is, it is computable. Furthermore, it is decidable whether a pair of terms can be unified. Sometimes ξ is called the *most general unifier* (mgu) — instead of the most general common instance — of χ_1 and χ_3.

We assumed (as a hypothesis of the lemma) that Z is typable, and we started with a derivation that results in a type assignment to Z, which guarantees typability. The typability of Z implies that Z_1 and Z_2, as well as $\mathsf{Z}_1\mathsf{Z}_2$ are typable; therefore, it follows that χ_1 and χ_3 have a common instance. Let us assume that ξ' (distinct from ξ) is also an mgci for χ_1 and χ_3. Then $s_3(\xi) = \xi'$ and $s_4(\xi') = \xi$. Then ξ and ξ' are each other's relabelings, because no implication has an atomic type as its instance and no substitution can distinguish two occurrences of a type variable. The fact that two mgci's are necessarily each other's reletterings justifies the use of the definite article in talking about mgci's.

Having found ξ, we can proceed in two ways to construct the derivation of the principal type schema for Z. If we suppose the earlier formulation of $\mathrm{CL}_{\rightarrow}$, then we can add substitution steps into the derivation. If we have type schemas in the combinatory axioms, then we can apply the substitutions s_1 and s_2 to the types in the subtrees ending in $\vdash \mathsf{Z}_1 : \chi_1 \to \chi_2$ and in $\vdash \mathsf{Z}_2 : \chi_3$, respectively. Either way we have a derivation of $\vdash \mathsf{Z}_1\mathsf{Z}_2 : \chi_2'$. Since ξ is an mgci, χ_2' is at least as general as the type of $\mathsf{Z}_1\mathsf{Z}_2$ in the original derivation. qed

By the previous lemma, we know that if Z can be assigned a type at all, then there are *infinitely many types* that can be assigned to Z. However, there is a type τ such that all those types are instances of τ, which justifies talking about *the principal type schema* of a combinator. Given a pair of formulas, *the most general common instance* is unique up to renaming atomic types, and the principal type schema of all the combinators — save S and K — are most general common instances of a pair of types. The close connection between these two notions explains that they are both unique (if exist) up to relettering.

Example 9.1.26. Let us consider the term SS. We might have a derivation that shows that the term is typable. Let π_1 and π_2 abbreviate $(\alpha \to \alpha \to \beta) \to (\alpha \to \alpha) \to \alpha \to \beta$ and $((\alpha \to \alpha \to \beta) \to \alpha \to \alpha) \to (\alpha \to \alpha \to \beta) \to \alpha \to \beta$, respectively.

$$\frac{\vdash \mathsf{S} : \pi_1 \to \pi_2 \qquad\qquad \vdash \mathsf{S} : \pi_1}{\vdash \mathsf{SS} : ((\alpha \to \alpha \to \beta) \to \alpha \to \alpha) \to (\alpha \to \alpha \to \beta) \to \alpha \to \beta}$$

This derivation shows that the term SS is typable. However, π_2 is *not* the principal type schema of this term.

Now let us take $(\chi_1 \to \chi_2 \to \chi_3) \to (\chi_1 \to \chi_2) \to \chi_1 \to \chi_3$ and $(\sigma_1 \to \sigma_2 \to \sigma_3) \to (\sigma_1 \to \sigma_2) \to \sigma_1 \to \sigma_3$ as instances of the type schema for S. In order to detach the second one from the first, χ_1 should be $\sigma_1 \to \sigma_2 \to \sigma_3$, χ_2 should be $\sigma_1 \to \sigma_2$ and χ_3 should be $\sigma_1 \to \sigma_3$. These conditions can be thought of as *equations* expressing *constraints*. The three conditions (taken together) do not force us to identify, for instance, σ_1 and σ_2. Thus the principal type schema that we get for SS is

$$((\sigma_1 \to \sigma_2 \to \sigma_3) \to \sigma_1 \to \sigma_2) \to (\sigma_1 \to \sigma_2 \to \sigma_3) \to \sigma_1 \to \sigma_3,$$

where the σ's may be pairwise distinct. An instance of the $\mathrm{pt}(\mathsf{SS})$ is the type

$$((\sigma_1 \to \sigma_2 \to \sigma_4 \to \sigma_4) \to \sigma_1 \to \sigma_2) \to (\sigma_1 \to \sigma_2 \to \sigma_4 \to \sigma_4) \to \sigma_1 \to \sigma_4 \to \sigma_4.$$

This type schema contains three variables, just like the previous one, however, the substitution of $\sigma_4 \to \sigma_4$ for σ_3 is not warranted by the unification of the two types of S (with χ's and with σ's).

LEMMA 9.1.27. *Given a combinator over the basis* $\{\mathsf{S},\mathsf{K}\}$, *it is* decidable *if the term is typable.*

Proof: From lemma 9.1.8, we know that if the term is typable (hence, there is a derivation of a type assignment for that term), then the formation tree of the term is isomorphic to the derivation tree of the type assignment (in the substitution-free version of $\mathrm{CL}_{\to}$). Performing unification at each step (if possible), as in the proof of the previous lemma, yields the principal type schema of the combinator, therefore, demonstrates typability. On the other hand, if unification fails at a step, then that demonstrates the untypability of the combinator. qed

The next claim is a straightforward consequence of the two preceding lemmas; nonetheless, it is useful to state it explicitly.

COROLLARY 9.1.27.1. *Given a term M and a type τ, it is decidable if* $\tau = \mathrm{pt}(M)$.

Finding types for terms, perhaps types with special properties, is one of the concerns we have. Often, especially when we think of the types as formulas, we are interested in looking at the type assignment relationship from the other direction, so to speak, that is, whether the type can be assigned to a combinator.

DEFINITION 9.1.28. If there is a combinator M, such that $\vdash M : \tau$, then M *inhabits* the type τ.

The $\{\mathsf{S},\mathsf{K}\}$ basis is very well behaved, because it is not only combinatorially complete and typability is decidable, but inhabitation is decidable too.

THEOREM 9.1.29. *Given a type τ, it is* decidable *whether τ is inhabited by a combinator over the base* $\{\mathsf{S},\mathsf{K}\}$.

Proof: We have seen in lemma 9.1.13 that all the types that can be assigned to a combinator in $\mathrm{CL}_{\rightarrow}$ are theorems of $H_{\rightarrow}$. The term-free part of $\mathrm{CL}_{\rightarrow}$ is in fact the *natural deduction* calculus for $H_{\rightarrow}$ (without the $I \rightarrow$ rule), which is, in turn, equivalent to the implicational fragments of Gentzen's *LJ* and Curry's $LA_m(\mathfrak{D})$.[9] Both calculi are decidable, hence so is $H_{\rightarrow}$. The decision procedures essentially depend on the existence of an upper bound for the search space for proofs that follows from the sufficiency of reduced sequents. qed

A combinator Z that inhabits a type τ can be viewed as a *blueprint for a proof* of the type (as a formula) in $H_{\rightarrow}$. The implicational fragment of intuitionistic logic can be presented as *an axiom system* by taking the two type schemas that appear in the axioms for S and K together with the rule of detachment. The definition of a proof in an axiomatic calculus is rather flexible, and we use that flexibility to restrict the class of proofs.

DEFINITION 9.1.30. The implicational fragment of intuitionistic logic, $H_{\rightarrow}$, comprises (A1)–(A2) and (R1).

(A1) $A \rightarrow B \rightarrow A$

(A2) $(A \rightarrow B \rightarrow C) \rightarrow (A \rightarrow B) \rightarrow A \rightarrow C$

(R1) $A \rightarrow B,\ A \Rightarrow B$

[9] See Gentzen [70, pp. 204–205] and Curry [52, p. 235].

A finite (nonempty) sequence of wff's is *a proof* iff each wff is either an instance of (A1) or (A2), or is obtained from preceding wff's in the sequence by (R1).

Note that inserting finitely many instances of axioms into a proof creates another proof. Also, typically, some wff's in a proof may be permuted, because (R1) does not stipulate — perhaps, despite appearances — that $A \to B$ must come before A. In what follows, proofs are considered as sequences of wff's arranged vertically with one wff on each line. This is a usual way to write out proofs, and it will make easier for us to describe proofs without introducing ambiguity or imprecision.

DEFINITION 9.1.31. We define recursively *a proof of* τ *in* $H_{\to}$, given a derivation of $\vdash \mathsf{Z}\colon \tau$ in $\mathrm{CL}_{\to}$ (with axiom schemas for S and K).[10]

(1) If the node is a leaf in the derivation tree, then we insert the type (as a formula) onto the next line of the proof (which is justified as an instance of an axiom schema).

(2) If the node is not a leaf, then the node is the result of an application of the $E\to$ rule, and we insert the lines obtained from the subtree rooted in the left premise, followed by the lines obtained from the subtree rooted in the right premise, followed by the type in the conclusion.

(3) No other formulas are included into the proof.

Notice that the sequence of wff's resulting from this definition is a proof in $H_{\to}$, because wff's by (1) are instances of axioms, and wff's by (2) are preceded by their premises in detachment. (3) guarantees that no loose wff's are inserted into the sequence. The latter condition is stronger than what is absolutely necessary to ensure that the sequence is a proof, as we will see below.

Example 9.1.32. Let us consider the derivation of a type for $\mathsf{S(KS)K}$. We start with the tree

$$\dfrac{\dfrac{\vdash \mathsf{S}\colon \pi_1 \qquad \dfrac{\vdash \mathsf{K}\colon [\alpha/\mathrm{pt}(\mathsf{S}), \beta/\beta\to\gamma]\,\mathrm{pt}(\mathsf{K}) \qquad \vdash \mathsf{S}\colon \mathrm{pt}(\mathsf{S})}{\vdash \mathsf{KS}\colon (\beta\to\gamma)\to\mathrm{pt}(\mathsf{S})}}{\vdash \mathsf{S(KS)}\colon \pi_2} \qquad \vdash \mathsf{K}\colon (\beta\to\gamma)\to\alpha\to\beta\to\gamma}{\vdash \mathsf{S(KS)K}\colon (\beta\to\gamma)\to(\alpha\to\beta)\to\alpha\to\gamma},$$

where π_1 is shorthand notation for

$$((\beta\to\gamma)\to\mathrm{pt}(\mathsf{S}))\to((\beta\to\gamma)\to\alpha\to\beta\to\gamma)\to(\beta\to\gamma)\to(\alpha\to\beta)\to\alpha\to\gamma,$$

and π_2 abbreviates the type

$$((\beta\to\gamma)\to\alpha\to\beta\to\gamma)\to(\beta\to\gamma)\to(\alpha\to\beta)\to\alpha\to\gamma,$$

[10] We use the second formulation of $\mathrm{CL}_{\to}$, where we have type *schemas* in $\vdash \mathsf{K}$ and $\vdash \mathsf{S}$, and there is no rule of substitution.

and further, $\mathrm{pt}(\mathsf{S})$ stands for the principal type schema of S with α, β and γ, similarly, $\mathrm{pt}(\mathsf{K})$ is the principal type schema of K with α and β, and the substitution notation indicates modifications to these particular instances of the pt's, for the sake of concreteness.

Let us denote by $\mathrm{pr}(\tau)$ the proof of τ in $H_{\rightarrow}$ obtained from the above definition. Starting at the root, we get by (1) and (2):

1–5. $\mathrm{pr}(\pi_2)$

6. $\mathrm{pr}((\beta \rightarrow \gamma) \rightarrow \alpha \rightarrow \beta \rightarrow \gamma)$

7. $(\beta \rightarrow \gamma) \rightarrow (\alpha \rightarrow \beta) \rightarrow \alpha \rightarrow \gamma$

(The line numbers are not known at this point; however, for the sake of transparency, we already use here the line numbers that we will have in the final proof. The line numbers mainly reinforce that the vertical series of wff's is a finite sequence.)

$\mathrm{pr}((\beta \rightarrow \gamma) \rightarrow \alpha \rightarrow \beta \rightarrow \gamma)$ is the formula itself by (1), because the corresponding node is a leaf, but $\mathrm{pr}(\pi_2)$ has to be expanded further. We obtain

1. $\mathrm{pr}(\pi_1)$

2–4. $\mathrm{pr}((\beta \rightarrow \gamma) \rightarrow \mathrm{pt}(\mathsf{S}))$

5. π_2

6. $(\beta \rightarrow \gamma) \rightarrow \alpha \rightarrow \beta \rightarrow \gamma$

7. $(\beta \rightarrow \gamma) \rightarrow (\alpha \rightarrow \beta) \rightarrow \alpha \rightarrow \gamma$

After a couple of other steps, we get the following completed proof in $H_{\rightarrow}$, which we annotate both with combinatory terms and the usual justifications often appended to proofs in axiom systems.

1. π_1 $[\mathsf{S}; (\mathrm{A2})]$

2. $\mathrm{pt}(\mathsf{S}) \rightarrow (\beta \rightarrow \gamma) \rightarrow \mathrm{pt}(\mathsf{S})$ $[\mathsf{K}; (\mathrm{A1})]$

3. $\mathrm{pt}(\mathsf{S})$ $[\mathsf{S}; (\mathrm{A2})]$

4. $(\beta \rightarrow \gamma) \rightarrow \mathrm{pt}(\mathsf{S})$ $[\mathsf{SK}; (\mathrm{R1})\ 2, 3]$

5. π_2 $[\mathsf{S}(\mathsf{KS}); (\mathrm{R1})\ 1, 4]$

6. $(\beta \rightarrow \gamma) \rightarrow \alpha \rightarrow \beta \rightarrow \gamma$ $[\mathsf{K}; (\mathrm{A1})]$

7. $(\beta \rightarrow \gamma) \rightarrow (\alpha \rightarrow \beta) \rightarrow \alpha \rightarrow \gamma$ $[\mathsf{S}(\mathsf{KS})\mathsf{K}; (\mathrm{R1})\ 5, 6]$

The above proof is of a rather *special shape*. A major premise for an application of detachment precedes the minor premise, and if the minor premise itself is obtained via detachment, then the major premise precedes all the lines that constitute the minor premise. There are no shortcuts in the proof in the sense of using a formula as a premise for two applications of detachment. Also, there are no spurious formulas in this proof either: if a formula is not used in obtaining the bottom most wff, then it is not inserted into the proof despite the formula being an instance of an axiom.

DEFINITION 9.1.33. A proof in $H_{\rightarrow}$ is *special* iff the proof can be annotated to satisfy (1)–(3).

(1) Each wff preceding the last wff is used in obtaining the last wff.

(2) Each wff preceding the last wff is used in no more that one application of (R1).

(3) If the wff on line n is by (R1), then there is an i $(i < n)$ such that the wff's on lines $1-(i-1)$ and on lines $i-(n-1)$ form special proofs of the major and minor premises, respectively.

LEMMA 9.1.34. *Every theorem of* $H_{\rightarrow}$ *has a* special proof.

Proof: Every theorem has a proof, by the definition of what a theorem is. A proof is a finite object, and we may assume that an annotation with line numbers is assigned to the formulas. Working from the bottom to the top, the annotation permits us to find wff's in the proof that are not used; deleting those formulas does not destroy the proof. If there is a wff in the proof that is cited more than once as a premise of (R1), then we copy that wff (together with its proof) sufficiently many times to achieve single citations, and then we readjust the annotation. After this step, we can clearly perform sufficiently many permutations (if needed at all), to ensure that each major premise precedes the whole derivation of a minor premise. The flexibility of the usual definition of a proof in an axiomatic system ensures that switching the minor and major premises preserves the property of being a proof. qed

LEMMA 9.1.35. (CURRY–HOWARD ISOMORPHISM) *There is a* 1–1 correspondence *between special proofs of theorems in* $H_{\rightarrow}$ *and proofs of type assignments in* $\mathsf{CL}_{\rightarrow}$.

Proof: Definition 9.1.31 associates a (unique) proof in $H_{\rightarrow}$ to each proof in $\mathsf{CL}_{\rightarrow}$. It is easy to see that by definition 9.1.33 these proofs are special. On the other hand, given a special proof, the construction of the $\mathsf{CL}_{\rightarrow}$ proof is straightforward: axioms are axioms, and applications of (R1) are applications of $E\rightarrow$, lastly, the tree structure of the $\mathsf{CL}_{\rightarrow}$ derivation is built-in into condition (3) in the definition of special proofs. It is also true that for any $H_{\rightarrow}$ proof, there is a proof of the same formula as a type for a combinator in $\mathsf{CL}_{\rightarrow}$; however, from lemma 9.1.34, it is obvious that several proofs in $H_{\rightarrow}$ may lead — via conversion into special proofs — to one and the same $\mathsf{CL}_{\rightarrow}$ proof. qed

The Curry–Howard isomorphism can be extended to other logics and other combinatory bases. *Nonclassical logics* have been introduced independently from combinatory logic and λ-calculi. An intriguing coincidence though is that the implicational fragment of the logic of relevant implication $R_{\rightarrow}$ was introduced by Church in [45], who gave a certain preference to the $\lambda\mathsf{I}$-calculus over the $\lambda\mathsf{K}$-calculus (both of which were invented by him).

We mention four implicational logics: $R_\to$, $E_\to$, $T_\to$ and $BCI_\to$. R is the logic of *relevant implication*, E is the logic of *entailment*, T is the logic of *ticket entailment* and $BCI_\to$ is the implicational fragment of *linear logic*.[11]

DEFINITION 9.1.36. The rule of detachment (R1), together with a selection of axiom schemas from the following list, defines the implicational logics mentioned.

(I) $A \to A$

(B) $(A \to B) \to (C \to A) \to C \to B$

(C) $(A \to B \to C) \to B \to A \to C$

(B′) $(A \to B) \to (B \to C) \to A \to C$

(2) $(A \to (B \to D) \to C) \to (B \to D) \to A \to C$

(W) $(A \to A \to B) \to A \to B$

(R1) $A \to B, A \Rightarrow B$

$BCI_\to$	$T_\to$	$E_\to$	$R_\to$
(B), (C), (I)	(B), (B′), (I), (W)	(B), (2), (I), (W)	(B), (C), (I), (W)

Except (2), all the other wff's are principal type schemas of *proper* combinators. There is no axiomatization for $E_\to$ using principal types of proper combinators only. $E_\to$ is located between $T_\to$ and $R_\to$. $BCI_\to$ does not include contraction, and it is included only in $R_\to$. On the other hand, $E_\to$ and $T_\to$ are not included in $BCI_\to$, because (C) is not a theorem of either of those logics.

For a stronger version of the Curry–Howard isomorphism, we now define rule **D**.

DEFINITION 9.1.37. The rule of *condensed detachment*, denoted by **D**, allows the conclusion of $s_1(B)$ from $A \to B$ and C, when (i)–(iii) hold.

(i) there are substitutions s_1 and s_2 such that $s_1(A) = s_2(C)$,

(ii) the number of propositional variables in $s_1(B)$ is minimal (among all the substitutions s_1 satisfying (i)),

(iii) the number of distinct propositional variables in $s_1(B)$ is maximal (among all the substitutions s_1 satisfying (ii)).

[11] For detailed treatments of these logics and for further information about them, see [3], [4], [129], [29], [106] and [150]. (See also section A.9.)

Condition (i) stipulates that the wff's A and C have to be unified. The two other conditions make precise the idea that the unification is carried out so as to produce an mgci of A and C, hence to yield a B as general as possible. That is, substitutions introduce more complex formulas and identify variables only if that is necessary in order to find a common instance of A and C.

It is immediate that a logic that is defined by axiom schemas and detachment (or by axioms, modus ponens and substitution) *admits* rule **D**. However, it is not obvious (and for some logics it is false) that axioms together with rule **D** suffice.

DEFINITION 9.1.38. A *condensed logic* is a formalization of a logic that contains axioms and the rule **D**.

Of course, the idea is that the axioms are closely related to the axiom schemas that formalize a logic. For example, if the propositional variables are $p,q,r,\ldots$, then axiom (I) looks like $p \to p$, or if we use basic types, then (I) is $\alpha \to \alpha$. For the sake of definiteness, we can stipulate that the meta-variables $A,B,C,\ldots$ are replaced by $p,q,r,\ldots$ in the axiom schemas for a logic, and detachment is replaced by condensed detachment yielding a *condensed version* of a logic. We will denote the condensed version of a logic by prefixing **D** to its label. Then, we can ask whether the two calculi have the same set of theorems.

THEOREM 9.1.39. *The logics* $\mathbf{D}T_{\to}$, $\mathbf{D}E_{\to}$ *and* $\mathbf{D}R_{\to}$ *each has the same set of theorems, respectively, as* $T_{\to}$, $E_{\to}$ *and* $R_{\to}$ *does, that is, each of these logics is* **D**-complete.

Proof: The proof for $T_{\to}$ is due to Meyer (see [111]), and an adaptation may be found in [86] too. The core ideas behind the proof are as follows. (1) Self-implication is a theorem of the condensed logic. (2) Generalized instances of prefixing and suffixing are also theorems. (3) If a wff is a theorem, then the identification of any two propositional variables occurring once is a theorem too. (4) The result of a substitution in a theorem is a theorem.

The truth of the claim for $E_{\to}$ and $R_{\to}$ follows from the **D**-completeness of $T_{\to}$ together with $p \to p$ being a theorem. qed

LEMMA 9.1.40. $\mathbf{D}BCI_{\to}$ *is* **D**-incomplete.

Proof: The wff

$$(p \to q \to r) \to (p \to s \to t) \to p \to p \to (r \to s) \to q \to t$$

is a theorem of $BCI_{\to}$ logic, however, it is not a theorem of $\mathbf{D}BCI_{\to}$. The proof of the formula is not difficult in a natural deduction calculus for $BCI_{\to}$. To prove that $\mathbf{D}BCI_{\to}$ cannot have this wff as a theorem, it is useful to prove that all theorems of the condensed version of $BCI_{\to}$ have exactly two occurrences of each propositional variable. (Informally, the absence of contraction or of self-distribution of implication means that variables cannot be identified via an application of rule **D**.) qed

Exercise 9.1.15. Consider a natural deduction system for $BCI_{\rightarrow}$ with $E\rightarrow$ and $I\rightarrow$ in which premises may be permuted and all premises must be used exactly once. Prove that the wff displayed above is provable.

Exercise* 9.1.16. Prove that if A is a theorem of $\mathbf{D}BCI_{\rightarrow}$, and p occurs in A, then p occurs exactly twice in A.

Rule **D** was invented to streamline proofs in propositional logic by delimiting potential proof steps. Thus we should not be surprised that a logic (thought of as a set of theorems) may have more than one condensed version, depending on the set of axiom schemas from which the axioms are lifted. A condensed version of a logic may not be equivalent to the original logic (like in the case of $BCI_{\rightarrow}$). Moreover, two condensed versions of one and the same logic may not be equivalent either.

LEMMA 9.1.41. *The condensed logic comprising axioms* (1)–(4) *and rule* **D** *is* equivalent *to the* $BCI_{\rightarrow}$ *logic.*

(1) $p \rightarrow p$

(2) $(p \rightarrow q) \rightarrow (r \rightarrow p) \rightarrow r \rightarrow q$

(3) $(p \rightarrow q \rightarrow r) \rightarrow q \rightarrow p \rightarrow r$

(4) $(p \rightarrow p \rightarrow q) \rightarrow p \rightarrow p \rightarrow q$

Proof: The proof follows the proof of the **D**-completeness of $\mathbf{D}T_{\rightarrow}$ with the modification that one has to first derive in the condensed logic the wff's that are used in that proof. Clearly, axiom (4), which is an instance of pt(I) with the antecedent of the pt(W), is the crucial addition that make the proof possible.[12] qed

THEOREM 9.1.42. (STRENGTHENED CURRY–HOWARD ISOMORPHISM) *Every theorem of* $H_{\rightarrow}$ *is* the principal type *of a combinator over the basis* $\{\mathsf{S},\mathsf{K}\}$.

Proof: The truth of the claim follows from the **D**-completeness of $T_{\rightarrow}$, once we note that $H_{\rightarrow}$ contains (under that identity translation) all the theorems of $T_{\rightarrow}$, including $p \rightarrow p$. qed

The theorem means that there is a tight connection between the typable combinators built from S and K, and the theorems of $H_{\rightarrow}$. The combinators *encode* proofs in a precise sense.

[12]For details of the proof, see [104].

9.2 Intersection types for combinators

Simple types reflect the idea that a function has a domain and a codomain. However, simple types are not the only types that have been considered in typing CL-terms. A function taking functions as its argument can be "uncurried," which makes it seem plausible to introduce a type constructor $\times$ for *Cartesian product.* Then projections on pairs might be naturally considered and so forth. In typed functional programming languages, a wide range of type constructors have been introduced, typically closely matching the programming constructs. In view of these sources and developments in typing, typing with simple types only may seem impoverished; however, typing with $\rightarrow$ has its own advantages.

Simple typed terms *strongly normalize.* This means that a term that is typable has a nf, moreover, *every* reduction sequence is finite, but strong normalization may seem too strong. If a term normalizes (at all), then it is possible to define a strategy that will guarantee that the nf of the term is actually reached. In other words, weak normalization can be thought of as sufficiently safe or reliable from a certain point of view of computation. The set of terms in which all reduction sequences are finite is an important set. However, the set of terms, which have an nf, is also interesting, and it is desirable to have a characterization of this set too.

Last but not least, the Curry–Howard isomorphism entices us to think about type constructors beyond $\rightarrow$. Logics often, perhaps even typically, include multiple connectives, even if the implication connective and the implicational fragment of a logic take a central place in the study of that logic. Classical sentential logic — the simplest logic — cannot be fully captured or described by implication alone or by its implicational fragment. Of course, we know from the Curry–Howard isomorphism that classical logic is not sufficiently sophisticated to provide the simple types; however, the previous remark about classical logic is a fortiori true concerning intuitionistic logic.

Intersection types have been introduced and investigated for both λ-calculi and CL. Because of their theoretical significance and their connection to nonclassical logics, we focus on this ramification of typing here.

DEFINITION 9.2.1. Let $\mathbb{P}$ be a nonempty set of *type variables.* Let ω be a *type constant,* and let $\rightarrow$ and $\wedge$ be two binary *type constructors.* The set of intersection types (i-types, for short) is inductively defined by (1)–(3).

(1) If $\alpha \in \mathbb{P}$, then α is an i-type;

(2) ω is an i-type;

(3) if π and σ are i-types, then so are $(\pi \rightarrow \sigma)$ and $(\pi \wedge \sigma)$.

Our notational convention for omitting parentheses will be the one for simple types — with the addition that $\wedge$ binds stronger than $\rightarrow$. The informal interpretation of $\alpha \wedge \beta$ as a type for a term M is that M is of both types α and β. This way of thinking about $\wedge$ is in parallel with the meaning of $x \in A \cap B$ as $x \in A$ and $x \in B$. That is, $M: \alpha \wedge \beta$ is like saying that $x \in A \cap B$.[13]

Example 9.2.2. Recall that M does not have a simple type, even though Mx has an nf, xx (which is, of course, not typable with simple types either). However, if x could be assigned the types $\alpha \to \beta$ and α at the same time, then xx could be assigned the type β, and we would have $\mathsf{M}\colon (\alpha \to \beta) \wedge \alpha \to \beta$.

Of course, allowing a term to have an arbitrary type in addition to its simple type would result in a too broad (and most likely useless) system. Therefore, type assignments in an intersection type assignment system will be governed by axioms and rules — like those in the simply typed calculi — together with separate rules for *type inference*. The latter kind of rules will be connected with the rest of the type assignment system via a rule that allows changing the type assigned to a term without a separate derivation of that type assignment. The type inference axioms and rules amount to the logic that is known as the implication–conjunction fragment of the relevance logic B with the constant T (for truth), that is, the *relevance logic* $B^T_\wedge$. (See, for example, Meyer [108].)

DEFINITION 9.2.3. The natural deduction-style calculus for the *intersection type assignment system* for combinatory logic, denoted by $\mathrm{CL}_\wedge$, contains the type inference axioms and rules:

$$\tau \le \tau \qquad \tau \le \omega \qquad \omega \le \omega \to \omega$$

$$\tau \le \tau \wedge \tau \qquad \tau \wedge \sigma \le \tau \qquad \tau \wedge \sigma \le \sigma$$

$$(\tau \to \sigma) \wedge (\tau \to \chi) \le \tau \to \sigma \wedge \chi$$

$$\frac{\tau \le \sigma \qquad \sigma \le \chi}{\tau \le \chi} \qquad \frac{\tau_1 \le \tau_2 \qquad \sigma_1 \le \sigma_2}{\tau_1 \wedge \sigma_1 \le \tau_2 \wedge \sigma_2} \qquad \frac{\tau_1 \le \tau_2 \qquad \sigma_1 \le \sigma_2}{\tau_2 \to \sigma_1 \le \tau_1 \to \sigma_2}.$$

The type assignment axioms and rules are the following ones. (All the assumption sets are required to be univalent.)

$$\vdash \mathsf{S}\colon (\tau_1 \to \tau_2 \to \tau_3) \to (\tau_4 \to \tau_2) \to \tau_1 \wedge \tau_4 \to \tau_3 \ \vdash\!\mathsf{S} \qquad \vdash \mathsf{K}\colon \tau_1 \to \tau_2 \to \tau_1 \ \vdash\!\mathsf{K}$$

$$\vdash \mathsf{I}\colon \tau \to \tau \ \vdash\!\mathsf{I} \qquad x\colon \tau \vdash x\colon \tau \ \text{id} \qquad \vdash M\colon \omega \ \vdash\!\omega$$

$$\frac{\Delta_1 \vdash M\colon \sigma \to \tau \qquad \Delta_2 \vdash N\colon \sigma}{\Delta_1, \Delta_2 \vdash MN\colon \tau}\ E\!\to$$

$$\frac{\Delta \vdash M\colon \tau \qquad \Delta \vdash M\colon \sigma}{\Delta \vdash M\colon \tau \wedge \sigma}\ I\wedge \qquad \frac{\Delta \vdash M\colon \tau \wedge \sigma}{\Delta \vdash M\colon \tau}\ E\wedge \qquad \frac{\Delta \vdash M\colon \tau \wedge \sigma}{\Delta \vdash M\colon \sigma}\ E\wedge$$

[13] We use $\wedge$ to emphasize the connection to certain nonclassical logics; however, $\cap$ is also used in the literature, for instance, in [9].

The following two rules ensure monotonicity and connect type inferences to type assignments, respectively.

$$\frac{\Delta_1 \vdash M\colon \tau}{\Delta_1, \Delta_2 \vdash M\colon \tau}\,\text{mon} \qquad\qquad \frac{\Delta \vdash M\colon \tau \qquad \tau \leq \chi}{\Delta \vdash M\colon \chi}\,\text{up}$$

In the case of simple types, we required that the assumption sets were univalent. The idea of intersection types is to allow terms to have *more than one type* simultaneously as we hinted at in example 9.2.2. That is, not only combinators, but also other terms too can have multiple types within one derivation. Thus it might seem that the requirement of the univalence of the set of assumptions should be, perhaps, removed. However, if multiple type assignments to variables would be allowed, then we would lose control over the relationship of those types to each other. Therefore, the requirement of *univalence* stays in place. The notion of a *derivation* is as before.

The axioms and rules for type inference may be divided into two parts. The axioms and rules that do not contain $\rightarrow$ define a meet semilattice with a top element. The rest of the axioms and rules characterize the interaction between the type constructor $\rightarrow$ and the semilattice. $\omega \leq \omega \rightarrow \omega$ expresses that $\rightarrow$ applied to ω behaves classically. The rule introducing $\rightarrow$'s expresses that implication is monotone (covariant) in its second and antitone (contravariant) in its first argument place. By the other axiom for $\rightarrow$, $\rightarrow$ distributes over $\wedge$ in its second argument place. (The other inequality, i.e., $\tau \rightarrow \sigma \wedge \chi \leq (\tau \rightarrow \sigma) \wedge (\tau \rightarrow \chi)$ follows from $\rightarrow$'s tonicity and $\wedge$'s being inf (greatest lower bound).)

Example 9.2.4. Recall that M is definable as SII. The next derivation shows that given this definition, $\mathsf{M}\colon (\alpha \rightarrow \beta) \wedge \alpha \rightarrow \beta$ is provable. (We abbreviate $\beta \rightarrow \gamma$ by π to shorten some types.)

$$\cfrac{\cfrac{\vdash \mathsf{S}\colon (\pi \rightarrow \pi) \rightarrow (\beta \rightarrow \beta) \rightarrow \pi \wedge \beta \rightarrow \gamma \qquad \vdash \mathsf{I}\colon \pi \rightarrow \pi}{\vdash \mathsf{SI}\colon (\beta \rightarrow \beta) \rightarrow \pi \wedge \beta \rightarrow \gamma} \qquad \vdash \mathsf{I}\colon \beta \rightarrow \beta}{\vdash \mathsf{SII}\colon (\beta \rightarrow \gamma) \wedge \beta \rightarrow \gamma}$$

Given that now we can type a combinator that was not typable by simple types, we should be curious how we get (if we get) the type assignment $\vdash xx\colon \gamma$ when $\mathsf{M}x\colon \gamma$. Of course, the application of M should work with any term, and so we will consider a slightly more complicated term that x.

Example 9.2.5. In the (Church- or Curry-style) simple type theory no self-application of a variable (or of a term comprising variables only) is typable. Here is a derivation showing that $xy(xy)$ is typable by i-types. (We abbreviate $\sigma \rightarrow (\tau_1 \rightarrow \tau_2) \wedge \tau_1$ by χ in some of the type assignments to save space.)

$$\cfrac{\cfrac{\cfrac{\{x\colon \chi\} \vdash x\colon \chi \qquad \{y\colon \sigma\} \vdash y\colon \sigma}{\{x\colon \chi,\, y\colon \sigma\} \vdash xy\colon (\tau_1 \rightarrow \tau_2) \wedge \tau_1}}{\{x\colon \chi,\, y\colon \sigma\} \vdash xy\colon \tau_1 \rightarrow \tau_2} \qquad \cfrac{\cfrac{\{x\colon \chi\} \vdash x\colon \chi \qquad \{y\colon \sigma\} \vdash y\colon \sigma}{\{x\colon \chi,\, y\colon \sigma\} \vdash xy\colon (\tau_1 \rightarrow \tau_2) \wedge \tau_1}}{\{x\colon \chi,\, y\colon \sigma\} \vdash xy\colon \tau_1}}{\{x\colon \sigma \rightarrow (\tau_1 \rightarrow \tau_2) \wedge \tau_1,\, y\colon \sigma\} \vdash xy(xy)\colon \tau_2}$$

The example shows that the effect of having two types for a term (that has no occurrences of combinators) is achieved by starting with a term that contains a conjunction in its type, which is then eliminated.

Exercise 9.2.1. Find suitable assumption sets to type xx and xxx, then write out the derivations that show that these terms are typable by i-types.

Exercise* 9.2.2. Recall that I is definable as SKK. We included an axiom into $\mathrm{CL}_\wedge$ for I. Can the type given in the $\vdash \mathsf{I}$ axiom be derived from the types for S and K in $\mathrm{CL}_\wedge$?

The following claim is almost trivial still it is worth to make it explicit. (We do not change the notion of typability from definition 9.1.6 except that $\mathrm{CL}_\to$ is replaced by $\mathrm{CL}_\wedge$.)

LEMMA 9.2.6. *If M is a* CL*-term, then M is typable in* $\mathrm{CL}_\wedge$.

Proof: An arbitrary term can be typed by ω. qed

LEMMA 9.2.7. *For every term M and for every type τ, if $\Delta \vdash M\colon \tau$ in $\mathrm{CL}_\to$, where Δ assigns a type to x iff $x \in \mathrm{fv}(M)$, then $\Delta \vdash M\colon \tau$ in $\mathrm{CL}_\wedge$.*

Proof: The type of K is the same in $\mathrm{CL}_\to$ and $\mathrm{CL}_\wedge$. The type of I is the type that can be derived in $\mathrm{CL}_\to$ for SKK, which defines I. The detachment rule is the same in both calculi. Thus we have to show only that $\mathrm{pt}(\mathsf{S})$ in $\mathrm{CL}_\to$ is derivable in $\mathrm{CL}_\wedge$.

$\tau \wedge \tau = \tau$, since $\tau \leq \tau \wedge \tau$ and $\tau \wedge \sigma \leq \tau$ (with arbitrary σ including τ).[14] Then $\tau_1 \wedge \tau_1 \to \tau_3 = \tau_1 \to \tau_3$, by one of the type inference rules. After applying the same rule two more times, we get two further equations. The inequality that we need is

$$(\tau_1 \to \tau_2 \to \tau_3) \to (\tau_1 \to \tau_2) \to \tau_1 \wedge \tau_1 \to \tau_3 \leq (\tau_1 \to \tau_2 \to \tau_3) \to (\tau_1 \to \tau_2) \to \tau_1 \to \tau_3,$$

and it is half of the second equations. (Let us abbreviate by π_1 and π_2 the types on the left- and on the right-hand sides of the $\leq$.) We have that

$$\frac{\vdash \mathsf{S}\colon \pi_1 \qquad \pi_1 \leq \pi_2}{\vdash \mathsf{S}\colon \pi_2}.$$

Clearly, any derivation of a type assignment in $\mathrm{CL}_\to$ can be reproduced in $\mathrm{CL}_\wedge$ by replacing a leaf that is an instance of the S axiom by a suitable instance of the tree described above. qed

The notion of principal types does not carry over from the simply typed system to i-types. For example, it is easy to see — given the above lemma — that the simple type of S is not a substitution instance of its intersection type, and S's i-type is not a substitution instance of its simple type. However, it is possible to define principal types for the i-type system together with operations on types that ensure that the principal types play a similar role as in the case of simple types.[15]

[14] The equality sign is not in our official language; however, we use it only to express two inequalities at once. This usage is unproblematic.

[15] See [126] for a detailed exposition of the new notion of principal types for the i-typed λ-calculus.

The i-types allow us to delimit the set of normalizable terms. We use the notion of a λ-translation of a CL-term, which is defined in chapter 4.

LEMMA 9.2.8. *If $\Delta \vdash M\colon \tau$ is provable in $\mathrm{CL}_\wedge$ and there is no occurrence of ω in τ or in the types assigned to variables in Δ, then M has a weak nf.*

If M_λ is in β nf, then there is a proof of $\Delta \vdash M\colon \tau$ in $\mathrm{CL}_\wedge$ such that ω does not occur in τ or in the types assigned to variables in Δ.

We will not prove this lemma here. A proof may be constructed from the similar result for the i-typed λ-calculus (where in both cases β nf's and the i-typed λ-calculus appear in the statement of the claim) together with properties of the λ-translation of CL-terms. We recall from chapter 4 that if M_λ is in β nf, then M is in wnf, but not necessarily the other way around.

Exercise** **9.2.3.** Prove the previous lemma. (Hint: You might find it useful to consult [9], [8], [59] and [60] for proofs of similar results.)

We mentioned in the previous section that all terms typable in $\mathrm{CL}_\rightarrow$ strongly normalize. In $\mathrm{CL}_\wedge$ — with a small restriction on derivations — we get a characterization of the set of strongly normalizable terms. It is trivially decidable if a derivation contains any types that have an occurrence of ω, because a derivation is finite and it can be directly inspected. Therefore, we could consider a slightly restricted version of $\mathrm{CL}_\wedge$ in which ω is not an i-type, hence the two type inference axioms concerning ω and the $\vdash \omega$ axiom in the type assignment system are omitted. (Of course, none of the other axioms or rules are allowed to be instantiated with a type containing ω either, since that is not an i-type now.) Let us denote this modification of $\mathrm{CL}_\wedge$ by $\mathrm{CL}_\wedge^{-\omega}$.

LEMMA 9.2.9. *If $\Delta \vdash M\colon \tau$ is a provable in $\mathrm{CL}_\wedge^{-\omega}$, then M has no infinite reduction sequences.*

If M_λ strongly normalizes, then there is a derivation of $\Delta \vdash M\colon \tau$ in $\mathrm{CL}_\wedge^{-\omega}$ with some Δ and τ.

Proof: We sketch the proof here.[16] As we saw in chapter 6, type-free combinatory logic has various kinds of models — including term models. A motivation for the type assignment systems is that a combinator may have many types, as for example, I can be typed by $\alpha \rightarrow \alpha$ as well as by $(\alpha \rightarrow \beta) \rightarrow \alpha \rightarrow \beta$. This idea can be realized in a model where combinators are sets of wff's. In a "dual manner," it is also possible to interpret types as sets of terms.

In order to interpret a type assignment $M\colon \tau$, both M and τ have to be interpreted together with a relation between those two objects. It is natural to interpret a term as a term, and then take $:$ to be $\in$. With the interpretation of τ defined suitably, we can have that $\vdash M\colon \tau$ implies $[M] \in I(\tau)$ (where $[\ ]$ and I are the components of the interpretation for terms and for types, respectively).

[16]For detailed expositions of the proof, see [8, §4.3] or [87, 11J].

A basic type α is interpreted by I as the set of *strongly normalizable* terms, and I is extended to $\sigma \to \pi$ recursively by

$$I(\sigma \to \pi) = \{ M \colon \forall N. N \in I(\sigma) \Rightarrow MN \in I(\tau) \}.$$

Next, we can prove that this interpretation assigns to a simple type a subset of strongly normalizable terms. (This step may be carried out via the notion of saturated sets of terms or strongly computable terms.)

It remains to show that the function I can be complemented by a valuation function v on variables, which ensures that the natural deduction calculus $\mathrm{CL}_{\to}$ is *sound*. For example, $v(x) = x$ is a suitable valuation. Informally, the valuation function can be thought of as a substitution function, because a term is interpreted as itself modulo the application of v, which is the substitution of $v(x)$ for x in all occurrences of x.

The proof of soundness shows, by induction on the structure of derivations, that if for each $x\colon \tau$ in the assumption set Δ, $v(x) \in I(\tau)$, then if $\Delta \vdash M\colon \chi$ is provable in $\mathrm{CL}_{\to}$, then $[\vec{x}/\overrightarrow{v(x)}]M \in I(\chi)$. The set of terms $I(\chi)$ is a subset of the set of strongly normalizable terms, therefore, M strongly normalizes too. qed

Exercise 9.2.4.** The two references ([8] and [87]) in the proof deal with simply typed λ-calculus and simply typed CL, whereas the lemma above concerns $\mathrm{CL}_{\wedge}$. Adapt those proofs to $\mathrm{CL}_{\wedge}$, and write out in detail the proof of the lemma.

The i-type assignment system leads to further models of CL. Terms can be interpreted as *sets of types*, and in the presence of conjunction $\wedge$, a desirable property of those sets means closure under conjunction. Together with upward closure (with respect to the $\leq$ relation on types), the sets of types that are suitable for the interpretation of terms are *filters* on the set of types. (See section 6.5 for the details of this model.) In the terminology of section 6.2, this kind of interpretation gives an *operational semantics* both for typed and type-free CL.

The set of type constructors may be further extended. In the design and study of programming languages, type systems for functional languages often contain a whole range of type constructors — some of which we have mentioned above. Given *intersection types*, it is straightforward to look for *union types*. The connection between $\mathrm{CL}_{\wedge}$ and $B_{\wedge}^{T}$ also suggests that the effect of adding *disjunction* as a binary type constructor may be useful.

Disjunction is denoted by $\vee$, and we assume that the set of types is inductively defined from a set of type variables and ω with an extra clause for $\vee$. That is, definition 9.2.1 may be extended by (4) — with the obvious modification that the set that is defined now is called "union types" (or u-types).

(4) If π and σ are u-types, then so is $(\pi \vee \sigma)$.

This extension adds new rules to the intersection type assignment system $\mathrm{CL}_{\wedge}$ (which is in definition 9.2.3).

DEFINITION 9.2.10. The type assignment system CL_+ is defined by the axioms and rules of $\mathrm{CL}_\wedge$ together with the following type assignment rules.

$$\frac{\Delta \vdash M\colon A}{\Delta \vdash M\colon A \vee B}\ I\vee \qquad \frac{\Delta \vdash M\colon B}{\Delta \vdash M\colon A \vee B}\ I\vee \qquad \frac{\Delta \vdash M\colon A \vee B \quad A \leq C \quad B \leq C}{\Delta \vdash M\colon C}\ E\vee$$

The set of axioms for type inference is expanded by the following four axioms.

$$\tau \leq \tau \vee \sigma \qquad \tau \leq \sigma \vee \tau \qquad (\tau \to \sigma) \wedge (\chi \to \sigma) \leq \tau \vee \chi \to \sigma$$

$$\tau \wedge (\sigma \vee \chi) \leq (\tau \wedge \sigma) \vee (\tau \wedge \chi)$$

An informal interpretation of union types is that a term that is assigned type $A \vee B$ is assigned the type A or the type B.

The set of typable terms, obviously, cannot change by the extension of the type assignment system: $\mathrm{CL}_\wedge$ is completely included in CL_+, thus the set of typable terms cannot shrink. $\mathrm{CL}_\wedge$ already allowed all CL-terms to be typed, thus the set of typable terms cannot grow. The addition of $\vee$ has at least two advantages. First, the connection between the minimal relevance logic B and CL (or the λ-calculus) is reinforced by this addition. Second, more elaborate models of CL and typed CL can be defined based on u-types.[17] The latter enhances our understanding of CL.

[17]Such a model is described at the end of section 6.5.

Appendix

Combinatory logic originated from *philosophical concerns* about classical first-order logic. *Logicians* and *mathematicians* developed CL with the idea of creating a framework for the whole of mathematics. Later on, with the appearance of computers and then of *computer science*, CL came to be seen primarily as a formalization of the notion of computable functions. With its many ties and aspects, CL assumes familiarity with a wide variety of abstract notions and ideas.

The aim of this appendix is to provide a *brief summary* of some of the concepts that are introduced informally or used without a definition in the preceding chapters. Some of these concepts are quite elementary, and repeating them here mainly serves the purpose of fixing notation. Some of the concepts are less well-known, and their inclusion here is intended to help to make the book more self-contained. There are also claims, lemmas and theorems here; however, we typically do not give a proof for them.[1] The proofs are often easy, or may be found in multiple sources.

The appendix is organized into sections paralleling the structure of the book. Considered separately, the appendix might look like jumping around between various subject areas. We do not intend to and cannot provide a systematic development of all the topics touched upon here, if for no other reason, because these topics are subsidiary to the main subject of the book. The present arrangement is hopefully practical and helpful in using the book together with the *list of symbols* and the *index*.

A.1 Elements of combinatory logic

Classes of objects are often defined using *induction*. The natural numbers may be defined inductively as the least set generated from the singleton set $\{0\}$ by the "$+1$" (i.e., the *successor*) function. An "official" definition of $\mathbb{N}$ (the set of *natural numbers*) may look like:

(1) $\{0\} \subseteq \mathbb{N}$;

(2) if $x \in \mathbb{N}$ then $x+1 \in \mathbb{N}$;

[1] Some exceptions are claims A.1.1 and A.1.2 that are included as examples of proofs by induction.

(3) $\mathbb{N}$ is the least set generated by clauses (1) and (2).[2]

To prove that all the elements of an inductively defined set possess some property, we use a *proof technique* that is also called *induction*. To prove that the sum of the natural numbers up to n is $n(n+1)/2$, we can use *weak mathematical induction*. We give a detailed proof of this claim to illustrate the structure of proofs by mathematical induction.

CLAIM A.1.1. *The sum of the natural numbers from* 0 *to* n *is* $\frac{n(n+1)}{2}$, *that is,*

$$\sum_{i=0}^{n} i = \frac{n(n+1)}{2}.$$

Proof: The proof is by induction on n.
Base case. If $n=0$, then $\sum_{i=0}^{0} i = 0$. Also, $0(0+1)/2 = 0$. $0 = 0$, that is the claim holds when $n = 0$.
Inductive case. The inductive hypothesis (IH) is that the claim is true for n. We will show that — given the hypothesis — the claim is true for $n+1$.

$\sum_{i=0}^{n+1} i = (\sum_{i=0}^{n} i) + (n+1)$. By (IH), $\sum_{i=0}^{n+1} i = \frac{n(n+1)}{2} + (n+1)$, which further equals to $\frac{n^2+n+2n+2}{2}$. On the other hand, if we substitute $n+1$ into the fraction on the right in the claim, then we get $\frac{(n+1)(n+2)}{2}$, and then $\frac{n^2+3n+2}{2}$. Thus $\sum_{i=0}^{n+1} i = \frac{(n+1)(n+2)}{2}$, which shows that the claim holds for $n+1$. qed

The set of natural numbers exhibits a very special structure. Other sets of objects may be defined by induction too, but starting with more that one object, and generating objects in several ways. Here we take as an example the inductive definition of *well-formed formulas* (wff's, for short). Informally, $\mathbb{P} = \{p_0, p_1, p_2, \ldots\}$ is the set of *propositional variables*, and the connectives are $\sim$, $\wedge$, $\vee$ and $\rightarrow$.

(1) $\mathbb{P} \subseteq \text{wff}$;

(2) if $A \in \text{wff}$, then $\sim A \in \text{wff}$;

(3) if $A, B \in \text{wff}$, then $(A \wedge B), (A \vee B), (A \rightarrow B) \in \text{wff}$.

Note that we suppressed the "least set" clause.

The definition may be rendered in English by saying that a propositional variable is a formula, and the negation of a formula, as well as the conjunction, the disjunction and the implication of a pair of formulas is a formula. Furthermore, nothing else is a formula.

Obviously, $\text{card}(\text{wff}) = \aleph_0$ when $\text{card}(\mathbb{P}) = \aleph_0$, that is, there are denumerably many well-formed formulas — just like natural numbers. However, there is no "$+1$"-like operation that generates all the wff's from some wff; rather there are several operations ($\sim$, $\wedge$, $\vee$ and $\rightarrow$) and some of them take *two arguments*. The

[2]The closure clause (3) is often omitted when the definition has been specified to be inductive.

proof method that accompanies an inductive definition like that of wff's is called *structural induction*. To illustrate this way of proving that a property holds of all the objects in the inductively defined set, we prove a (very simple) claim in detail.

CLAIM A.1.2. *Every well-formed formula contains an* equal number *of* (*'s and*) *'s.*

Proof: The proof is by induction on the structure of formulas.
Base case. The base set in the definition is $\mathbb{P}$ by clause (1). If the formula is a propositional variable, that is, p_i for some i, then it has 0 ('s and 0)'s. Therefore, the claim is obviously true.
Inductive case. We consider separately those formulas in which the main connective is negation, and those formulas in which the main connective is binary.
1. The hypothesis of the induction is that the number of ('s and)'s equals in A. The symbol $\sim$ is not a parenthesis, and $\sim A$, which is generated according to (2), contains exactly the ('s and)'s that A does. Therefore, by the (IH), the claim holds for $\sim A$.
2. The hypothesis of the induction is that A has as many ('s as)'s, and B has as many ('s as)'s. Let us say these numbers are, respectively, m and n. $(A \wedge B)$ contains the (at the beginning of the wff, and the ('s from A and from B, that amounts to $1+m+n$. The number of)'s is $m+n+1$ — counting from left to right again. $1+m+n = m+n+1$, hence the claim is true for $(A \wedge B)$. The subcases for $(A \vee B)$ and $(A \rightarrow B)$ are similar (hence, omitted). qed

→→→→→ ❄ ←←←← ❄ →→→→ ❄ ←←←←←

DEFINITION A.1.3. (LEFT-ASSOCIATED TERMS) The set of *left-associated terms* is inductively defined by (1)–(3).

(1) If x is a variable, then x is a left-associated term;

(2) if Z is a constant, then x is a left-associated term;

(3) if M is a left-associated term, then (Mx) and $(M\mathsf{Z})$ are left-associated terms, where x is a variable and Z is a constant.

Left-associated terms appear in the axioms of combinators on the left-hand side of the $\triangleright$. The dual notion is the set of *right-associated terms*, which appear in the axioms of dual combinators in chapter 7.

DEFINITION A.1.4. (RIGHT-ASSOCIATED TERMS) The set of *right-associated terms* is inductively defined by (1)–(3).

(1) If x is a variable, then x is a right-associated term;

(2) if Z is a constant, then x is a right-associated term;

(3) if M is a right-associated term, then (xM) and $(\mathsf{Z}M)$ are right-associated terms, where x is a variable and Z is a constant.

Not all terms are left- or right-associated, though some CL-terms are both. Left- and right-associated terms are singled out because it is easy to specify their shape and they are linked to axioms (though some other CL-terms fall into one of these two categories too). Also, by convention, parentheses may be omitted from left-associated terms.

CL-terms may contain grouping, that is, there might be ('s and) 's in a term. The structure of a CL-term is naturally depicted as a tree (rather than a sequence of symbols). Nonetheless, it is useful to be able to assign a number to a term, which shows how many atomic terms (more precisely, term occurrences) make up the term.

DEFINITION A.1.5. (LENGTH OF A TERM) The length of a CL-term is a *positive integer*. The *length function*, denoted by ℓ, assigns a number to a term according to (1)–(3).

(1) $\ell(x) = 1$, when x is a variable;

(2) $\ell(\mathsf{Z}) = 1$, when Z is a constant;

(3) $\ell(MN) = m+n$, when $\ell(M) = m$ and $\ell(N) = n$.

ℓ is *well-defined*, that is, ℓ is a function from the set of CL-terms into $\mathbb{Z}^+$ (the set of positive integers), because each CL-term matches the shape of exactly one term from (1)–(3). Obviously, ℓ is surjective, but not injective.

Terms may be depicted as *trees*, more precisely, as nonempty ordered trees with branching factor two.[3] The *leaves* of the tree are the atomic terms, and the branchings correspond to the binary application operation. The *root* of the tree is the term itself, and the *interior nodes* are proper subterms of that term. The latter foreshadows that in the case of leaves and interior nodes it is more accurate to talk about *occurrences* of subterms than subterms per se.

Example A.1.6. The terms x, $x(y(zv))$, $\mathsf{SKK}x$ and $\mathsf{C}(\mathsf{I}x)(y(\mathsf{KK}))$ are depicted by the trees below.

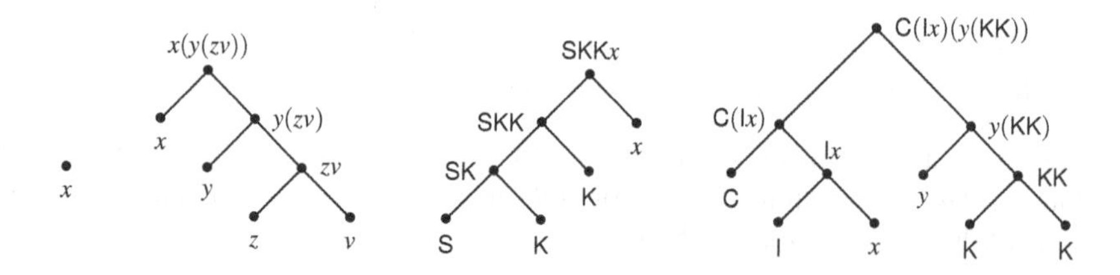

In the first two terms, each subterm has just one occurrence. In the latter two terms, K has two occurrences. Complex subterms may have multiple occurrences too, though there is no such subterm in these particular examples. Given a term and its *formation tree*, we may uniquely identify the occurrences of subterms.

[3]The formation trees of CL-terms are not the so-called Böhm-trees for λ-terms; that is, the two kinds of trees for a term are not isomorphic.

DEFINITION A.1.7. Given a term M, the *occurrences of subterms* in M are identified by assigning an *index* to them inductively by (1) starting with M_0.

(1) If M_i is (NP), then the subterm occurrences are $(N_{0i}P_{1i})$.

Example A.1.8. We continue with the terms displayed above. The indices we get are: x_0; $(x_{00}(y_{010}(z_{0110}v_{1110})_{110})_{10})_0$; $(((\mathsf{S}_{0000}\mathsf{K}_{1000})_{000}\mathsf{K}_{100})_{00}x_{10})_0$; and lastly, $((\mathsf{C}_{000}(\mathsf{I}_{0100}x_{1100})_{100})_{00}(y_{010}(\mathsf{K}_{0110}\mathsf{K}_{1110})_{110})_{10})_0$. (We have restored all the parentheses for the placement of the indices.) Admittedly, with all the parentheses included and all the indices added, the terms became rather bulky. Still, it should be clear that each subterm occurrence has *a unique* index, hence distinct occurrences of a subterm have distinct indices. For example, the K's in the third term have indices 1000 and 100; in the fourth term their indices are 0110 and 1110.

DEFINITION A.1.9. A term M *has an occurrence* in the term N iff M is a subterm of N as defined in 1.1.4. A *particular occurrence* of M in N is M_i, where $_i$ is the index assigned to an M in N_0 by definition A.1.7.

With each subterm occurrence uniquely indexed, the replacement of an occurrence can be viewed as follows.

DEFINITION A.1.10. (REPLACEMENT OF A SUBTERM) If M_i is an occurrence of M in N, then $[P_i/M_i]N$ is *the replacement* of that occurrence of M by P. If P is a complex term, then the subterms of P are assigned indices by (1) in definition A.1.7 starting with P_i.

Visually, the replacement of a subterm occurrence is the *replacement of a subtree* of the formation tree of the whole term with another subtree (which is itself the formation tree of a term). A replacement of an occurrence of a subterm does not affect the indices of occurrences of subterms that are not subterm occurrences within the replaced occurrence of a subterm.[4]

Variables are one kind of atomic terms in CL.

DEFINITION A.1.11. (VARIABLE OCCURRENCES) A variable x *has an occurrence* in M iff x is a subterm of N. *A particular occurrence of a variable* x in M is x_i, where $_i$ is the index that occurrence gets in M_0.

All occurrences of variables in all CL-terms are *free*.

Replacement and *substitution* are different ways to alter terms, and they play different roles in theories. However, in CL, the substitution of a term M for the variable x in the term N may be viewed — equivalently — as the replacement of all the occurrences of x in N by M.

It is usual to define the sets of free and bound variables in λ-calculi (and in logics with quantifiers or other variable binding operators). (See definitions A.4.1 and A.4.2 in section A.4.) The definitions of those sets take a much simpler form in CL.

[4]The clumsiness of this sentence, and the bulkiness of the indexed terms are the reasons to opt for a slightly more sloppy way of talking about CL-terms — once it has been understood that the sloppiness can be eliminated if necessary, and when there is no danger of confusion.

DEFINITION A.1.12. (FREE AND BOUND VARIABLES OF A CL-TERM) The set of *free* variables of M is denoted by $\text{fv}(M)$, similarly, the set of *bound* variables is $\text{bv}(M)$. $x \in \text{fv}(M)$ iff x is a subterm of M, and $\text{bv}(M) = \emptyset$.

Sometimes we want to emphasize that a term is a subterm of a term, but not identical to it. The two clauses (1) and (2) in the next definition give the same notion.

DEFINITION A.1.13. (PROPER SUBTERMS)

(1) M is *a proper subterm of* N iff M is a subterm of N and M is not the same term as N.

(2) M is *a proper subterm of* N iff M_i is an occurrence of M in N_0 and $i \neq 0$.

Substitutions are *total maps* from the set of *variables* into the set of CL-*terms* with (typically) only finitely many variables changed. It is customary, however, to start with a notion of substitution as a *one-point modification* of the identity function. In chapter 1, we used the notation M_x^N for the substitution of the CL-term N for all the (free) occurrences of the variable x in the CL-term M.

DEFINITION A.1.14. (SIMULTANEOUS SUBSTITUTION) Let $x_1,\ldots,x_n$ be n variables and let $N_1,\ldots,N_n$ be CL-terms. The *simultaneous substitution* of $N_1,\ldots,N_n$ for $x_1,\ldots,x_n$ in the CL-term M, that is, $[x_1/N_1,\ldots,x_n/N_n]M$, is recursively defined by (1)–(4).

(1) If M is x_i, where $1 \leq i \leq n$, then $[x_1/N_1,\ldots,x_n/N_n]x_i$ is N_i;

(2) if M is y, where y is a variable distinct from each of $x_1,\ldots,x_n$, then $[x_1/N_1,\ldots,x_n/N_n]y$ is y;

(3) if M is Z, a constant, then $[x_1/N_1,\ldots,x_n/N_n]\mathsf{Z}$ is Z;

(4) if M is (P_1P_2), then $[x_1/N_1,\ldots,x_n/N_n](P_1P_2)$ is $([x_1/N_1,\ldots,x_n/N_n]P_1\,[x_1/N_1,\ldots,x_n/N_n]P_2)$.

This definition is very similar to definition 1.2.2. A difference is that clause (1) in the above definition can be viewed as a variable is "selecting" the term that is to be substituted for that variable. Note that we have not stipulated that the $N_1,\ldots,N_n$ are distinct from each other or from the variables $x_1,\ldots,x_n$.

Example A.1.15. For comparison, let us consider x in place of the CL-term in which we substitute. $[z/x][y/z][x/y]x$ indicates three *successive* substitutions. $[x/y]x$ yields y, the next substitution results in z, and lastly, $[z/x]z$ is x. $[z/x,y/z,x/y]x$ indicates a simultaneous substitution for three variables, including a substitution (for z) of the same variable x as the term in which the substitution is performed. Since x is atomic, and it is a variable, only one application of clause (1) from definition A.1.14 is possible. The result of $[z/x,y/z,x/y]x$ is y, the same term that would result from $[x/y]x$.

PROPOSITION A.1.16. (EQUIVALENCE OF SUBSTITUTIONS) *Every simultaneous substitution can be simulated by* finitely many single substitutions. *Single and simultaneous substitutions are essentially* equivalent.

A single substitution is a simultaneous one (with $n = 1$). The above example shows that a simultaneous substitution is not necessarily equivalent to a series of single substitutions where the simultaneous substitution is chopped up into single steps. However, sequentializing works with a bit of refinement.

PROPOSITION A.1.17. *Let* $[x_1/N_1, \ldots, x_n/N_n]M$ *be a simultaneous substitution, and let* $y_1, \ldots, y_{n-1}$ *be variables distinct from each other and from the* x_i*'s, as well as from each variable in* $\mathrm{fv}(N_1 \ldots N_n M)$. *For each* N_i *(where* $1 < i \leq n$*), let* N_i' *be* $[x_1/y_1, \ldots, x_{i-1}/y_{i-1}]N_i$. *Then the effect of the simultaneous substitution* $[x_1/N_1, \ldots, x_n/N_n]M$ *can be obtained by* $2n-1$ consecutive single substitutions:

$$[y_1/x_1] \ldots [y_{n-1}/x_{n-1}][x_1/N_1'] \ldots [x_n/N_n']M.$$

An eagle-eyed reader might object that we used simultaneous substitution in specifying the N_i''s, and so there is a "circularity" in the proposition. However, all the x_i's and y_i's are distinct, hence $[x_1/y_1, \ldots, x_{i-1}/y_{i-1}]N_i$ can always be emulated by $[x_1/y_1] \ldots [x_{i-1}/y_{i-1}]N_i$. (Incidentally, there are always sufficiently many y's available for our purposes, because $n \in \mathbb{N}$ and each term has a finite length.)

We prove the proposition (as we promised in chapter 1 and unlike we said we typically do in the introduction to the Appendix).

Proof: We show by induction on n that if $\{x_1, \ldots, x_{i-1}\} \cap \mathrm{fv}(N_i) = \emptyset$, then $[x_1/N_1, \ldots, x_{n-1}/N_{n-1}, x_n/N_n]M$ is the same CL-term as $[x_1/N_1] \ldots [x_{n-1}/N_{n-1}][x_n/N_n]M$. If $n = 2$, then we have $[x_1/N_1, x_2/N_2]M$ and $x_1 \notin \mathrm{fv}(N_2)$, by assumption. Then the term $[x_2/N_2]M$ contains the same occurrences of x_1 as M does. (They are the same in the sense of having the same numerical indices.) Thus the terms $[x_1/N_1, x_2/N_2]M$ and $[x_1/N_1][x_2/N_2]M$ are the same.

Let us assume that $[x_1/N_1, \ldots, x_{n-1}/N_{n-1}, x_n/N_n]M$ (with any M) satisfies the claim. Suppose additionally, that $\{x_1, \ldots, x_n\} \cap \mathrm{fv}(N_{n+1}) = \emptyset$. Then all the occurrences of the variables $x_1, \ldots, x_n$ in M coincide with those in $[x_{n+1}/N_{n+1}]M$. Thus $[x_1/N_1, \ldots, x_n/N_n, x_{n+1}/N_{n+1}]M$ and $[x_1/N_1, \ldots, x_n/N_n][x_{n+1}/N_{n+1}]M$ are the same term, and then by the inductive hypothesis $[x_1/N_1, \ldots, x_n/N_n, x_{n+1}/N_{n+1}]M$ is $[x_1/N_1] \ldots [x_n/N_n][x_{n+1}/N_{n+1}]M$.

It only remains to observe that the local disidentifications of variables yielding the N_i''s are reversed by the initial segment $[y_1/x_1] \ldots [y_{n-1}/x_{n-1}]$ in the sequence of substitutions in the term in the proposition. qed

→→→→→ ❄ ←←←← ❄ →→→→ ❄ ←←←←←

CL-terms may be viewed as sequences of symbols. In formal language theory and in theoretical computer science, *strings* (also called *words*) and their sets, *formal languages* are the primary objects. We provide a brief comparison of strings and CL-terms.

DEFINITION A.1.18. (STRINGS) Let $\mathbb{A}$ be a (finite) set $\{a_1,\ldots,a_n\}$, *an alphabet.*[5] The set of *nonempty strings* over $\mathbb{A}$ is denoted by $\mathbb{A}^+$, and inductively defined by (1) and (2).

(1) $\mathbb{A} \subseteq \mathbb{A}^+$;

(2) if $s_1 \in \mathbb{A}^+$ and $s_2 \in \mathbb{A}^+$, then so is $s_1 s_2$.

The set of *strings* over $\mathbb{A}$ is denoted by $\mathbb{A}^*$, and defined by (3) (given the definition of $\mathbb{A}^+$).

(3) $\mathbb{A}^* = \mathbb{A}^+ \cup \{\varepsilon\}$.

Notice that no parentheses surround $s_1 s_2$ in clause (2), and this is not sloppiness. Juxtaposition in strings denotes a binary operation that is called *concatenation.* The lack of parentheses means that concatenation is *associative.*

$\mathbb{A}^*$ differs from $\mathbb{A}^+$ by exactly one string: ε is the *empty string.* ε is called empty, because it does not contain any letters (i.e., elements of the alphabet).[6]

LEMMA A.1.19. *If* $\mathrm{card}(\mathbb{A}) \leq \aleph_0$, *then* $\mathrm{card}(\mathbb{A}^+) = \aleph_0$.

Strings can be thought of as finite sequences, or n-tuples. Accordingly, the identity of a pair of strings can be determined by comparing them letter by letter: $s_1 = s_2$ iff there is an $a_i \in \mathbb{A}$ such that $s_1 = a_i s_1'$ and $s_2 = a_i s_2'$, plus $s_1' = s_2'$, or both s_1 and s_2 are ε.

For CL-terms we stipulated the existence of a denumerable set of variables. The latter set may be generated from a finite set via indexing, thus in sum each CL-term is a string over a finite alphabet. However, it is also interesting to consider a set of terms that can be generated from a finite set of starting terms. For example, when we are interested only in combinators, then we are interested in a set of CL-terms that can be generated from a combinatory basis alone.

DEFINITION A.1.20. (TERMS OVER $\mathbb{T}$) Let $\mathbb{T}$ be a (finite) set of starting elements $\{t_1,\ldots,t_n\}$. The set of CL-*terms over* $\mathbb{T}$ is denoted by $\mathbb{T}^\oplus$, and inductively defined by (1) and (2).

(1) $\mathbb{T} \subseteq \mathbb{T}^\oplus$;

(2) if $M_1 \in \mathbb{T}^\oplus$ and $M_2 \in \mathbb{T}^\oplus$, then so is $(M_1 M_2)$.[7]

We deliberately made this definition to look as similar as possible to definition A.1.18. Therefore, the differences should be salient. The juxtaposition is now thought of as *function application* and $t_1 t_2$ is enclosed into a pair of parentheses.

[5] Sometimes it is useful to allow infinite sets as alphabets.

[6] The empty string has been denoted by a wide variety of symbols from o, Φ, $\emptyset$, to Λ. Nowadays, ε seems to be the most commonly used notation.

[7] As always, we assume that the delimiters are not elements of the set $\mathbb{T}$.

We do not have a counterpart to ε, that is, there is no empty CL-term (hence, the notation $^{\oplus}$). If we allow $\mathbb{T}$ to be infinite, and stipulate that the set of variables and the set of constants together are $\mathbb{T}$, then definition A.1.20 gives the same set of CL-terms as definition 1.1.1 does.

For readers who are familiar with *context-free grammars*, we give two CFGs.

Let $\mathbb{A}$ and $\mathbb{T}$ be as in definitions A.1.18 and A.1.20. The set of *strings* over the alphabet $\mathbb{A}$ is defined by the following CFG.[8]

$$W := a_1 \mid \ldots \mid a_n \mid WW \mid \varepsilon$$

The set of CL-*terms over* $\mathbb{T}$ is defined by the following CFG.

$$M := t_1 \mid \ldots \mid t_n \mid (MM)$$

Terms, obviously, can be viewed as *strings* (which is demonstrated by the second CFG). This aspect of CL-terms leads to considerations of the number of parentheses, and if they are well-balanced. These issues are important for the determination of whether a string of symbols is a CL-term as well as for the convention of omitting parentheses. However, the parentheses never form a subterm by themselves; they are *auxiliary* symbols.

Given n letters in an alphabet $\mathbb{A}$, it is quite easy to calculate the number of strings of length m (where the length is simply the number of letter occurrences). The strings of length m comprise all the *variations* of n elements with repetitions: n^m. An informal way to deduce this number is by considering the possible letters that can occupy a position in a series with m spots. There are n letters for the first, and for the second, for the third, etc., places — m times.

If we have just one letter, then there is *one string* of length m, because $m^1 = m$. However, CL-terms have more structure because function application is not an associative operation unlike concatenation is. The number of ways to parenthesize a string into a CL-term is given by the Catalan numbers.

DEFINITION A.1.21. (CATALAN NUMBERS) The *Catalan numbers* are denoted by C_n, and they are defined, for $n \in \mathbb{N}$, as

$$\frac{1}{n+1}\binom{2n}{n}.$$

The first seven Catalan numbers are

$$C_0 = 1,\ C_1 = 1,\ C_2 = 2,\ C_3 = 5,\ C_4 = 14,\ C_5 = 42,\ C_6 = 132.$$

In general, the numbers of distinct CL-terms that consist of n occurrences of x (or of another atomic term) is C_{n-1}.

CLAIM A.1.22. *The Catalan numbers can be* equivalently defined, *for* $n \geq 1$, *as*

$$\binom{2n}{n} - \binom{2n}{n-1}.$$

[8]The ... indicate possible additional rewrite rules if there are more that two elements in $\mathbb{A}$ or $\mathbb{T}$. Because of the lack of concrete alphabets, these sets of rules may be better viewed as meta-CFGs.

A.2 Main theorems

A subterm M of a term N may *overlap* with another subterm P. The notion of overlapping may be made rigorous using the identification of occurrences of subterms introduced in definition A.1.7.

DEFINITION A.2.1. (OVERLAPPING OCCURRENCES) If M_i is a subterm of a term N, then P_j, a subterm of N, *overlaps* with M_i iff either i is kj or j is li, where i and j are binary strings, and k or l may be the empty string.

Example A.2.2. Let us consider the term $x\mathsf{W}(\mathsf{I}(\mathsf{B}x)(yy))$. The fully indexed term is

$$((x_{000}\,\mathsf{W}_{100})_{00}\,((\mathsf{I}_{0010}\,(\mathsf{B}_{01010}\,x_{11010})_{1010})_{010}\,(y_{0110}\,y_{1110})_{110})_{10})_0.$$

First of all, if M_i and P_j are identical, then M_i and P_j overlap. For instance, $(x\mathsf{W})_{00}$, overlaps with $(x\mathsf{W})_{00}$. If M_i is x_{11010} and N_j is $(\mathsf{I}(\mathsf{B}x))_{010}$, then i is $11j$, that is, the string 11 concatenated with the binary string j. Symmetrically, if M_i is $(yy)_{110}$, then both occurrences of y, y_{0110} and y_{1110} overlap with M_i. j is either $0i$ or $1i$.

On the other hand, the two occurrences of x in this term do not overlap. The indices of x_{000} and x_{11010} are such that neither is the *final segment* of the other.

The indexing now may be seen also as a tool to turn questions about overlapping between subterms into questions of coincidence of final segments of binary strings. The indexing makes precise the notion of overlapping, and reinforces the visual idea.

LEMMA A.2.3. *Let $\mathsf{Z}^n M_1 \ldots M_n$ be a redex in the term N_0. If $(\mathsf{Z}M_1 \ldots M_n)_i$ and $(\mathsf{Z}M_1 \ldots M_n)_j$ are subterms of N, then the occurrences* overlap *iff $i = j$.*[9]

In the definition of $\triangleright_{p1}$ in chapter 2, we are interested in *occurrences of redexes.* However, the following more general statement is also true.

LEMMA A.2.4. *If M_i and M_j are subterm occurrences of M in N_0, then M_i and M_j are* two disjoint *(i.e.,* nonoverlapping*) occurrences of M just in case $i \neq j$.*

Another way to generalize lemma A.2.3 — without requiring that the two redex occurrences are occurrences of one and the same redex — is the following lemma.

LEMMA A.2.5. *Let $(\mathsf{Z}_1^n M_1 \ldots M_n)_i$ and $(\mathsf{Z}_2^m N_1 \ldots N_m)_j$ be overlapping redex occurrences in P_0. If $i \neq j$, then either $(\mathsf{Z}_1^n M_1 \ldots M_n)_i$ is* a subterm occurrence in *N_k (where $1 \leq k \leq m$), or $(\mathsf{Z}_2^m N_1 \ldots N_m)_j$ is* a subterm occurrence in *M_l (where $1 \leq l \leq n$).*[10]

[9] The indices of the M's are used in the usual way; only i, j and 0 are indices indicating subterm occurrences.

[10] Only the three indices i, j and 0 identify subterm occurrences in this statement.

Example A.2.6. Let Z_1 be the first S in the term $\mathsf{S}(\mathsf{SKK}x)(\mathsf{K}yz)(\mathsf{S}(\mathsf{SKK}x)(\mathsf{K}yz)\mathsf{S})$, which contains four distinct redexes and six redex occurrences. The whole term itself is a redex, and all the other redex occurrences are proper subterms of this term.

The first occurrence of $\mathsf{SKK}x$, which gets index $_{1000}$, comprises the first argument of the first S (which is S_{0000}). The second occurrence of the same subterm is $(\mathsf{SKK}x)_{10010}$, and this is a proper subterm of the third argument of S_{0000}.

If a redex is a proper subterm of another redex, then the former is completely confined to an argument of the latter. This result does not follow merely from the notions of subterms and occurrences. If we were not assuming that redexes are left-associated terms, then redexes with distinct heads could turn out to be just one occurrence of one subterm. We will see in chapter 7 that this can happen if dual and symmetric combinators are included, and then proper subterms of a term that form a redex are not confined inside of an argument.

Various operations on sets of objects fall under the concept of a *closure operator.*

DEFINITION A.2.7. (CLOSURE OPERATOR) Let $\mathfrak{c}$ be of type $\mathfrak{c}\colon \mathcal{P}(X) \longrightarrow \mathcal{P}(X)$. $\mathfrak{c}$ is a *closure operator* iff it satisfies (1)–(3) for all $Y, Z \subseteq X$.

(1) $Y \subseteq \mathfrak{c}(Y)$;

(2) $Y \subseteq Z$ implies $\mathfrak{c}(Y) \subseteq \mathfrak{c}(Z)$;

(3) $\mathfrak{c}(\mathfrak{c}(Y)) \subseteq \mathfrak{c}(Y)$.

It is obvious that $\mathfrak{c}(\mathfrak{c}(Y)) = \mathfrak{c}(Y)$ once (1) and (3) are combined. The idea of "closure" is expressed by (1).[11] Property (2) is called *monotonicity*, and (3) is the *idempotence* of $\mathfrak{c}$.

Concretely, we consider the reflexive, the transitive and the symmetric closure of binary relations. (Various properties of binary relations such as reflexivity and transitivity are defined in section A.6.)

DEFINITION A.2.8. (REFLEXIVE CLOSURE) Let R be a binary relation on the set X. The *reflexive closure* of R, denoted by R^{id}, is $R \cup \{ \langle x,x \rangle : x \in X \}$.

The reflexive closure of R simply expands R to the least reflexive relation (which includes R). It may be useful to note that the identity relation (i.e., $\{ \langle x,x \rangle : x \in X \}$) *does not depend* on R itself, rather it depends on X, the set that is the range of the relation R.

LEMMA A.2.9. ^{id} *is a* closure operator *on binary relations.*

DEFINITION A.2.10. (TRANSITIVE CLOSURE) Let R be a binary relation on the set X. The *transitive closure* of R, denoted by R^{+}, is defined by (1)–(2).

[11] In topology, the closure of a set is the dual of the interior of a set. The closure may add points, whereas the interior may omit points, so to speak.

(1) $R \subseteq R^+$;

(2) R^+xz if there are $y_1, \ldots, y_n$ ($n \in \mathbb{Z}^+$) such that $Rxy_1 \wedge \ldots \wedge Ry_iy_{i+1} \wedge \ldots \wedge Ry_nz$, where $1 \leq i \leq n-1$.

(1) makes salient that R is included in its transitive closure. (1) and (2) may be combined into (3).

(3) R^+xz if there are $y_1, \ldots, y_n$ ($n \in \mathbb{N}$) such that $Rxy_1 \wedge \ldots \wedge Ry_iy_{i+1} \wedge \ldots \wedge Ry_nz$, where $1 \leq i \leq n-1$.

The case $n = 0$ is understood as R^+xz if Rxz. In either formulation, (2) or (3), the transitive closure adds "shortcuts" between points that are connected by a *finite* number of R edges.

LEMMA A.2.11. $^+$ *is a* closure operator *on binary relations.*

DEFINITION A.2.12. Let R be a binary relation on the set X. The *reflexive transitive closure* of R, denoted by R^*, is $(R^+)^{\text{id}}$.

It is not difficult to prove that these two closure operators *commute*.

LEMMA A.2.13. *For any binary relation R, $(R^+)^{\text{id}} = (R^{\text{id}})^+$. That is, the reflexive transitive closure of a relation* coincides *with its transitive reflexive closure.*

DEFINITION A.2.14. (SYMMETRIC CLOSURE) Let R be a binary relation on the set X. The *symmetric closure* of R, denoted by $R^{\leftrightarrow}$, is $R \cup \{ \langle x,y \rangle : \langle y,x \rangle \in R \}$, that is, $R \cup R^{\cup}$.

The relation $R^{\cup}$ is the *converse* of R, which contains the pairs in R with their order reversed. Thus $R^{\leftrightarrow}$ depends on R itself, not merely on the range of R.

LEMMA A.2.15. $^{\leftrightarrow}$ *is a* closure operator *on binary relations.*

LEMMA A.2.16. *For any binary relation R, $(R^{\leftrightarrow})^{\text{id}} = (R^{\text{id}})^{\leftrightarrow}$. That is, the reflexive symmetric and the symmetric reflexive closures of a relation* coincide.

The smallest relation we considered on the set of CL-terms (i.e., $\triangleright_1$) is neither reflexive, nor symmetric. However, we did not consider the reflexive and the symmetric closures in the absence of transitivity.

LEMMA A.2.17. *The closure operators* $^{\leftrightarrow}$ *and* $^+$ do not commute. *That is, there is a binary relation R such that $(R^+)^{\leftrightarrow} \neq (R^{\leftrightarrow})^+$. However, $(R^+)^{\leftrightarrow} \subseteq (R^{\leftrightarrow})^+$.*

Example A.2.18. Let us consider the terms $\mathsf{K}(\mathsf{K}xy)z$ and $\mathsf{K}(\mathsf{K}xv)w$. $\mathsf{K}(\mathsf{K}xy)z \triangleright_1 \mathsf{K}xy \triangleright_1 x$ and $\mathsf{K}(\mathsf{K}xv)w \triangleright_1 \mathsf{K}xw \triangleright_1 x$. Of course, $\mathsf{K}(\mathsf{K}xy)z \triangleright_1^+ x$ and $\mathsf{K}(\mathsf{K}xv)w \triangleright_1^+ x$. However, $\langle \mathsf{K}(\mathsf{K}xy)z, \mathsf{K}(\mathsf{K}xv)w \rangle \notin (\triangleright_1^+)^{\leftrightarrow}$ (indeed, $\langle \mathsf{K}(\mathsf{K}xy)z, \mathsf{K}(\mathsf{K}xv)w \rangle \notin (\triangleright_{\mathrm{w}})^{\leftrightarrow}$). But $x \triangleright_1^{\leftrightarrow} \mathsf{K}xw \triangleright_1^{\leftrightarrow} \mathsf{K}(\mathsf{K}xv)w$, hence

$$\mathsf{K}(\mathsf{K}xy)z \triangleright_1^{\leftrightarrow} \mathsf{K}xy \triangleright_1^{\leftrightarrow} x \triangleright_1^{\leftrightarrow} \mathsf{K}xw \triangleright_1^{\leftrightarrow} \mathsf{K}(\mathsf{K}xv)w,$$

which means that $\langle \mathsf{K}(\mathsf{K}xy)z, \mathsf{K}(\mathsf{K}xv)w \rangle \in (\triangleright_1^{\leftrightarrow})^+$.

→→→→→❄←←←←❄→→→→❄←←←←←

A function may have *no*, *one*, *two*, ..., $n \in \mathbb{N}$, or *infinitely many* fixed points. We gave some examples in the text; here are some further examples: $f(x) = \log_2 x$ has no fixed points; $f(x) = x^3$ has three fixed points, 0, $+1$ and -1; $f(x) = x/1$ has as many fixed points as the cardinality of its domain, potentially including $\aleph_0$ and $2^{\aleph_0}$.

If we think about unary numerical functions, then it is easy to visualize the fixed points, because they are simply the points where the curve representing the function in the coordinate system intersects with the linear function $f(x) = x$. The numerical functions evaluate to a number, thus it is not clear (without further conventions or encoding) what the fixed point of a function of several arguments should be.[12]

CL-terms applied to CL-terms always result in a CL-term, which means that it makes sense to consider fixed points for any CL-term.

→→→→→❄←←←←❄→→→→❄←←←←←

A *lattice* is a particular kind of algebra with two operations that are often called *meet* and *join*.[13]

DEFINITION A.2.19. $\mathfrak{A} = \langle A, \wedge, \vee \rangle$ is a *lattice* iff equations (1)–(8) hold. ($a, b, c \in A$, and the equations are — tacitly — universally quantified.)

(1) $a \wedge a = a$	(5) $a \vee a = a$
(2) $a \wedge (b \wedge c) = (a \wedge b) \wedge c$	(6) $a \vee (b \vee c) = (a \vee b) \vee c$
(3) $a \wedge b = b \wedge a$	(7) $a \vee b = b \vee a$
(4) $a \wedge (b \vee a) = a$	(8) $a \vee (b \wedge a) = a$

An alternative, but completely equivalent definition of lattices starts with *partially ordered* sets (i.e., posets). (Order relations are introduced in definition A.6.2 below.)

DEFINITION A.2.20. (BOUNDS) Let $\mathfrak{X} = \langle X, \leq \rangle$ be a partially ordered set, and let $Y \subseteq X$.

(1) $z \in X$ is an *upper bound* of Y iff for all $y \in Y$, $y \leq z$.

(2) $z \in X$ is *the least upper bound* of Y, denoted as $\sup(Y)$ or $\mathrm{lub}(Y)$, iff z is an upper bound of Y, and for any z' that is an upper bound of Y, $z \leq z'$.

(3) $z \in X$ is a *lower bound* of Y iff for all $y \in Y$, $z \leq y$.

(4) $z \in X$ is *the greatest lower bound* of Y, denoted as $\inf(Y)$ or $\mathrm{glb}(Y)$, iff z is a lower bound of Y, and for any z' that is a lower bound of Y, $z' \leq z$.

---

[12] In chapter 6, we will see that this difficulty can be overcome, and there is a model for CL comprising functions on natural numbers.

[13] A classic compendium on lattice theory is Birkhoff [27].

The upper bound of a subset need not be an element of that particular subset. Moreover, no upper bound has to exist, in general. (E.g., $\mathbb{N}$ has no upper bound with respect to $\leq$, the usual ordering of $\mathbb{N}$.) On the other hand, a subset may have several upper bounds, and there may be no *least* one among the upper bounds. However, if a poset has an upper bound, then that is the least upper bound of the set. Because of antisymmetry, the least upper bound is unique, which explains the definite article *the*. Similar remarks apply *dually* to lower bounds.

DEFINITION A.2.21. Let $\mathfrak{A} = \langle A, \leq \rangle$ be a partially ordered set.

(1) $\mathfrak{A}$ is a *lattice* iff nonempty finite subsets of A have sup's and inf's.

(2) $\mathfrak{A}$ is a *bounded lattice* iff every finite subset of A has a sup and an inf.

(3) $\mathfrak{A}$ is a *complete lattice* iff every subset of A has a sup and an inf.

The definitions of lattices in A.2.19 and A.2.21 are *equivalent* when we amend them, respectively, by a definition of a partial order and of two operations. Given a lattice $\mathfrak{A} = \langle A, \wedge, \vee \rangle$, a partial order $\leq$ may be defined by $a \leq b$ iff $a \wedge b = a$, or $a \leq b$ iff $a \vee b = b$, and these two orders coincide. Given a lattice $\mathfrak{A} = \langle A, \leq \rangle$, meet and join are defined as inf and sup, respectively, of subsets of cardinality 1 or 2.

Adding sup($\emptyset$) amounts to adding $\bot$ to $\mathfrak{A}$ as an algebra, whereas inf($\emptyset$) is tantamount to adding $\top$. Thus the equivalent algebraic formulation of (2) from definition A.2.21 is $\mathfrak{A} = \langle A, \wedge, \vee, \bot, \top \rangle$. According to the usual understanding of algebras as structures comprising *finitary operations* only, there is no algebra corresponding to (3) from definition A.2.21. However, sometimes $\bigwedge$ and $\bigvee$ are used in place of inf and sup, which resemble $\wedge$ and $\vee$. If $\mathfrak{A}$ is a complete lattice, then $\bigwedge$ and $\bigvee$ are functions of type $\mathcal{P}(A) \longrightarrow A$, which justifies viewing $\bigwedge$ and $\bigvee$ as polyadic operations that might have infinitely many arguments. Thus a complete lattice is an algebra in a wider sense.

DEFINITION A.2.22. Let $\mathfrak{A} = \langle A, \wedge, \vee, \circledast \rangle$ be a lattice with a unary operation $\circledast$ on A. $\circledast$ is *monotone* (or *increasing*) when $a \leq b$ implies $\circledast(a) \leq \circledast(b)$.

A.3 Recursive functions and arithmetic

Turing machines, TMs are one of the most popular models of computation. (We give a brief description of them here.) They were originally introduced in Turing [152] (reprinted in Davis [56]). TMs are abstract machines, but they radiate the feel of concreteness, which may be the reason why many variants of TMs have been developed. The class of TM-computable functions is the same as the class of partial recursive functions, but the values of functions are calculated via different steps in the two models.

DEFINITION A.3.1. A *Turing machine* $\mathfrak{T}$ is a quadruple $\langle Q,A,M,P\rangle$, where the elements satisfy (1)–(4).

(1) $Q = \{q_i \colon i \in \mathbb{Z}^+\}$ is the set of *internal states* of $\mathfrak{T}$;[14]

(2) $A = \{a_i \colon i \in \mathbb{Z}^+\}$ is the set of *symbols* that can appear on the tape;

(3) $M = \{L,R\}$ is the set of allowed *moves* for the head of the TM;

(4) P, the *program* of the TM, is a finite set of quadruples of the form $\langle a_i,q_j,a_k,m\rangle$, where $a_i,a_k \in A$, $q_j \in Q$ and $m \in M$.

If P is a function with respect to its first two arguments, then $\mathfrak{T}$ is *deterministic*. If there can be two quadruples of the form $\langle a_i,q_j,a_k,m\rangle$ and $\langle a_i,q_j,a'_k,m'\rangle$, where $a_k \neq a'_k$ or $m \neq m'$, then $\mathfrak{T}$ is *nondeterministic*.

A TM should be imagined as a read-write head moving above a two-way infinite tape, which is divided into squares, each containing exactly one symbol. The head contains the program, and it is always in one of the possible internal states. A quadruple $\langle a_i,q_j,a_k,m\rangle$ in the program of $\mathfrak{T}$ means that, when the head is in state q_j over a square containing a_i, then $\mathfrak{T}$ replaces the symbol a_i by a_k, and moves one square to the left or to the right, depending on whether m is L or R.

Natural numbers may be encoded in unary notation as sequences of a symbol, for instance, as sequences of a_1. Then a function f of one argument is TM-*computable* iff there is a TM such that started in q_0 over the leftmost symbol of the sequence representing the number, the TM stops (after finitely many steps) over the leftmost symbol of the sequence representing the number that is the value of the function. (It is usually assumed that there is a symbol, called "blank," which fills in the rest of the square on the tape at the start and at the end of the computation.) For n-ary functions, the inputs may be stipulated to be separated with exactly one blank.

THEOREM A.3.2. *A function f on natural numbers is* TM-computable *iff f is* partial recursive.

We do not describe further models, rather we simply list them (together with our favorite expositions of them). *Register machines* are detailed in Boolos and Jeffrey [28, Ch. 6], where they are called "abaci." *Markov algorithms* are described in Curry [52, Ch. 2, §E] and in Mendelson [105, §5.5]. *Post systems* (and the closely related *semi-Thue systems* are detailed in Davis [57, Ch. 6]. Some further models include *Büchi machines*, 0-*type grammars* and *abstract state machines*.

→→→→→❄←←←←❄→→→→❄←←←←←

In chapter 3, Exercises 3.1.1 and 3.1.2 concern *multiplication* and *factorial*. We assume that these two functions have been defined, and it is known that they are primitive recursive. The following is a short list of some other functions that are primitive recursive. We use the abbreviation $\mathrm{pr}(f_1,f_2)$ when f_1 and f_2 are the two

[14] Since i is a positive integer, Q is a *finite* set.

functions that were denoted by f and g in definition 3.1.4. These two functions completely characterize primitive recursive equations.

(1) $\exp = \mathrm{pr}(\circ(\mathfrak{s},\mathfrak{z}),\circ(\times,\pi_1^3,\pi_3^3))$ (where $\times$ is multiplication); $\exp = \lambda xy.x^y$, that is, exp is *exponentiation*;

(2) $\mathrm{pred} = \mathrm{pr}(0,\pi_1^2)$; $\mathrm{pred}(0) = 0$ and $\mathrm{pred}(\mathfrak{s}(x)) = x$, that is, pred is the *predecessor* function;

(3) $\dot{-} = \mathrm{pr}(\pi_1^1,\circ(\mathrm{pred},\pi_3^3))$; $x \dot{-} y = x - y$ if $x \geq y$, and 0 otherwise, that is, $\dot{-}$ is *total truncated subtraction*.

The following functions are also useful to have:

(1) $\mathrm{sg} = \circ(\dot{-},\pi_1^1,\mathrm{pred})$ is the *sign* function; that is, $\mathrm{sg}(0) = 0$ and $\mathrm{sg}(\mathfrak{s}(x)) = 1$;

(2) $\overline{\mathrm{sg}} = \circ(\dot{-},\circ(\mathfrak{s},\mathfrak{z}),\mathrm{sg})$ is the *inverted sign* function; that is, $\overline{\mathrm{sg}}(0) = 1$ and $\overline{\mathrm{sg}}(\mathfrak{s}(x)) = 0$;

(3) $|\ -\ | = \circ(+,\dot{-},\circ(\dot{-},\pi_2^2,\pi_1^2))$ is the *absolute value of the difference*; that is, $|\,x-y\,| = x \dot{-} y$ is $x \geq y$, and $y \dot{-} x$ otherwise;

(4) $\min = \circ(\dot{-},\pi_1^2,\dot{-})$ is the *minimum* function; that is, $\min(x,y) = x$ if $x \leq y$, and y otherwise;

(5) $\max = \circ(+,\pi_2^2,\dot{-})$ is the *maximum* function; that is, $\max(x,y) = x$ if $x \geq y$, and y otherwise.

Notice that $\overline{\mathrm{sg}}$ is like *negation* if 0 and 1 are thought of as the truth values "false" and "true." Also, for all x, $\overline{\mathrm{sg}}(\mathrm{sg}(x)) = \overline{\mathrm{sg}}(x)$. This means that *conjunction* and *disjunction* from classical logic may be defined by (6) and (7).

(6) $\wedge = \circ(\times,\circ(\mathrm{sg},\pi_1^2),\circ(\mathrm{sg},\pi_2^2))$ is *conjunction*, that is, $x \wedge y = 1$ if $x = 1 = y$, and 0 otherwise;

(7) $\vee = \circ(\overline{\mathrm{sg}},\circ(\overline{\mathrm{sg}},\pi_1^2),\circ(\overline{\mathrm{sg}},\pi_2^2))$ is *disjunction*, that is, $x \vee y = 0$ if $x = 0 = y$, and 1 otherwise.

→→→→→❄←←←←❄→→→→❄←←←←←

The first-order theory of arithmetic is usually called *Peano arithmetic*, and abbreviated as PA. The language of PA is a first-order language that includes a *name constant* 0, a *unary function symbol* $\mathfrak{s}$, as well as two *binary function symbols* $+$ and $\times$. Informally, 0 is the number zero, $\mathfrak{s}$ is the successor function, and $+$ and $\times$ are addition and multiplication. The axioms of PA are as follows.[15]

(A1) $\forall x \forall y.\, \mathfrak{s}(x) = \mathfrak{s}(y) \Rightarrow x = y$

[15]By saying that PA is a first-order theory, we mean that the axioms listed here are added to a (sound and complete) formalization of *classical first-order logic with identity*.

(A2) $\forall x. \neg 0 = \mathfrak{s}(x)$

(A3) $\forall x. x + 0 = x$

(A4) $\forall x \forall y. x + \mathfrak{s}(y) = \mathfrak{s}(x + y)$

(A5) $\forall x. x \times 0 = 0$

(A6) $\forall x \forall y. x \times \mathfrak{s}(y) = (x \times y) + x$

(A7) $(A(0) \wedge \forall x. A(x) \Rightarrow A(\mathfrak{s}(x))) \Rightarrow \forall x. A(x)$

Axiom (A7) is the first-order induction axiom; that is, A is a one-place predicate expressible in the language of PA (rather than a tacitly quantified predicate variable).

The axioms do not look very complicated, and so it may be unexpected that this first-order theory is remarkably complicated and powerful. The next two theorems, in effect, are Gödel's theorems for PA.[16] They are often labeled as "the first" and "the second" incompleteness theorems, respectively.

THEOREM A.3.3. (FIRST INCOMPLETENESS THEOREM) There is a sentence *in the language of* PA *such that neither the sentence nor its negation is provable in* PA.

The notion of incompleteness in this theorem is *syntactic*, though it is closely related to semantic incompleteness.

THEOREM A.3.4. (SEMANTIC INCOMPLETENESS) *Given any interpretation of* PA*, which makes all the theorems of* PA *true, there is a sentence that is true in that interpretation but unprovable in* PA.

Both theorems might seem to suggest that we simply need to find a "better" theory, let us say, PA$'$, that appropriately extends PA. In other words, this purported theory PA$'$ should contain all the axioms of PA and more.

THEOREM A.3.5. (INCOMPLETABILITY) *There is no extension of* PA *that is axiomatizable, consistent and complete.*

Axiomatizability means having a *recursive* (i.e., *decidable*) set of sentences. The semantic incompleteness theorem can then be phrased as "the set of true sentences in the language of PA is not recursive."

Gödel's aim, in the spirit of the formalist program for the foundations of mathematics, was to *prove consistency*. PA is sufficiently expressive to be able to formalize its own consistency, for instance, in the form "$0 = 1$ is not provable."

THEOREM A.3.6. (SECOND INCOMPLETENESS THEOREM) PA *cannot prove its own consistency, provided that* PA *is consistent.*

[16] Gödel [76] is the original "incompleteness paper."

If PA is *in*consistent, then it proves any formula in its language, including all the sentences expressing the consistency of PA — as well as those expressing the inconsistency of PA. In view of this, the theorem says that, if there is at least one formula that is not provable in PA, then a formula expressing the consistency of PA is not provable either.[17]

The second incompleteness theorem is not "absolute" in the sense that it does not state or imply that the consistency of PA cannot be proved at all. (See Gentzen [71] for the first proof.) There are "stronger" formal systems, for example, ZF in which the consistency of PA can be proved. The latter means that PA is *consistent relative to* ZF. On the other hand, PA cannot prove the consistency of ZF; neither can ZF prove its own consistency.

A.4 Connections to λ-calculi

A complication in Λ is the λ *operator* which explicitly binds a variable. The next two definitions introduce the same notions for Λ that are introduced in definition A.1.12 for CL.

DEFINITION A.4.1. The set of *free variables* of a term M is inductively defined by (1)–(3).

(1) $\mathrm{fv}(x) = \{x\}$;

(2) $\mathrm{fv}(MN) = \mathrm{fv}(M) \cup \mathrm{fv}(N)$;

(3) $\mathrm{fv}(\lambda x.M) = \mathrm{fv}(M) - \{x\}$.

$\mathrm{fv}(M)$ is always a *proper* subset of the set of variables. Since fv is a set, the number of occurrences of variables are not counted. If $\mathrm{fv}(M) = \emptyset$, then M is called a *closed* Λ-term.

DEFINITION A.4.2. The set of *bound variables* of a term M is inductively defined by (1)–(3).

(1) $\mathrm{bv}(x) = \emptyset$;

(2) $\mathrm{bv}(MN) = \mathrm{bv}(M) \cup \mathrm{bv}(N)$;

(3) $\mathrm{bv}(\lambda x.M) = \mathrm{bv}(M) \cup \{x\}$.

[17] There are detailed formal presentations of theorems of this sort for PA and for other systems like Q and ZF. See, for example, Boolos and Jeffrey [28].

$\text{bv}(M)$ is also a *proper* subset of the set of variables. In order to make the parsing of Λ-terms easier, it may be useful to use two different infinite sequences of variables. For instance, x's can be used for bound and v's for free variables. This practice yields Λ-terms that are easier to read, but it leads to more complicated definitions of Λ-terms, of substitution and of other related concepts. We do not adopt this convention here. Then, there are Λ-terms that contain a variable both as a bound and as a free variable, that is, $\text{fv}(M) \cap \text{bv}(M) \neq \emptyset$ is not excluded.

Example A.4.3. $x(\lambda xyz.\, yx)y$ is a Λ-term by definition 4.1.1. $\text{fv}(x(\lambda xyz.\, yx)y) = \{x, y\}$ and $\text{bv}(x(\lambda xyz.\, yx)y) = \{x, y, z\}$. Notice that $z \notin \text{fv}(yx)$, but it follows a λ. Beyond the fact that $\lambda xyz.\, yx$ is a ternary function (because it is $\lambda x \lambda y \lambda z.\, yx$), it is necessary to list z in $\text{bv}(\lambda xyz.\, yx)$ when it comes to substituting Λ-terms for x or y.

Definition A.1.7 gave a convenient way to identify occurrences of subterms in CL-terms. The same indexing cannot be adopted for Λ-terms without changes, because of the possibility to form λ-abstracts. We do not extend the indexing for CL-terms to Λ-terms here (even though this can be done rigorously). Rather, we assume that in the few cases when we need, we can rely on the informal notion of occurrences.

DEFINITION A.4.4. (FREE AND BOUND OCCURRENCES OF VARIABLES) An occurrence of the variable x in M is *free* iff no subterm of M of the form $\lambda x.N$ contains that occurrence. Otherwise, the occurrence is *bound*.

Each occurrence of a variable is free or bound, but never both. The occurrence of a variable that comes immediately after a λ, is always bound. A variable cannot be bound more that once, that is, by two or more distinct λ's. Thus it is important, for instance, for β-reduction to determine *binding* relationships between λ's and variables in Λ-terms.

DEFINITION A.4.5. A variable occurrence x that is bound in M, but is not preceded by a λ, is bound by the λ in $\lambda x.N$ if the occurrence falls into N, which has no proper subterm $\lambda x.P$ containing that occurrence of x.

More concisely, though a bit sloppily, a variable occurrence is bound by the λ (with a matching variable) that has the *narrowest scope*.

→→→→→❄←←←←❄→→→→❄←←←←←

Famously, Church gave a preference to the λI-calculus over the λK-calculus. In the λK-calculus, a λ may bind just one occurrence of a variable, which is immediately after the λ.

DEFINITION A.4.6. (λI-TERMS) The set of λI*-terms* is defined by (1)–(3).

(1) If x is a variable, then x is a λI-term;

(2) if M and N are λI-terms, then (MN) is a λI-term;

(3) if x is a variable and M is a term such that $x \in \text{fv}(M)$, then $\lambda x.M$ is a λI-term.

Clause (3) does not allow the possibility of forming a *vacuous* λ-abstraction. λI-terms are connected to the *logic of relevant implication.*

Linear λ-terms are named after linear logic, which is an extension of the implicational *BCI* logic. Linear λ-terms constitute a proper subset of the λI-terms.

DEFINITION A.4.7. (LINEAR λ-TERMS) The set of *linear λ-terms* is defined by (1)–(3).

(1) If x is a variable, then x is a linear λ-term;

(2) if M and N are linear λ-terms and $\text{fv}(M) \cap \text{fv}(N) = \emptyset$, then (MN) is a linear λ-term;

(3) if x is a variable, and M is a linear λ-term such that $x \in \text{fv}(M)$, then $\lambda x.M$ is a linear λ-term.

A.5 (In)equational combinatory logic

Equations play an important role in defining classes of algebras. *Equational* and *inequational* logics were invented to axiomatize reasoning that is used in algebras that are defined by sets of equations or inequations.

Let us assume that there is an indexed set of *operations* in an algebra, $\circledast_{i \in I}$. We need to single out an argument place of each operation, and for this purpose we use the notation $\circledast_i(\vec{a}, [b]_j)$ to indicate that there are some a's that fill out the argument places of $\circledast_i$ save the jth place, which has b in it. $\vec{a}$ can be thought of as a sequence or vector of some elements. (We always assume that j is within the range of 1 to n if $\circledast_i$ is an n-ary operation.)

Terms in an algebra are formed in the usual way from variables and operations; and *formulas* are always of the form $t_1 = t_2$, where t_1 and t_2 are terms.

The *equational calculus* comprises an *axiom* and rules. There are three rules that are not specific for the particular set of operations. For each operation $\circledast_i$, there are as many rules as the arity of the operation.

$$x = x \quad \text{id}$$

$$\frac{x = y \quad y = z}{x = z}\ \text{tr} \qquad \frac{x = y}{y = x}\ \text{sym}$$

$$\frac{r = s}{[x_1/t_1, \ldots, x_n/t_n]r = [x_1/t_1, \ldots, x_n/t_n]s}\ \text{sub}$$

$$\frac{t = s}{\circledast_i(\vec{r}, [t]_j) = \circledast_i(\vec{r}, [s]_j)}\ \text{rep}_i^j$$

The axiom is called *identity*, and the next two rules make $=$ *transitive* and *symmetric*. (See section A.6 for definitions of properties of binary relations.) The rule sub permits a substitution of terms, for instance, of $t_1,\ldots,t_n$ for variables, for instance, for $x_1,\ldots,x_n$ in a pair of terms r and s with their identity being preserved. The last rule should be understood as a *rule schema* in the sense that for an n-ary $\circledast_i$, it abbreviates n rules. That is, rep is a rule for each j, where $1 \leq j \leq n$.

The language of an algebra is always assumed to contain $=$ as a binary predicate, but it may not contain $\leq$ as another binary predicate. Some algebras allow a *partial order* to be defined relying on the operations in the algebra. For instance, lattices and Boolean algebras have a partial order standardly associated to them. (See definitions A.2.19, A.2.21, A.9.7 and A.6.6.)

An algebra may contain a *partial order*, denoted by $\leq$, as a primitive — in addition to an indexed set of operations $\circledast_{i \in I}$. Then $\leq$ can be used similarly to $=$ as a binary predicate. The resulting formulas are called *inequations* or *inequalities*.

DEFINITION A.5.1. (TONICITY) An operation $\circledast$ is *monotone* in its jth argument place iff for all $\vec{a}$, and for all b and c, $b \leq c$ implies $\circledast(\vec{a},[b]_j) \leq \circledast(\vec{a},[c]_j)$. An operation $\circledast$ is *antitone* in its jth argument place iff for all $\vec{a}$, and for all b and c, $b \leq c$ implies $\circledast(\vec{a},[c]_j) \leq \circledast(\vec{a},[b]_j)$.

Example A.5.2. A Boolean algebra may be defined using only two operations, $-$ and $\vee$, which are *complementation* and *join*. $a \leq c$ implies $a \vee b \leq a \vee c$ as well as $b \vee a \leq b \vee c$, that is, join is monotone in both argument places. Complementation, on the other hand, is antitone, because $a \leq b$ implies $-b \leq -a$.

The *ineqational* calculus contains an *axiom* and three general rules, together with a set of rules specific to the operations of the algebra.

$$x \leq x \ \text{id}$$

$$\frac{x \leq y \quad y \leq z}{x \leq z}\ \text{tr} \qquad \frac{x \leq y \quad y \leq x}{x = y}\ \text{antisym}$$

$$\frac{r \leq s}{[x_1/t_1,\ldots,x_n/t_n]r \leq [x_1/t_1,\ldots,x_n/t_n]s}\ \text{sub}$$

$$\frac{t \leq s}{\circledast_i(\vec{r},[t]_j) \leq \circledast_i(\vec{r},[s]_j)}\ \text{mono}_i^j$$

$$\frac{t \leq s}{\circledast_i(\vec{r},[s]_k) \leq \circledast_i(\vec{r},[t]_k)}\ \text{anti}_i^k$$

The rules tr and sub are similar to those in the equational calculus. antisym ensures that $\leq$ is *antisymmetric*. The mono and anti rules are postulated for an operation and an argument place, depending on the *tonicity* of the operation in that argument place. Typically, an operation is not both monotone and antitone for an argument place, and it even may not be either. That is, the mono and anti rules together give at most as many rules as rep gives in the equational calculus.

In chapter 5, and then in chapter 7, we use extensions and modifications of the equational and inequational calculi given in this section.

First of all, combinators are constants (or nullary operations), and they require the addition of axioms, in which the combinators interact with the function application operation, because the only non-zeroary operation in CL is function application.

The inequational calculi that we use to axiomatize systems of CL do not contain the rule antisym, because $\triangleright_w$ is not antisymmetric. Perhaps, another name such as "pre-equational calculi" would be more faithful to the preorder on CL-terms that is $\triangleright_w$. Instead of introducing the new term "pre-equation," we opted for using "inequation" with a proviso.

A.6 Models

Binary relations appear in many areas; $\triangleright_w$ and $=_w$ are the leading examples in CL. The following is a (fairly) standard list of properties and names for kinds of binary relations. To suggest general applicability, the binary relation is denoted by R. The arguments are written without parentheses and commas (which is one of the commonly used notations).

DEFINITION A.6.1. (PROPERTIES OF RELATIONS) Let R be a binary relation on a set X, that is, $R \subseteq X \times X$.

(1) R is *reflexive* iff for all $x \in X$, Rxx.

(2) R is *transitive* iff for all $x, y, z \in X$, Rxy and Ryz imply Rxz.

(3) R is *symmetric* iff for all $x, y \in X$, Rxy implies Ryx.

(4) R is *asymmetric* iff for all $x, y \in X$, Rxy implies $\neg Ryx$, that is, Rxy implies that Ryx does not hold.

(5) R is *antisymmetric* iff for all $x, y \in X$, Rxy and Ryx imply $x = y$.

(6) R is *connected* iff for all $x, y \in X$, Rxy or Ryx.

(7) R is the *identity relation* iff for all $x, y \in X$, Rxy just in case $x = y$.

(8) R is the *total relation* iff for all $x, y \in X$, Rxy.

DEFINITION A.6.2. (ORDER RELATIONS) Let R be a binary relation on the set X.

(1) R is a *preorder* (or *quasi-order*) iff R is reflexive and transitive.

(2) R is a *partial order* iff R is an antisymmetric preorder.

(3) R is a *strict partial order* iff R is irreflexive and transitive.

(4) R is a *linear order* iff R is a connected partial order.

(5) R is a *strict linear order* iff R is a connected strict partial order.

Sometimes the strict variants of these order relations are considered more fundamental, in which case (2) and (4) are called *weak partial order* and *weak linear order*, respectively. Linear orders have been called *total orders* as well as *simple orders*.

Equivalence relations could also be considered among order relations, but it is perhaps, useful to separate them into their own group.

DEFINITION A.6.3. (EQUIVALENCE RELATIONS AND EQUIVALENCE CLASSES) Let R be a binary relation on the set X.

(1) R is an *equivalence relation* iff R is a symmetric partial order. In other words, an equivalence relation is *reflexive*, *transitive* and *symmetric*.

Let R be an equivalence relation on X. A subset Z of X is an *equivalence class* with respect to R just in case Z is $[x]$ for some $x \in X$, where

(2) $[x] = \{y \in X : Rxy\}$.

Sets have subsets, and the collection of all the subsets of a set X is the *power set* of X. Certain other collections of subsets of a set have important features.

DEFINITION A.6.4. (PARTITIONS) Let X be a set, and $\mathbb{Y} = \{Y_i : i \in I \wedge Y_i \subseteq X\}$ be a collection of nonempty subsets of X. $\mathbb{Y}$ is a *partition* of X when (1) and (2) hold.

(1) $\forall x \in X\ \exists Y_i \in \mathbb{Y}.\, x \in Y_i$;

(2) $\forall x \in X.\, x \in Y_i \Rightarrow Y_i \neq Y_j \Rightarrow x \notin Y_j$.

$\mathbb{Y}$ is a partition of X when $\mathbb{Y}$ comprises nonempty subsets of X that are pairwise disjoint and together cover all the elements of X. Thus, alternatively, we may define $\mathbb{Y}$ to be a *partition* of X when (3) and (4) hold.

(3) $\forall i, j.\, i \neq j \Rightarrow Y_i \cap Y_j = \emptyset$;

(4) $\bigcup_{i \in I} Y_i = X$.

Partitions are easy to visualize, and they are intimately linked to equivalence relations.

LEMMA A.6.5. (EQUIVALENCE RELATIONS AND PARTITIONS) *Let E be the set of all equivalence relations on X and let P be the set of all partitions of X. There is a* 1–1 correspondence *between the elements of E and P.*

The idea for a proof of this lemma is to use equivalence classes in place of the subsets figuring into a partition. (It is easy to show that each equivalence relation gives rise to a set of equivalence classes, and the other way around.) The *coarsest partition* of a set comprises just one subset, namely, the set itself. The set of singletons — where each element of a set is an element of a singleton — is the *finest partition* of a set. These two extreme partitions correspond to the *total* relation and the *identity* relation on the set, respectively.

→→→→→ ❄ ←←←← ❄ →→→→ ❄ ←←←←←

Boolean algebras are very well understood and extremely important operational structures in universal algebra.

DEFINITION A.6.6. (BOOLEAN ALGEBRAS) The algebra $\mathfrak{B} = \langle B, \wedge, \vee, -, 0, 1\rangle$ of similarity typc $\langle 2,2,1,0,0\rangle$ is *a Boolean algebra* when the following equations hold.

$$
\begin{array}{cc}
a \wedge a = a & a \vee a = a \\
(a \wedge b) \wedge c = a \wedge (b \wedge c) & (a \vee b) \vee c = a \vee (b \vee c) \\
a \wedge b = b \wedge a & a \vee b = b \vee a \\
a \wedge (a \vee b) = a & a \vee (a \wedge b) = a \\
a \wedge (b \vee c) = (a \wedge b) \vee (a \wedge c) & a \vee (b \wedge c) = (a \vee b) \wedge (a \vee c) \\
a \wedge -a = 0 & a \vee -a = 1 \\
a \wedge 1 = a & a \vee 0 = a
\end{array}
$$

The *carrier set* of an algebra (e.g., B above) is always assumed to be closed under the operations of the algebra.[18] The *similarity type* of an algebra is a listing of the arity of its operations. (Zeroary operations are elements of the carrier set, and they are typically stipulated to satisfy some equations or conditions.)

Algebras come with an identity relation, and replacement of identical elements is always permitted. Equations, like the ones above, are understood to be *universally quantified*. For example, $a \wedge b = b \wedge a$ means that for all $a \in B$, and for all $b \in B$, the elements $a \wedge b$ and $b \wedge a$ are identical.

The operations of $\mathfrak{B}$ are sometimes called *meet* ($\wedge$), *join* ($\vee$) and (Boolean or ortho-)*complement* ($-$). 0 and 1 are the *bottom* and *top* elements, respectively; alternatively, they are the *identity elements* for join and meet, and the *null elements* for meet and join.

Boolean algebras may be specified by a set of equations (e.g., as above), hence they form a *variety*. Varieties have been extensively studied, and they possess properties — such as closure under *homomorphic images* — that makes them an easy topic for investigations.

The smallest Boolean algebra is $\langle \{a\}, \wedge, \vee, -, 1, 0\rangle$. All algebras with just one element are called *trivial*, and as it often happens to extremal elements, trivial algebras are excluded or included, depending on the aims of the investigations. We consider some nontrivial Boolean algebras as examples.

[18] An exception is when so-called *partial algebras* are considered too.

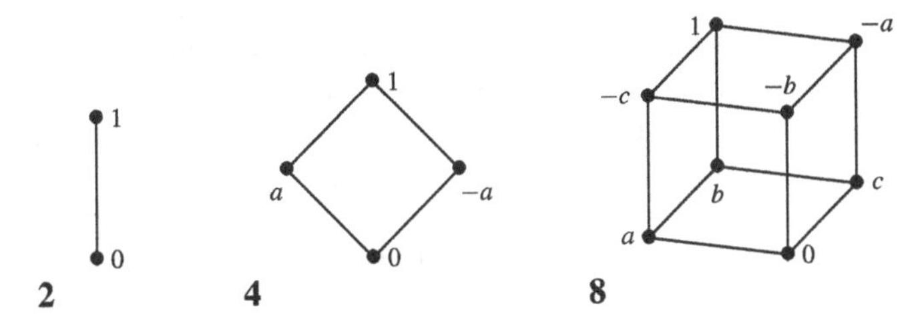

2 is theonly *irreducible* Boolean algebra, that is, it is not (isomorphic to) a product of Boolean algebras. **4** is $\mathbf{2} \times \mathbf{2}$, and **8** is $\mathbf{4} \times \mathbf{2}$. These Boolean algebras have 1, 2 and 3 *atoms*, respectively (which are 1; a and $-a$; and a, b and c). Every finite Boolean algebra has 2^n many elements when it has n atoms.

A prototypical example of a Boolean algebra is the *power set* of a set. The above examples can be easily seen to be isomorphic to the power set algebras of a singleton, of a doubleton and of a three element set. Despite the importance of these kinds of Boolean algebras, not all Boolean algebras are isomorphic to a power set algebra.

→→→→→ ❄ ←←←← ❄ →→→→ ❄ ←←←←←

Topologies are often useful to select a subset of the powerset of a set.

DEFINITION A.6.7. (TOPOLOGY) Let X be a set. $\mathfrak{T} = \langle X, \mathfrak{O} \rangle$ is *a topology* on X when (1)–(3) hold.

(1) $X, \emptyset \in \mathfrak{O}$;

(2) $O_1, O_2 \in \mathfrak{O}$ imply $O_1 \cap O_2 \in \mathfrak{O}$;

(3) $O_{i \in I} \in \mathfrak{O}$ implies $\bigcup_{i \in I} O_i \in \mathfrak{O}$.

The elements of $\mathfrak{O}$ are called *open sets*.

The powerset of a set is a topology, but often interesting collections of subsets are topologies without comprising all the subsets. Conditions (2) and (3) are closure conditions: $\mathfrak{O}$ is closed under *finite intersections* and *arbitrary unions*.

DEFINITION A.6.8. (BASIS AND SUBBASIS FOR A TOPOLOGY) Let $\mathfrak{T} = \langle X, \mathfrak{O} \rangle$ be a topology on X. $\mathfrak{B}$ is *a basis* for $\mathfrak{T}$ if (1)–(3) are true.

(1) $\mathfrak{B} \subseteq \mathfrak{O}$;

(2) $B_1, B_2 \in \mathfrak{B}$ imply $B_1 \cap B_2 \in \mathfrak{B}$;

(3) $O \in \mathfrak{O}$ implies $O = \bigcup C$ for some $C \subseteq \mathfrak{B}$.

$\mathfrak{S}$ is *a subbasis* for $\mathfrak{T}$ if (1)–(2) are true.

(1) $\mathfrak{S} \subseteq \mathfrak{O}$;

(2) $\{B \colon B = \bigcap C \wedge C \subseteq_{\text{fin}} \mathfrak{S}\}$ is a basis for $\mathfrak{T}$.

A basis generates a topology. If only a basis $\mathfrak{B}$ is given, then we denote *the topology generated* by the basis as $\tau(\mathfrak{B})$. A subbasis $\mathfrak{S}$ also generates a topology, which is denoted by $\tau(\mathfrak{S})$. Given $\mathfrak{S}$, all sets obtainable by finite intersections yield a basis $\mathfrak{B}$, then arbitrary unions are formed, which results in $\mathfrak{O}$.

DEFINITION A.6.9. (CONTINUOUS FUNCTIONS) Let $\mathfrak{T}_1 = \langle X_1, \mathfrak{O}_1 \rangle$ and $\mathfrak{T}_2 = \langle X_2, \mathfrak{O}_2 \rangle$ be topologies, and let f be a function of type $X_1 \longrightarrow X_2$. f is *continuous* when for $O \in \mathfrak{O}_2$, $f^{-1}[O] \in \mathfrak{O}_1$.

f^{-1} followed by [ ] is the type-raised *inverse function* of f. $f^{-1}\colon \mathcal{P}(X_2) \longrightarrow \mathcal{P}(X_1)$ is the type of f^{-1}. $f^{-1}[O] = \{x\colon \exists y \in O. f(x) = y\}$, that is, the *existential image* operation of the inverse of f. The inverse of f may or may not be a function. However, f^{-1} collects into a subset some elements and always yields a function from the powerset of f's codomain into the powerset of f's domain. If f is injective (and onto), then $f^{-1}[\{x\}] = \{y\}$, for all $x \in \mathrm{cod}(f)$.

A function f is *continuous* in the topological sense iff the inverse image of f maps an open set into an open set.

THEOREM A.6.10. (CONTINUOUS FUNCTIONS ARE CONTINUOUS) *A function $f\colon \mathcal{P}(\mathbb{N}) \longrightarrow \mathcal{P}(\mathbb{N})$ is* continuous *in the sense of definition 6.3.2 just in case f is* continuous *in the sense of definition A.6.9 when $\mathbb{N}$ is equipped with the Scott topology from definition 6.3.1.*

→→→→→ ❄ ←←←← ❄ →→→→ ❄ ←←←←←

Binary strings are finite sequences of characters each of which is a 0 or a 1. In section A.1, we gave a definition of strings (or words) over an alphabet using a context-free grammar. For binary strings, the alphabet is $\{0,1\}$.

It is not difficult to see that there are $\aleph_0$-many binary strings. Natural numbers have a natural ordering, namely, $\leq$. Binary strings may be viewed as natural numbers in base-two notation. Strings, in general, need not contain any digits (i.e., numerical characters), for instance, if the alphabet does not have any. A *lexicographic order* is a strict linear ordering on a set of strings induced by the linear ordering of the alphabet. The term "lexicographic ordering" is used — somewhat confusingly — to refer to two different orderings.

DEFINITION A.6.11. (LEXICOGRAPHIC ORDERING, I) Let the linear ordering on the alphabet $\mathbb{A} = \{a_0, \ldots, a_n\}$ be $a_0 < \ldots < a_i < \ldots < a_i' < \ldots < a_n$. The *lexicographic ordering* on $\mathbb{A}^*$ is the least relation satisfying (1)–(2).[19]

(1) $w_1 < w_2$ if $\ell(w_1) < \ell(w_2)$;

(2) $w_1 < w_2$ if $\ell(w_1) = \ell(w_2)$ and $w_1 = va_iu_1$ and $w_2 = va_i'u_2$.

[19] w's, v's and u's range over strings. The length of a string w, denoted by $\ell(w)$, is the number of letters w contains.

Example A.6.12. According to the definition, ε precedes all the nonempty words, because ε is the shortest word of length 0. Let $a_0 < a_1 < a_2 < a_3 < a_4$ be the linear order on the alphabet. $a_1a_0a_2a_2 < a_1a_0a_2a_4$, because $\ell(a_1a_0a_2a_2) = \ell(a_1a_0a_2a_3) = 4$, and v in clause (2) of the definition may be taken to be $a_1a_0a_2$, whereas $u_1 = u_2 = \varepsilon$. $a_0a_0a_2a_2 < a_0a_4a_1a_3$ by clause (2) too. $a_0 < a_4$ in the linear ordering on $\mathbb{A}$, and $u_1 = a_2a_2$, whereas $u_2 = a_1a_3$.

If the set of strings $\mathbb{S}$ is a proper subset of $\mathbb{A}^*$, then the clauses from the above definition are still applicable, and the ordering generated by (1)–(2) is the same on $\mathbb{S}$ as $<\restriction \mathbb{S}\times\mathbb{S}$.

The lexicographic ordering of nonempty binary strings starting with 1 coincides with $<$ on $\mathbb{Z}^+$, when a binary string is viewed as base-2 notation for a number.

We mention for the sake of comparison the other sense in which the term "lexicographic order" is used. A mnemonic label for this ordering is "phone-book ordering," because the names in a phone book are ordered this way. (Of course, real phone books rarely contain strings of length, perhaps, 50 or greater; that is, they contain a small finite subset of $\mathbb{A}^*$, which is lexicographically ordered.)

DEFINITION A.6.13. (LEXICOGRAPHIC ORDERING, II) Let the linear ordering on the alphabet $\mathbb{A} = \{a_0, \ldots, a_n\}$ be $a_0 < \ldots < a_i < \ldots < a'_i < \ldots < a_n$. The *lexicographic ordering* on $\mathbb{A}^*$ is the least relation satisfying (1)–(3).

(1) $\varepsilon < w$ if $\ell(w) > 0$;

(2) $w_1 < w_2$ if $w_1 = va_iu_1$ and $w_2 = va'_iu_2$;

(3) $w_1 < w_2$ if $w_2 = w_1a_i$.

The strings that we considered in example A.6.12, will get the same ordering according to the second definition. However, in general, the orderings determined by the two definitions differ — because of the absence of any reference to the length of the strings in (2) in definition A.6.13.

Example A.6.14. Assuming the same linear order for the alphabet as in the previous example, $a_1a_2 < a_3$ in the second lexicographic ordering — despite the fact that a_3 is shorter than a_1a_2. If $\ell(w_1) = \ell(w_2)$, then $w_1 <_{\mathrm{I}} w_2$ iff $w_1 <_{\mathrm{II}} w_2$. However, $a_0a_1a_2$ is immediately followed by $a_0a_1a_3$ in the first, but not in the second ordering.

If there is more than one letter in $\mathbb{A}$, and $\mathbb{A}^*$ is the set of strings ordered, then there are *infinitely* many strings preceding a_1 (the second letter).

If the set $\mathbb{S}$ is a proper subset of $\mathbb{A}^*$, then the clauses from definition A.6.13 are still applicable, and $<$ is the same relation as $<\restriction \mathbb{S}\times\mathbb{S}$.

Given $\mathbb{A}^*$, where $\mathrm{card}(\mathbb{A}) \geq 2$, the second lexicographic ordering *does not* yield *an enumeration* of $\mathbb{A}^*$.

LEMMA A.6.15. (ENUMERATION OF STRINGS) *The lexicographic ordering in definition A.6.11 is* an enumeration of *the set* $\mathbb{A}^*$.

Real phone books do not run into problems with enumerating strings simply because they contain finitely many items. The following table illustrates the *difference* between the two orderings on a small set of binary strings. The numerical values of strings — in decimal notation — are given below the binary strings. (The boldface strings in the $<_{\mathrm{I}}$ ordering are the binary strings that start with 1.)

$<_{\mathrm{I}}$	0	**1**	00	01	**10**	**11**	000	001	010	011	**100**	**101**	**110**	**111**
#	0	1	0	1	2	3	0	1	2	3	4	5	6	7

$<_{\mathrm{II}}$	0	00	000	001	01	010	011	1	10	100	101	11	110	111
#	0	0	0	1	1	2	3	1	2	4	5	3	6	7

A.7 Dual and symmetric combinatory logic

Programming languages come in various styles, for instance, there are functional, declarative and logic programming languages. They also vary widely in the *syntactic conventions* they use.

We mentioned the programming language APL as an example in which some operators apply from the *right*. We cannot give a comprehensive introduction to this language. However, we want to illustrate the possibility of having a language that contains functions and that has a more subtle way of forming expressions than simple strings like $f(x_1,\ldots,x_n)$.

A feature of APL is that the main data structure comprises *matrices* of finite dimensions. This naturally leads to the polymorphism of the operations, which are typically performed *pointwise*, depending on the parameters characterizing the dimensions of the argument. Incidentally, this can be thought of as an example of *an argument* being *applied to a function* and selecting the function from a range of similar ones that applies to the argument.

Another feature of APL is that expressions are parsed and executed *from right to left*. This means that long (unparenthesized) strings are viewed as right-associated terms. Some operations can take arguments from both sides, but the parsing order of strings means that they work as functions applied from the right, that is, like dual combinators do.

We give a small program in APL as an illustration. To utilize the default data structure, we define a function that gives a one-dimensional matrix as the result. A one-dimensional matrix is sometimes called a *vector*, an *array* or a *list* in other programming languages.

Recall that the *factorial* function is denoted by !. The values of the factorial function for the first five natural numbers are 1, 1, 2, 6 and 24. The APL program that we define is called FAC, takes a positive natural number n, and returns a matrix comprising $\langle 0!,\ldots,n!\rangle$. The program comprises three lines.

∇ FAC n

[1.] $A \leftarrow 1$

[2.] $\rightarrow 2 \times n > -1 + \rho A \leftarrow A, \rho A \times {}^{-1} \uparrow A \nabla$

The ∇'s delineate the *definition of the function*, and the bracketed numbers allow references to lines of code. The first line could be written informally as FAC $=$ $\lambda n.\ldots$, or in some versions of LISP as **`(define FAC (lambda (n)....`**

Line 1 says that A is set to be the matrix with the one element 1.

The execution of line 2 starts with taking the last element of A (${}^{-1} \uparrow A$). The next step is to multiply this number ($\times {}^{-1} \uparrow A$) by the length of A (ρA). We used ℓ as the function for length of strings and CL-terms, and ρ for a one-dimensional matrix exactly like ℓ. The resulting number is concatenated to A. This operation is like **`append`** in versions of LISP; here this operation is denoted by $,$. The operations so far are accomplished by the segment of the string $A, \rho A \times {}^{-1} \uparrow A$. The next step is to replace the old A by the new A, that is, to set A to be $A, \rho A \times {}^{-1} \uparrow A$. The $\leftarrow$ functions as $:=$ or **`set!`** in some programming languages. The ρ operation returns the length of the matrix, but it is applied after the assignment; that is, ρ is applied to A after all the operations mentioned so far have been carried out. Then we decrease the number, which is the length of A, by 1 ($-1 + \rho A \ldots$). This accounts for the fact that we start with 0! in the matrix, and the length of the matrix is the next natural number with which we multiply the last number in the matrix. The result is compared to the input n. A convenient "type-free" feature of $>$ is that if $x > y$ is false, then the returned value is 0, otherwise, it is 1. In other words, there is no separate type "Boolean," rather the truth values are treated as numbers. The $\rightarrow$ is a **`go to`** statement, which makes clear the role of the numbering of the lines in the program. If the instruction is **`go to 0`**, then the program terminates, because there is no line labeled by 0. If the input number is greater than the length of the matrix, then the instructions on line 2 are repeated again.

The separation of the definition of the program into a block by the ∇'s is like defining functions or subroutines. Once the function is defined, the factorials up to n can be calculated by FAC n, which places the results into a matrix of factorials $\langle 0!, \ldots, n! \rangle$.

→→→→→❄←←←←❄→→→→❄←←←←←

Constructing classifications of objects like the *genus–species* hierarchy has a long history. As the previous example from APL illustrates, there are similarities between operations applied to different kinds of objects. Of course, it is somewhat arbitrary where we draw the line of difference between this object and that object, and it is also up to us to view certain operations as the same operation rather than different ones. For example, depending on how two-dimensional matrices are implemented, they may or may not be quite similar to a vector (i.e., a one-dimensional matrix). Or, we might think about addition on integers to be the same operation as addition on reals, but perhaps, the latter can be only represented as floating point numbers and the actual addition algorithms are different.

Object-oriented programming (OOP) takes these observations one step further. Rather than to leave it to chance which abstract data types are similar and which operations are similar it allows the user to introduce similarities where they are desired. In a sense, object-oriented programming is an approach in which the user can decide *how to view* the world; that is, what the classes of objects are and what the ways to manipulate them are.

Example A.7.1. Let us assume that a university has a database that contains information about those who are affiliated with the university. There may be a class of objects each called a **`person`**. Let us assume that the database can be queried, and one of the possible query functions is **`affiliation`** which takes a last name of a person as its argument. The result may be a list of people with matching last names together with their affiliation. The latter is not going to be simply the university (which would be uninformative), but either the department in which they work, or the program in which they are enrolled. The latter difference can be achieved by having two subclasses of objects: **`faculty`** and **`student`**. Another possible query function may return the university provided e-mail address of the person, and can function in the same way for the two subclasses of the class **`person`** to which the object belongs.

In this setting, the final result of the application of a function to an *object* is a more complicated process than adding 1 plus 2, because the function may be a class of functions called a *method*, which contains suitable specifications for the subclasses of objects. The analogy to which we alluded takes *methods* to be combinators and *objects* (or the feature of an object that it belongs to a class) to be dual combinators. An application of a method to some arguments may give one result, but if the object selects a certain method first, that may lead to a different result. If the lack of confluence does not happen in OOP language, then one of the reasons might be that the syntax of the language is so specified that the order of evaluation on analogues of overlapping redexes is fully determined.

→→→→→ ❄ ←←←← ❄ →→→→ ❄ ←←←←←

The *first sequent calculus* was introduced by Gentzen in [69]. For the sake of comparison with structurally free logics, we briefly recall the sentential part of the calculus LK. (We do not use the original symbols everywhere; rather, notationally, we bring LK closer to LC^*.)

The *language* contains $\wedge$, $\vee$, $\supset$ and $\neg$ as connectives, and propositional variables. Formulas are defined in the usual way, and $A, B, C, \ldots$ are meta-variables for formulas. Capital Greek letters refer to *sequences* of formulas, and the $,$ indicates joining sequences together. (Sequences of formulas are n-tuples of formulas, without the $\langle\ \rangle$ brackets.) For instance, B, Δ is the sequence in which B is concatenated to the beginning of Δ; Γ, Δ is the concatenation of the sequences Γ and Δ in that order. A sequence of formulas may be *empty*.

A *sequent* consists of an *antecedent*, a *succedent* and $\vdash$. (*Sequences* and *sequents* are not the same in LK, and the two words are not spelling variants of each other.) Both the antecedent and the succedent are sequences of formulas, and either one may be the empty sequence.

The *rules* are divided into three groups: structural, cut and connective rules. *LK* contains only *one axiom*: $A \vdash A$.

The *structural rules* are:

$$\frac{\Gamma \vdash \Theta}{A, \Gamma \vdash \Theta}\ \text{thin-A} \qquad \frac{\Gamma \vdash \Theta}{\Gamma \vdash \Theta, A}\ \text{thin-S}$$

$$\frac{A, A, \Gamma \vdash \Theta}{A, \Gamma \vdash \Theta}\ \text{con-A} \qquad \frac{\Gamma \vdash \Theta, A, A}{\Gamma \vdash \Theta, A}\ \text{con-S}$$

$$\frac{\Gamma, A, B, \Delta \vdash \Theta}{\Gamma, B, A, \Delta \vdash \Theta}\ \text{int-A} \qquad \frac{\Gamma \vdash \Theta, A, B, \Delta}{\Gamma \vdash \Theta, B, A, \Delta}\ \text{int-S.}$$

The abbreviations are deciphered as *thinning*, *contraction* and *interchange*. The attached A or S indicate the place of the applicability of the structural rule, namely, either in the antecedent or in the succedent.

The *cut rule* is formulated as

$$\frac{\Gamma \vdash \Theta, A \qquad A, \Delta \vdash \Xi}{\Gamma, \Delta \vdash \Theta, \Xi}\ \text{cut.}$$

This is the so-called *single cut rule*, however, the formulation differs from the cut rule in *LC**, because the succedent may contain more than one formula (and in *LC**, the antecedent and the succedent have more structure than a string has).

The *connective rules* are:

$$\frac{A, \Gamma \vdash \Theta \quad B, \Gamma \vdash \Theta}{A \vee B \vdash \Theta}\ \vee\vdash \qquad \frac{\Gamma \vdash \Theta, A}{\Gamma \vdash \Theta, A \vee B}\ \vdash\vee \qquad \frac{\Gamma \vdash \Theta, B}{\Gamma \vdash \Theta, A \vee B}\ \vdash\vee$$

$$\frac{A, \Gamma \vdash \Theta}{A \wedge B, \Gamma \vdash \Theta}\ \wedge\vdash \qquad \frac{B, \Gamma \vdash \Theta}{A \wedge B, \Gamma \vdash \Theta}\ \wedge\vdash \qquad \frac{\Gamma \vdash \Theta, A \quad \Gamma \vdash \Theta, B}{\Gamma \vdash \Theta, A \wedge B}\ \vdash\wedge$$

$$\frac{\Gamma \vdash \Theta, A}{\neg A, \Gamma \vdash \Theta}\ \neg\vdash \qquad \frac{A, \Gamma \vdash \Theta}{\Gamma \vdash \Theta, \neg A}\ \vdash\neg$$

$$\frac{\Gamma \vdash \Theta, A \quad B, \Delta \vdash \Xi}{A \supset B, \Gamma, \Delta \vdash \Theta, \Xi}\ \supset\vdash \qquad \frac{A, \Gamma \vdash \Theta, B}{\Gamma \vdash \Theta, A \supset B}\ \vdash\supset.$$

The definitions of a *derivation* and of a *proof* are usual. There is no need in *LK* for a special constant like $\boldsymbol{t}$ in *LC**, because the antecedent is permitted to be empty.

The structural connective , may occur both on the left- and on the right-hand side of the $\vdash$ symbol. The rules for $\wedge$ and $\vee$ though do not involve the structural connective, because the occurrences of the , only *situate* the formula which becomes a subformula of a conjunction or disjunction within the antecedent or succedent. Despite the lack of an apparent connection, the structural connective can be interpreted as $\wedge$ on the left- and as $\vee$ on the right-hand side of the $\vdash$. The structural rules on the two sides of the $\vdash$ make the structural connective commutative and idempotent. In addition, , is associative to start with. These are some of the properties that conjunction and disjunction have in classical logic. The following two rules are *derivable* in *LK*.

$$\frac{\Gamma, A, B, \Delta \vdash \Theta}{\Gamma, A \wedge B, \Delta \vdash \Theta} \qquad \frac{\Gamma \vdash \Delta, A, B, \Theta}{\Gamma \vdash \Delta, A \vee B, \Theta}$$

The original proof of the *admissibility of the cut rule* (or if included, then its eliminability) was proved via the admissibility of the mix rule, which can be proved to be equivalent to the cut rule in *LK*. The importance of the cut theorem is underscored by its original name, "main theorem" (Hauptsatz, in German). The *mix rule* is the following rule in which * has a special meaning.

$$\frac{\Gamma \vdash \Theta \qquad \Delta \vdash \Xi}{\Gamma, \Delta^* \vdash \Theta^*, \Xi} \text{ mix}$$

Δ^* and Θ^* are like their nonstarred versions except that each of Δ and Θ must contain at least one occurrence of a formula M, called *the mix formula*, and the starred versions are void of all occurrences of M.

The proof of the cut theorem requires careful definitions, and it has a double inductive structure.[20] Instead of getting into the details — no matter how exciting those are — we only briefly sketch the idea behind the proof. (We take for granted that the equivalence of the cut and mix rules has been proved.)

If a proof contains an application of the mix rule, then either (1) that mix rule is completely eliminated from the proof (when it involves the axiom), (2) that mix rule is moved toward the top of the proof by permuting the mix with the rule immediately above or (3) that mix rule is replaced by one or more mix rules in which the mix formula is simpler (i.e., contains fewer connectives). In each case, it is shown that the whole proof can be transformed accordingly into a proof.

In LC^* and many other nonclassical logics, mix is *not equivalent* to the cut rule because of the absence of some of the structural rules. On the other hand, the proof of the admissibility of the single cut rule with double induction runs into a difficulty if there is a *contraction-like rule* in the calculus. The following rule is called *multiple cut rule*, because it allows several occurrences of the cut formula in the right premise to be replaced by the antecedent of the left premise.

$$\frac{\Gamma \vdash A \qquad \Delta\langle A \rangle \vdash \Theta}{\Delta\langle \Gamma \rangle \vdash \Theta} \text{ cut}$$

Here $\langle\,\rangle$ indicates that there is at least one occurrence of A in Δ that is selected, and in the lower sequent, all those selected occurrences are replaced by Γ.

The single cut rule is equivalent to the multiple cut rule in *LK*, and with Γ, Δ and Θ reinterpreted as structures, also in LC^*. This is an equivalence with respect to provable sequents, and it does not imply that the rules are interchangeable within a proof of the Hauptsatz too.

The LC^* *systems* differ from *LK* in the set of connectives their languages include, and also in their notion of a sequent, which comprises a structure, $\vdash$ and a formula.

[20]For a direct inductive proof of the admissibility of the single cut (not of the mix) in *LK*, see [22].

Structures are introduced in definition 7.3.2; they cannot be empty, hence, a sequent always contains at least one formula on the left- and one formula on the right-hand side of the $\vdash$. Structures have many similar properties to those that CL-terms have — and this is intended.

DEFINITION A.7.2. (SUBSTRUCTURES) The *set of substructures* of a structure $\mathfrak{A}$, denoted by $\mathrm{subst}(\mathfrak{A})$, is defined recursively by (1)–(2).

(1) If $\mathfrak{A}$ is A, a wff, then $A \in \mathrm{subst}(\mathfrak{A})$;

(2) if $\mathfrak{A}$ is $(\mathfrak{B};\mathfrak{C})$, then $\mathrm{subst}(\mathfrak{A}) = \{\mathfrak{A}\} \cup \mathrm{subst}(\mathfrak{B}) \cup \mathrm{subst}(\mathfrak{C})$.

A structure $\mathfrak{A}$ has an occurrence in $\mathfrak{B}$ just in case $\mathfrak{A} \in \mathrm{subst}(\mathfrak{B})$.

DEFINITION A.7.3. (FORMATION TREES FOR STRUCTURES) The *formation tree* of a structure $\mathfrak{A}$ is a binary ordered tree that is defined by (1)–(2).

(1) If $\mathfrak{A}$ is a wff (i.e., an A), then the formation tree of $\mathfrak{A}$ comprises the node that is the root and is labeled by A;

(2) if $\mathfrak{A}$ is $(\mathfrak{B};\mathfrak{C})$, then the formation tree comprises all the nodes in the trees of $\mathfrak{B}$ (on the left) and $\mathfrak{C}$ (on the right), together with a new node for root, which is labeled by $(\mathfrak{B};\mathfrak{C})$.

We do not want to identify the nodes in a formation tree with their labels, because a structure may have identically looking substructures in different locations. Given the formation trees, we can define *indexing* similarly to the indexing of CL-terms in definition A.1.7.

DEFINITION A.7.4. Let $\mathfrak{A}$ be a structure. *Indexes* are assigned to structures that appear as labels of the nodes in the formation tree of $\mathfrak{A}$ according to clauses (1) and (2).

(1) The root is labeled by 0;

(2) if a node is labeled by $(\mathfrak{B};\mathfrak{C})$ with index $_i$, then the nodes $\mathfrak{B}$ and $\mathfrak{C}$ are indexed with $_{i0}$ and $_{i1}$, respectively.

A structure has a *unique* formation tree, and the nodes in the formation tree receive a *unique* label via the indexing. This permits the occurrences of a structure in a structure to be identified. Each node in the formation tree of a structure is the root of a subtree.

Replacement then simply means replacing a subtree with a subtree together with reindexing the new subtree. (The root of the new subtree is indexed as the root of the old subtree, and then (2) is applied as long as it is applicable.) The leaves of the formation tree of a structure are formulas, and so the replacement of an occurrence of a formula by a structure is a special case of the replacement of a structure by a structure.

DEFINITION A.7.5. A structure comprising $\mathfrak{B}_1,\ldots,\mathfrak{B}_n$ in this order is a *left-associated* structure whenever it is of the form $(\ldots(\mathfrak{B}_1;\mathfrak{B}_2);\ldots;\mathfrak{B}_n)$. Similarly, the structure of the same components is a *right-associated* structure whenever it is of the form $(\mathfrak{B}_1;\ldots;(\mathfrak{B}_{n-1};\mathfrak{B}_n)\ldots)$.

As a notational convention, we omit parentheses from left-associated structures.

We rarely need to make precise the internal structure of structures. It is convenient to use *meta-variables* to describe a CL-term or a wff to a certain level of detail. In this streak, definition A.1.20 allowed us to generate a set of CL-terms, given a set of terms, which were taken to be the atoms with respect to the newly generated terms.

DEFINITION A.7.6. (STRUCTURES OVER $\mathbb{B}$) Let $\mathbb{B}$ be a (finite) set of structures $\{\mathfrak{A}_1,\ldots,\mathfrak{A}_n\}$. The set of *structures over* $\mathbb{B}$ is denoted by $\mathbb{B}^\oplus$, and this set is inductively defined by (1) and (2).

(1) $\mathbb{B}\subseteq\mathbb{B}^\oplus$;

(2) if $\mathfrak{C}\in\mathbb{B}^\oplus$ and $\mathfrak{D}\in\mathbb{B}^\oplus$, then $(\mathfrak{C};\mathfrak{D})\in\mathbb{B}^\oplus$.

A.8 Applied combinatory logic

First-order logic may be defined based on a language from a range of languages that share certain properties, but some other components in the language are optional. We start with a "minimal" language.

DEFINITION A.8.1. (FIRST-ORDER LANGUAGE, I) A *first-order language* contains (1)–(4).

(1) A denumerable set of *variables*, $\{x_i : i\in\mathbb{N}\}$;

(2) at least one *predicate symbol*, for instance, P;

(3) a functionally complete set of *connectives*, for instance, $\{\neg,\wedge,\vee\}$;

(4) a quantifier, for instance, $\forall$.

The set of *well-formed formulas*, denoted by *wff*, is inductively defined by (5)–(7).

(5) If P is an n-place predicate symbol, and $y_1,\ldots,y_n\in\{x_i\}_{i\in\mathbb{N}}$, then $P(y_1,\ldots,y_n)$ is in wff;

(6) if $A,B\in$ wff, then $\neg A,(A\wedge B),(A\vee B)\in$ wff;

(7) if $A\in$ wff and $y\in\{x_i\}_{i\in\mathbb{N}}$, then $\forall yA\in$ wff.

$\forall$ is the *universal quantifier*. From a syntactic point of view, it is similar to λ: it binds the variable that comes immediately after it, as well as all the free occurrences of the same variable in A. Accordingly, the notions of *free* and *bound occurrences* of variables, of sets of free and bound variables, of closed formulas may be defined similarly to those notions for CL-terms and Λ-terms. (See section A.1, and especially, section A.4.)

Often a "richer" language is used — if for no other reason, for convenience. The following definition of a first-order language includes further kinds of expressions.

DEFINITION A.8.2. (FIRST-ORDER LANGUAGE, II) A *first-order language* contains (1)–(6).

(1) A denumerable set of *variables*, $\{x_i \colon i \in \mathbb{N}\}$;

(2) a denumerable set of *name constants*, $\{a_i \colon i \in \mathbb{N}\}$;

(3) a denumerable set of *function symbols*, $\{f_i \colon i \in \mathbb{N}\}$;

(4) a denumerable set of *predicate symbols*, $\{P_i \colon i \in \mathbb{N}\}$, containing a two-place predicate symbol called *identity*, which may be denoted also by $=$;

(5) a functionally complete set of *connectives*, for instance, $\{\neg, \wedge, \vee, \supset, \equiv\}$;

(6) two quantifiers, $\forall$ and $\exists$.

The set of *terms*, denoted by *trm*, is inductively defined by (7)–(9).

(7) If $y \in \{x_i\}_{i \in \mathbb{N}}$, then $y \in \text{trm}$;

(8) if $b \in \{a_i\}_{i \in \mathbb{N}}$, then $b \in \text{trm}$;

(9) if $g \in \{f_i\}_{i \in \mathbb{N}}$ an n-ary function symbol and $t_1, \ldots, t_n \in \text{trm}$, then $f(t_1, \ldots, t_n) \in \text{trm}$.

The set of *well-formed formulas*, denoted by *wff*, is inductively defined by (10)–(12).

(10) If P is an n-ary predicate symbol and $t_1, \ldots, t_n \in \text{trm}$, then $P(t_1, \ldots, t_n) \in \text{wff}$;

(11) if $A, B \in \text{wff}$, then $\neg A, (A \wedge B), (A \vee B), (A \supset B), (A \equiv B) \in \text{wff}$;

(12) if $y \in \{x_i\}_{i \in \mathbb{N}}$ and $A \in \text{wff}$, then $\forall y A, \exists y A \in \text{wff}$.

It is not feasible to define a "maximal" first-order language in the sense that all other first-order languages are subsets of this purported maximal language. There are infinitely many connectives and operators that are definable in first-order logic, and any of them could be added to the language. The connectives and the variable binding operators are *logical constants*, that is, they have a *fixed interpretation*. We can specify a handful of them using their (more or less) standard notation; however, denoting large numbers of them would require a notation that would, in effect, give the interpretation of each of those logical constants.

In the late 19th century, *Charles S. Peirce* discovered the truth-functional interpretation of classical sentential logic. He also discovered two binary truth functions each of which is sufficient alone to express all others.

Peirce first considered *joint denial.* Let $A \dagger B$ mean that neither A, nor B is true. Using our notation, $A \dagger B$ has the same meaning as $\neg(A \vee B)$. In computer science, this connective is usually called NOR, and we gave its truth matrix at the end of section 3.2.

The Boolean dual of NOR is NAND, which is usually denoted by $|$, and called "Sheffer's stroke." $A \mid B$ means that either not A or not B is true. The names of these connectives as NOR and NAND are mnemonic, when N abbreviates negation. $A \mid B$ may be expressed by other connectives as $\neg(A \wedge B)$.

A set of connectives is *functionally complete* when any 2-valued truth function can be expressed by a wff that contains propositional variables and connectives from the set. This statement glides over the distinction between a connective and the truth function that interprets that connective. However, the connectives correspond straightforwardly to truth functions, and so hopefully, no confusion arises.

Truth tables provide a systematic way to represent truth functions. Given a truth table with reference columns for $P_1, \ldots, P_n$, the truth function represented by the truth table may be easily described by a wff in *disjunctive normal form* using $\neg, \wedge$ and $\vee$.[21] The complete truth table with reference columns for n propositional variables contains 2^n situations (i.e., rows). These situations are *mutually exclusive* and *collectively exhaustive*. If the truth function is true in one of these situations, then a conjunction of true propositional variables with the negation of false propositional variables is added to the disjunction that expresses the truth function.

Example A.8.3. The truth table for $A \supset B$ may be described by the formula $(A \wedge B) \vee (\neg A \wedge B) \vee (\neg A \wedge \neg B)$.[22] The three conjunctions in the disjunction stand for the three situations where $A \supset B$ is true.

The procedure we described and then illustrated by an example is applicable to any truth functions of n variables except in one case. The exception is when the function represented by the truth table never takes the value true. However, this function is definable uniformly by $(P_1 \wedge \neg P_1) \vee \ldots \vee (P_n \wedge \neg P_n)$. In sum, all functions given by a truth table can be described by a formula that contains only propositional variables (on which the truth function depends) and some of the connectives $\neg, \wedge$ and $\vee$. Definition A.8.1 prognosticated the following lemma.

LEMMA A.8.4. *The set of connectives* $\{\neg, \wedge, \vee\}$ *is* functionally complete.

The connective $\dagger$ has a fixed interpretation as a truth function. We described this interpretation utilizing $\neg$ and $\vee$.

Example A.8.5. The connective $\dagger$ itself is sufficient to express $\neg$ and $\vee$ (without using the other). $\neg A$ is true exactly when A is false. $A \dagger A$ is true when both

[21] Propositional variables may be viewed as zeroary predicate symbols.

[22] A truth matrix defining the truth function $\supset$ may be found at the end of section 3.2.

arguments of $\dagger$ are false; but both of those are A, that is, $A \dagger A$ is true just in case A is false.

If $A \dagger B$ is $\neg(A \vee B)$, then $\neg(A \dagger B)$ is $\neg\neg(A \vee B)$. However, in classical logic, double negations are eliminable, which means that the latter formula is logically equivalent to $A \vee B$. Then $A \vee B$ is definable as $(A \dagger B) \dagger (A \dagger B)$.

The connective $|$ is sufficient to express $\neg$ and $\wedge$ similarly. The so-called *De Morgan laws* — (1) and (2) below — may be used to show that $\wedge$ or $\vee$, but not both can be omitted from $\{\neg, \wedge, \vee\}$ without reducing the expressive power of the language. (3) is *double negation* elimination and introduction.

(1) $\neg(A \vee B)$ is equivalent to $\neg A \wedge \neg B$;

(2) $\neg(A \wedge B)$ is equivalent to $\neg(A \vee B)$;

(3) $\neg\neg A$ is equivalent to A.

LEMMA A.8.6. *The sets of connectives* $\{\dagger\}$ *and* $\{|\}$ *are* functionally complete.

It is well-known that the quantifiers cannot be eliminated in favor of connectives. The domain of quantification may contain elements that are *not named* by any name constant in the language. For instance, definition A.8.1 does not require that any name constants are included in the language. Even if a first-order language would contain sufficiently many names for each object in the domain of quantification, the domain may be infinite, which would require forming infinitely long formulas in lieu of quantified ones.

Example A.8.7. To express that all natural numbers less than 3 have property P, we can use the wff $\forall x\,(x < 3 \supset P(x))$, where $<$ is a two-place predicate and P is a one-place predicate (and we assume that all objects in the domain are natural numbers). We may have 0, 1 and 2 as name constants in the language with the intended interpretation that they denote 0, 1 and 2, respectively. Then the universally quantified sentence can be replaced by

$$(0 < 3 \supset P(0)) \wedge (1 < 3 \supset P(1)) \wedge (2 < 3 \supset P(2)),$$

which may be further simplified to $P(0) \wedge P(1) \wedge P(2)$.

On the other hand, to express that all natural numbers not less than (i.e., greater than or equal to) 3 have property P, we can only use the universally quantified formula $\forall x\,(\neg x < 3 \supset P(x))$. We could consider starting the wff with

$$(\neg 3 < 3 \supset P(3)) \wedge (\neg 4 < 3 \supset P(4)) \wedge (\neg 5 < 3 \supset P(5)) \wedge (\neg 6 < 3 \supset P(6))\ldots$$

However, the ellipsis indicates infinitely many subformulas of the form $(\neg a < 3 \supset P(a))$ — even if we assume that we have a name constant for each natural number.

A.9 Typed combinatory logic

Mathematical objects are abstract entities. Representation theorems and axiomatic theories underscore the idea that intended interpretations and preferred views do not characterize objects uniquely. To give a simple example, we may have a somewhat informal idea of ordered pairs. We also have the notation $\langle a,b\rangle$ for the ordered pair comprising a and b in that order. The ordered pairs $\langle a_1,b_1\rangle$ and $\langle a_2,b_2\rangle$ are *identical* whenever $a_1 = a_2$ and $b_1 = b_2$. Various sets can function as the ordered pair $\langle a,b\rangle$, for instance, $\{\{a\},\{a,b\}\}$ or $\{\{a,0\},\{b,1\}\}$. Treating ordered pairs in one or another way may be convenient or aesthetically pleasing, but there is *nothing inherent* in ordered pairs that would allow us to choose "the one true way" to think of ordered pairs as sets. The usual expression "set-theoretical representation" reflects this point of view deftly.

Mathematical objects can be viewed "even more abstractly," so to speak, as being characterized by the *maps* or *transformations* (and the properties of the latter) that are admissible for them. For example, a line in an affine geometry is an object of which it can be stated that it is parallel to another line in the plane when they have no common points, whereas a line in a projective geometry always intersects all other lines in the plane. The objects in affine versus projective geometry are characterized by the sets of affine and projective transformations — notwithstanding what they are called or how they are visualized.

Category theory takes as its primary objects of study maps or functions.[23]

DEFINITION A.9.1. (CATEGORY) A *category* comprises $\mathbb{M}$, a set of *maps*, and $\mathbb{O}$, a set of *objects* and a partial *operation* $\circ$ that satisfy (1)–(5).

(1) For any $f \in \mathbb{M}$, there are exactly one $A \in \mathbb{O}$ and exactly one $B \in \mathbb{O}$ such that $f\colon A \longrightarrow B$;

(2) for any $A \in \mathbb{O}$, there is an $i_A \in \mathbb{M}$;

(3) for any $f,g \in \mathbb{M}$, if $f\colon A \longrightarrow B$ and $g\colon B \longrightarrow C$, then $g \circ f \in \mathbb{M}$ and $g\colon A \longrightarrow C$;

(4) for any $f \in \mathbb{M}$, if $f\colon A \longrightarrow B$, then $f \circ i_A = f$ and $i_B \circ f = f$;

(5) for any $f,g,h \in \mathbb{M}$, if $h \circ (g \circ f) \in \mathbb{M}$, then so is $(h \circ g) \circ f$, and vice versa; further, $h \circ (g \circ f) = (h \circ g) \circ f$.

We used a bit of set notation in the definition, but maps should not be thought of as sets of ordered pairs — indeed such an understanding is (strongly) discouraged. *Maps* are functions or operations, and each has a pair of objects assigned to them.

[23]The term "object" is used for the labels attached to "arrows." Nonetheless, the main interest lies with studying the maps.

The latter are the *domain* and *codomain* of the map, what we intended to suggest by the notation $\longrightarrow$ in clause (1). Note that the domain and the codomain are unique for each map. When a function is a set of ordered pairs, it makes perfect sense to extend the domain of a function to a larger set, perhaps, changing a total function into a partial one. In a category, we are not given an order relation on $\mathbb{O}$, which means that a map simply has a domain and a codomain.[24]

Each object A has an *identity map* i_A. That is, the set of objects could be recovered without loss from the set of maps by looking at the identity maps.

Maps with matching codomains and domains *compose*. The *composition operation* is denoted by $\circ$, which is the usual notation for function composition. Clause (5) stipulates that this operation is *associative*, and (4) means that there are left and right identities for this operation, namely, the maps that are introduced in (2).

Example A.9.2. Let $\mathcal{M} = \langle \{A\}, \{f_k\colon k \in K\} \rangle$ be a category. Since $\{A\}$ is a singleton, $\forall f_{k\in K}. f_k\colon A \longrightarrow A$. $\{f_k\colon k \in K\}$ may *not* be a singleton. $\{f_k\colon k \in K\} \neq \emptyset$, because i_A is an element. We can see that i_A is *unique*: $i_A = i_A \circ i'_A = i'_A$ by (4), applied twice. However, in general, $f_k \circ f_l \neq f_k$, and $f_k \circ f_l \neq f_l$ either.

The category $\mathcal{M}$ is quite special, because it has only one object, but it shows how a *monoid* can be viewed as a category. Examples of monoids include the set of strings over an alphabet with ε, the empty string being the identity element (see definition A.1.18), and meet semilattices with $\top$, the top being the identity element (see definition A.9.7).

DEFINITION A.9.3. (MONOIDS) The algebra $\mathfrak{A} = \langle A, \circ, 1 \rangle$ of similarity type $\langle 2, 0 \rangle$ is a *monoid* iff (1) and (2) hold for all elements a, b and c of A.

(1) $(a \circ b) \circ c = a \circ (b \circ c)$

(2) $1 \circ a = a = a \circ 1$

Universal algebra studies monoids as a variety, that is, an equational class of algebras, where the equations are those in (1) and (2) above. The *category of monoids* is the set of all monoids with *monoid homomorphisms* as maps. A *homomorphism* preserves the operations; a monoid homomorphism, in particular, preserves $\circ$ and 1. Given $\mathfrak{A}_1 = \langle A_1, \circ_1, 1_1 \rangle$, $\mathfrak{A}_2 = \langle A_2, \circ_2, 1_2 \rangle$ and $h\colon A_1 \longrightarrow A_2$, $h(a \circ_1 b) = ha \circ_2 hb$ and $h1_1 = 1_2$ should hold for h to be a monoid homomorphism.

Another example of a category is *partially ordered sets* with *monotone maps*. The last two examples illustrate the idea that the sort of maps that are included in the category typify the objects. Then, unsurprisingly, the category of *topological spaces* contains *continuous functions* as maps.

Categories themselves can be viewed as objects. If maps can be found between categories, then this may allow back-and-forth moves between categories. *Duality*

[24]This does not contradict that if there are special maps in the category of sets then, for instance, the subset relation can be emulated.

theorems show the usefulness of such moves, when a problem is recast as a question about other sorts of objects.

We have put a great emphasis on the role of maps in categories. Then it may bewilder us why category theory is not combinatory logic or what is the difference between them. In type-free CL, there are no two separate sorts of entities, whereas the (co)domains and the maps are disjoint sets in a category. CL has function application, which is a binary operation, whereas a category has another binary operation, function composition. Their similarities end with both operations being binary; function application is certainly not associative, and function composition is not assumed to be a total operation on the set of maps. (The category $\mathcal{M}$ in example A.9.2 was exceptional in this respect.) Maps in a category are always *unary*; combinators, on the other hand, have a *fixed positive* arity.

Set theory gives a different view of functions. An n-ary function is a set of $n+1$-tuples. If f is a function of type $A_1 \times \ldots \times A_n \longrightarrow A_{n+1}$, then $f \subseteq A_1 \times \ldots \times A_n \times A_{n+1}$, where $f(a_1, \ldots, a_n) = b$ and $f(a_1, \ldots, a_n) = c$ imply that $b = c$.

A *domain* and a *codomain* (also called *image*) may be defined, respectively, as $\mathrm{dom}(f) = \{\langle a_1, \ldots, a_n \rangle \colon \exists b. f(a_1, \ldots, a_n) = b\}$ and $\mathrm{cod}(f) = \{b \colon \exists \langle a_1, \ldots, a_n \rangle . f(a_1, \ldots, a_n) = b\}$. Alternatively, the domain and codomain may be stated, as we did above, by specifying the type of the function.

If dom and cod are defined from the set of tuples, then the function is *total* and *onto* (i.e., *surjective*) for the sets obtained. If the type of the function is simply given, then it can naturally emerge that the function is not total or onto.

Example A.9.4. The function $/$ (i.e., division) is conveniently described as of type $/ \colon \mathbb{R} \times \mathbb{R} \longrightarrow \mathbb{R}$, but the second argument should come from $\mathbb{R} - \{0\}$. Or $f(x) = x^2$ (i.e., $\lambda x.x \cdot x$) can be viewed as a function of type $\mathbb{Z} \longrightarrow \mathbb{Z}$, but more narrowly, the image can be taken as $\mathbb{N}$.

A partial function can be turned into a total one not only by delineating a subset of its domain, but also by introducing a *new* object into the codomain as the value for all elements for which the function was undefined previously. Fixing up a non-surjective function might require a bigger addition to the domain: for each element of the declared codomain that is not a value for any element of the domain, a new element has to be included into the domain. The reason why it is preferable to deal with $/$ as a *partial function* rather than to have an extra element added to $\mathbb{R}$ is, probably, the widespread use of $\mathbb{R}$ as it is — without exotic intruders.

Sets come with $\subseteq$, the subset relation, which is straightforwardly definable given the extensional understanding of sets. Then it becomes justifiable to blur functions or to select the domain and codomain of a function according to needs.

Example A.9.5. $+_{\mathbb{N}}$, $+_{\mathbb{Z}}$ and $+_{\mathbb{Q}}$ are alike, because $+_{\mathbb{Q}} \restriction \mathbb{Z} = +_{\mathbb{Z}}$ and $+_{\mathbb{Z}} \restriction \mathbb{N} = +_{\mathbb{N}}$. Thus we often simply use $+$.

$+_{\mathbb{N}}$ and $+_{\mathrm{mod}\,2}$ on $\mathbb{Z}$ cannot be blended together similarly, because $1 +_{\mathbb{N}} 3 = 4$ but $1 +_{\mathrm{mod}\,2} 3 = 0$.

The more relaxed approach to the type of a function contrasts functions in set theory with maps in categories. The existence of a type (or domain–codomain pair) for a function contrasts functions in set theory with functions in CL. Another difference between functions in set theory and CL is the lack of self-application in the former — at least in the more commonly used systems of set theory where a set cannot be its own member. The axiom of foundation excludes this possibility, thereby, excluding ordered pairs $\langle x,y \rangle$ where x is itself $\langle x,y \rangle$.

→→→→→❄←←←←❄→→→→❄←←←←←

Natural deduction systems for classical and intuitionistic logic were introduced by Gentzen in [69]. Natural deduction systems are intended to mimic the natural steps of reasoning that a mathematician uses.

There are many variations on natural deduction systems. As an example, we describe briefly a formulation for propositional *intuitionistic* logic.

The *language* contains four logical constants and a denumerable set of *propositional variables*. Let us assume that $\mathbb{P}$ is a nonterminal symbol that can be rewritten as a propositional variable. Then the following context-free grammar defines the set of formulas (as the set of strings generated).

$$A ::= \mathbb{P} \mid \bot \mid (A \wedge A) \mid (A \vee A) \mid (A \supset A),$$

where $\bot$ is *contradiction*, a nullary logical constant; $\wedge$ and $\vee$ are *conjunction* and *disjunction*, and $\supset$ is *implication*. (Negation is defined as: $\neg A$ is $(A \supset \bot)$.)

The system comprises rules that split into *introduction* and *elimination* rules. The introduction rules are the following four. (I abbreviates introduction.)

$$\frac{A \quad B}{A \wedge B}\,\wedge I \qquad \frac{A}{A \vee B}\,\vee I \qquad \frac{B}{A \vee B}\,\vee I \qquad \frac{\begin{array}{c}[A]\\ \vdots \\ B\end{array}}{A \supset B}\,\supset I$$

The bracketing of A shows that the assumption A has been *discharged* or *canceled*. *Multiple occurrences* of an assumption may be canceled at once, and *vacuous* discharges are permitted too. There is no introduction rule for $\bot$. The elimination rules are as follows. (E stands for elimination.)

$$\frac{A \wedge B}{A}\,\wedge E \qquad \frac{A \wedge B}{B}\,\wedge E \qquad \frac{A \vee B \quad \begin{array}{c}[A]\\ \vdots \\ C\end{array} \quad \begin{array}{c}[B]\\ \vdots \\ C\end{array}}{C}\,\vee E \qquad \frac{A \quad A \supset B}{B}\,\supset E \qquad \frac{\bot}{A}\,\bot E$$

A *derivation* is a tree where the leaves are the assumptions, and the formulas in the other nodes of the tree are obtained according to the rules. A derivation is a *proof* when all the assumptions have been discharged.

<u>Example</u> A.9.6. The proof of $(A \supset A \supset B) \supset A \supset B$ shows that the last but one step

cancels both A's.

$$\dfrac{\dfrac{\dfrac{\dfrac{[A \supset A \supset B] \qquad [A]}{A \supset B} \qquad [A]}{B}}{A \supset B}}{(A \supset A \supset B) \supset A \supset B}$$

A vacuous cancellation is illustrated by the next proof, where the second formula results by discharging B.

$$\dfrac{\dfrac{[A]}{B \supset A}}{A \supset B \supset A}$$

Lately, natural deduction systems are more often presented in a "horizontal" style. This rendering makes a natural deduction system superficially look like a sequent calculus. Nonetheless, the two types of calculi are different; the rules in a natural deduction system are elimination and introduction rules rather than left- and right-introduction rules. ($\Gamma, \Delta, \ldots$ are sets of wff's. Γ, Δ abbreviates $\Gamma \cup \Delta$, and Γ, A is $\Gamma \cup \{A\}$.)

$$\dfrac{\Gamma \vdash A \quad \Delta \vdash B}{\Gamma, \Delta \vdash A \wedge B} \wedge I \qquad \dfrac{\Gamma \vdash A}{\Gamma \vdash A \vee B} \vee I \qquad \dfrac{\Gamma \vdash B}{\Gamma \vdash A \vee B} \vee I \qquad \dfrac{\Gamma \vdash B}{\Gamma - \{A\} \vdash A \supset B} \supset I$$

$$\dfrac{\Gamma \vdash A \wedge B}{\Gamma \vdash A} \wedge E \qquad \dfrac{\Gamma \vdash A \wedge B}{\Gamma \vdash B} \wedge E \qquad \dfrac{\Gamma \vdash A \vee B \qquad \Delta, A \vdash C \qquad \Theta, B \vdash C}{\Gamma, \Delta, \Theta \vdash C} \vee E$$

$$\dfrac{\Gamma \vdash A \supset B \qquad \Delta \vdash A}{\Gamma, \Delta \vdash B} \supset E \qquad \dfrac{\Gamma \vdash \bot}{\Gamma \vdash A} \bot E$$

→→→→→ ❄ ←←←← ❄ →→→→ ❄ ←←←←←

Classical sentential logic can be presented in many ways because the *two-valued* truth functional connectives are often interdefinable. But even in classical logic, the conditional connective by itself cannot express all the other connectives, which means that the implicational fragment of classical logic differs from classical logic as a whole. In *nonclassical* logics, the connectives are typically "even less" interdefinable. For the sake of easy comparison between the logics that we mentioned in chapter 9 (and passim in other chapters), we give a list of implicational formulas various selections from which give various implicational logics. (Strictly speaking, each logic has its own language, but for simplicity, we assume the identity translation.) On the far right, we also give one of the popular names for each wff.

(A1) $((A \to B) \to A) \to A$ [Peirce's law]

(A2) $A \to B \to A$ [positive paradox]

(A3) $(A \to B \to C) \to (A \to B) \to A \to C$ [self-distribution of $\to$ on the major]

(A4) $A \to A$ [self-implication]

(A5) $(A \to B \to C) \to B \to A \to C$ [permutation]

(A6) $(A \to A \to B) \to A \to B$ [contraction]

(A7) $(A \to B) \to (B \to C) \to A \to C$ [suffixing]

(A8) $(A \to B) \to (C \to A) \to C \to B$ [prefixing]

(A9) $(A \to B) \to (A \to B \to C) \to A \to C$ [self-distribution of $\to$ on the minor]

(A10) $(A \to B) \to ((A \to B) \to C) \to C$ [restricted assertion]

(A11) $(A \to (B \to C) \to D) \to (B \to C) \to A \to D$ [restricted permutation]

(A12) $(A \to B) \to (C \to (A \to B) \to D) \to C \to D$ [restricted conditional MP]

(A13) $((A \to A) \to B) \to B$ [specialized assertion]

Implicational fragments of various logics can be axiomatized by the following sets of axioms — together with the rule of *modus ponens* (or *detachment*): A and $A \to B$ imply B.

Classical (two-valued) logic, $TV_{\to}$**.** (A1), (A2) and (A3). All the other formulas listed are its theorems. An interesting alternative axiomatization is (A1), (A2) and (A7), which at once shows that (A1) and (A2) are very productive axioms, and (A7) is rather powerful. (A7) is in fact more powerful than the closely related (A8) is.

Intuitionistic logic, $H_{\to}$**.** (A2) and (A3). All the other formulas save (A1) are its theorems.

Logic of relevant implication, $R_{\to}$**.** (A4), (A5), (A6) and (A8). All the other formulas except (A1) and (A2) are its theorems. Notice that the axioms that are not theorems are exactly the two formulas that we previously called very productive. (Alternatively, we could have called them wildly permissive.)

Entailment, $E_{\to}$**.** (A3), (A4), (A7) and (A10). The nontheorems of $E_{\to}$ are (A1), (A2) and (A5). The difference between (A5) and (A11) is that the latter requires the second antecedent in the antecedent of the wff to be an implicational formula, that is, an entailment; hence (A11) is the restricted version of (A5).

Logic of ticket entailment, $T_{\to}$**.** (A3), (A4), (A7), (A8) and (A9). The only other wff listed that is a theorem of $T_{\to}$ is (A6). The latter wff can be taken as an axiom in lieu of (A3) and (A9).

***BCI* logic,** $BCI_{\to}$**.** (A4), (A5) and (A8). The only other formula that is a theorem of $BCI_{\to}$ from among those listed is (A7).

Minimal relevant logic, $B_{\to}$**.** (A1) together with rules obtainable from (A7) and (A8). (That is, infer $(B \to C) \to A \to C$ from $A \to B$, and similarly, infer $(C \to A) \to C \to B$ from $A \to B$.)

The relationships between the sets of theorems of these logics can be pictured as shown below, where the $\to$'s indicate the $\supseteq$ relations on sets of theorems.

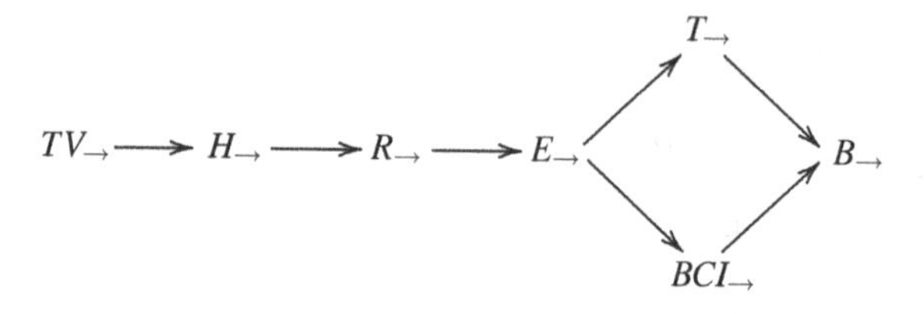

In section A.2 we gave characterizations of lattices. See definitions A.2.19 and A.2.21. In chapter 9, *meet semilattices* become important because intersection is a prototypical meet-like operation on sets.

DEFINITION A.9.7. (MEET SEMILATTICES) The algebra $\mathfrak{A}=\langle A,\wedge\rangle$ where $\wedge$ is a binary operation on A is a *meet semilattice* iff the equations (1)–(3) hold for all elements of A.

(1) $a\wedge a=a$

(2) $a\wedge b=b\wedge a$

(3) $a\wedge(b\wedge c)=(a\wedge b)\wedge c$

A meet semilattice taken in itself is completely indistinguishable from a join semilattice. We tend to read $\wedge$ as meet, but if we were to take $\wedge$ to be join, then $\mathfrak{A}$ would be a join semilattice.

The usual order relation associated with a meet semilattice is defined as $a\leq b$ iff $a\wedge b=a$. $\leq$ is a *partial order.*

Example A.9.8. A meet semilattice may *not* be bounded from below or from above. The set $\mathbb{Z}$ with min (as a binary operation) is a meet semilattice, but $\mathbb{Z}$, of course, does not have a least or a greatest element.

If card$(A)\in\mathbb{N}$, then $\mathfrak{A}$ is bounded from below. To see this, let $A=\{a_0,\ldots,a_n\}$. For any a_i,a_j, $a_i\wedge a_j\leq a_i$. Omitting parentheses — by appealing to (3) — $a_0\wedge\ldots\wedge a_n\leq a_i$ for $0\leq i\leq n$, which means that $a_0\wedge\ldots\wedge a_n$ is the least element.

A finite meet semilattice need not be bounded from above, let alone should a meet semilattice with an infinite carrier set have a top element.

DEFINITION A.9.9. (BOUNDED MEET SEMILATTICE) The algebra $\mathfrak{A}=\langle A,\wedge,\top,\bot\rangle$ of similarity type $\langle 2,0,0\rangle$ is a *bounded meet semilattice* when $\langle A,\wedge\rangle$ is a meet semilattice (i.e., (1)–(3) from the previous definition hold), and (4) and (5) are also true for any $a\in A$.

(4) $a\wedge\top=a$

(5) $\bot\wedge a=\bot$

If one of $\top$ or $\bot$ is dropped (together with (4) or (5), as appropriate), then $\mathfrak{A}$ is a meet semilattice with top or with bottom (only).

The meet operation is the binary version of the *greatest lower bound*, which is denoted by glb, inf or $\bigwedge$. (See definition A.2.20 above.) The next definition parallels — and should be compared to — definition A.2.21.

DEFINITION A.9.10. Let $\mathfrak{A} = \langle A, \leq \rangle$ be a partially ordered set.

(1) $\mathfrak{A}$ is a *meet semilattice* iff nonempty finite subsets of A have inf's.

(2) $\mathfrak{A}$ is a *meet semilattice with top* iff every finite subset of A has a inf.

(3) $\mathfrak{A}$ is a *complete lattice*, hence also a *complete meet semilattice* iff every subset of A has an inf.

Some subsets of the carrier set of a (semi)lattice are of particular importance.

DEFINITION A.9.11. (UPWARD-CLOSED SUBSETS) Let $\mathfrak{A} = \langle A, \wedge \rangle$ be a meet semilattice, and let $B \subseteq A$. B is an *upward-closed subset* of A iff

(1) $a \leq b$ and $a \in B$ imply $b \in B$.

Upward closed sets reappear in many places, essentially, as soon as there is an order relation on a set. They have been labeled variously as *cones*, *upsets* and *increasing sets*. In chapter 6, definition 6.6.5 introduced downward-directed cones, which are special cones on *arbitrary* partially ordered sets.

DEFINITION A.9.12. (DOWNWARD-DIRECTED SUBSETS) Let A be a partially ordered set, and let $B \subseteq A$. B is a *downward-directed subset* of A when (1) and (2) are true.

(1) $B \neq \emptyset$;

(2) $a, b \in B$ implies that there is a $c \in B$ such that $c \leq a$ and $c \leq b$.

If A is the carrier set of a meet semilattice, then $c \leq a$ means that $c \wedge a = c$ and $c \leq b$ means that $c \wedge b = c$. That is, B is a downward-directed subset of a meet semilattice whenever B is *closed under meet*.

DEFINITION A.9.13. (FILTERS) Let $\mathfrak{A}$ be a meet semilattice. $B \subseteq A$ is a *filter* iff (1)–(3) hold.

(1) $B \neq \emptyset$;

(2) $a \in B$ and $a \wedge b = a$ imply $b \in B$;

(3) $a, b \in B$ imply $a \wedge b \in B$.

Filters are exactly the downward-directed cones, when the partially ordered set is a meet semilattice. We may note that occasionally $\emptyset$, the empty set is counted as a filter — depending on the purpose of the definition — in which case filters are downward-directed cones or the empty set.

DEFINITION A.9.14. (PRINCIPAL CONES AND FILTERS) Let $\mathfrak{A}$ be a meet semilattice. The set $[a)$ is defined as $[a) = \{b\colon a \leq b\}$, and it is *the principal cone* generated by a. $[a)$ is a filter too, and it is called *the principal filter* generated by a.

Not all cones are filters, and not all cones are principal. In general, not all filters are principal, but in a finite meet semilattice, all (nonempty) filters are principal.

Bibliography

1. Roberto M. Amadio and Pierre-Louis Curien. *Domains and Lambda-Calculi*. Number 46 in Cambridge Tracts in Computer Science. Cambridge University Press, Cambridge, 1998.

2. Alan R. Anderson. Entailment shorn of modality (abstract). *Journal of Symbolic Logic*, 25:388, 1960.

3. Alan R. Anderson and Nuel D. Belnap. *Entailment. The Logic of Relevance and Necessity*, volume 1. Princeton University Press, Princeton, NJ, 1975.

4. Alan R. Anderson, Nuel D. Belnap, and J. Michael Dunn. *Entailment. The Logic of Relevance and Necessity*, volume 2. Princeton University Press, Princeton, NJ, 1992.

5. Franco Barbanera and Stefano Berardi. A symmetric lambda calculus for classical program extraction. *Information and Computation*, 125:103–117, 1996.

6. Franco Barbanera, Mariangiola Dezani-Ciancaglini, and Ugo de'Liguoro. Intersection and union types: syntax and semantics. *Information and Computation*, 119:202–230, 1995.

7. Franco Barbanera, Stefano Berardi, and Massimo Schivalocchi. "Classical" programming-with-proofs in $\lambda_{\mathbf{PA}}^{Sym}$: an analysis of non-confluence. In Martin Abadi and Takayasu Ito, editors, *Proceedings of the 3rd Symposium (TACS'97) held at Tohuku University, September 23–26, 1997*, number 1281 in Lecture Notes in Computer Science, pages 365–390, Berlin, 1997. Springer.

8. Henk Barendregt. Lambda calculi with types. In S. Abramsky, D. M. Gabbay, and T. S. E. Maibaum, editors, *Handbook of Logic in Computer Science. Background: Computational Structures*, volume 2, pages 117–309. Oxford University Press, New York, NY, 1992.

9. Henk Barendregt, Mario Coppo, and Mariangiola Dezani-Ciancaglini. A filter lambda model and the completeness of type assignment. *Journal of Symbolic Logic*, 48:931–940, 1983.

10. Henk P. Barendregt. *The Lambda Calculus. Its Syntax and Semantics*, volume 103 of *Studies in Logic and the Foundations of Mathematics*. North-Holland, Amsterdam, 1981.

11. Nuel D. Belnap. Display logic. *Journal of Philosophical Logic*, 11:375–417, 1982.

12. Nuel D. Belnap and John R. Wallace. A decision procedure for the system $E_{\overline{I}}$ of entailment with negation. *Zeitschrift für mathematische Logik und Grundlagen der Mathematik*, 11:277–289, 1965.

13. Katalin Bimbó. Investigation into combinatory systems with dual combinators. *Studia*

Logica, 66:285–296, 2000.

14. Katalin Bimbó. Semantics for structurally free logics $LC+$. *Logic Journal of IGPL*, 9: 525–539, 2001.

15. Katalin Bimbó. The Church-Rosser property in dual combinatory logic. *Journal of Symbolic Logic*, 68:132–152, 2003.

16. Katalin Bimbó. Semantics for dual and symmetric combinatory calculi. *Journal of Philosophical Logic*, 33:125–153, 2004.

17. Katalin Bimbó. The Church-Rosser property in symmetric combinatory logic. *Journal of Symbolic Logic*, 70:536–556, 2005.

18. Katalin Bimbó. Types of I-free hereditary right maximal terms. *Journal of Philosophical Logic*, 34:607–620, 2005.

19. Katalin Bimbó. Admissibility of cut in LC with fixed point combinator. *Studia Logica*, 81:399–423, 2005.

20. Katalin Bimbó. Curry-type paradoxes. *Logique et Analyse*, 49:227–240, 2006.

21. Katalin Bimbó. Relevance logics. In Dale Jacquette, editor, *Philosophy of Logic*, volume 5 of *Handbook of the Philosophy of Science* (D. Gabbay, P. Thagard and J. Woods, eds.), pages 723–789. Elsevier (North-Holland), Amsterdam, 2007.

22. Katalin Bimbó. $LE^{t}_{\rightarrow}$, $LR^{\circ}_{\underset{\sim}{\wedge}}$, LK and cutfree proofs. *Journal of Philosophical Logic*, 36:557–570, 2007.

23. Katalin Bimbó. Combinatory logic. In E. Zalta, editor, *Stanford Encyclopedia of Philosophy*, pages 1–54. CSLI, Stanford, CA, URL: `plato.stanford.edu/entries/logic-combinatory`, 2008.

24. Katalin Bimbó. Schönfinkel-type operators for classical logic. *Studia Logica*, 95:355–378, 2010.

25. Katalin Bimbó and J. Michael Dunn. *Generalized Galois Logics. Relational Semantics of Nonclassical Logical Calculi*, volume 188 of *CSLI Lecture Notes*. CSLI Publications, Stanford, CA, 2008.

26. Katalin Bimbó and J. Michael Dunn. Two extensions of the structurally free logic LC. *Logic Journal of IGPL*, 6:403–424, 1998.

27. Garrett Birkhoff. *Lattice Theory*, volume 25 of *AMS Colloquium Publications*. American Mathematical Society, Providence, RI, 3rd edition, 1967.

28. George S. Boolos and Richard C. Jeffrey. *Computability and Logic*. Cambridge University Press, Cambridge, UK, 3rd edition, 1992.

29. Ross T. Brady, editor. *Relevant Logics and Their Rivals. A Continuation of the Work of R. Sylvan, R. Meyer, V. Plumwood and R. Brady*, volume II. Ashgate, Burlington, VT, 2003.

30. Sabine Broda and Luis Damas. On principal types of combinators. *Theoretical Computer Science*, 247:277–290, 2000.

31. Maarten Bunder. *Set Theory Based on Combinatory Logic*. PhD thesis, University of

Amsterdam, 1969.

32. Martin Bunder. A weak absolute consistency proof for some systems of illative combinatory logic. *Journal of Symbolic Logic*, 48:771–776, 1983.

33. Martin Bunder. A simplified form of condensed detachment. *Journal of Logic, Language and Information*, 4:169–173, 1995.

34. Martin W. Bunder. Expedited Broda–Damas bracket abstraction. *Journal of Symbolic Logic*, 65:1850–1857, 2000.

35. Martin W. Bunder. Proof finding algorithms for implicational logics. *Theoretical Computer Science*, 232:165–186, 2000.

36. Martin W. Bunder. Intersection types for lambda-terms and combinators and their logics. *Logic Journal of IGPL*, 10:357–378, 2002.

37. Martin W. Bunder. Intersection type systems and logics related to the Meyer–Routley system B^+. *Australasian Journal of Logic*, 1:43–55, 2003.

38. Martin W. Bunder. Corrections to some results for **BCK** logics and algebras. *Logique et Analyse*, 31:115–122, 1988.

39. Martin W. Bunder. Combinatory logic and lambda calculus with classical types. *Logique et Analyse*, 137–138:69–79, 1992.

40. Martin W. Bunder. Theorems in classical logic are instances of theorems in condensed BCI logic. In K. Došen and Schroeder-Heister P., editors, *Substructural Logics*, pages 48–62. Clarendon, Oxford, UK, 1993.

41. Martin W. Bunder. Lambda terms definable as combinators. *Theoretical Computer Science*, 169:3–21, 1996.

42. Martin W. Bunder and Robert K. Meyer. A result for combinators, BCK logics and BCK algebras. *Logique et Analyse*, 28:33–40, 1985.

43. Felice Cardone and J. Roger Hindley. History of lambda-calculus and combinatory logic. In Dov M. Gabbay and John Woods, editors, *Logic from Russell to Church*, volume 5 of *Handbook of the History of Logic*, pages 732–617. Elsevier, Amsterdam, 2009.

44. Alonzo Church. *The Calculi of Lambda-Conversion*. Princeton University Press, Princeton, NJ, 1st edition, 1941.

45. Alonzo Church. The weak theory of implication. In A. Menne, A. Wilhelmy, and H. Angsil, editors, *Kontrolliertes Denken, Untersuchungen zum Logikkalkül und zur Logik der Einzelwissenschaften*, pages 22–37. Komissions-Verlag Karl Alber, Munich, 1951.

46. Alonzo Church. *Introduction to Mathematical Logic*. Princeton University Press, Princeton, NJ, revised and enlarged edition, 1996.

47. Mario Coppo and Mariangiola Dezani-Ciancaglini. An extension of the basic functionality theory for the λ-calculus. *Notre Dame Journal of Formal Logic*, 21:685–693, 1980.

48. Mario Coppo, Mariangiola Dezani-Ciancaglini, and Betti Venneri. Principal type schemes and λ-calculus semantics. In J. R. Hindley and J. P. Seldin, editors, *To H. B. Curry*, pages 535–560. Academic Press, London, UK, 1980.

49. Pierre-Louis Curien and Hugo Herberlin. The duality of computation. In *ICFP'00 Proceedings of the 5th ACM SIGPLAN International Conference on Functional Programming*, pages 1–11. ACM, 2000.

50. Haskell B. Curry. Grundlagen der kombinatorischen Logik. *American Journal of Mathematics*, 52:509–536, 789–834, 1930.

51. Haskell B. Curry. Basic verifiability in the combinatory theory of restricted generality. In Y. Bar-Hillel, E. T. J. Poznanski, M. O. Rabin, and A. Robinson, editors, *Essays on the Foundations of Mathematics. Dedicated to A. A. Fraenkel on his seventieth anniversary*, pages 165–189. Magnes Press, Hebrew University, Jerusalem, 1961.

52. Haskell B. Curry. *Foundations of Mathematical Logic*. McGraw-Hill Book Company, New York, NY, 1963. (Dover, New York, NY, 1977).

53. Haskell B. Curry and Robert Feys. *Combinatory Logic, vol. 1*. Studies in Logic and the Foundations of Mathematics. North-Holland, Amsterdam, 1st edition, 1958.

54. Haskell B. Curry, J. Roger Hindley, and Jonathan P. Seldin. *Combinatory Logic, vol. 2*. Studies in Logic and the Foundations of Mathematics. North-Holland, Amsterdam, 1972.

55. Brian A. Davey and Hilary A. Priestley. *Introduction to Lattices and Order*. Cambridge University Press, Cambridge, UK, 2nd edition, 2002.

56. Martin Davis, editor. *The Undecidable. Basic Papers on Undecidable Propositions, Unsolvable Problems and Computable Functions*. Dover Publications, Mineola, NY, 2004.

57. Martin Davis. *Computability and Unsolvability*. Dover Publications, New York, NY, enlarged edition, 1982.

58. Klaus Denecke and Shelly L. Wismath. *Universal Algebra and Applications in Theoretical Computer Science*. Chapman & Hall/CRC, Boca Raton, FL, 2002.

59. Mariangiola Dezani-Ciancaglini and J. Roger Hindley. Intersection types for combinatory logic. *Theoretical Computer Science*, 100:303–324, 1992.

60. Mariangiola Dezani-Ciancaglini, Elio Giovannetti, and Ugo de'Liguoro. Intersection types, λ-models, and Böhm trees. In Masako Takahashi, Mitsuhiro Okada, and Mariangiola Dezani-Ciancaglini, editors, *Theories of Types and Proofs*, volume 2 of *MSJ Memoirs*, pages 45–97. Mathematical Society of Japan, Tokyo, 1998.

61. Mariangiola Dezani-Ciancaglini, Robert K. Meyer, and Yoko Motohama. The semantics of entailment omega. *Notre Dame Journal of Formal Logic*, 43:129–145, 2003.

62. J. Michael Dunn. Ternary relational semantics and beyond: programs as data and programs as instructions. *Logical Studies*, 7:1–20, 2001.

63. J. Michael Dunn. A 'Gentzen system' for positive relevant implication (abstract). *Journal of Symbolic Logic*, 38:356–357, 1973.

64. J. Michael Dunn. Relevance logic and entailment. In D. Gabbay and F. Guenthner, editors, *Handbook of Philosophical Logic*, volume 3, pages 117–229. D. Reidel, Dordrecht, 1st edition, 1986.

65. J. Michael Dunn and Gary M. Hardegree. *Algebraic Methods in Philosophical Logic*, volume 41 of *Oxford Logic Guides*. Oxford University Press, Oxford, UK, 2001.

66. J. Michael Dunn and Robert K. Meyer. Combinators and structurally free logic. *Logic Journal of IGPL*, 5:505–537, 1997.

67. Erwin Engeler. Algebras and combinators. *Algebra Universalis*, 13:389–392, 1981.

68. Andrzej Filinski. Declarative continuations: an investigation of duality in programming language semantics. In D. H. Pitt, D. E. Rydeheard, P. Dybjer, A. M. Pitts, and A. Poigné, editors, *Category Theory and Computer Science. Proceedings of the Third Biennial Conference Held in Manchester*, number 389 in Lecture Notes in Computer Science, pages 224–249, Springer, Berlin, 1989.

69. Gerhard Gentzen. Untersuchungen über das logische Schließen. *Mathematische Zeitschrift*, 39:176–210, 1935.

70. Gerhard Gentzen. Untersuchungen über das logische Schließen, II. *Mathematische Zeitschrift*, 39:405–431, 1935.

71. Gerhard Gentzen. Die Widerspruchsfreiheit der reinen Zahlentheorie. *Mathematische Annalen*, 112:493–565, 1936.

72. Pietro Di Gianantonio, Furio Honsell, and Luigi Liquori. A lambda calculus of objects with self-inflicted extension. In *Proceedings of ACM-SIGPLAN OOPSLA-98 International Symposium on Object Oriented Programming, System Languages and Applications*, pages 1–13. ACM Press, New York, NY, 1998.

73. Gerhard Gierz, Karl H. Hofmann, Klaus Keimel, Jimmie D. Lawson, Michael W. Mislove, and Dana S. Scott. *Continuous Lattices and Domains*, volume 93 of *Encyclopedia of Mathematics and Its Applications*. Cambridge University Press, Cambridge, UK, 2003.

74. Jean-Yves Girard. Linear logic. *Theoretical Computer Science*, 50:1–102, 1987.

75. Lou Goble. Combinator logic. *Studia Logica*, 76:17–66, 2004.

76. Kurt Gödel. Über formal unentscheidbare Sätze der *Principia mathematica* und verwandter Systeme I. In Solomon Feferman, editor, *Collected Works*, volume 1, pages 144–195. Oxford University Press and Clarendon Press, New York, NY, and Oxford, UK, 1986, 1931.

77. Kurt Gödel. On undecidable propositions of formal mathematical systems. In Solomon Feferman, editor, *Collected Works*, volume 1, pages 346–371. Oxford University Press and Clarendon Press, New York, NY, and Oxford, UK, 1986, 1934.

78. George Grätzer. *Universal Algebra*. Springer-Verlag, New York, NY, 2nd edition, 1979.

79. Chris Hankin. *Lambda Calculi. A Guide for Computer Scientists*, volume 3 of *Graduate Texts in Computer Science*. Clarendon Press, Oxford, UK, 1994.

80. David Harel. *Algorithmics. The Spirit of Computing*. Addison-Wesley Publishing Co.,

Wokingham, UK, 1987.

81. J. Roger Hindley. Axioms for strong reduction in combinatory logic. *Journal of Symbolic Logic*, 32:224–236, 1967.

82. J. Roger Hindley. An abstract form of the Church-Rosser theorem, I. *Journal of Symbolic Logic*, 34:545–560, 1969.

83. J. Roger Hindley. Lambda-calculus models and extensionality. *Zeitschrift für mathematische Logik und Grundlagen der Mathematik*, 26:289–310, 1980.

84. J. Roger Hindley. The simple semantics for Coppo–Dezani–Sallé types. In M. Dezani-Ciancaglini and U. Montanari, editors, *International Symposium on Programming*, number 137 in Lecture Notes in Computer Science, pages 212–226, Springer, Berlin, 1982.

85. J. Roger Hindley. Coppo–Dezani types do not correspond to propositional logic. *Theoretical Computer Science*, 28:235–236, 1984.

86. J. Roger Hindley. *Basic Simple Type Theory*, volume 42 of *Cambridge Tracts in Theoretical Computer Science*. Cambridge University Press, Cambridge, UK, 1997.

87. J. Roger Hindley and Jonathan P. Seldin. *Lambda-Calculus and Combinators, an Introduction*. Cambridge University Press, Cambridge, UK, 2008.

88. J. Roger Hindley and Jonathan P. Seldin, editors. *To H. B. Curry*. Academic Press, London, UK, 1980.

89. J. Roger Hindley and Jonathan P. Seldin. *Introduction to Combinators and* λ*-Calculus*. Cambridge University Press, Cambridge, UK, 1986.

90. W. A. Howard. The formulae-as-types notion of construction. In J. R. Hindley and J. P. Seldin, editors, *To H. B. Curry*, pages 479–490. Academic Press, London, UK, 1980.

91. Stephen C. Kleene. *Introduction to Metamathematics*. Van Nostrand, New York, NY, 1952.

92. Jan W. Klop. *Combinatory reduction systems*, volume 127 of *Mathematical Centre Tracts*. Mathematisch Centrum, Amsterdam, 1980.

93. Jan W. Klop. Term rewriting systems. In S. Abramsky, D. M. Gabbay, and T. S. E. Maibaum, editors, *Handbook of Logic in Computer Science. Background: Computational Structures*, volume 2, pages 1–116. Oxford University Press, New York, NY, 1992.

94. C. P. J. Koymans. Models of the lambda calculus. *Information and Control*, 52:306–332, 1982.

95. Joachim Lambek. The mathematics of sentence structure. *American Mathematical Monthly*, 65:154–169, 1958.

96. Joachim Lambek. On the calculus of syntactic types. In R. Jacobson, editor, *Structure of Language and Its Mathematical Aspects*, pages 166–178. American Mathematical Society, Providence, RI, 1961.

97. Edward J. Lemmon, Carew A. Meredith, David Meredith, Arthur N. Prior, and Ivo

Thomas. Calculi of pure strict implication. In J. W. Davis, D. J. Hockney, and W. K. Wilson, editors, *Philosophical Logic*, pages 215–250. D. Reidel, Dordrecht, 1969.

98. Bruce Lercher. Strong reduction and normal form in combinatory logic. *Journal of Symbolic Logic*, 32:213–223, 1967.

99. Bruce Lercher. The decidability of Hindley's axioms for strong reduction. *Journal of Symbolic Logic*, 32:237–239, 1967.

100. G. Longo. Set-theoretical models of λ-calculus: theories, expansions, isomorphisms. *Annals of Pure and Applied Logic*, 24:153–188, 1983.

101. Guilo Manzonetto and Antonino Salibra. Applying universal algebra to lambda calculus. *Journal of Logic and Computation*, 20:877–915, 2010.

102. Edwin D. Mares. Relevance logic. In Dale Jacquette, editor, *A Companion to Philosophical Logic*, pages 609–627. Blackwell, Madden, MA, 2002.

103. Edwin D. Mares and Robert K. Meyer. Relevant logics. In Lou Goble, editor, *The Blackwell Guide to Philosophical Logic*, Blackwell Philosophy Guides, pages 280–308. Blackwell Publishers, Oxford, UK, 2001.

104. Norman D. Megill and Martin W. Bunder. Weaker **D**-complete logics. *Journal of IGPL*, 4:215–225, 1996.

105. Elliott Mendelson. *Introduction to Mathematical Logic*. Chapman & Hall/CRC, Boca Raton, FL, 4th edition, 1997.

106. Carew A. Meredith and Arthur N. Prior. Notes on the axiomatics of the propositional calculus. *Notre Dame Journal of Formal Logic*, 4:171–187, 1963.

107. Albert R. Meyer. What is a model of the lambda calculus? *Information and Control*, 52:87–122, 1982.

108. Robert K. Meyer. What entailment can do for type theory. *(preprint)*, 2000.

109. Robert K. Meyer. A general Gentzen system for implicational calculi. *Relevance Logic Newsletter*, 1:189–201, 1976.

110. Robert K. Meyer and Martin W. Bunder. Condensed detachment and combinators. Technical Report TR–ARP–8188, Australian National University, Canberra, Australia, 1988.

111. Robert K. Meyer and Martin W. Bunder. The D-completeness of $T_{\rightarrow}$. Technical report, Australian National University, Canberra, Australia, 1998.

112. Robert K. Meyer and Richard Routley. Algebraic analysis of entailment I. *Logique et Analyse*, 15:407–428, 1972.

113. Robert K. Meyer, Richard Routley, and J. Michael Dunn. Curry's paradox. *Analysis (n.s.)*, 39:124–128, 1979.

114. Robert K. Meyer, Martin W. Bunder, and Lawrence Powers. Implementing the "fool's model" of combinatory logic. *Journal of Automated Reasoning*, 7:597–630, 1991.

115. Koushik Pal and Robert K. Meyer. Basic relevant theories for combinators at levels I and II. *Australasian Journal of Logic*, 3:14–32, 2005.

116. Benjamin C. Pierce. *Types and Programming Languages*. MIT Press, Cambridge, MA, 2002.

117. A. Piperno. Abstraction problems in combinatory logic: a composite approach. *Theoretical Computer Science*, 66:27–43, 1989.

118. Gordon Plotkin. A semantics for static type inference. *Information and Computation*, 109:256–299, 1994.

119. Garrel Pottinger. A type assignment for the strongly normalizable λ-terms. In J. R. Hindley and J. P. Seldin, editors, *To H. B. Curry*, pages 561–577. Academic Press, London, UK, 1980.

120. Garrel Pottinger. A tour of the multivariate lambda calculus. In J. Michael Dunn and Anil Gupta, editors, *Truth or Consequences*, pages 209–229. Kluwer Academic Publishers, Amsterdam, 1990.

121. W. V. Quine. Predicate functors revisited. *Journal of Symbolic Logic*, 46:649–652, 1981.

122. Willard V. O. Quine. Predicate-functor logic. In J. E. Fenstad, editor, *Proceedings of the Second Scandinavian Logic Symposium*, volume 63 of *Studies in Logic and the Foundations of Mathematics*, pages 309–315, Amsterdam, 1971. North-Holland.

123. Willard Van Orman Quine. *Selected Logic Papers*. Harvard University Press, Cambridge, MA, enlarged edition, 1995.

124. György E. Révész. *Lambda-Calculus, Combinators and Functional Programming*. Cambridge University Press, Cambridge, UK, 1988.

125. Hartley Rogers. *Theory of Recursive Functions and Effective Computability*. MIT Press, Cambridge, MA, 1987.

126. Simona Ronchi Della Rocca and Betti Venneri. Principal type schemes for an extended type theory. *Theoretical Computer Science*, 28:151–169, 1984.

127. Barry K. Rosen. Tree-manipulating systems and Church-Rosser theorems. *Journal of the Association for Computing Machinery*, 20:160–187, 1973.

128. J. Barkley Rosser. A mathematical logic without variables. I. *Annals of Mathematics*, 36:127–150, 1935.

129. Richard Routley, Robert K. Meyer, Val Plumwood, and Ross T. Brady. *Relevant Logics and Their Rivals*, volume 1. Ridgeview Publishing Company, Atascadero, CA, 1982.

130. Moses Schönfinkel. On the building blocks of mathematical logic. In Jean van Heijenoort, editor, *From Frege to Gödel. A Source Book in Mathematical Logic*, pages 355–366. Harvard University Press, Cambridge, MA, 1967, 1924.

131. Dana Scott. Outline of a mathematical theory of computation. Technical report, Oxford University Computing Laboratory Programming Research Group, 1970.

132. Dana Scott. A type-theoretical alternative to ISWIM, CUCH, OWHY. *Theoretical Computer Science*, 121:411–440, 1993.

133. Jonathan P. Seldin. The logic of Curry and Church. In Dov M. Gabbay and John

Woods, editors, *Handbook of the History of Logic*, volume 5, pages 819–873. Elsevier, Amsterdam, 2009.

134. Natarajan Shankar. *Metamathematics, Machines, and Gödel's Proof*, volume 38 of *Cambridge Tracts in Theoretical Computer Science*. Cambridge University Press, Cambridge, UK, 1997.

135. Michael Sipser. *Introduction to the Theory of Computation*. PWS Publishing Company, Boston, MA, 1997.

136. Raymond M. Smullyan. *Theory of Formal Systems*. Princeton University Press, Princeton, NJ, revised edition, 1962.

137. Raymond M. Smullyan. *To Mock a Mockingbird. And Other Logic Puzzles Including an Amazing Adventure in Combinatory Logic*. Alfred A. Knopf, New York, NY, 1985.

138. Raymond M. Smullyan. *Diagonalization and Self-Reference*. Clarendon, Oxford, UK, 1994.

139. Morten H. Sørensen and Pawel Urzyczyn. *Lectures on the Curry–Howard Isomorphism*. Number 149 in Studies in Logic and the Foundations of Mathematics. Elsevier, Amsterdam, 2006.

140. John Staples. Church-Rosser theorems for replacement systems. In J. N. Crossley, editor, *Algebra and Logic*, volume 450 of *Lecture Notes in Mathematics*, pages 291–307. Springer, Berlin, 1975.

141. Sören Stenlund. *Combinators, λ-Terms and Proof Theory*. D. Reidel, Dordrecht, 1972.

142. Joseph E. Stoy. *Denotational Semantics: The Scott–Strachey Approach to Programming Language Theory*. MIT Press, Cambridge, MA, 1977.

143. William Tait. Intensional interpretations of functionals of finite type I. *Journal of Symbolic Logic*, 32:198–212, 1967.

144. William Tait. A realizability interpretation of the theory of species. In R. Parikh, editor, *Logic Colloquium*, volume 453 of *Lecture Notes in Mathematics*, pages 240–251, Boston, 1975. Springer.

145. William W. Tait. Normal form theorem for bar recursive functions of finite type. In J. E. Fenstad, editor, *Proceedings of the Second Scandinavian Logic Symposium*, number 63 in Studies in Logic and the Foundations of Mathematics, pages 353–367, Amsterdam, 1971. North-Holland.

146. Alfred Tarski. Contributions to the theory of models, III. *Indagationes Mathematicae*, 17:56–64, 1955.

147. Alfred Tarski. A lattice-theoretical fixpoint theorem and its applications. *Pacific Journal of Mathematics*, 5:285–309, 1955.

148. Terese. *Term Rewriting Systems*, volume 55 of *Cambridge Tracts in Theoretical Computer Science*. Cambridge University Press, Cambridge, UK, 2003.

149. Peter Trigg, J. Roger Hindley, and Martin W. Bunder. Combinatory abstraction using B, B' and friends. *Theoretical Computer Science*, 135:405–422, 1994.

150. Anne S. Troelstra. *Lectures on Linear Logic*, volume 29 of *CSLI Lecture Notes*. CSLI Publications, Stanford, CA, 1992.

151. Anne S. Troelstra and H. Schwichtenberg. *Basic Proof Theory*, volume 43 of *Cambridge Tracts in Theoretical Computer Science*. Cambridge University Press, Cambridge, UK, 2nd edition, 2000.

152. Alan M. Turing. On computable numbers, with an application to the Entscheidungsproblem. *Proceedings of the London Mathematical Society*, 42:230–265, 1936–37.

153. David A. Turner. Another algorithm for bracket abstraction. *Journal of Symbolic Logic*, 44:267–270, 1979.

154. Hirofumi Yokouchi. Syntax and semantics of type assignment systems. In M. Takahashi, M. Okada, and M. Dezani-Ciancaglini, editors, *Theories of Types and Proofs*, number 2 in MSJ Memoirs, pages 99–141. Mathematical Society of Japan, Tokyo, 1998.

List of symbols

Index

G

H

I